ISNM 86:
International Series of Numerical Mathematics
Internationale Schriftenreihe zur Numerischen Mathematik
Série Internationale d'Analyse Numérique
Vol. 86

ISNM 98:
International Series of Numerical Mathematics
Internationale Schriftenreihe zur Numerischen Mathematik
Série Internationale d'Analyse Numérique
Vol. 98

Edited by
K.-H. Hoffmann, Augsburg; H. D. Mittelmann, Tempe;
J. Todd, Pasadena

Birkhäuser Verlag
Basel · Boston · Berlin

Multigrid Methods III

Edited by

W. Hackbusch
U. Trottenberg

1991

Birkhäuser Verlag
Basel · Boston · Berlin

Editors

W. Hackbusch
Universität Kiel
Institut für Informatik
und Praktische Mathematik
Olshausenstrasse 40
D–2300 Kiel 1

U. Trottenberg
Gesellschaft für Mathematik
und Datenverarbeitung mbH
Postfach 1240
Schloss Birlinghoven
D–5205 Sankt Augustin 1

Deutsche Bibliothek Cataloging-in-Publication Data

Multigrid methods III / ed. by W. Hackbusch; U. Trottenberg.
– Basel ; Boston ; Berlin : Birkhäuser, 1991
(International series of numerical mathematics ; Vol. 98)
ISBN 3-7643-2632-8 (Basel ...)
ISBN 0-8176-2632-8 (Boston)
NE: Hackbusch, Wolfgang [Hrsg.]; European Multigrid Conference <03, 1990, Bonn>; GT

Printed from the authors' camera-ready manuscripts on acid-free paper in Germany
ISBN 3-7643-2632-8
ISBN 0-8176-2632-8

PREFACE

These proceedings contain a selection of papers presented at the

Third European Conference on Multigrid Methods

which was held in Bonn on October 1– 4, 1990.

Following conferences in 1981 and 1985, a platform for the presentation of new Multigrid results was provided for a third time. Multigrid methods no longer have problems being accepted by numerical analysts and users of numerical methods; on the contrary, they have been further developed in such a successful way that they have penetrated a variety of new fields of application. The high number of 154 participants from 18 countries and 76 presented papers show the need to continue the series of the European Multigrid Conferences.

The papers of this volume give a survey on the current Multigrid situation; in particular, they correspond to those fields where new developments can be observed. For example, several papers study the appropriate treatment of time dependent problems. Improvements can also be noticed in the Multigrid approach for semiconductor equations. The field of parallel Multigrid variants, having been started at the second European Multigrid Conference, is now at the centre of interest.

Whereas the papers of this volume are assumed to be of general interest for a large number of scientists in academic as well as industrial research, those more specialized papers presented at the conference will be published in the series GMD-Studien, volume no. 189: *Multigrid Methods: Special Topics and Applications II.* (These papers are listed at the end of this volume.)

At this point, we would like to thank again all participants and especially the lecturers and those having presented posters for their contributions to the success of the conference.

The conference was initiated by

- Gesellschaft für Mathematik und Datenverarbeitung mbH (GMD), St. Augustin,
- GAMM–Fachausschuß 'Effiziente numerische Verfahren für partielle Differentialgleichungen'.

The scientific organization was excellently conceived by Dr. Ute Gärtel, Institut für Methodische Grundlagen of the GMD. Together with Christine Harms, Conference Secretariat of the GMD, she was also responsible for the practical realization. Furthermore, several co-workers, the secretary and the students of the Institut für Methodische Grundlagen provided substantial assistance to the organization of the conference. We thank all persons involved.

We would like to express our gratitude to the Executive Board of the GMD for their support of the conference. Furthermore, we thank the industrial sponsors (Bayer AG, Dornier GmbH, Krupp Atlas Elektronik GmbH, Stardent Computer GmbH, Sun Microsystems GmbH, SUPRENUM GmbH) for their interest in the topics of the conference.

Finally, we would like to thank Birkhäuser Verlag for their friendly cooperation.

Kiel / St. Augustin, March 1991

Wolfgang Hackbusch
Ulrich Trottenberg

CONTENTS

PART I: INVITED PAPERS

PART II: CONTRIBUTED PAPERS

Part I
Invited Papers

MULTIGRID METHODS FOR STEADY EULER- AND NAVIER-STOKES EQUATIONS BASED ON POLYNOMIAL FLUX-DIFFERENCE SPLITTING

by

Erik DICK

ABSTRACT. *Steady Euler- and Navier-Stokes equations are discretized with a vertex-centered finite volume method. For the definition of the convective fluxes, a flux-difference splitting upwind method, based on the polynomial character of the flux-vectors with respect to the primitive variables is used. The diffusive fluxes are defined in the classical central way.*
In the first order formulation, flux-difference splitting leads to a discretization of vector-positive type. This allows a solution by classical relaxation methods in multigrid form.
By the use of the flux-difference splitting concept a consistent discrete formulation of boundary conditions, including the treatment of diffusive fluxes, is possible for all types of boundaries : in- and outflow and solid boundaries. A second order formulation is achieved by using a limited flux-difference extrapolation according to the method suggested by Chakravarthy and Osher. The minmod-limiter is used. In second order form, direct relaxation of the discrete equations is not possible anymore due to the loss of positivity. To solve the second order system, defect-correction is used.

1. INTRODUCTION

Multigrid methods for calculation of steady compressible flows are now well established. Several classes of methods can be distinguished. The methods orgininated by Jameson [1] use explicit time-stepping of Runge-Kutta type. The stabilization of these time-stepping methods is based on the use of artificial viscosity terms which give the space discretization some upwind appearance. An example for unstructured grids is given by Mavriplis and Jameson [2]. In these methods, the Runge-Kutta coefficients can be altered to obtain good smoothing rates, improving the multigrid performance. Examples of optimization of these coefficients are given by Jameson [1], by van Leer, Tai and Powell [3], and by Catalano and Deconinck [4]. Similar Runge-Kutta time-stepping schemes, but using upwind discretizations of flux-difference type were developed by Dervieux et al. [5] and by Lallemand and Dervieux [6]. Their flux-difference splitting can be considered as an inconsistent simplification of Roe's flux-difference splitting [7]. Very similar work, but using the original Roe flux-difference splitting was done by von Levante et al. [8].

A second class of methods uses upwind discritizations in order to allow relaxation-type solution methods. These upwind methods are based on flux-vector splitting or flux-difference splitting. The most popular splitting techniques are the flux-vector splitting of van Leer [9] and the flux-difference splitting of Osher and Chakravarthy [10]. For steady equations, only for first order accuracy, these upwind techniques lead to discrete equations that can be solved by relaxation methods in multigrid form. In order to reach second order accuracy, two approaches are possible. The first way consists of adding a correction to the first order form by a defect-correction procedure. The second way consists in using an implicit time-stepping procedure as relaxation step. Both techniques are comparable to each other in efficiency. The defect-correction approach was first illustrated using the Osher-Chakravarthy flux-difference splitting by Hemker [11] and further developed by Hemker, Spekreijse and Koren [12,13,14]. Examples of implicit time-stepping procedures using van Leer flux-vector splitting were given by Mulder [15] and by Schröder and Hänel [16]. In the cited multigrid methods based on relaxation schemes, either van Leer splitting or Osher-Chakravarthy splitting is used because of their differentiability, allowing Newton-linearization. The quadratic convergence associated to the Newton-linearization is generally believed to be important to reach good efficiency in relaxation-type multigrid methods. This explains why the non-differentiable Roe flux-difference splitting in general is only used in multigrid methods based on time-stepping.

Flux-vector splitting is known to be much less accurate than flux-difference splitting, especially for use with Navier-Stokes equations, since flux-vector splitting cannot recognize shear layers. To make flux-vector splitting acceptable, at least a modification of the basic van Leer splitting is necessary. An example of such a modification is given by Schröder and Hänel [16]. Despite these possible modifications, flux-difference splitting still has the advantage of giving a much more consistent definition of the numerical flux. This has particular advantages with respect to treatment of boundary conditions.

Multigrid methods based on relaxation schemes are much more efficient than multigrid methods based on time-stepping schemes due to their better smoothing properties. This holds at least in terms of work units, where the work unit is the computational time for one basic relaxation or time-stepping operation on the fine grid. The computational work involved in doing one relaxation with the differentiable Osher-Chakravarthy flux-difference splitting is however much larger than the work involved in doing one Runge-Kutta time step with the non-differentiable Roe flux-difference splitting. From these considerations, the necessity follows to devise simple flux-difference splitting methods that allow relaxation. The foregoing statements about efficiency have to be somewhat relativated for advanced computer architectures allowing vectorization and parallelization. Time-stepping procedures allow easy vectorization and parallelization. Relaxation procedures usually are highly sequential. We do not discuss these aspects here.

In this paper, it is demonstrated that a simple flux-difference splitter can be sufficient to construct relaxation-type multigrid methods. The splitter is of Roe-type, i.e. it satisfies the requirements formulated by Roe to construct a flux-difference splitting [7]. It is however much simpler. Its simplicity follows from dropping the secondary requirement of having a unique definition of average flow variables. This secondary requirement defines the original Roe flux-difference splitting within the class of splittings allowed by the primary requirements. This secondary requirement however is not necessary.

The flux-difference splitting was formulated by the author in [17]. It is called the polynomial flux-difference splitting. Its construction is based on the earlier work of Lombard et al.[18]. A multigrid method based on a first order accurate dicretization was presented by the author in [19]. Second order formulations based on defect-correction were given in [20] and [21]. In the cited references, slightly different variants of the method are used. Since these publications, the method has undergone simplifications. The opportunity is taken here to describe the method in its latest version, with sufficient detail so that interested readers can implement it. Several aspects of simplicity are discussed.

It is not generally recognized that similar methods also can be applied to incompressible flow. Some recent applications of flux-difference splitting methods in incompressible flow are due to Hartwich and Hsu [22] and to Gorski [23], using Roe's splitting. These methods however resort to the concept of artificial compressibility in order to construct, through time integration, a solution of the steady incompressible Navier-Stokes equations. The method is then very similar to the time dependent approach for the compressible equations.

The relaxation methodology however also can be applied to incompressible fluids. This basically was demonstrated by the author, using non-conservative flux-matrix splitting, in [24] and [25]. The conservative formulation using the polynomial flux-difference splitting and second order formulation by flux-limiting and defect-correction was treated in [26]. For incompressible fluids, the polynomial flux-difference splitting is identical to Roe flux-difference splitting. The principles of the application to incompressible fluids are also discussed in this paper. The method formulated in [26] has a boundary treatment that is slightly different from the boundary treatment used for the Euler equations. In this paper, the boundary treatment for the incompressible Navier-Stokes equations is formulated in the same way as for the compressible Euler equations. This brings the method in a unified form, equally well applicable to compressible and incompressible, viscous and inviscid flow problems.

2. POLYNOMIAL FLUX-DIFFERENCE SPLITTING FOR COMPRESSIBLE FLOW

Steady Euler equations, in two dimensions, take the form

$$\frac{\partial f}{\partial x} + \frac{\partial g}{\partial y} = 0 ,$$

where the flux-vectors are

$$f^T = \{\rho u,\ \rho uu+p,\ \rho uv,\ \rho Hu\} , \qquad g^T = \{\rho v,\ \rho uv,\ \rho vv+p,\ \rho Hv\} ,$$

and where ρ is density, u and v are Cartesian velocity components, p is pressure, $H = \gamma p/(\gamma-1)\rho + \frac{1}{2} u^2 + \frac{1}{2} v^2$ is total enthalpy and γ is the adiabatic constant.

Since the components of the flux-vectors form polynomials with respect to the primitive variables ρ, u, v and p, components of flux-differences can be written as follows :

$$\Delta \rho u = \bar{u}\ \Delta\rho + \bar{\rho}\ \Delta u ,$$

$$\Delta(\rho uu+p) = \overline{\rho u}\ \Delta u + \bar{u}\ \Delta\rho u + \Delta p = \bar{u}^2\ \Delta\rho + (\overline{\rho u} + \bar{\rho}\ \bar{u})\Delta u + \Delta p\ ,$$

$$\begin{aligned}\Delta\rho Hu &= \overline{\rho u}(\tfrac{1}{2}\ \Delta u^2 + \tfrac{1}{2}\ \Delta v^2) + \tfrac{1}{2}\ (\overline{u^2} + \overline{v^2})\Delta\rho u + \frac{\gamma}{\gamma-1}\ \Delta pu \\ &= \tfrac{1}{2}(\overline{u^2}+\overline{v^2})\bar{u}\ \Delta\rho + \tfrac{1}{2}(\overline{u^2}+\overline{v^2})\bar{\rho}\ \Delta u + \overline{\rho u}\ \bar{u}\ \Delta u \\ &\quad + \frac{\gamma}{\gamma-1}\ \bar{p}\ \Delta u + \overline{\rho u}\ \bar{v}\ \Delta v + \frac{\gamma}{\gamma-1}\ \bar{u}\ \Delta p\ ,\end{aligned}$$

etc., where the bar denotes mean value.

With the definition of $\bar{q} = \frac{1}{2}\ (\overline{u^2} + \overline{v^2})$, the flux-difference Δf can be written as

$$\Delta f = \begin{pmatrix} \bar{u} & \bar{\rho} & 0 & 0 \\ \bar{u}^2 & \bar{\rho}\ \bar{u} + \overline{\rho u} & 0 & 1 \\ \bar{u}\ \bar{v} & \bar{\rho}\ \bar{v} & \overline{\rho u} & 0 \\ \bar{q}\ \bar{u} & \bar{q}\ \bar{\rho} + \overline{\rho u}\ \bar{u} + \frac{\gamma}{\gamma-1}\ \bar{p} & \overline{\rho u}\ \bar{v} & \frac{\gamma}{\gamma-1}\ \bar{u} \end{pmatrix} \Delta\xi\ ,$$

where ξ is the vector of primitive variables

$$\xi^T = \{\rho,\ u,\ v,\ p\}.$$

With the definition of $\bar{\bar{u}}$ by $\bar{\rho}\ \bar{\bar{u}} = \overline{\rho u}$, the flux-difference Δf is given by

$$\Delta f = \begin{pmatrix} 1 & 0 & 0 & 0 \\ \bar{u} & \bar{\rho} & 0 & 0 \\ \bar{v} & 0 & \bar{\rho} & 0 \\ \bar{q} & \bar{\rho}\ \bar{u} & \bar{\rho}\ \bar{v} & 1/\gamma-1 \end{pmatrix} \begin{pmatrix} \bar{u} & \bar{\rho} & 0 & 0 \\ 0 & \bar{\bar{u}} & 0 & 1/\bar{\rho} \\ 0 & 0 & \bar{\bar{u}} & 0 \\ 0 & \gamma\bar{p} & 0 & \bar{u} \end{pmatrix} \Delta\xi\ . \qquad (1)$$

By denoting the first matrix in (1) by T, it is easily seen that the flux-difference Δg can be written in a similar way as

$$\Delta g = T \begin{pmatrix} \bar{v} & 0 & \bar{\rho} & 0 \\ 0 & \bar{\bar{v}} & 0 & 0 \\ 0 & 0 & \bar{\bar{v}} & 1/\bar{\rho} \\ 0 & 0 & \gamma\bar{p} & \bar{v} \end{pmatrix} \Delta\xi\ ,$$

where $\bar{\rho}\ \bar{\bar{v}} = \overline{\rho v}$.

Any linear combination of Δf and Δg can be written as

$$\Delta\phi = n_x\ \Delta f + n_y\ \Delta g = \bar{A}\ \Delta\xi = T\ \tilde{A}\ \Delta\xi\ , \qquad (2)$$

where

$$\tilde{A} = \begin{pmatrix} \bar{w} & n_x \bar{\rho} & n_y \bar{\rho} & 0 \\ 0 & \bar{\bar{w}} & 0 & n_x/\bar{\rho} \\ 0 & 0 & \bar{\bar{w}} & n_y/\bar{\rho} \\ 0 & n_x \gamma\bar{p} & n_y \gamma\bar{p} & \bar{w} \end{pmatrix} , \qquad (3)$$

with $\bar{w} = n_x \bar{u} + n_y \bar{v}$, $\bar{\bar{w}} = n_x \bar{\bar{u}} + n_y \bar{\bar{v}}$.

It is easy to verify that the matrix $\tilde{A}$ has real eigenvalues and a complete set of eigenvectors.

For $n_x^2 + n_y^2 = 1$, the eigenvalues are given by

$$\lambda_1 = \bar{w}, \quad \lambda_2 = \bar{\bar{w}}, \quad \lambda_3 = \tilde{w} + \bar{c} \quad \text{and} \quad \lambda_4 = \tilde{w} - \bar{c} , \qquad (4)$$

where $\tilde{w} = (\bar{w} + \bar{\bar{w}})/2$, $\bar{c}^2 = \gamma\bar{p}/\bar{\rho} + (\bar{w} - \bar{\bar{w}})^2/4$.

The matrix $\tilde{A}$ can be split into positive and negative parts by

$$\tilde{A}^+ = R\ \Lambda^+ L , \qquad \tilde{A}^- = R\ \Lambda^- L ,$$

where R and L denote right and left eigenvector matrices, in orthonormal form and where

$$\Lambda^+ = \mathrm{diag}(\lambda_1^+, \lambda_2^+, \lambda_3^+, \lambda_4^+) , \quad \Lambda^- = \mathrm{diag}(\lambda_1^-, \lambda_2^-, \lambda_3^-, \lambda_4^-) ,$$

with $\lambda_i^+ = \max(\lambda_i, 0)$, $\lambda_i^- = \min(\lambda_i, 0)$.

With positive and negative matrices, matrices with respectively non-negative and non-positive eigenvalues are meant.

The left eigenvector matrix of $\tilde{A}$ is

$$L = \begin{pmatrix} 1 & 0 & 0 & -\bar{\rho}/\gamma\bar{p} \\ 0 & n_y & -n_x & 0 \\ 0 & n_x & n_y & (1+\delta)\bar{c}/\gamma\bar{p} \\ 0 & n_x & n_y & -(1-\delta)\bar{c}/\gamma\bar{p} \end{pmatrix} ,$$

where $\delta = (\bar{w} - \bar{\bar{w}})/2\bar{c}$.

The matrix T is the transformation matrix between differences of conservative variables and differences of primitive variables, i.e.

$$\Delta\varsigma = T\ \Delta\xi , \qquad (5)$$

where ς is the vector of conservative variables

$$\varsigma^T = \{\rho, \rho u, \rho v, \rho E\} ,$$

with the total energy

$$E = p/(\gamma-1)\rho + \tfrac{1}{2}\,u^2 + \tfrac{1}{2}\,v^2 \ .$$

Combination of (2) and (5) gives

$$\Delta\phi = T\,\tilde{A}\,T^{-1}\,\Delta\zeta = T\,R\,\Lambda\,L\,T^{-1}\,\Delta\zeta = A\,\Delta\zeta \ . \tag{6}$$

The left eigenvector matrix of A can be written as

$$LT^{-1} = \begin{pmatrix} 1 - \dfrac{\bar\rho\,\beta\,\bar{\bar q}}{\bar c} & \dfrac{\bar\rho\,\beta\,\bar u}{\bar c} & \dfrac{\bar\rho\,\beta\,\bar v}{\bar c} & -\dfrac{\bar\rho\,\beta}{\bar c} \\ \bar t/\bar\rho & n_y/\bar\rho & -n_x/\bar\rho & 0 \\ -\dfrac{\bar w}{\bar\rho} + (1+\delta)\beta\,\bar{\bar q} & \dfrac{n_x}{\bar\rho} - (1+\delta)\beta\,\bar u & \dfrac{n_y}{\bar\rho} - (1+\delta)\beta\,\bar v & (1+\delta)\beta \\ -\dfrac{\bar w}{\bar\rho} - (1-\delta)\beta\,\bar{\bar q} & \dfrac{n_x}{\bar\rho} + (1-\delta)\beta\,\bar u & \dfrac{n_y}{\bar\rho} + (1-\delta)\beta\,\bar v & -(1-\delta)\beta \end{pmatrix}, \tag{7}$$

where $\quad \bar{\bar q} = \bar u^2 + \bar v^2 - \bar q \ , \quad \bar t = n_x\,\bar v - n_y\,\bar u \quad$ and $\quad \beta = \dfrac{\gamma-1}{\gamma}\dfrac{\bar c}{\bar p}$.

Furthermore, we have

$$A^+ = T\,R\,\Lambda^+\,L\,T^{-1} \ , \qquad A^- = T\,R\,\Lambda^-\,L\,T^{-1} \ . \tag{8}$$

This allows a splitting of the flux-difference (2) by

$$\Delta\phi = A^+\Delta\zeta + A^-\Delta\zeta \ . \tag{9}$$

3. VERTEX-CENTERED FINITE VOLUME FORMULATION

Figure 1 shows the control volume centered around the node (i,j).

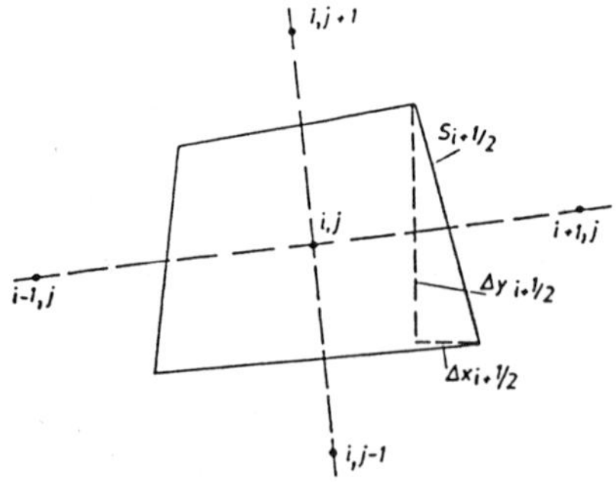

Fig. 1. Control volume around node (i,j)

With piecewise constant interpolation of variables, the flux-difference over the surface $S_{i+\frac{1}{2}}$ of the control volume can be written as

$$\Delta F_{i,i+1} = \Delta y_{i+\frac{1}{2}}\, \Delta f_{i,i+1} + \Delta x_{i+\frac{1}{2}}\, \Delta g_{i,i+1}$$
$$= \Delta s_{i+\frac{1}{2}}\, (n_x\, \Delta f_{i,i+1} + n_y\, \Delta g_{i,i+1}) \ ,$$

where $n_x = \Delta y_{i+\frac{1}{2}}/\Delta s_{i+\frac{1}{2}}$, $n_y = \Delta x_{i+\frac{1}{2}}/\Delta s_{i+\frac{1}{2}}$, $\Delta s^2_{i+\frac{1}{2}} = \Delta x^2_{i+\frac{1}{2}} + \Delta y^2_{i+\frac{1}{2}}$,

and where n_x and n_y denote the components of the unit outgoing normal to the control surface.

With the notation of the previous section, the flux-difference is

$$\Delta F_{i,i+1} = F_{i+1} - F_i = \Delta s_{i+\frac{1}{2}}\, A_{i,i+1}\, \Delta\zeta_{i,i+1} \ . \tag{10}$$

Furthermore, the matrix $A_{i,i+1}$ can be split into positive and negative parts.

This allows the definition of the absolute value of the flux-difference by

$$|\Delta F_{i,i+1}| = \Delta s_{i+\frac{1}{2}}\, (A^+_{i,i+1} - A^-_{i,i+1})\Delta\zeta_{i,i+1} \ . \tag{11}$$

Based on (11) an upwind definition of the flux is

$$F_{i+\frac{1}{2}} = \tfrac{1}{2}[F_i + F_{i+1} - |\Delta F_{i,i+1}|] \ . \tag{12}$$

That (12) represents an upwind flux can be verified by writing it in either of the two following equivalent ways

$$F_{i+\frac{1}{2}} = F_i + \tfrac{1}{2}\, \Delta F_{i,i+1} - \tfrac{1}{2}\, |\Delta F_{i,i+1}|$$
$$= F_i + \Delta s_{i+\frac{1}{2}}\, A^-_{i,i+1}\, \Delta\zeta_{i,i+1} \ ; \tag{13}$$

$$F_{i+\frac{1}{2}} = F_{i+1} - \tfrac{1}{2}\, \Delta F_{i,i+1} - \tfrac{1}{2}\, |\Delta F_{i,i+1}|$$
$$= F_{i+1} - \Delta s_{i+\frac{1}{2}}\, A^+_{i,i+1}\, \Delta\zeta_{i,i+1} \ . \tag{14}$$

Indeed, when $A_{i,i+1}$ only has positive eigenvalues, the flux $F_{i+\frac{1}{2}}$ is taken to be F_i and when $A_{i,i+1}$ only has negative eigenvalues, the flux $F_{i+\frac{1}{2}}$ is taken to be F_{i+1}.

The expression for the flux to be preferred for implementation is (13). A similar expression can be written for all surfaces of the control volume, by replacing $i+1$ by the outside node and using the corresponding length of the surface and the corresponding unit outgoing normal. The flux balance then symbolically can be written as

$$\sum_k \Delta s_{i,j,k}\, A^-_{i,j,k}\, (\zeta_{i,j,k} - \zeta_{i,j}) = 0 \ , \tag{15}$$

where k is an index indicating the nodes surrounding the node (i,j).

The discrete set of equations (15) is a so-called vector-positive set, since the matrix-coefficients of the surrounding nodes are negative and the matrix-coefficient of the node (i,j) is minus the sum of the coefficients of the surrounding nodes. As a consequence of the positivity, a solution can be obtained by a collective variant of any scalar relaxation method. By a collective variant it is meant that in each node, all components of the vector of dependent variables ζ are relaxed simultaneously.

In practice, however, it is not necessary to solve for the conservative variables ζ. The set of equations directly can be solved for the primitive variables ξ. This means that in (15), the matrices A according to (6) can be replaced by the matrices $\tilde{A}$ according to (2) when the variables ζ are replaced by the variables ξ. This simplifies the construction of the discrete equations considerably. The negative part of the matrices $\tilde{A}$ can be written explicitely. The negative part of the matrices A is then obtained by premultiplication with the matrices T. This makes the flux-difference splitting procedure extremely simple. It is simpler than the original Roe flux-difference splitting and it is much simpler than the Osher-Chakravarthy flux-difference splitting.

4. BOUNDARY CONDITIONS

Figure 2 shows the half-volumes centered around a node at inlet and around a node at a solid boundary.

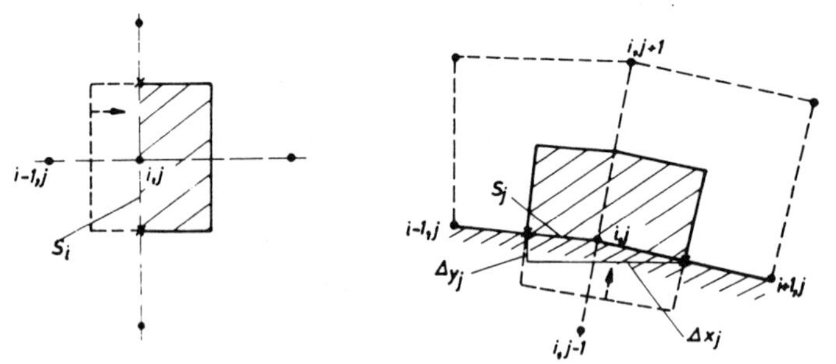

Fig. 2. Control volumes at inlet and at a solid boundary

These half-volumes can be seen as the limit of complete volumes in which one of the sides tends to the boundary.

As a consequence, the flux on the side S_i of the inlet control volume can be expressed according to (13) by

$$F_i + \Delta s_i \, A^-_{i,j}(\zeta_{i-1} - \zeta_i) \;, \tag{16}$$

where the matrix $A_{i,j}$ is calculated in the node (i,j).

Similarly, the flux on the side S_j of the controle volume at the solid boundary can be expressed by

$$F_j + \Delta s_j \, A^-_{i,j}(\zeta_{j-1} - \zeta_j) \;, \tag{17}$$

where again the matrix $A^-_{i,j}$ is calculated in the node (i,j).

With the definitions (16) and (17), the flux balance on the control volumes at boundaries takes the form (15) in which a node outside the domain comes in. This node, however, can be eliminated.

For subsonic inflow, one of the eigenvalues of $A^-_{i,j}$ is zero. So there exists one combination of the equations in the set by which the node i-1 is eliminated. The combination is given by the fourth eigenvector in (7). The resulting equation is to be supplemented with three boundary conditions : stagnation temperature, stagnation pressure and flow direction.

At the outflow boundary, the set of equations is similar, now involving a node i+1. For subsonic outflow, the first three eigenvalues of $A^-_{i,j}$ are now equal to zero. As a a consequence the three combinations corresponding to the first three eigenvectors in (7) eliminate the node i+1. The resulting equations are to be supplemented by one boundary condition. This can be the specification of the Mach number or the pressure.

At a solid boundary, according to figure 2, the condition of impermiability is

$$n_x u_{i,j} + n_y v_{i,j} = 0 \; . \qquad (18)$$

As a consequence, three eigenvalues in the matrix $A^-_{i,j}$ are now zero. Indeed using the notation of the previous sections one has now

$$\bar{w} = \bar{\bar{w}} = \tilde{w} = 0 \; ,$$

so that $\lambda_1 = 0$, $\lambda_2 = 0$, $\lambda_3 = c$, $\lambda_4 = -c$.

The first three eigenvalues used to construct A^- are zero. This means that the combinations corresponding to the first three eigenvectors in (7) eliminate the outside node. These combinations are proportional to (dropping the index i,j) :

$$\ell_1 = (\frac{\gamma}{\gamma-1}\frac{p}{\rho} - q, \quad u, \quad v, \quad -1) \; ,$$

$$\ell_2 = (t, \quad n_y, \quad -n_x, \quad 0) \; ,$$

$$\ell_3 = (q, \quad \frac{n_x}{\rho\beta} - u, \quad \frac{n_y}{\rho\beta} - v, \quad 1) \; ,$$

where $\beta = \frac{\gamma-1}{\gamma}\frac{c}{p}$.

Further ℓ_1 and ℓ_3 can be combined into

$$(c, \; n_x, \; n_y, \; 0) \; .$$

Due to (18), ℓ_2 is proportional to

$$(-u^2-v^2, \; u, \; v, \; 0) \; .$$

Combining this with ℓ_1 gives

$$(H, \; 0, \; 0, \; -1) \; .$$

So, the following combinations can be used :

(1) as mass equation :

$$(1, \quad n_x/c, \quad n_y/c, \quad 0) \; ;$$

(2) as tangential momentum equation :

$$(t, \; n_y, \; -n_x, \; 0) \; ;$$

(3) as energy equation :

$$(-H, \; 0, \; 0, \; 1) \; .$$

These three combinations are to be supplemented by the kinematic condition of tangentiality (18), which serves as normal momentum equation.

The foregoing procedure defines so-called consistent or characteristic boundary equations at all boundaries. It is certainly advisable to use conditions of this type at solid boundaries. At inflow and at outflow, however, simpler boundary equations can be used. In particular at inflow, the consistent boundary treatment is difficult to apply. Indeed one quasi-linear equation is given as result of characteristic combinations. Further flow direction and stagnation conditions are to be added. The stagnation conditions are highly non-linear combinations of the primitive variables. As a consequence many local iterations are necessary to impose these conditions. Therefore, it is better to treat inflow and outflow in the more classical way by using so-called algebraic boundary equations based on extrapolation. In this technique no (half) control volumes are formed around nodes at inlet and outlet. These nodes are considered as auxiliary nodes and information to be obtained from the flow field is extrapolated. Often extrapolations based on local one-dimensional characteristics are used. In the sequel we use a simpler approach. At subsonic inflow, Mach number is extrapolated along the given flow direction. Together with the given stagnation conditions, this determines density, velocity and pressure in an explicit way. At subsonic outflow, stagnation values and flow direction are extrapolated. Together with an imposed Mach number, again this determines all variables in an explicit way. At supersonic inflow, all variables are imposed. At supersonic outflow, all variables are extrapolated.

The boundary treatment at solid boundaries is equivalent to the so-called image point treatment in the finite-difference method.

5. SECOND ORDER FORMULATION

In order to obtain second order accuracy, the definition of the flux (13) is to be modified.

First we remark that, using (8), the flux-difference (10) can be written as

$$\Delta F_{i,i+1} = \Delta s_{i+\frac{1}{2}} \sum_n r^n_{i+\frac{1}{2}} \lambda^n_{i+\frac{1}{2}} \ell^n_{i+\frac{1}{2}} \Delta\zeta_{i,i+1} \, , \tag{19}$$

where the superscript n refers to the nth eigenvalue and where r^n and ℓ^n denote the nth right and left eigenvectors. r^n and ℓ^n are components of TR and LT^{-1}. By denoting the projection of $\Delta\zeta_{i,i+1}$ on the nth eigenvector by

$$\sigma^n_{i+\frac{1}{2}} = \ell^n_{i+\frac{1}{2}} \Delta\zeta_{i,i+1} \, ,$$

(19) can be written as

$$\Delta F_{i,i+1} = \sum_n \Delta F^n_{i,i+1} = \Delta s_{i+\frac{1}{2}} \sum_n r^n_{i+\frac{1}{2}} \lambda^n_{i+\frac{1}{2}} \sigma^n_{i+\frac{1}{2}} = \Delta s_{i+\frac{1}{2}} \sum_n r^n_{i+\frac{1}{2}} \gamma^n_{i+\frac{1}{2}} \, , \tag{20}$$

where $\Delta F^n_{i,i+1}$ is the component of the flux-difference associated to the nth eigenvalue and $r^n_{i+\frac{1}{2}}$ is the projection of the flux-difference on the nth eigenvector.

Using (20), the first order flux (12) can be written as

$$F_{i+\frac{1}{2}} = \frac{1}{2}(F_i + F_{i+1}) - \frac{1}{2}\sum_n \Delta F^{n+}_{i,i+1} + \frac{1}{2}\sum_n \Delta F^{n-}_{i,i+1} , \tag{21}$$

where the + and - superscripts denote the positive and negative parts of the components of the flux-difference, i.e. the parts obtained by taking the positive and negative parts of the eigenvalues.

According to Chakravarthy and Osher [27], assuming a structured sufficiently smooth grid, a second order flux corresponding to (21) can be defined by

$$\begin{aligned} F_{i+\frac{1}{2}} = \frac{1}{2}(F_i + F_{i+1}) &- \frac{1}{2}\sum_n \Delta F^{n+}_{i,i+1} + \frac{1}{2}\sum_n \Delta F^{n-}_{i,i+1} \\ &+ \frac{1}{2}\sum_n \Delta\tilde{F}^{n+}_{i-1,i} - \frac{1}{2}\sum_n \Delta\tilde{F}^{n-}_{i+1,i+2} , \end{aligned} \tag{22}$$

where

$$\begin{aligned} \Delta\tilde{F}^{n+}_{i-1,i} &= \Delta s_{i+\frac{1}{2}}\, r^n_{i+\frac{1}{2}}\, \lambda^n_{i+\frac{1}{2}}\, \ell^n_{i+\frac{1}{2}}\, \Delta\varsigma_{i-1,i} \\ &= \Delta s_{i+\frac{1}{2}}\, r^n_{i+\frac{1}{2}}\, \lambda^n_{i+\frac{1}{2}}\, \tilde{\sigma}^n_{i-\frac{1}{2}} , \end{aligned} \tag{23}$$

with a similar definition for $\Delta\tilde{F}^{n-}_{i+1,i+2}$.

Clearly (23) is constructed by considering a flux-difference over the surface $S_{i+\frac{1}{2}}$, i.e. using the geometry of this surface, with data shifted in the negative i-direction. $\tilde{\sigma}^n_{i-\frac{1}{2}}$ represents the projection of the shifted difference of the dependent variables on the nth eigenvector of the original flux-difference. The second order correction also could be defined using the r-variables, i.e. the projections of the flux-difference. This would mean that also the eigenvalue in (23) is shifted. In practice, there is little difference between the results of both formulations. In the sequel we only use (23). The calculation of the σ-values is further simplified by taking the projection of the difference of primitive variables on the left eigenvectors of $\tilde{A}$.

The definition (22) corresponds to a second order upwind flux. This easily can be seen by considering the case where all eigenvalues have the same sign. Second order accuracy also can be reached by taking a central definition of the flux vector

$$\bar{F}_{i+\frac{1}{2}} = \frac{1}{2}(F_i + F_{i+1}) . \tag{24}$$

As is well known, using either (22) or (24) leads to a scheme which is not monotonicity preserving so that wiggles in the solution become possible. Following the theory of the flux limiters [28], a combination of (22) and (24) is to be taken. This has the form

$$\begin{aligned} F_{i+\frac{1}{2}} = \frac{1}{2}(F_i + F_{i+1}) &- \frac{1}{2}\sum_n \Delta F^{n+}_{i,i+1} + \frac{1}{2}\sum_n \Delta F^{n-}_{i,i+1} \\ &+ \frac{1}{2}\sum_n \Delta\tilde{\tilde{F}}^{n+}_{i-1,i} - \frac{1}{2}\sum_n \Delta\tilde{\tilde{F}}^{n-}_{i+1,i+2} , \end{aligned} \tag{25}$$

with

$$\tilde{\tilde{\Delta F}}^{n+}_{i-1,i} = \mathrm{Lim}(\tilde{\Delta F}^{n+}_{i-1,i}, \Delta F^{n+}_{i,i+1}) \ , \tag{26}$$

$$\tilde{\tilde{\Delta F}}^{n-}_{i+1,i+2} = \mathrm{Lim}(\tilde{\Delta F}^{n-}_{i+1,i+2}, \Delta F^{n-}_{i,i+1}) \ , \tag{27}$$

where Lim denotes some limited combination of both arguments. We choose here the simplest possible form of a limiter, i.e. Lim = MinMod, where the function MinMod returns the argument with minimum absolute value if both arguments have the same sign and returns zero otherwise. By the use of the limiter to the vectors (26)(27) it is meant that the limiter is used per σ-component.

In the vicinity of boundaries, some components of flux-differences in (26) or (27) do not exist. For these components, the limiter then returns a zero. This does not degrade the second order accuracy since due to the characteristic boundary treatment these components do not enter the boundary equations.

The foregoing second order correction procedure is called the flux-extrapolation technique. In contrast to the more common MUSCL-technique [12-16], it gives the second order correction in an explicit way. Flux-extrapolation is therefore much simpler to use with defect-correction.

6. MULTIGRID DEFECT-CORRECTION FORMULATION

Since for the discretization obtained by the second order formulation, the positivity is not guaranteed, a relaxation solution is impossible. Therefore as solution procedure a defect-correction formulation is used. By denoting symbolically the first order and second order formulation on the finest grid by

$$\overset{1}{L}_h = \overset{1}{r}_h \ , \tag{28}$$

$$\overset{2}{L}_h = \overset{2}{r}_h \ , \tag{29}$$

a defect correction means that (28) is replaced by

$$\overset{1}{L}_h = \overset{1}{r}_h + [(\overset{1}{L}_h - \overset{1}{r}_h) - (\overset{2}{L}_h - \overset{2}{r}_h)] \ , \tag{30}$$

where L and r indicate left and right hand sides.

In (30) the difference of the defects of the first and second order discretization is added to the right hand side. The defect-correction is only performed on the finest grid so that a multigrid formulation on the first order discretization can be used.

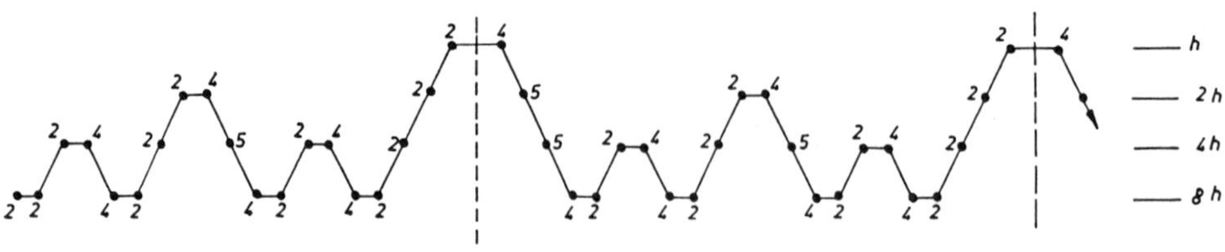

Fig. 3. The multigrid cycle

Figure 3 shows the cycle-structure of the multigrid method. Both the starting cycle and the repeated cycle have W-form. A full approximation scheme (FAS) on the non-linear equations (15) is used. The relaxation algorithm is Gauss-Seidel in lexicographic order, but alternately starting in the lower left corner going up in j-direction and in the upper left corner going down in j-direction. Three relaxations are done per level. In relaxing the set of equations (15), the coefficients are formed with the latest available information. This means for instance that the coefficient $A^{-}_{i,i-1}$ is evaluated with the function values in node (i,j) on the old level, but with the function values in node (i-1,j) on the new level. After determination of the new values in node (i,j), no updates of coefficients and no extra iterations are done. This means that the set of equations (15) is treated as a quasi-linear set and that the multigrid procedure corresponds to a Picard-iteration and not to a Newton-iteration. As restriction operator for residuals, full weighting in the flow field and injection at the boundaries is used. The restriction for function values is injection. The prolongation operator is bilinear interpolation.

In figure 3, the operation count is indicated. A relaxation on the current grid is taken as one local work unit. A residual evaluation plus the associated grid transfer is also taken as one local work unit. Hence, the 5 in figure 3, in going down, stands for the construction of the right hand side in the FAS-formulation, three relaxations and one residual evaluation. With this way of evaluating the work, the cost of the repeated cycle is 13.0625 work units on the finest level. The cost of the starting cycle is 7.5 work units.

7. COMPUTATIONAL EXAMPLES

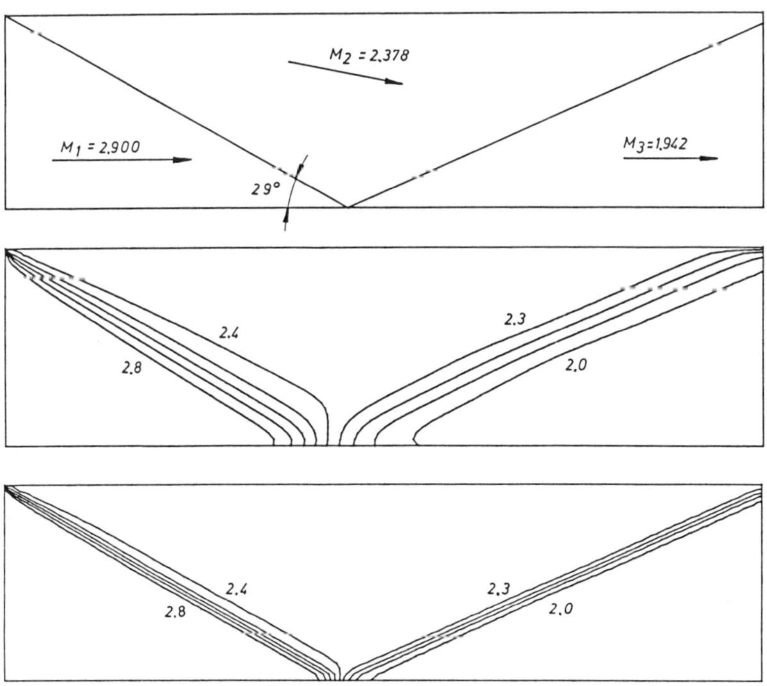

Fig. 4. First order and second order solution of Harten's shock reflection problem (iso-Machlines per 0.1)

Figure 4 shows first order and second order iso-Machline results for Harten's well known shock reflection problem, using a rectangular 97 x 33 grid. The low quality of the first order result is obvious.

Figure 5a shows the convergence behaviour of the first order and second order multigrid methods. The residual shown is the maximum residual over all equations and all points. The defect-correction was used from the first cycle. The calculation starts from a uniform flow with Machnumber 2.9 on the coarsest grid. In calculating the work units in figure 5, the work involved in the defect correction was taken to be one work unit. The residual reduction in the first order formulation is 0.690 per work unit. For the second order formulation it degrades to about 0.944 per work unit in the first phase of the convergence (up to about 120 work units). This is 0.471 per cycle. This is not an excellent but a well acceptable multigrid performance.

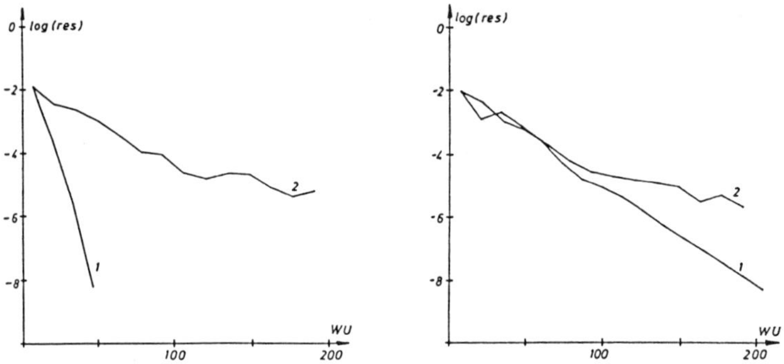

Fig. 5. Convergence behaviour of first order (1) and second order (2) formulation for Harten's shock reflection problem (left). Convergence behaviour of first order (1) and second order (2) formulation for the GAMM-testcase (right)

Figure 6 shows the well known GAMM-testcase for transonic flows. The finest grid has 97 x 33 grid points. In the figure, every fourth line is shown. Figure 7 shows the first and second order iso-Machline results. There is almost no difference between both solutions. This can be explained by the alignment of the shock with the grid lines. Figure 5b shows the convergence behaviour of the first order and second order multigrid methods.

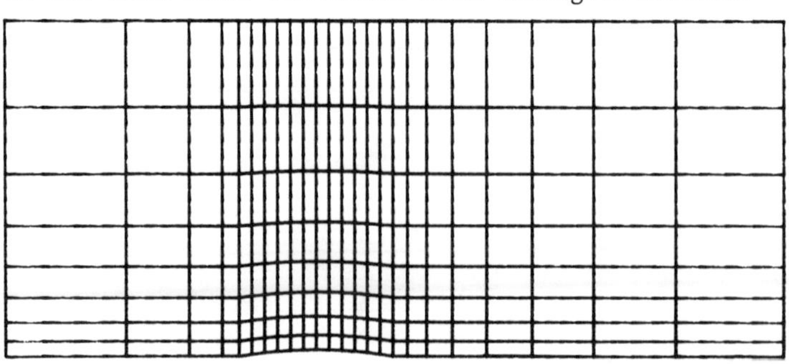

Fig. 6. The GAMM-testcase for transonic flows

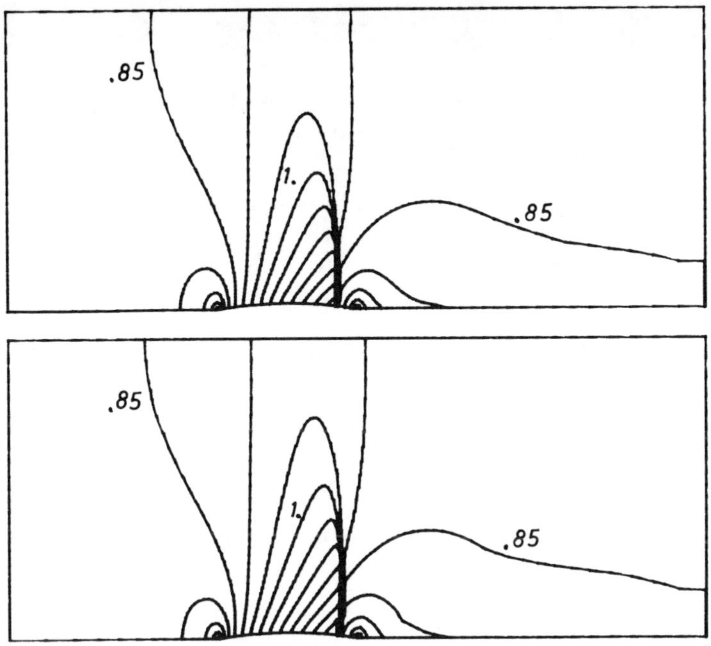

Fig. 7. First order (upper) and second order (lower) results for the GAMM-testcase (iso-Machlines per 0.05)

The calculation starts now from a uniform flow with Machnumber 0.85 on the coarsest grid. The outlet boundary condition is M = 0.85. The residual reduction of the first order formulation is about 0.927 per work unit. For the second order formulation, it is 0.940 per work unit in the first phase of the convergence (up to about 120 work units). This is 0.444 per cycle. The second order performance of both test problems is very similar. It is better than the performance reported by Koren and Spekreijse [14] for the differentiable Osher-Chakravarthy flux-difference splitting, using a similar defect-correction procedure. This shows that differentiability of the splitting is not a necessity to reach good performance. Like for any defect-correction procedure, the convergence behaviour shown in figure 5 has some saturation. The solution obtained at a convergence level of 10^{-4} however cannot be distinguished from the fully converged solution. So, in practice, the saturation has not much meaning.

8. FLUX-DIFFERENCE SPLITTING FOR INCOMPRESSIBLE FLOW

The steady Navier-Stokes equations for an incompressible fluid, in conservative form, are

$$\frac{\partial}{\partial x} u^2 + \frac{\partial}{\partial y} uv + \frac{\partial}{\partial x} p = \nu\left(\frac{\partial^2 u}{\partial x^2} + \frac{\partial^2 u}{\partial y^2}\right) , \tag{31}$$

$$\frac{\partial}{\partial x} uv + \frac{\partial}{\partial y} v^2 + \frac{\partial}{\partial y} p = \nu\left(\frac{\partial^2 v}{\partial x^2} + \frac{\partial^2 v}{\partial y^2}\right) , \tag{32}$$

$$c^2(\frac{\partial u}{\partial x} + \frac{\partial v}{\partial y}) = 0 \ . \tag{33}$$

Here, c is a reference velocity introduced to homogenize the eigenvalues of the system matrices defined in the sequel, p is pressure divided by density and ν is the kinematic viscosity coefficient.

The set of equations can be written in system form by

$$\frac{\partial f}{\partial x} + \frac{\partial g}{\partial y} = \frac{\partial f_v}{\partial x} + \frac{\partial g_v}{\partial y} \ , \tag{34}$$

where f and g are the convective fluxes while f_v and g_v are the viscous fluxes. These are

$$f = \begin{pmatrix} u^2+p \\ uv \\ c^2u \end{pmatrix} , \quad g = \begin{pmatrix} uv \\ v^2+p \\ c^2v \end{pmatrix} , \quad f_v = \begin{pmatrix} \nu \frac{\partial u}{\partial x} \\ \nu \frac{\partial v}{\partial x} \\ 0 \end{pmatrix} , \quad g_v = \begin{pmatrix} \nu \frac{\partial v}{\partial x} \\ \nu \frac{\partial v}{\partial y} \\ 0 \end{pmatrix} . \tag{35}$$

Differences of the convective fluxes can be written in algebraically exact form as follows :

$$\Delta f = \begin{pmatrix} 2\bar{u} & 0 & 1 \\ \bar{v} & \bar{u} & 0 \\ c^2 & 0 & 0 \end{pmatrix} \Delta \begin{pmatrix} u \\ v \\ p \end{pmatrix} , \qquad \Delta g = \begin{pmatrix} \bar{v} & \bar{u} & 0 \\ 0 & 2\bar{v} & 1 \\ 0 & c^2 & 0 \end{pmatrix} \Delta \begin{pmatrix} u \\ v \\ p \end{pmatrix} , \tag{36}$$

where the bar means the algebraic mean of the differenced variables.

The matrices defined by (36) are discrete Jacobians. Any linear combination of these Jacobians has the form

$$A = n_x A_1 + n_y A_2 = \begin{pmatrix} n_x\bar{u}+\bar{w} & n_y\bar{u} & n_x \\ n_x\bar{v} & n_y\bar{v}+\bar{w} & n_y \\ c^2 n_x & c^2 n_y & 0 \end{pmatrix} ,$$

where $\bar{w} = n_x\bar{u} + n_y\bar{v}$.

For $n_x^2 + n_y^2 = 1$, the eigenvalues of the Jacobian A are

$$\lambda_1 = \bar{w} \ , \qquad \lambda_{2,3} = \bar{w} \pm a \ ,$$

with $a = \sqrt{\bar{w}^2+c^2}$.

The corresponding left and right eigenvector matrices are given by

$$L = \begin{pmatrix} \frac{\bar{v}\,\bar{w} + n_y c^2}{a^2} & -\frac{\bar{u}\,\bar{w} + n_x c^2}{a^2} & \frac{n_x\bar{v} - n_y\bar{u}}{a^2} \\ \frac{n_x}{2}(\frac{\bar{w}}{a} + 1) & \frac{n_y}{2}(\frac{\bar{w}}{a} + 1) & \frac{1}{2a} \\ \frac{n_x}{2}(\frac{\bar{w}}{a} - 1) & \frac{n_y}{2}(\frac{\bar{w}}{a} - 1) & \frac{1}{2a} \end{pmatrix} , \tag{37}$$

$$R = \begin{pmatrix} n_y & \frac{\bar u}{a} - n_x(\frac{\bar w}{a} - 1) & \frac{\bar u}{a} - n_x(\frac{\bar w}{a} + 1) \\ -n_x & \frac{\bar v}{a} - n_y(\frac{\bar w}{a} - 1) & \frac{\bar v}{a} - n_y(\frac{\bar w}{a} + 1) \\ 0 & a - \bar w & a + \bar w \end{pmatrix} , \quad (38)$$

where $R = L^{-1}$.

The inviscid fluxes can now further be treated in the same way as was discussed in the previous sections for Euler equations, including the second order corrections. The viscous fluxes are to be determined in a central way. The usual procedure is based on Peyret-control volumes. The structure of the discrete viscous part, when transferred to the left hand side in (34) is positive in the sense of the previous sections. Adding the viscous flux-balance to the inviscid flux-balance enforces the positivity.

9. VISCOUS BOUNDARY CONDITIONS

Using the left and right eigenvector matrices (37-38), the general expression for A^- is found to be

$$A^- = \begin{pmatrix} n_y \bar w^- \beta + n_x \lambda^2 \hat u/2 & -n_y \bar w^- \alpha + n_y \lambda^2 \hat u/2 & (n_y \bar w^- \gamma + \lambda \hat u/2)/a \\ -n_x \bar w^- \beta + n_x \lambda^2 \hat v/2 & n_x \bar w^- \alpha + n_y \lambda^2 \hat v/2 & (-n_x \bar w^- \gamma + \lambda \hat v/2)/a \\ -n_x c^2 \lambda/2 & -n_y c^2 \lambda/2 & -c^2/2a \end{pmatrix} , \quad (39)$$

where $\hat u = \bar u - n_x(\bar w + a)$, $\hat v = \bar v - n_y(\bar w + a)$, $\lambda = (\bar w - a)/a$,

$\alpha = (\bar u\ \bar w + n_x c^2)/a^2$, $\beta = (v\ w + n_y c^2)/a^2$, $\gamma = (n_x \bar v - n_y \bar u)/a$.

Clearly on a solid boundary, due to the condition of tangentiality $\bar w = n_x \bar u + n_y \bar v = 0$, (39) simplifies to (even more simplifications are possible)

$$A^- = \begin{pmatrix} n_x \lambda^2 \hat u/2 & n_y \lambda^2 \hat u/2 & \lambda \hat u/2a \\ n_x \lambda^2 \hat v/2 & n_y \lambda^2 \hat v/2 & \lambda \hat v/2a \\ -n_x c^2 \lambda/2 & -n_y c^2 \lambda/2 & -c^2/2a \end{pmatrix} . \quad (40)$$

Because rank $(A^-) = 1$, two independent combinations of the discrete set of equations exist, eliminating the outside node. These correspond to the first two eigenvectors of L in (37). The boundary conditions are

$$u = 0 , \qquad v = 0 . \quad (41)$$

As a consequence, we only need one boundary equation. The first eigenvector in (37) contains the tangential velocity. Existence of tangential velocity at a solid boundary is not consistent with the viscous equations. Therefore the second eigenvector is to be taken. This eigenvector is proportional to

$$(n_x c\ ,\ n_y c,\ 1) . \quad (42)$$

This combination is used to determine pressure at solid walls.

With (42), the viscous terms are combined on the solid boundary in a term proportional to

$$n_x \nabla^2 u + n_y \nabla^2 v = \nabla^2 w .$$

Using the finite volume integration gives

$$\int_\Omega \nabla.(\nabla w)d\Omega = \int_S \nabla w.\vec{n}\, dS = \int_S \frac{\partial w}{\partial n} dS .$$

On the part of the surface coinciding with the boundary, w stands for the outward normal component of the velocity. Using the mass equation in a coordinate system aligned with the boundary gives

$$\frac{\partial w}{\partial n} + \frac{\partial t}{\partial s} = 0 ,$$

where t is the tangential velocity component and s is the tangential direction. Since obviously $\partial t/\partial s = 0$, the result is

$$\frac{\partial w}{\partial n} = 0 .$$

This means that for the viscous terms, there is no contribution from the boundary in the combination given by (42).

A similar reasoning applies to the other boundaries.

At outflow $\overline{w}^- = 0$. This leads also to the simplification of A^- to the form (40). The combinations eliminating the outside node are again given by the first two eigenvectors of L in (37). It can immediately be verified on the expression of A^- given by (40) that these eigenvectors can be combined to give

$$(1,\ 0,\ \lambda\hat{u}/c^2) ,$$

$$(0,\ 1,\ \lambda\hat{v}/c^2) .$$

These determine u- and v-equations. The boundary condition to be added can be the prescription of pressure ($p = 0$).

For fully developed outflow $\partial u/\partial n = 0$ and $\partial v/\partial n = 0$. So again there is no contribution in the viscous terms at the boundary.

At inflow, the expression for A^- given by (39) cannot be simplified. However, the combination (42) still holds. This gives the pressure. Further boundary conditions are prescription of u and v. For fully developed inflow again there is no contribution in the viscous terms at the boundary

The boundary traitment outlined here for the incompressible Navier-Stokes equations, also can be used with minor modifications for the compressible Navier-Stokes equations.

ACKNOWLEDGEMENT

The research reported in this paper was partly granted by the Belgian National Science Foundation (N.F.W.O.).

REFERENCES

1. Jameson A., Solution of the Euler equations for two dimensional transonic flow by a multigrid method, Appl. Math. Comp. 13 (1983), 327-355.

2. Mavriplis D.and Jameson A., Multigrid solution of the Euler equations on unstructured and adaptive meshes, Lecture Notes in Pure and Applied Mathematics 110 (1988), 413-429.

3. van Leer B., Tai C.H. and Powell K.G., Design of optimally smoothing multi-stage schemes for the Euler equations, AIAA-paper 89-1983, 1989.

4. Catalano L.A. and Deconinck H., Two-dimensional optimization of smoothing properties of multi-stage schemes applied to hyperbolic equations. Third European Conference on Multigrid Methods, Bonn, October 1990.

5. Dervieux A., Desideri J.A., Fezoui F., Palmerio B., Rosenblum J.P. and Stoufflet B., Euler calculations by upwind finite element methods and adaptive mesh algorithm, Notes on Numerical Fluid Mechanics 26 (1986).

6. Lallemand M.H. and Dervieux A., A multigrid finite element method for solving the two-dimensional Euler equations, Lecture Notes in Pure and Applied Mathematics 110 (1988), 337-363.

7. Roe P.L., Approximate Riemann solvers, parameter vectors and difference schemes, J. Comp. Phys. 43 (1981), 357-372.

8. von Levante E., El-Miligui A., Cannizzaro F.E. and Warda H.A., Simple explicit upwind schemes for solving compressible flows, Notes on Numerical Fluid Mechanics 29 (1990), 293-302.

9. Van Leer B., Flux-vector splitting for the Euler equations, Lecture Notes in Physics 170 (1982), 507-512.

10. Osher S. and Chakravarthy S.R., Upwind schemes and boundary conditions with applications to Euler equations in general geometries, J. Comp. Phys. 50 (1983), 447-481.

11. Hemker P.W., Defect correction and higher order schemes for the multigrid solution of the steady Euler equations, Lecture Notes in Mathematics 1228 (1986), 149-165.

12. Hemker P.W. and Spekreijse S., Multiple grid and Osher's scheme for the efficient solution of the steady Euler equations, Appl. Num. Math. 2 (1986), 475-493.

13. Koren B., Defect correction and multigrid for an efficient and accurate computation of airfoil flows, J. Comp. Phys. 77 (1988), 183-206.

14 Koren B. and Spekreijse S., Solution of the steady Euler equations by a multigrid method, Lecture Notes in Pure and Applied Mathematics 110 (1988), 323-336.

15. Mulder W.A., Multigrid relaxation for the Euler equations, J. Comp. Phys. 60 (1985), 235-252.

16. Schröder W. and Hänel D., An unfactored implicit scheme with multigrid acceleration for the solution of the Navier-Stokes equations, Comp. and Fluids 15 (1987), 313-336.

17. Dick E., A flux-difference splitting method for steady Euler equations, J. Comp. Phys. 76 (1988), 19-32.

18. Lombard C.K., Oliger J. and Yang J.Y., A natural conservative flux difference splitting for the hyperbolic systems of gas dynamics, AIAA paper 82-0976, 1982.

19. DICK E., A multigrid method for steady Euler equations based on flux-difference splitting with respect to primitive variables, Notes on Numerical Fluid Mechanics 23 (1988), 69-85.

20. DICK E., A multigrid flux-difference splitting method for steady Euler equations, Proc. 4th Copper Mountain Conference on Multigrid Methods, SIAM (1989), 117-129.

21. DICK E., Multigrid formulation of polynomial flux-difference splitting for steady Euler equations, J. Comp. Phys. 91 (1990), 161-173.

22. Hartwich P.M. and Hsu C.H., High resolution upwind schemes for the three-dimensional, incompressible Navier-Stokes equations, AIAA paper 87-0547, 1987.

23. Gorski J.J., Solutions of the incompressible Navier-Stokes equations using an upwind-differenced TVD scheme, Lecture notes in physics 323 (1989), 278-282.

24. Dick E., A flux-vector splitting method for steady Navier-Stokes equations, Int. J. Num. Meth. Fl. 8 (1988), 317-326.

25. Dick E., A multigrid method for steady incompressible Navier-Stokes equations based on partial flux-splitting, Int. J. Num. Meth. Fl. 9 (1989), 113-120.

26. Dick E. and Linden J., A multigrid flux-difference splitting method for steady incompressible Navier-Stokes equations, Notes on Numerical Fluid Mechanics 29 (1990), 99-108.

27. Chakravarthy S.R. and Osher S., A new class of high accuracy TVD schemes for hyperbolic conservation laws, AIAA paper 85-0363, 1985.

28. Sweby P.K., High resolution schemes using flux limiters for hyperbolic conservation laws, SIAM J. Num. Anal. 21 (1984), 995-1011.

Erik DICK
Department of Machinery
State University of Ghent
Sint Pietersnieuwstraat 41
B-9000 GENT
BELGIUM

RECENT DEVELOPMENTS FOR THE PSMG MULTISCALE METHOD

Paul O. Frederickson [1]
RIACS,
NASA-Ames Research Center
Moffet Field, CA 94035

Oliver A. McBryan [2]
Dept. of Computer Science
University of Colorado
Boulder, CO 80309-0430

In this paper we discuss new developments for the PSMG multiscale method, which we have introduced previously as an efficient PDE solver for massively parallel architectures.

After an overview of the algorithm we introduce the fundamental multiscale recursion relation, as well as appropriate Fourier space notation. We derive the multiscale recursion as a single functional equation without reference to grids. We prove a sequence of rigorous convergence rate bounds which provide increasingly accurate estimates of the convergence rate for translation invariant problems. We show that in constant coefficient situations the convergence rates for the method may be derived to arbitrary precision, and we develop an efficient numerical scheme for computing such rates. Convergence rates are shown to be faster than reported previously. We present estimates for the normalized work involved in PSMG solution: the number of parallel arithmetic and communication operations required per digit of error reduction. The work estimates show that the algorithm is highly efficient.

---

1. Work supported by the NAS Systems Division of NASA and the Defense Advanced Research Project Agency via Cooperative Agreement NCC 2-387 between NASA and the University Space Research Association (USRA).
2. Work supported by the Air Force Office of Scientific Research, under grant AFOSR-89-0422

1. OVERVIEW

In many situations the most efficient algorithms for the numerical solution of large sparse elliptic problems are the various multigrid algorithms[1-3]. Usually these methods are able to compute a solution with N unknowns in $O(N)$ operations, asymptotically faster than most other algorithms. Efficient parallel implementations of multigrid algorithms have been reported on both SIMD and MIMD parallel computers[4-14]. Most of these methods are designed for moderately parallel systems, where each processor will hold many degrees of freedom.

The PSMG method, introduced in[15], provides a parallel algorithm appropriate to the case where the number of processors is comparable to the number of degrees of freedom. Related ideas have been introduced by Hackbusch[16]. As remarked in [17], PSMG is an example of an *intrinsically parallel algorithm.* It is highly efficient if sufficient processors are available, but is extremely inefficient on serial or low-parallelism computers. In situations where there are substantially more fine grid points than processors, an efficient approach might use a hybrid algorithm---using standard multigrid on the finest grids, but switching to PSMG on grid levels where the number of processors approximates or exceeds the number of grid points.

A brief but complete description of the PSMG algorithm is presented in Sections 2 and 3 below. Section 4 introduces the underlying multiscale recurrence relation which is used throughout the paper. Section 5 specializes to translation invariant problems and introduces notation in Fourier transform space. In section 6 we show that a single functional equation on the unit square is equivalent to the PSMG recursion formula across grid levels, greatly simplifying the analysis of PSMG convergence. Section 7 presents proofs of basic convergence rate bounds which show that the PSMG method actually converges, uniformly in the grid size. Section 8 discusses the numerical evaluation of convergence rates and develops an $O(N)$ algorithm for such evaluations. In section 9 we discuss methods for estimating the parallel work required by PSMG. Section 10 discusses actual PSMG performance, comparing the work per digit of convergence achieved across several different algorithms.

2. THE BASIC IDEA

Consider a simple discretization problem on a 1-dimensional grid. Standard multigrid techniques work with a series of coarser grids, each obtained by eliminating the odd numbered points of the previous grid. The error equation for the fine grid is then projected to the coarse grid at even numbered points, the coarse grid equation is solved approximately, and the error is interpolated back to the fine grid and added to the solution there. Finally a smoothing operation is performed on the fine grid. Recursive application of this procedure defines the complete multigrid procedure[1, 3].

The basic idea behind PSMG is the observation that for each fine grid there are two natural coarse grids - the even and odd points of the fine grid. (For simplicity we assume that periodic boundary conditions are enforced). Either of these coarse grids could be used at any point to construct the coarse grid solution, and both would presumably provide approximately equivalent quality solutions. Thus it ought to be possible to find a combination of the two solutions that is significantly better than either separately. It would follow immediately that such a scheme would converge faster (fewer iterations) than the corresponding standard multigrid scheme. Note that on a massively parallel machine the two coarse grid solutions may be solved simultaneously, in the same time as one of them would take - we assume here that the number of processors is comparable to the number of fine grid points. Both coarse grid problems are solved using the same set of machine instructions. Consequently the algorithm is well suited to SIMD parallel computers, as well as to MIMD machines.

The idea outlined above extends naturally to multi-dimensional problems. In d dimensions, 2^d coarse grids are obtained from a fine grid by selecting either the even or the odd points in each of the d coordinate directions. The fine grid solution is then defined by performing a suitable linear interpolation of all 2^d coarse grid points.

3. THE PSMG ALGORITHM

The PSMG algorithm works with a single grid of points $G^{(L)}$ of size $n = n(L) = 2^L$ in each dimension (called the level L grid, or the *fine* grid), but utilizes operators with different scales $l \leq L$ on that grid. Thus the algorithm is strictly speaking multiscale rather than multigrid. There are three basic operators: a finite difference operator A, an interpolation operator

Q and a smoothing operator $\mathbf{S} = I - ZA$. All operators are periodic on the grid in each coordinate direction. The PSMG algorithm extends naturally to both Neumann and Dirichlet boundary conditions, with no increase in convergence rate. The simplest approach to implementing Neumann or Dirichlet boundary conditions is to use reflection or anti-reflection boundary conditions and an extended grid.

The operators at scale level l, denoted $A^{(l)}$, $Q^{(l)}$, and $Z^{(l)}$, couple points at a distance $d_l \equiv 2^{L-l}$. Each level l operator is defined at all points of the grid $G^{(L)}$. The basic steps involved at level l, $0<l\le L$, for the solution of $A^{(L)}U = f$, starting with an initial guess u, with error e and residual r, are described by:

Algorithm PSMG(l,u,f):

1. Compute residual: $r = A^{(l)}e = f - A^{(l)}u$
2. Project residual to coarse grid: $r = r$ (trivial injection).
3. Solve coarse grid residual equation using PSMG: $e' = \mathbf{PSMG}(l-1, 0, r)$
4. Interpolate to fine grid: $e'' = Q^{(l)}e'$
5. Apply a relaxation: $e''' = (I - Z^{(l)}A^{(l)})e'' + Z^{(l)}r$
6. Compute and return the new solution: $u''' = u + e'''$

An exact solver is utilized instead of **PSMG** on the coarsest grid in step 3. As indicated in step 3, the initial guess for all coarse grids is taken to be 0. Consequently steps 1 and 6 are required only on the finest grid. The process as described depends on an explicit choice for $Q^{(l)}$ and $Z^{(l)}$. The PSMG strategy is to choose $Q^{(l)}$ and $Z^{(l)}$ as functions of $A^{(l)}$ in such a way as to optimize the convergence rate of the above algorithm. Explicit choices for $Q^{(l)}$ and $Z^{(l)}$ are given in[15, 18], and in Section 10, for the cases where $A^{(l)}$ represents either the standard 5-point or Mehrstellen discretizations of the Laplacian.

4. THE PSMG RECURRENCE EQUATION

We note that a two-grid PSMG algorithm may be described in the form: $r^{(T)} = \mathbf{T}^{(L)} r$, where the two-grid residual reduction operator $\mathbf{T}^{(L)}$ is determined by:

$$\mathbf{T}^{(L)} = \mathbf{S}^{(L)}(I - A^{(L)} Q^{(L)} A^{(L-1)-1}) \ .$$

We define the *two-grid convergence rate* τ of this iteration procedure as the quantity: $\tau = \sup_L ||\mathbf{T}^{(L)}||$. Clearly τ provides a bound on the convergence rate of the two-grid method on any grid.

We obtain an equation for the residual reduction operator of the full PSMG algorithm by recursive application of the two-grid algorithm described above. The corresponding residual reduction then takes the form: $r^{(M)} = \mathbf{M}^{(l)} r$, where the multi-grid iteration operator $\mathbf{M}^{(l)}$ is determined by:

$$\mathbf{M}^{(l)} = \mathbf{T}^{(l)} + (\mathbf{S}^{(l)} - \mathbf{T}^{(l)})\, \mathbf{M}^{(l-1)} \ , \quad l = L, \cdots, 1 \ ,$$

with $\mathbf{M}^{(0)} \equiv 0$. We define the *multigrid convergence rate* of this procedure as the quantity:

$$\mu = \sup_L \sup_{l \le L} ||\mathbf{M}^{(l)}|| \ .$$

Clearly μ provides a bound on the convergence rate of PSMG on any grid $G^{(L)}$.

5. FOURIER MODE ANALYSIS

In order to complete the description of the PSMG algorithm it is essential to define the operators $Q^{(l)}$ and $Z^{(l)}$ used for interpolation and smoothing. In this section, we describe suitable classes of $Q^{(l)}$ and $Z^{(l)}$ for the special case of an operator which has translation invariant coefficients. We will illustrate the ideas for the Poisson equation discretized on a periodic rectangular grid $G^{(L)}$ of $N = n \times n$ points, $n = 2^L$, which we label with the index $i = (i_1, i_2)$, $0 \le i_1, i_2 < n$. We will use two discretizations of the negative Laplacian $-\Delta$. The first of these is the standard five-point discretization defined by

$$(A_5^{(l)} u)_i = h_l^{-2} \,(4u_i - u_{i-e_1^l} - u_{i+e_1^l} - u_{i-e_2^l} - u_{i+e_2^l}) \ ,$$

where e_i^l are integer vectors of length $d_l \equiv 2^{L-l}$ in the coordinate directions in index space, or alternatively by the familiar five-point star notation:

$$A_5^{(l)} = h_l^{-2} \begin{bmatrix} & -1 & \\ -1 & 4 & -1 \\ & -1 & \end{bmatrix} .$$

The second discretization we will study is the more accurate *Mehrstellen* discretization represented by the nine-point star

$$A_9^{(l)} = (6h_l^2)^{-1} \begin{bmatrix} -1 & -4 & -1 \\ -4 & 20 & -4 \\ -1 & -4 & -1 \end{bmatrix} .$$

Similarly, we will choose the operators $Q^{(l)}$ and $Z^{(l)}$ to be defined by simple symmetric three parameter nine-point star operators (with appropriate scale length depending on l):

$$Q^{(l)} = \begin{bmatrix} q_{11} & q_1 & q_{11} \\ q_1 & q_0 & q_1 \\ q_{11} & q_1 & q_{11} \end{bmatrix} , \quad Z^{(l)} = h_l^2 \begin{bmatrix} z_{11} & z_1 & z_{11} \\ z_1 & z_0 & z_1 \\ z_{11} & z_1 & z_{11} \end{bmatrix} ,$$

or equivalently in operator notation:

$$(Q^{(l)}u)_i = q_0 u_i + q_1 (u_{i+e_1^l} + u_{i-e_1^l} + u_{i+e_2^l} + u_{i-e_2^l}) +$$

$$q_{11} (u_{i+e_1^l+e_2^l} + u_{i-e_1^l+e_2^l} + u_{i+e_1^l-e_2^l} + u_{i-e_1^l-e_2^l}) ,$$

with a similar expression for $Z^{(l)}u$. For simplicity, we take the parameters q_i and z_i to be independent of the scale parameter l.

Since all of these operators are translation invariant, they are diagonalized by the discrete Fourier transform. The analysis of the PSMG algorithm then becomes particularly convenient. In the following we work entirely in Fourier transform space, where each of the operators $A^{(l)}$, $Q^{(l)}$ and $Z^{(l)}$ will reduce to multiplication by a trigonometric function. In terms of the two-dimensional discrete Fourier transform on $G^{(L)}$:

$$\hat{u}_k = n^{-1} \sum_{j_1, j_2=0}^{n-1} e^{i\, 2\pi j \cdot k/n} u_j , \quad 0 \le k_1, k_2 < n ,$$

and the notation $x_i^{(l)} = \cos(2\pi k_i d_l/n)$, $i = 1,2$, the operators $A^{(l)}$ and $Q^{(l)}$ reduce to multiplication by the trigonometric polynomials:

$$A_5^{(l)} = h_l^{-2} (4 - 2(x_1^{(l)} + x_2^{(l)})) ,$$

$$A_9^{(l)} = (6h_l^2)^{-1} (20 - 8(x_1^{(l)} + x_2^{(l)}) - 4x_1^{(l)} x_2^{(l)}) ,$$

$$Q^{(l)} = q_0 + 2q_1(x_1^{(l)} + x_2^{(l)}) + 4q_{11}x_1^{(l)}x_2^{(l)} \ ,$$

with a similar expression for $Z^{(l)}$ in terms of parameters z_0, z_1 and z_{11}. Application of the two-grid iteration operator $\mathbf{T}^{(l)}$ over the grid $G^{(L)}$ reduces to multiplication by the function:

$$\mathbf{T}^{(l)}{}_k = \mathbf{S}^{(l)}{}_k \, \mathbf{C}^{(l)}{}_k \ ,$$

where $\mathbf{S}^{(l)}{}_k \equiv 1 - Z^{(l)}{}_k A^{(l)}{}_k$ and $\mathbf{C}^{(l)}{}_k \equiv 1 - Q^{(l)}{}_k A^{(l-1)}{}_k{}^{-1} A^{(l)}{}_k$ are the Fourier representations of the smoothing and coarse-scale correction operators, respectively.

For both the Laplace and Mehrstellen case, $\mathbf{C}^{(l)}{}_k$ has apparent poles at the four points $|x_1^{(l)}| = |x_2^{(l)}| = 1$, which are the zeroes of the coarse grid difference operator. The pole at $x_1^{(l)} = x_2^{(l)} = 1$ is canceled by a corresponding zero in the numerator, but this is not so for the other three poles. We cancel the remaining zeroes in the denominator by carefully choosing the three parameters q_i. It is easily checked that the two conditions:

$$q_1 = q_0/2 \ , \quad q_{11} = q_0/4 \ ,$$

suffice, leaving one free parameter q_0 in the Q operator to be chosen later.

A further constraint is necessary if the multigrid iteration operator $\mathbf{M}^{(l)}$ is to have a bound independent of l, namely the constraint that the coarse grid correction operator $\mathbf{C}^{(l)}$ vanish at the origin in frequency space. This constraint leads to the condition q_0=.25 on the interpolation operator $Q^{(l)}$ which is therefore uniquely determined. The resulting form of $\mathbf{C}^{(l)}$ is then:

$$\mathbf{C}^{(l)}{}_k = 1 - .5q_0\,(1+x_1^{(l)})(1+x_2^{(l)})\,\frac{2-x_1^{(l)}-x_2^{(l)}}{2-x_1^{(l)2}-x_2^{(l)2}} \ ,$$

for the five-point operator, with a similar expression for the nine-point operator. Note that the same restrictions on Q are required for the five and nine-point operators.

To get an improved convergence rate we have also used a 25-point star operator for Q:

$$Q = \begin{bmatrix} q_{22} & q_{12} & q_2 & q_{12} & q_{22} \\ q_{12} & q_{11} & q_1 & q_{11} & q_{12} \\ q_2 & q_1 & q_0 & q_1 & q_2 \\ q_{12} & q_{11} & q_1 & q_{11} & q_{12} \\ q_{22} & q_{12} & q_2 & q_{12} & q_{22} \end{bmatrix} .$$

Again we compute an explicit rational function expression for the coarse-scale operator $\mathbf{C}_k^{(l)}$ as a function of the trigonometric variables $x_i^{(l)}$. As before, there are three poles of the denominator in $\mathbf{C}_k^{(l)}$ which we cancel by careful choice of Q, leading to the two constraints:

$$0 = q_0 - 4q_1 + 4q_{11} + 4q_2 + 4q_{22} - 8q_{12} ,$$

$$0 = q_0 - 4q_{11} + 4q_2 + 4q_{22} .$$

We must add the constraint

$$1 = q_0 + 4q_1 + 4q_{11} + 4q_2 + 4q_{22} + 8q_{12} ,$$

required to ensure that the coarse grid correction operator $\mathbf{C}^{(l)}$ vanishes at the origin in frequency space. It follows that there are 3 independent coefficients of Q, along with the 3 parameters of Z, that are available to minimize μ in the multigrid analysis.

In the translation invariant case all of the operators in the recurrence equations for the multigrid residual reduction operator $\mathbf{M}^{(l)}$ introduced in the previous section commute. Furthermore $\mathbf{M}^{(l)}$ is also then the error reduction operator. $\mathbf{M}_k^{(l)}$ is therefore determined recursively as a sum of products of low-degree rational functions in the $x_i^{(l)}$. Since $\mathbf{M}^{(l)}$ is a multiplication operator in the translation invariant case, its norm is the maximum value of $|\mathbf{M}_k^{(l)}|$ evaluated over all relevant frequencies k_i, or equivalently, evaluated over those discrete points in the square $-1 \le x_1^{(l)}, x_2^{(l)} \le 1$ that correspond to Fourier frequencies on the grid $G^{(L)}$.

Translation invariant problems will have an inherent singularity if constant functions are in the null space of the differential operator A. In Fourier space this results in bad behavior at the origin k=(0,0). The singularity must be dealt with. One solution is to omit constant functions and zero frequency from the domain of interest. Alternatively one may include these functions by using the Moore-Penrose pseudo-inverse[19, 20] to extend operators from the space of non-constant data. This latter choice is equivalent to defining $\mathbf{M}^{(l)}(0,0) = 1$ in the basic recurrence relation.

6. A FUNCTIONAL EQUATION FOR M

A remarkable feature of PSMG for translation invariant PDE without lower order terms (such as the Poisson equation) is that we can phrase the defining equation for the iteration in a form completely independent of grids or grid levels. The resulting equation is a simple functional equation on the space of trigonometric polynomials. The key point in deriving this relationship is to make a transition from unbounded frequency variables (k) to uniformly bounded angle (θ) or cosine ($x = \cos(\theta)$) variables.

In previous sections, all operators have been explicitly shown as dependent on the grid $G^{(L)}$ and the level l, as well as on frequency k_i. We begin with a lemma which shows that the kernels of these operators are actually independent of L and need not be computed at even frequencies k.

Lemma 1: Each of the operators $O = \mathbf{S}^{(l)}, \mathbf{T}^{(l)}, \mathbf{M}^{(l)}$ defined on $G^{(L)}$ has a kernel $O_k^{(l)}$ independent of L, and satisfying $O_{2k}^{(l)} = O_k^{(l-1)}$.

Proof: From section 5 we note that the only explicit l or L dependence in the operators **S**, **T** is in the trigonometric functions $x_k^{(l)} = \cos(2\pi k/2^l)$. Factors of h_l involved in the operators A and Z always cancel out provided there are no first order or constant terms in the difference operator. Thus the operator kernels depend explicitly on l but not on L. Furthermore we note that $x_{2k}^{(l)} = \cos(2\pi\, 2k/2^l) = x_k^{(l-1)}$, and consequently we have the result of the lemma for **S** and **T**.

We prove the result for **M** by induction on l. For l=1 we apply the recurrence along with $\mathbf{M}^{(0)} \equiv 1$, obtaining $\mathbf{M}^{(1)} = \mathbf{S}^{(1)}$, and so the lemma applies for l=1. Now assume the lemma applies to $\mathbf{M}^{(l-1)}$. Then

$$\begin{aligned}
\mathbf{M}_{2k}^{(l)} &= \mathbf{T}_{2k}^{(l)} + (\mathbf{S}_{2k}^{(l)} - \mathbf{T}_{2k}^{(l)})\mathbf{M}_{2k}^{(l-1)} \\
&= \mathbf{T}_k^{(l-1)} + (\mathbf{S}_k^{(l-1)} - \mathbf{T}_k^{(l-1)})\mathbf{M}_k^{(l-2)} \\
&= \mathbf{M}_k^{(l-1)} .
\end{aligned}$$

This completes the induction proof.

As a consequence of the lemma, we may write the recurrence as a single-level functional equation:

$$\mathbf{M}_k^{(l)} = \mathbf{T}_k^{(l)} + (\mathbf{S}_k^{(l)} - \mathbf{T}_k^{(l)})\mathbf{M}_{2k}^{(l)} \ .$$

We now regard $\mathbf{M}^{(l)}$ as a function of the normalized variables $x_i = \cos(2\pi k_i/n)$ rather than of k_i. These variables span the range $[-1,1]$ as k_i and n vary. Alternatively we may introduce the variables $\theta_i \equiv 2\pi k_i/n$ and regard $\mathbf{M}$ as a function $\mathbf{M}(\theta)$ on $[0,2\pi]\times[0,2\pi]$. We denote by D_θ the set of values (θ_1,θ_2) where $0 \le k_i < n$, for any n of the form 2^l, and where the origin, $k_1=k_2=0$, is omitted. We denote the corresponding set of values of x by D_x. Thus all of the values of $\mathbf{M}_k^{(l)}$, for all grids $G^{(L)}$, are values of a single function $\mathbf{M}(\theta)$, $\theta=(\theta_1,\theta_2)$ in D_θ, or equivalently of a single function $\mathbf{M}(x)$, $x=(x_1,x_2)$ in D_x. Similarly the smoothing and two grid kernels extend into functions $\mathbf{S}(\theta)$ and $\mathbf{T}(\theta)$ defined over D_θ, or $\mathbf{S}(x)$ and $\mathbf{T}(x)$ defined over D_x, and the multiscale recursion relation reduces to the form:

$$\mathbf{M}(\theta) = \mathbf{T}(\theta) + (\mathbf{S}(\theta)-\mathbf{T}(\theta))\,\mathbf{M}(2\theta) \, , \quad \mathbf{M}(0) = 1 \, ,$$

or equivalently

$$\mathbf{M}(x) = \mathbf{T}(x) + (\mathbf{S}(x)-\mathbf{T}(x))\,\mathbf{M}(2x^2-1) \, , \quad \mathbf{M}(1) = 1 \, ,$$

where we have used the fact that $\cos(2\theta) = 2\cos^2(\theta) - 1$. We will use these functional equations in the following section. The choice of $\mathbf{M}(1) = 1$ was explained at the end of the previous section, and effectively defines the Moore-Penrose pseudo-inverse on constant functions.

7. CONVERGENCE RATE BOUNDS: STATEMENT AND PROOFS

In this section we will derive rigorous upper bounds on the multigrid convergence rate of the PSMG algorithm in the translation invariant case described in the previous section.

Theorem 1: Define $\mu_1 \equiv \sup_x |\mathbf{T}(x)| \, / \, (1 - |\mathbf{S}(x) - \mathbf{T}(x)|)$. Assume that the smoother satisfies $|\mathbf{S}| \le \mu_1$ at the three high frequency points $(-1,1)$, $(1,-1)$ and $(-1,-1)$. Then the multigrid convergence rate μ satisfies the bound $\mu \le \mu_1$.

Proof: Let P_3 denote the set $\{(1,-1), (-1,1), (-1,-1)\}$. We will use a mapping $g: R^2 \to R^2$ defined by $g(x_1,x_2) = (2x_1^2-1, 2x_2^2-1)$. We decompose D_x into a union of disjoint subsets D_n, $n \geq 0$, where we define recursively:

$$D_n = \{x \in D_x : x \textit{ not } \in D_{n-1} \text{ and } g(x) \in D_{n-1}\} \ , \quad D_0 \equiv P_3 \, .$$

D_n consists of the points of D_x which correspond to angles in D_θ of the form $2\pi k/2^n$ with k odd. We will prove Theorem 1 holds for the sets D_n by induction on n.

The induction hypothesis holds for $D_0 \equiv P_3$. To see this we note that for each point $p \in P_3$, $g(p) = (1,1)$ and since $M(1,1) = 1$, the recursion relation reduces at these points to

$$\mathbf{M}(p) = \mathbf{T}(p) + (\mathbf{S}(p) - \mathbf{T}(p)) \cdot 1 = \mathbf{S}(p) \ , \quad p \in P_3 \, .$$

Thus by the assumption of the theorem that $|\mathbf{S}(p)| \leq \mu_1$ for $p \in P_3$, we conclude that $|\mathbf{M}(p)| \leq \mu_1$, and thus that the induction hypothesis holds for D_0.

Now assume the result is true for D_{n-1} and let $x \in D_n$. Since $g(x) \in D_{n-1}$, it follows that:

$$\begin{aligned} |\mathbf{M}(x)| &\leq |\mathbf{T}(x)| + |\mathbf{S}(x) - \mathbf{T}(x)| \, |\mathbf{M}(2x^2-1)| \\ &\leq (1 - |\mathbf{S}(x) - \mathbf{T}(x)|)\mu_1 + |\mathbf{S}(x) - \mathbf{T}(x)| \mu_1 \\ &= \mu_1 \, . \end{aligned}$$

which completes the proof of the induction hypothesis for D_n.

Remarks: The assumption on **S** in the statement of Theorem 1 is a requirement that **S** be small at high frequencies, which is essential anyway if **S** is to be a good smoother. Thus it is not a major restriction. However the assumption can be avoided by terminating the PSMG iteration at the level 1 grid rather than at level 0. In that case an exact solution is performed on level 1, so that **M** is 0 on the set P_3. We also remark that the theorem does not require that the differential operator be homogeneous, as required to derive the grid independent recursion relation. The theorem is easily extended by the same proof to the case where $\mathbf{M}(x)$ is dependent on the level l. The theorem does however require translation invariance.

The bound described above was used in[15] to design good choices for Q and Z. In fact the bound μ_1 requires only that we find the suprenum of a rational function on the unit square. One can estimate μ_1 by evaluation of the suprenum over a large grid of values in the unit square. The resulting quantity is then used as an objective function and minimized with respect to the parameters defining Q and Z.

Sharper convergence criteria result if one first applies the iteration formula for $\mathbf{M}(x)$ several times. It is notationally convenient to write $\mathbf{M}$ as a function of the angles $0 \le \theta_i \equiv 2\pi k_i / n < 2\pi$. We also define $\mathbf{R}(\theta) \equiv \mathbf{S}(\theta) - \mathbf{T}(\theta)$. The recurrence relation then reduces to:

$$\mathbf{M}(\theta) = \mathbf{T}(\theta) + \mathbf{R}(\theta)\mathbf{M}(2\theta), \quad \mathbf{M}(\mathbf{0}) = 1 .$$

Iterating $\nu - 1$ times we arrive at:

$$\begin{aligned}\mathbf{M}(\theta) &= \mathbf{T}(\theta) + \mathbf{R}(\theta)\mathbf{T}(2\theta) + \mathbf{R}(\theta)\mathbf{R}(2\theta)\mathbf{T}(4\theta) + \cdots \\ &\quad + \mathbf{R}(\theta)\mathbf{R}(2\theta)\cdots\mathbf{R}(2^{\nu-1}\theta)\mathbf{T}(2^{\nu}\theta) + \mathbf{R}(\theta)\mathbf{R}(2\theta)\cdots\mathbf{R}(2^{\nu}\theta)\mathbf{M}(2^{\nu+1}\theta) \\ &\equiv \mathbf{T}_{\nu}(\theta) + \mathbf{R}_{\nu}(\theta)\mathbf{M}(2^{\nu+1}\theta) .\end{aligned}$$

This again has a similar form to the original recurrence equation, and we prove exactly as in the previous theorem:

Theorem 2: Define $\mu_\nu \equiv \sup_\theta |\mathbf{T}_\nu(\theta)| / (1 - |\mathbf{R}_\nu(\theta)|)$. Assume that the smoother satisfies $|\mathbf{S}| \le \mu_\nu$ at the three high frequency points (–1,1), (1,–1) and (–1,–1). Then the convergence rate satisfies the bound $\mu \le \mu_\nu$.

It is important to take care in evaluating the bounds μ_ν. Even if $\mathbf{R}_\nu(\theta)$ approaches 1 at some points, the bound may still be useful provided the numerator $\mathbf{T}_\nu(\theta)$ also vanishes at the same points. Indeed this situation occurs at the origin even for μ_1. In evaluating $|\mathbf{T}_\nu|$ it is also important to take the absolute value only after the various terms have been added together.

These convergence results may be extended to provide various bounds on derivatives of $\mathbf{M}(\theta)$, and in particular to show that with appropriate conditions imposed, μ may be approximated arbitrarily closely by taking the suprenum of $|\mathbf{M}(\theta)|$ on a suitably fine mesh of points θ. For further details we refer to[21].

For illustration we provide here results obtained on grids with edges of size $n(L)$ from 4 through 1024 using each of the bounds $\mu_1, .., \mu_5$. The final column is the exact norm as computed using the full iteration formula. The results are for the case of a Mehrstellen operator with 9-point Q and Z stencils - the coefficients of Q and Z are given in the last section of the paper under PSMG9-9.

TABLE 1: CONVERGENCE BOUNDS μ_v

n(L)	μ_1	μ_2	μ_3	μ_4	μ_5	μ
4	.0009	.0049	.0049	.0049	.0049	.0049
8	.0248	.0174	.0174	.0174	.0174	.0174
16	.0264	.0216	.0216	.0216	.0216	.0217
32	.0291	.0230	.0216	.0216	.0216	.0217
64	.0291	.0241	.0216	.0216	.0216	.0217
128	.0292	.0241	.0225	.0216	.0216	.0217
256	.0292	.0243	.0224	.0220	.0217	.0217
512	.0292	.0243	.0226	.0220	.0219	.0217
1024	.0292	.0243	.0226	.0221	.0219	.0217

8. COMPUTATION OF CONVERGENCE RATES

In our paper[15] we have used the first theorem of the previous section to provide upper bounds on PSMG convergence rates and to guide thereby, optimization processes. Basing an optimization process for a rate on optimization of a possibly crude upper bound for that rate is certainly not ideal. More recently[18], we have improved our knowledge of PSMG convergence rates by *exactly* computing the spectral radius $\mu^{(L)}$ of the self-adjoint PSMG multigrid error reduction operator $\mathbf{M}^{(L)}$ for all grids $G^{(L)}$ with L ranging from 0 to 11. Here $G^{(L)}$ is a square grid with $n \equiv 2^L$ points on a side. The verification is based on the iterative formula:

$$\mathbf{M}^{(l)} = \mathbf{T}^{(l)} + (\mathbf{S}^{(l)} - \mathbf{T}^{(l)})\,\mathbf{M}^{(l-1)}, \quad 1 \le l \le L,$$

for the multigrid operator $\mathbf{M}^{(l)}$. The correct domain for the translation invariant Poisson equation is the set of grid functions orthogonal to constants. In Fourier space this means that points aliased to (0,0) are omitted from the computation. Equivalently one defines $\mathbf{M}^{(0)} \equiv 1$ in (1).

We compute $\mu^{(L)}$ by evaluating the recurrence above from l=0 to l=L in Fourier space for every frequency pair k_1, k_2 appropriate to $G^{(L)}$ - i.e. for $0 \le k_i < n$. The only approximation in this procedure is that the kernels are evaluated in double precision rather than infinite precision arithmetic.

At first sight it would appear that the evaluation of $\mu^{(L)}$ would require $O(n^2 \log L)$ operations - the number of grid points for level L times the cost $O(\log L)$ of an evaluation of the recurrence. Remarkably, in Fourier space only $O(n^2)$ operations are required. The idea is to use the previously derived fact that $\mathbf{M}_{2k}^{(l)} = \mathbf{M}_k^{(l-1)}$. The value of $\mathbf{M}_k^{(l)}$ is independent of which grid level l it is evaluated at (assuming frequency k exists at that level). Therefore we develop an inverse process where we first evaluate $\mathbf{M}$ at all coarse grid points, then evaluate $\mathbf{M}$ on the next finer grid but only at those points not in the coarse grid. Furthermore one can use symmetry in k_1, k_2 and between $k_i \le n/2$ and $k_i \ge n/2$ to reduce the work by a further factor of 8. We store all evaluated $\mathbf{M}_k$ on a given level, so that the recursion on the following level can be terminated after one step. It is this last fact that reduces the overall cost by $\log L$. Given that the result of a norm evaluation is a single data point in the optimization of the choice of Q and Z, the ability to have such a fast evaluation of the norm of $\mathbf{M}$ is the key to obtaining highly efficient PSMG processes in both 2D and 3D.

9. PARALLEL OPERATION COUNTS

In this section we discuss methods for estimating the work performed in a PSMG iteration. We use a model of parallel computation introduced in[22]. Thus we assume only nearest neighbor communication with four neighbors, purely SIMD communication and computation, and we consider separately each grid level. The operation counts will be those for intermediate grids and the communication unit is grid level dependent - the cost for communication between "nearest neighbors" at that grid level.

The PSMG algorithm was described as 6 separate steps in section 3 above to which we now refer. For intermediate grids ($0<l<L$) the initial guess is taken to be 0 so that step 1 is not needed. Similarly step 6 is relevant only to the top level grid L. Step 2 is free because PSMG uses injection, while step 3 is counted at level $l-1$ or lower. Thus only steps 4-5 are to be counted at level l. Apart from the 3 operators involved, we note that no communication is required in these steps while two computations are required in step 5. Therefore we conclude that at intermediate grid levels ($0<l<L$), the parallel communication and computation costs for the PSMG algorithm are given by:

$$comm(PSMG(l)) = comm(Q) + comm(A) + comm(Z) \quad ,$$

$$comp(PSMG(l)) = comp(Q) + comp(A) + comp(Z) + 2 \quad ,$$

where $comm(O)$ and $comp(O)$ are the number of parallel communication and computation operations required to execute operator O.

In Table 2 below we present the operation counts for each of the individual operators encountered in the four PSMG algorithms corresponding to choice of 5 or 9 point A and to 9 or 25 point Q stencils (often referred to as the PSMG5-9, PSMG9-9, PSMG5-25 and PSMG9-25 methods). For full details of the derivation of these operation counts we refer to[18].

TABLE 2: OPERATION COUNTS FOR A, Q, Z

Operator	Comp. Steps	Comm. Steps
5-pt A (Laplacian)	3	4
9-pt A (Laplacian)	5	4
9-pt Q (Interpolation)	4	4
25-pt Q (Interpolation)	12	8
9-pt Z (Relaxation)	5	4

10. PSMG PERFORMANCE

An effective measure of algorithm performance depends on several factors besides convergence rate. For parallel environments, measures of parallel work and of communication need to be incorporated. Furthermore one needs to be precise about the underlying model machine specification. As a model of parallel computation we have followed[22], as discussed in the previous section.

If the asymptotic convergence rate of a method is ρ and the method requires w parallel operations per iteration, the normalized operation count is defined as $w/\log_{10}\rho$, and measures the parallel work required per grid level to reduce the error by a factor of 10. We consider parallel communication and computation operations separately.

For several PSMG methods we present asymptotic convergence rates, the number of parallel arithmetic and communication operations required on each grid per iteration, and also the normalized operation count for arithmetic and communication. We summarize the results for several simple cases in Table 3. For each cycle we use a single relaxation with a 9-point Z stencil. The cases are labeled PSMG$a-q$ where a and q are the stencil size (5 or 9) of the difference operator a and of the q matrix (9 or 25) respectively.

TABLE 3: PSMG CONVERGENCE RATES

Method	Convergence Rate	Steps per Level Comp.	Steps per Level Comm.	Normalized Steps Comp.	Normalized Steps Comm.
PSMG 5-9	.08867	14	12	13.31	11.40
PSMG 5-25	.02504	22	16	13.74	9.99
PSMG 9-9	.02165	16	12	9.61	7.21
PSMG 9-25	.00165	24	16	8.62	5.75

The corresponding coefficients for the interpolation operator Q and the smoothing operator Z are (in the notation of[18]):

PSMG5-9:	q_0=.25	q_1=.125	q_{11}=.0625
	z_0=.278079	z_1=.0534577	z_{11}=.0125615
PSMG5-25:	q_0=.361017	q_1=.11458	q_{11}=.0625
	q_2=-.0309162	q_{12}=.00521024	q_{22}=.00316188
	z_0=.361452	z_1=.0891718	z_{11}=.0293793
PSMG9-9:	q_0=.25	q_1=.125	q_{11}=.0625
	z_0=.300589	z_1=.0432465	z_{11}=.0139994
PSMG9-25:	q_0=.34152	q_1=.0995677	q_{11}=.0625
	q_2=-.0199225	q_{12}=.0127161	q_{22}=-.00295755
	z_0=.283286	z_1=.0323815	z_{11}=.00835795

The convergence rates given in Table 3 are the maximum values of $\mu^{(L)}$ for $0 \le L \le 11$ and therefore bound the exact convergence rates for all grids up to size 4 million points. In practice we find that the convergence rates $\mu^{(L)}$ are unchanged to several digits of precision beyond about level L=6. We therefore are confident, that the convergence rates in Table 3 extend to arbitrary numbers of grid levels. As a final check we have solved the Poisson equation using PSMG, with zero right hand side and a random initial guess, and in each case verified the convergence rates of Table 3.

As a comparison point we note that the PSMG9-25 algorithm uses about half as much computation and close to one fifth as much communication as the best standard red-black algorithms[18, 22], assuming of course that there are about as many processors as fine grid points.

References

1. A. Brandt, "Multi-level adaptive solutions to boundary-value problems," *Math. Comp.*, vol. 31, pp. 333-390, 1977.
2. W. Hackbusch, "Convergence of multi-grid iterations applied to difference equations," *Math. Comp.*, vol. 34, pp. 425-440, 1980.
3. K. Stüben and U. Trottenberg, "Multigrid Methods: Fundamental algorithms, model problem analysis and applications," in *Multigrid Methods*, ed. U. Trottenberg, Lecture Notes in Mathematics 960., pp. 1-176, Springer-Verlag, Berlin, 1981,.
4. A. Brandt, "Multi-Grid Solvers on Parallel Computers," ICASE Technical Report 80-23, NASA Langley Research Center, Hampton Va., 1980.
5. D. Gannon and J. van Rosendale, "Highly Parallel Multi-Grid Solvers for Elliptic PDEs: An Experimental Analysis," ICASE Technical Report 82-36, NASA Langley Research Center, Hampton Va., 1982.
6. O. McBryan and E. Van de Velde, "Parallel Algorithms for Elliptic Equation Solution on the HEP Computer," *Proceedings of the Conference on Parallel Processing using the Heterogeneous Element Processor, March 1985*, University of Oklahoma, March 1985.
7. O. McBryan and E. Van de Velde, "Parallel Algorithms for Elliptic Equations," *Commun. Pure and Appl. Math.*, vol. 38, pp. 769-795, 1985.
8. C. Thole, "Experiments with Multigrid Methods on the Caltech Hypercube," GMD-Studien Nr. 103, Gesellschaft fur Mathematik und Datenvararbeitung, St. Augustin, West Germany, 1986.
9. O. McBryan and E. Van de Velde, "The Multigrid Method on Parallel Processors," in *Multigrid Methods II*, ed. W. Hackbusch and U. Trottenberg, Lecture Notes in Mathematics, vol. 1228, Springer-Verlag, Berlin, Heidelberg, 1986.
10. O. McBryan and E. Van de Velde, *Hypercube Algorithms and Implementations*, SIAM J. Sci. Stat. Comput., 8, pp. 227-287, 1987.
11. T. F. Chan, Y. Saad, and M. H. Schultz, "Solving elliptic partial differential equations on hypercubes," *Hypercube Multiprocessors 1986*, SIAM, Philadelphia, 1986.
12. P. O. Frederickson and M. W. Benson, "Fast parallel algorithms for the Moore-Penrose pseudo-inverse," *Hypercube Multiprocessors 1986*, SIAM, Philadelphia, 1986.
13. O. McBryan, "The Connection Machine: PDE Solution on 65536 Processors," *Parallel Computing*, vol. 9, pp. 1-24, North-Holland, 1988.
14. O. McBryan, P. Frederickson, J. Linden, A. Schuller, K. Solchenbach, K. Stüben, C.-A. Thole, and U. Trottenberg, "Multigrid Methods on Parallel Computers - a Survey on Recent Developments," *University of Colorado Tech. Report CU-CS-504-90, and*

IMPACT of Supercomputing, 1991, to appear.

15. P. O. Frederickson and O. McBryan, "Parallel Superconvergent Multigrid," in *Multigrid Methods: Theory, Applications and Supercomputing*, ed. S. McCormick, Math Applications Series, vol. 110, pp. 195-210, Marcel-Dekker Inc., New York, 1988.
16. W. Hackbusch, "A New Approach to Robust Multigrid Methods," Proceedings of the 1987 ICIAM Conference, 1987.
17. P. O. Frederickson and O. McBryan, "Intrinsically Parallel Multiscale Algorithms for Hypercubes," in *Proceedings of 3rd Conference on Hypercube Concurrent Computers and Applications, Vol II.*, ed. G. Fox, pp. 1726-1734, ACM Press, 1988.
18. P. Frederickson and O. McBryan, "Normalized Convergence Rates for the PSMG Method," *and SISSC*, vol. 12, no. 1, pp. 221-229, 1991.
19. E. H. Moore, "On the reciprocal of the general algebraic matrix," *Bull. Amer. Math. Soc.*, vol. 26, pp. 394-395, 1919.
20. R. Penrose, "A Generalized Inverse for Matrices," *Proc. Cambridge Philos. Soc.*, vol. 51, pp. 406-413, 1955.
21. P. Frederickson and O. McBryan, "Convergence Bounds for the PSMG Method," University of Colorado CS Dept Technical Report, 1991, to appear.
22. N. Decker, "A Note on the Parallel Efficiency of the Frederickson-McBryan Multigrid Algorithm," *SISSC*, vol. 12, no. 1, pp. 208-220, 1991.

An adaptive multigrid approach for the solution of the 2D semiconductor equations

P.W. Hemker and J. Molenaar

Abstract

An adaptive multigrid method is presented for the solution of the two-dimensional steady state Van Roosbroeck equations for semiconductor device modeling. The discretisation is based on the (hybrid) mixed finite element method on rectangles. The integrals involved are approximated by the trapezoidal rule. In this way, in the interior of the domain, the classical Scharfetter Gummel discretisation is retained. A 5-point collective Vanka-type relaxation procedure is used as a smoother.

The mixed finite elements give rise to a cell-centered multigrid method and the multigrid grid-transfer operators are chosen in agreement with the discretisation. The main difficulties are the proper use of very coarse grids and the construction of suitable initial estimates. In order to admit very coarse grids, it appears necessary to take special measures and to introduce local damping of the residual in the coarse grid correction.

It is shown that, under these conditions, a fast convergence can be obtained that seems to be independent of the grid size. Hence, in combination with nested iteration, an efficient procedure is obtained. Results are shown for a realistic two-dimensional transistor model.

1 Introduction

The usual mathematical model to describe the electric behaviour of semiconductor devices is the drift diffusion model, that was essentially proposed by Van Roosbroeck [18] in 1950. It is given by a nonlinear system of three second order partial differential equations. Let $\Omega \subset I\!R^n$, n=1,2,3, be an open bounded region with a piecewise smooth boundary $\partial\Omega$, then, scaled to dimensionless form, the equations are [11] [15]

$$\begin{aligned} &\text{div}\,(\lambda^2 \text{grad}\,\psi) = n - p - D, \\ &\tfrac{\partial n}{\partial t} = \text{div}\,(\mu_n(\text{grad}\,n - n\,\text{grad}\,\psi - n\,\text{grad}\,\log n_i)) - R, \\ &\tfrac{\partial p}{\partial t} = \text{div}\,(\mu_p(\text{grad}\,p + p\,\text{grad}\,\psi - p\,\text{grad}\,\log n_i)) - R. \end{aligned} \tag{1}$$

Here the dependent variables n and p, denote the local density of free electrons and holes in the device respectively, and ψ is the electrostatic potential. These variables are functions of $x \in \Omega$ and $t \geq 0$; λ^2 is associated with the dielectric constant; D(x), the net doping function, describes the location of the impurities that characterises the device. Also n_i, the effective intrinsic carrier density, is a function of x; μ_n and μ_p the carrier mobilities are functions of x, n, p, grad ψ, and the net recombination rate R, that models the recombination and generation of electrons and holes, is a function of n and p. The first equation in (1) is the

Poisson equation for the electric field, whereas the second and the third are the continuity equations for the electrons and holes.

The usual approach for the numerical solution of (1) is the application of a box method (finite volume method) for the discretisation, where the Scharfetter-Gummel exponential fitting scheme is used for the approximation of the fluxes between the control volumes. Damped Newton methods are generally used for the solution of the discrete nonlinear systems. For details about the equations and the techniques for their numerical solution, we refer to [11] and [15]. In this paper we investigate a nonlinear multigrid technique for the solution of eq.(1) in order to see whether it could be advantageously applied.

In order to reduce the large number of technical difficulties involved, we restrict ourselves to the computation of the steady state, and we assume the intrinsic carrier density n_i and the mobilities μ_n and μ_p to be constant. With these simplifications, still many of the essential difficulties remain. In the first place the equations are singularly perturbed because the parameter λ, that can be related with the Debye length of the device, is generally small compared with the size of the device. Moreover, a strong convective behaviour of the equations is caused by the possibly large coefficient grad ψ in the continuity equations.

In the past, several attempts have already been made to explore the possibilities of multigrid techniques for the solution of the equations (1). However, up to now the question of whether the multigrid technique is feasible for practical application for these equations is still open. It appears that the use of multigrid is not straightforward at all and that a number of difficulties are encountered with its application. In this paper we want to show some progress made towards an applicable MG method for semiconductor device modeling.

Adaptive grids and nested meshes

In this paper we apply the nonlinear multigrid approach to the solution of the discretised equation (1). The difficulties lie in the bad scaling, the strong nonlinearity and the singular perturbation character. Because of the singular perturbation behaviour, sharp shifts will appear in the solution, whereas in other regions the variables vary only gradually. This makes it unfeasible to represent the solution on a regular mesh. A priori some physical insight may be available about the location of the various regions, but the true solution is only known after numerical approximation. Therefore automatic adaptive local mesh refinements are introduced. This is done in a more or less straightforward sense, as basicly treated e.g in Brandt [2] or McCormick [12]. One difference is that the method is now applied in its cell-centered form. The solution is represented by its values at cell centers and cells are divided into smaller cells to obtain finer meshes. This approach is advantageous in the case conservation laws play a role and it has direct consequences for the transfer operators used.

The adaptive algorithm is flexible in the sense that it allows a completely arbitrary refinement of already existing cells. This is done by allowing any rectangular cell to be divided into four smaller rectangles of the same shape. The smaller cells are part of the next level of refinement in the sequence of discretisations used for the multigrid method. In this way all cells belong to a quad-tree structure and each cell has at most four neighbours on the same refinement level. The same quad-tree structure is used in the program to store the data. The domain of definition for the equations needs to be covered only by the very coarsest grid. Finer grids may cover the domain only partially. This freedom allows another (independent) algorithm to take full responsibility of the grid refinement procedure.

Multigrid

In the first place multigrid can be used for nested iterations, i.e. to obtain initial estimates for the solution on a coarser grid than the one that is required to give an accurate representation of the solution. The usual way of solving the equation for a set of boundary conditions, is by starting at a zero bias and incrementing the boundary values in small steps until the desired conditions are reached. Such a continuation process requires a number of intermediate calculations that are best made on a grid that is as coarse as possible. As soon as the problem has been approximately solved on the coarse mesh, the mesh can be refined to yield higher accuracy.

The iteration process to solve the problem on each level might be an approximate Newton method where multigrid is used to solve the linear systems. For this approach see e.g. [1]. A drawback of global linearisation is the time-consuming evaluation of Jacobian entries and the large memory requirements to store them. This can be avoided by the use of nonlinear multigrid where linearisations are made only locally. We are interested in such nonlinear multigrid techniques for the solution of the large nonlinear systems.

Because of the strong local variations in the solution, one difficulty to deal with is to know what information, available from a very coarse grid solution, may still be useful for the acceleration of the convergence on the fine grid. Such problems were also known from CFD problems with shocks, where -for the Euler equations- these difficulties could be solved by strict adherence to the discrete conservation laws. For the semiconductor equations this problem appears to be much harder because source terms, that are related to approximate truncation errors, may be so large that -without special measures- no longer a (positive) solution for the coarse grid problem can be guaranteed.

2 The equations

After the simplifications, mentioned in Section 1, the equations to be solved are

$$\begin{array}{lll} \text{div } J_\psi - n + p + D = 0, & J_\psi = \lambda^2 \text{grad } \psi, & \\ \text{div } J_n - R = 0, & J_n = +\mu_n(\text{grad } n - n\text{grad } \psi), & \\ \text{div } J_p + R = 0, & J_p = -\mu_p(\text{grad } p + p\text{grad } \psi). & \end{array} \tag{2}$$

In shorthand we write these equations also as $N(\psi, n, p) = 0$. For the recombination rate we assume the Shockley-Read-Hall model

$$R = \frac{np - 1}{\tau_p(n+1) + \tau_n(p+1)}. \tag{3}$$

In order to bring the variables to quantities of the same dimension, it is useful to introduce the quasi-Fermi potentials ϕ_n and ϕ_p by

$$\begin{aligned} n &= \exp(\psi - \phi_n), \\ p &= \exp(\phi_p - \psi). \end{aligned} \tag{4}$$

As a starting point for the discretisation, we use the Slotboom variables

$$\begin{aligned} \Phi_n &= \exp(-\phi_n), \\ \Phi_p &= \exp(+\phi_p), \end{aligned} \tag{5}$$

for which the equations appear in symmetric positive definite form:

$$\begin{aligned}
&-\text{div } (\mu_\psi \text{grad } \psi) + Q = D, \\
&-\text{div } (\mu_n \exp(+\psi) \text{grad } \Phi_n) + R = 0, \\
&-\text{div } (\mu_p \exp(-\psi) \text{grad } \Phi_p) + R = 0,
\end{aligned} \tag{6}$$

where we use the notation $\mu_\psi = \lambda^2$, $Q = e^\psi \Phi_n - e^{-\psi}\Phi_p$. The boundary conditions are of Dirichlet type (ψ, ϕ_n, ϕ_p given) at the Ohmic contacts, and homogeneous Neumann conditions ($J_\psi = J_n = J_p = 0$) at the remaining parts of the boundary.

3 The discretisation by Mixed Finite Elements

Each of the equations (6) can be cast in the form

$$\begin{array}{lcll}
\sigma & = & a \text{ grad } u & \left.\right\} \text{ on } \Omega, \\
\text{div } \sigma & = & f(u) & \\
u & = & g & \text{on } \Gamma_D, \\
\mathbf{n} \cdot \sigma & = & 0 & \text{on } \Gamma_N,
\end{array} \tag{7}$$

where Γ_D and Γ_N denote the parts of the boundary with Dirichlet or homogeneous Neumann boundary conditions, respectively. The sign is chosen such that $a(x) > 0$. As a starting point for the discretisation we use its variational form: find $\sigma \in \mathbf{H}^{BC}(\text{div }, \Omega)$ and $u \in L^2(\Omega)$ such that

$$\left\{ \begin{array}{lcll}
\int_\Omega a^{-1} \sigma \cdot \mathbf{v} \, d\Omega + \int_\Omega u \text{ div } \mathbf{v} \, d\Omega & = & \oint_{\Gamma_D} g \, \mathbf{v} \cdot \mathbf{n} \, ds \, , & \forall \mathbf{v} \in \mathbf{H}^{BC}(\text{div }, \Omega), \\
\int_\Omega \phi \text{ div } \sigma \, d\Omega & = & \int_\Omega \phi f(u) \, d\Omega, & \forall \phi \in L^2(\Omega),
\end{array} \right. \tag{8}$$

where $\mathbf{H}^{BC}(\text{div }, \Omega) = \{\mathbf{v} \in H(\text{div }, \Omega) \mid \mathbf{v} \cdot \mathbf{n} = 0 \text{ on } \Gamma_N\}$.

For the discretisation we assume that Ω can be divided by a regular partitioning in open disjoint rectangular cells Ω_i, $\overline{\Omega} = \cup \overline{\Omega}_i$. We denote by E_j the edges of the rectangles, by ϵ_i the characteristic function on Ω_i, and we use the notation

$$d_{ij} = \left\{ \begin{array}{ll}
+1 & \text{if } E_j \text{ is a N- or E-edge of } \Omega_i, \\
-1 & \text{if } E_j \text{ is a S- or W-edge of } \Omega_i, \\
0 & \text{if } E_j \text{ is not an edge of } \Omega_i.
\end{array} \right. \tag{9}$$

By $\mathbf{e}_j \in \mathbf{H}(\text{div }, \Omega)$ we denote the *tent function* for E_j, i.e. a vector function $\mathbf{e}_j$ of which each component is linear on each Ω_i and which satisfies $\mathbf{e}_j \cdot \mathbf{n}_k = \delta_{jk}$, where $\mathbf{n}_k$ is the unit normal on edge E_k (in the positive x- or y- direction); δ_{jk} is the Kronecker delta. We introduce $\overline{\epsilon}_\ell$, a function defined on all edges E_j, by

$$\overline{\epsilon}_\ell(x) = \left\{ \begin{array}{ll}
1 & \text{if } x \in E_\ell, \\
0 & \text{if } x \notin E_\ell,
\end{array} \right. \tag{10}$$

and the *half tent function* $\mathbf{e}_{ij}$ defined by $\mathbf{e}_{ij} = \mathbf{e}_j \epsilon_i$.

We define the discrete spaces

$$\begin{array}{ll}
L_h(\Omega) & = \text{Span}(\epsilon_i), \\
M_h & = \text{Span}(\overline{\epsilon}_\ell), \\
\mathbf{H}_h^{BC}(\text{div }, \Omega) & = \text{Span}(\mathbf{e}_i) \cap \mathbf{H}^{BC}(\text{div }, \Omega), \\
W_h(\Omega) & = \text{Span}(\mathbf{e}_{ij}).
\end{array} \tag{11}$$

The mixed finite element (MFE) discretisation of equation (6) reads:
find $\boldsymbol{\sigma}_h \in \mathbf{H}_h^{BC}(\text{div},\Omega)$ and $u_h \in L_h^2(\Omega)$ such that

$$\left\{\begin{array}{lll} \int_\Omega a^{-1}\boldsymbol{\sigma}_h\cdot\mathbf{v}_h\,d\Omega + \int_\Omega u_h \text{div}\,\mathbf{v}_h\,d\Omega & = \oint_{\Gamma_D} g\,\mathbf{v}_h\cdot\mathbf{n}, & \forall \mathbf{v}_h \in \mathbf{H}_h^{BC}(\text{div},\Omega), \\ \int_\Omega \phi_h \,\text{div}\,\boldsymbol{\sigma}_h\,d\Omega & = \int_\Omega \phi_h f(u)\,d\Omega, & \forall \phi \in L_h^2(\Omega). \end{array}\right. \tag{12}$$

We notice that f may be a nonlinear function of u, so linearisation yields a discrete linear system of the form

$$\begin{pmatrix} A & B \\ B^T & C \end{pmatrix}\begin{pmatrix} \sigma_j \\ u_i \end{pmatrix} = \begin{pmatrix} b \\ f \end{pmatrix}, \tag{13}$$

where

$$\begin{array}{lll} a_{k,j} & = \int_\Omega a^{-1}\mathbf{e}_k\cdot\mathbf{e}_j\,d\Omega, & \\ b_{k,i} & = \int_{\Omega_i} \text{div}\,\mathbf{e}_k\,d\Omega & = \sum_j d_{ij}h_j, \\ c_{m,i} & = -\delta_{m,i}\int_{\Omega_m}\frac{\partial f}{\partial u}\,d\Omega, & \\ b_k & = \int_{E_k} g\,\mathbf{e}_k\cdot\mathbf{n}\,d\Gamma & = \sum_i d_{ik}\int_{E_k} g\,d\Gamma, \\ f_m & = \int_{\Omega_m} f(u)\,d\Omega. & \end{array} \tag{14}$$

Here h_j denotes the length of E_j and σ_j and u_i are the coefficients in

$$\boldsymbol{\sigma}_h = \sum_j \sigma_j\,\mathbf{e}_j, \qquad u_h = \sum_i u_i\,\epsilon_i. \tag{15}$$

(Notice that $C=0$ in classical MFE theory.) In the usual stable case (no avalanche) the sign of f is such that $c_{i,i}\le 0$. One of the advantages of the mixed finite element method is that the second equation in its discrete form guarantees the property of discrete current conservation.

Lumping, Scharfetter-Gummel

Taking piecewise constant approximations for f and g, all entries in the system (13) are simple to evaluate, except $a_{k,j}$. This coefficient may give rise to problems because in the continuity equations $a(x)$ can be a rapidly varying exponential function. The quadrature used to approximate $a_{k,j}$ is the weighted trapezoidal rule for rectangles

$$\int_{\Omega_i} w(x)\,z(x)\,d\Omega \cong \sum_{\nu=1,2,3,4} z(x_\nu)\int_{\Omega_i^\nu} w(x)\,d\Omega,$$

where x_ν are the four vertices and Ω_i^ν are the four quarter rectangles, parts of Ω_i, associated with these vertices respectively. We use $w(x)=a^{-1}$, and $z(x)=\sum_i \mathbf{e}_{ij}\cdot\mathbf{e}_{ik}$. The use of this quadrature rule is called *lumping* because it makes the matrix A diagonal. This is seen e.g. in the case of constant a (Poisson equation), where exact quadrature would yield

$$\int_{\Omega_i} a^{-1}\mathbf{e}_{ij}\cdot\mathbf{e}_{jk}\,d\Omega = \left\{\begin{array}{ll} \frac{1}{3}a^{-1}a_i & \text{if } k=j, \\ \frac{1}{6}a^{-1}a_i & \text{if } E_k \text{ and } E_j \text{ are opposite neighbours}, \\ 0 & \text{otherwise}, \end{array}\right.$$

where $a_i = \text{area}(\Omega_i)$. In the case of the trapezoidal rule we obtain

$$\int_{\Omega_i} a^{-1}\mathbf{e}_{ij}\cdot\mathbf{e}_{jk}\,d\Omega = \left\{\begin{array}{ll} \frac{1}{2}a^{-1}a_i & \text{if } k=j, \\ 0 & \text{otherwise}. \end{array}\right.$$

It was shown by Schilders [16] that the trapezoidal rule is advantageous, because the lumped form of the discretisation still yields an M-matrix. In the non-lumped case it is easily shown that the matrix obtained after elimination of σ is not necessarily an M-matrix in the case of a non-zero matrix C. Hence stability problems may rise. (In the non-lumped linear one-dimensional case with constant coefficients, $f(u) = f'u$ with $f' > 0$, the matrix is an M-matrix only if $h^2 f'/a \leq 6$!) Therefore, in the remainder of this paper we restrict ourselves to the lumped case only.

By the trapezoidal rule we get the approximation

$$a_{j,k} = \sum_i \int_{\Omega_i} a^{-1}\mathbf{e}_{ij} \cdot \mathbf{e}_{ik}\, d\Omega \cong \sum_i \sum_{\nu=1,2,3,4} (\mathbf{e}_{ij} \cdot \mathbf{e}_{ik})(x_\nu) \int_{\Omega_i^\nu} a^{-1}(x)\, d\Omega =$$

$$= \sum_i \sum_{\nu=1,2,3,4} \delta_{jk}\bar{e}_k(x_\nu) \int_{\Omega_i^\nu} a^{-1}(x)\, d\Omega = \delta_{jk} \int_{\Omega^k} a^{-1}\, d\Omega,$$

where $\Omega^k = \bigcup_{\{i,\nu | \bar{\Omega}_i^\nu \cap E_k \neq \emptyset\}} \Omega_i^\nu$; i.e. Ω^k is the dual box related with the edge E_k. If we approximate ψ in Ω^k by a linear function, interpolating the values ψ_i from the neighbouring cell centers, then $a = \exp(\pm\psi)$, and we obtain

$$\int_{\Omega^k} a^{-1}\, d\Omega = \text{area}(\Omega_k)\, \text{Bexp}^{-1}(\mp\psi_{i_1}, \mp\psi_{i_2}),$$

where we introduced the function

$$\text{Bexp}(x, y) = \frac{x - y}{e^x - e^y}. \tag{16}$$

Thus we retain the well-known Scharfetter-Gummel scheme (cf. [3]). In [16] it was shown that currents may be computed more accurately by the present MFE method than by the classical box scheme.

We see that, after lumping, the variables σ_j may be eliminated to obtain a five-point difference scheme between the variables u_i. For the discretisation of (6), we apply the above scheme for $u = (\psi, \Phi_n \Phi_p)$, so that $\sigma = (J_\psi, J_n, J_p)$, and $a = (\mu_\psi, \mu_n \exp(\psi), \mu_p \exp(-\psi))$, to obtain the system

$$\begin{aligned}
&\textstyle\sum_j h_j d_{ij} J_{\psi j} = e^{\phi_{p_i} - \psi_i} - e^{\psi_i - \phi_{n_i}} + D(x_i), && J_{\psi j} = -\tfrac{h_j d_{ij}}{a_j} \mu_\psi (\psi_j - \psi_i), \\
&\textstyle\sum_j h_j d_{ij} J_{nj} = +R(\psi_i, \phi_{n_i}, \phi_{p_i}), && J_{nj} = -\tfrac{h_j d_{ij}}{a_j} \mu_n \frac{\text{Bexp}\,(-\psi_j, -\psi_i)}{\text{Bexp}\,(-\phi_{n_j}, -\phi_{n_i})} (\phi_{n_j} - \phi_{n_i}), \\
&\textstyle\sum_j h_j d_{ij} J_{pj} = -R(\psi_i, \phi_{n_i}, \phi_{p_i}), && J_{pj} = -\tfrac{h_j d_{ij}}{a_j} \mu_p \frac{\text{Bexp}\,(\psi_j, \psi_i)}{\text{Bexp}\,(\phi_{p_j}, \phi_{p_i})} (\phi_{p_j} - \phi_{p_i}),
\end{aligned} \tag{17}$$

where $a_j = \text{area}(\Omega_j)$.

Green boundaries

In the case of partially refined grids, green boundaries appear. Green boundaries are those boundaries of a fine grid that are not part of the boundary of the domain Ω. Such green boundaries (green edges) separate areas where finest cells have different mesh sizes. Here the finer mesh needs an additional boundary condition. Hence, values of the potentials u are needed at edges E_j, to serve as boundary conditions for discretisation on the fine mesh. These values are obtained from the coarse grid by the use of half tent functions as weighting functions in equation (12a), and by forcing sufficient continuity by the introduction of a Lagrange multiplier, as is usual for the hybrid mixed finite element method [3]. We denote these test functions by τ_h^*. The value of the potential at wall E_k, denoted by λ_k, is derived from the variational equation: find $(u_h^*, \sigma_h^*, \lambda_h) \in L_h(\Omega) \times W_h(\Omega) \times M_h(\Omega)$ such that

$$\begin{cases} \sum_i \int_{\Omega_i} a^{-1}\sigma_h^* \cdot \tau_h^* + \sum_i \int_{\Omega_i} u_h^* \mathrm{div}\ \tau_h^* & = \sum_i \oint_{\partial\Omega_i} \lambda_h \tau_h^* \cdot \mathbf{n}_i, \\ \sum_i \int_{\Omega_i} \phi_h \mathrm{div}\ \sigma_h^* & = \sum_i \int_{\Omega_i} \phi_h\ f(u)\ d\Omega, \\ \sum_i \oint_{\partial\Omega_i} \mu_h \sigma_h^* \cdot \mathbf{n}_i & = 0, \end{cases} \tag{18}$$

for all $(\phi_h, \tau_h^*, \mu_h) \in L_h(\Omega) \times W_h(\Omega) \times M_h(\Omega)$. Now the third equation guarantees that the fluxes in the solution satisfy $\sigma_h^* \in \mathbf{H}^{BC}(\mathrm{div}\ , \Omega)$. Hence, in the interior the solution of system (18) is the same as the solution of (12), and λ_h can be interpreted as the value of the potentials at the edges. The values λ_k are the coefficients in $\lambda_h = \sum_k \lambda_k \bar{e}_k$, with $\bar{e}_k$ the characteristic function on E_k, and the λ_k can be expressed as

$$\lambda_k = u_{i_1} \frac{\int a^{-1}\ d\Omega_{k,i_1}}{\int a^{-1}\ d\Omega_k} + u_{i_2} \frac{\int a^{-1}\ d\Omega_{k,i_2}}{\int a^{-1}\ d\Omega_k}, \tag{19}$$

where i_1 and i_2 denote adjacent cells. This actually comes down to linear interpolation for the Poisson equation, or exponential interpolation for the continuity equations as was used for the one-dimensional case in [5].

4 Vanka type relaxation

For the efficiency of the multigrid method the choice of a proper relaxation procedure is of prime importance. Several procedures are available to solve the system of equations that arises from the mixed finite element method. Blockwise relaxation with current conservation has been used by Schmidt and Jacobs [17] for the solution of a Poisson problem with Neumann boundary conditions, Maitre c.s. [10] give an analysis of Uzawa relaxation. Vanka [19] describes a block-implicit method applied to the incompressible Navier-Stokes equations. In that study the equations associated with the pressure in a cell and the velocities over the cell faces are solved in a coupled manner.

In the present study we use a method similar to the procedure used by Vanka. In our relaxation all cells on a given level are scanned in a predetermined order, either lexicographically or in a red-black ordering. When a cell is visited the variables related with that cell and the fluxes over its four edges are relaxed simultaneously. In this way 5 variables are relaxed for each equation in (6), and in the relaxation of a single cell 15 equations are solved simultaneously.

This system of equations for $(\psi_i, \phi_{n_i}, \phi_{p_i})$ and $(J_{\psi k}, J_{nk}, J_{pk})$, $k = N, E, S, W$, has the following form

$$
\begin{array}{ll}
(a1) & \sum_k d_{i,k} h_k J_{\psi k} = (e^{\phi_{p_i}-\psi_i} - e^{\psi_i-\phi_{n_i}}) + D(x_i), \\
(a2) & J_{\psi k} = -\frac{d_{i,k}h_k}{a_k}\mu_\psi(\psi_k - \psi_i), \\
(b1) & \sum_k d_{i,k} h_k J_{nk} = +R(\psi_i, \phi_{n_i}, \phi_{p_i}), \\
(b2) & J_{nk} = -\frac{d_{i,k}h_k}{a_k}\mu_n(\phi_{nk} - \phi_{n_i})\frac{\mathrm{Bexp}\,(-\psi_k,-\psi_i)}{\mathrm{Bexp}\,(-\phi_{n_k},-\phi_{n_i})}, \\
(c1) & \sum_k d_{i,k} h_k J_{pk} = -R(\psi_i, \phi_{n_i}, \phi_{p_i}), \\
(c2) & J_{pk} = -\frac{d_{i,k}h_k}{a_k}\mu_p(\phi_{p_k} - \phi_{p_i})\frac{\mathrm{Bexp}\,(\psi_k,\psi_i)}{\mathrm{Bexp}\,(\phi_{p_k},\phi_{p_i})}.
\end{array}
\tag{20}
$$

Due to the structure of the equations, the computational work in each cell is limited. We can exploit the linear appearance of J in the equations, and, as was the case in [19], the linearised form of the equations can be arranged in a block structure

$$
\begin{bmatrix}
b_N & 0 & 0 & 0 & +h_N \\
0 & b_E & 0 & 0 & +h_E \\
0 & 0 & b_S & 0 & -h_S \\
0 & 0 & 0 & b_W & -h_W \\
h_N & h_E & -h_S & -h_W & -h_N h_W f\prime(x_i)
\end{bmatrix}
\begin{bmatrix}
\sigma_N \\ \sigma_E \\ \sigma_S \\ \sigma_W \\ u_i
\end{bmatrix}
=
\begin{bmatrix}
S_N + u_N h_N \\ S_E + u_E h_E \\ S_S - u_S h_S \\ S_W - u_W h_W \\ S_i + h_N h_W f(x_i)
\end{bmatrix},
\tag{21}
$$

where $b_k = \mathrm{area}(\Omega_k)/\mu$, $k = N, E, S, W$, for the Poisson equation, or $b_k = \mathrm{area}(\Omega_k)\mu^{-1}\mathrm{Bexp}(\mp\psi_i, \mp\psi_k)$ for the continuity equations. S_k denotes a possible source term and u_k the potential in the neighbouring cells. The upper 4×4 block in this system is inverted analytically, which comes down to the local elimination of the fluxes.

Because the equations associated with the edges of a cell are satisfied as soon as that cell has been relaxed, it is a property of our 5-point Vanka relaxation that all equations related to the fluxes (i.e. eq. (17,a2,b2,c2)) are satisfied as soon as a complete relaxation sweep has been performed. (Notice that an over or under relaxation would spoil this property.) The residuals left are associated with the cells and describe the extent to which the conservation property is not satisfied.

Newton vs Gummel

What remains in the relaxation of a cell is the solution of the nonlinear part of the equations. For this we resort to two approaches (1) Newton's iteration, and (2) Gummel's iteration. (Notice that we apply these methods locally, in contrast with the usual approaches where these methods are used for all points in the grid simultaneously.) The advantage of Newton's method is its quadratic convergence in the neighbourhood of the solution. This well known phenomenon makes Newton's method efficient when good initial approximations are available. For practical problems it appeared that relaxation based on Newton's iteration took about 60% of the computing time needed by Gummel's iteration. (This figure depends on a number of factors, but it gives some qualitative impression.) For our equations, the problem with Newton's method is the strong nonlinearity in the potential variables and a possible lack of good initial approximations. Much of the nonlinearity is characterised by the fact that the variables ψ, ϕ_n and ϕ_p, appear as exponents in exponential functions.

Correction transformations and initial estimates

Because the equations are better linearised with respect to the variables n and p than to ϕ_n and ϕ_p (see Section 1) Schilders' correction transformation [15] is used, both in the case of Newton's and Gummel's method. This means that first the linearisation is made with respect to the potentials and the corresponding correction is computed. Then this correction is transformed to the correction that would have been obtained if a linearisation with respect to n or p were made. From equation (4) it follows that the corrections, expressed in the quasi-Fermi potentials, are related by

$$\Delta\phi_n^{\text{new}} = \Delta\psi^{\text{old}} - \log(1 - (\Delta\phi_n^{\text{old}} - \Delta\psi^{\text{old}})), \tag{22}$$

$$\Delta\phi_p^{\text{new}} = \Delta\psi^{\text{old}} + \log(1 + (\Delta\phi_p^{\text{old}} - \Delta\psi^{\text{old}})). \tag{23}$$

Such a transformation, introduced in [15] for the continuity equations, can also be used for the nonlinear Poisson equation. There we have to determine what part of the equation is dominating, the linear part or the nonlinear (exponential) part. We took the following strategy. Without loss of generality the Poisson equation (20 a) can be written as

$$a \sinh\psi + b\psi = 1. \tag{24}$$

If $|b| > |a\cosh\psi|$ we decide that the linear part is dominating and we apply the correction

$$\psi^{\text{new}} = \psi^{\text{old}} + \Delta\psi,$$

if $|b| < |a\cosh\psi|$ the nonlinear part is dominating and we take

$$\psi^{\text{new}} = \text{arsinh}(\sinh\psi^{\text{old}} + \Delta\psi\cosh\psi^{\text{old}}).$$

If we need an initial estimate for Gummel's method, we can also start from equation (24). Depending on the size of ψ, we can find two approximations for the solution: $\psi = 1/(a+b)$, or $\psi = \text{arsinh}(1/a)$. In order to decide which one is the more appropriate, we select the one for which the functional

$$G(\psi) = a\cosh\psi + \frac{1}{2}b\psi^2 - \psi$$

is minimal.

The convergence of pointwise Gummel iteration

Newton's method is used in the later stages of the local solution process, when good initial estimates are already available; in the absence of good initial estimates we use (locally) Gummel's iteration because it is more robust.

Little is known about the convergence of the pointwise Gummel iteration. Therefore we present here an analysis of the convergence of Gummel's decoupling method for the solution of the system (20). The objective is to obtain a more precise understanding of the convergence properties of this iterative scheme. The analysis presented predicts that the convergence of Gummel's method depends only on the difference in the values of ψ in the neighbouring control volumes, and not on the initial estimate or on the properties of the doping profile $D(x)$.

In the spirit of [8] we study the Gummel iteration as a fixed-point mapping $T : I\!R^2 \to I\!R^2$, that maps a pair (ϕ_n, ϕ_p) onto a pair $(\tilde{\phi}_n, \tilde{\phi}_p) = T(\phi_n, \phi_p)$. To compute $T(\phi_n, \phi_p)$, first the electric potential $\psi(\phi_n, \phi_p)$ is computed as an intermediate result by the solution of (20 a). The values $\tilde{\phi}_n$ and $\tilde{\phi}_p$ are obtained from this $\psi(\phi_n, \phi_p)$ by the solution of (20 b,c). Existence of a solution in $A \subset I\!R^2$ follows when T is a contraction mapping on A. Then Gummel's iteration converges and the contraction factor may give an indication of the convergence speed of the iteration. To measure the distance in $I\!R^2$ we use the max-norm:

$$\|(\phi_n^1, \phi_p^1) - (\phi_n^2, \phi_p^2)\| = \max(|\phi_n^1 - \phi_n^2|, |\phi_p^1 - \phi_p^2|). \tag{25}$$

In order to be able to be more specific, we restrict the analysis to the zero recombination case. This enables us to find explicit expressions for the iterates.

Theorem 1 *If the variation in the ψ-values in the four neighbouring points is sufficiently small* ($\max_k \psi_k - \min_k \psi_k < 12$), *then the operator T for the pointwise Gummel iteration is a contraction, i.e.*

$$\|T(\phi_n^1, \phi_p^1) - T(\phi_n^2, \phi_p^2)\| \leq C\|(\phi_n^1, \phi_p^1) - (\phi_n^2, \phi_p^2)\|, \tag{26}$$

with $C = \frac{1}{12}(\max_k \psi_k - \min_k \psi_k)$ *and for all* $(\phi_n^i, \phi_p^i) \in I\!R^2$, $i = 1, 2$.

Proof: The proof is given in two parts. We consider the iteration sequence

$$(\phi_n^i, \phi_p^i) \to \psi^i \to (\tilde{\phi}_n^i, \tilde{\phi}_p^i), \quad i = 1, 2, \tag{27}$$

so that $\psi^i = \psi(\phi_n^i, \phi_p^i)$ and $(\tilde{\phi}_n^i, \tilde{\phi}_p^i) = T(\phi_n^i, \phi_p^i)$. In the first part we prove

$$|\psi^1 - \psi^2| \leq \|(\phi_n^1, \phi_p^1) - (\phi_n^2, \phi_p^2)\|, \tag{28}$$

and in the second part we show

$$\|(\tilde{\phi}_n^1, \tilde{\phi}_p^1) - (\tilde{\phi}_n^2, \tilde{\phi}_p^2)\| \leq C|\psi^1 - \psi^2|. \tag{29}$$

In fact we show (29) only for ϕ_p,

$$|\tilde{\phi}_p^1 - \tilde{\phi}_p^2| \leq C|\psi^1 - \psi^2|, \tag{30}$$

because a similar result for ϕ_n follows by analogy, and both results together yield (29).

In order to prove equation (28) we consider (20a), which yields for $i = 1, 2$,

$$\sum_k w_k \mu_\psi(\psi_k - \psi^i) + (e^{\phi_p^i - \psi^i} - e^{\psi^i - \phi_n^i}) + D(x) = 0,$$

with $w_k = h_k^2/\text{area}(\Omega_k)$. By subtraction we obtain

$$\sum_k w_k \mu_\psi(\psi_2 - \psi_1) + (e^{\phi_p^1 - \psi^1} - e^{\psi^1 - \phi_n^1} - e^{\phi_p^2 - \psi^2} + e^{\psi^2 - \phi_n^2}) = 0$$

or

$$(\psi_1 - \psi_2)\mu_\psi \sum_k w_k = (e^{\psi^1 - \phi_n^1}(e^{(\phi_n^1 - \phi_n^2) - (\psi^1 - \psi^2)} - 1) + e^{\phi_p^2 - \psi^2}(e^{(\phi_p^1 - \phi_p^2) - (\psi^1 - \psi^2)} - 1)). \tag{31}$$

From this equality, the inequality (28) follows for the following reason.

Assume that (28) is *not* true, then we consider two cases: either $\psi^1 - \psi^2 > 0$ or $\psi^1 - \psi^2 < 0$. In the former case from the negation of (28) follows that $\psi^1 - \psi^2 \geq \phi_n^1 - \phi_n^2$ and $\psi^1 - \psi^2 \geq \phi_p^1 - \phi_p^2$. It follows that the left-hand side of the equality (31) is positive and the right-hand side is negative. This is a contradiction. Similarly, if $\psi^1 - \psi^2 < 0$ it follows that $\psi^1 - \psi^2 \leq \phi_n^1 - \phi_n^2$ and $\psi^1 - \psi^2 \leq \phi_p^1 - \phi_p^2$. Now it follows that the left-hand side of the equality (31) is negative and the right-hand side is positive. This also yields a contradiction. Because (28) is trivially satisfied for $\psi^1 = \psi^2$, we may conclude that (28) holds.

In order to prove the second part (30), we consider (20c). With zero recombination this yields for $i = 1, 2$, (dropping the subscript p)

$$\sum_k w_k(\phi_k - \phi^i)\frac{\text{Bexp }(\psi_k, \psi^i)}{\text{Bexp }(\phi_k, \phi^i)} = 0,$$

using the definition of Bexp for the denominators, we obtain

$$e^{\phi^i}\sum_k w_k \text{Bexp }(\psi_k, \psi^i) = \sum_k w_k\ e^{\phi_k}\ \text{Bexp }(\psi_k, \psi^i). \tag{32}$$

First we notice that all factors and terms in this expression are positive, and hence $\min_k e^{\phi_k} \leq e^{\phi^i} \leq \max_k e^{\phi_k}$, for $i = 1, 2$, which yields (without any restriction on ψ_k)

$$\min_k \phi_k \leq \phi^i \leq \max_k \phi_k, \quad \text{for } i = 1, 2,$$

and

$$\phi^1 - \phi^2 \leq |\max_k \phi_k - \min_k \phi_k|.$$

Further, from (32) we derive

$$e^{\phi^1 - \phi^2} = \frac{\sum_k w_k \text{Bexp }(\psi_k, \psi^2)}{\sum_k w_k \text{Bexp }(\psi_k, \psi^1)} \cdot \frac{\sum_k w_k \text{Bexp }(\psi_k, \psi^1)e^{\phi_k}}{\sum_k w_k \text{Bexp }(\psi_k, \psi^2)e^{\phi_k}}.$$

Now we define ψ_A to be the value of ψ_k for which

$$\frac{\text{Bexp }(\psi_A, \psi^2)}{\text{Bexp }(\psi_A, \psi^1)} \geq \frac{\text{Bexp }(\psi_k, \psi^2)}{\text{Bexp }(\psi_k, \psi^1)} \tag{33}$$

for all k, and similarly ψ_B such that

$$\frac{\text{Bexp }(\psi_B, \psi^1)}{\text{Bexp }(\psi_B, \psi^2)} \geq \frac{\text{Bexp }(\psi_k, \psi^1)}{\text{Bexp }(\psi_k, \psi^2)}$$

for all k, then

$$e^{\phi^1 - \phi^2} \leq \frac{\text{Bexp }(\psi_A, \psi^2)}{\text{Bexp }(\psi_A, \psi^1)} \cdot \frac{\text{Bexp }(\psi_B, \psi^1)}{\text{Bexp }(\psi_B, \psi^2)}. \tag{34}$$

Taking the logarithm and introducing the function $g(x) = \log\left(\frac{x}{e^x - 1}\right)$, we may write (34) as

$$\phi^1 - \phi^2 \leq g(\psi^2 - \psi_A) - g(\psi^1 - \psi_A) - g(\psi^2 - \psi_B) + g(\psi^1 - \psi_B)$$

or

$$\phi^1 - \phi^2 \leq \int_{-\psi_B}^{-\psi_A} \int_{\psi_2}^{\psi_1} (-g''(x+y))\, dx\, dy \;.$$

Since

$$g''(x) = \frac{1}{2\cosh(x) - 2} - \frac{1}{x^2}$$

we know that $0 < -g''(x) \leq 1/12$ and

$$\phi^1 - \phi^2 \leq \frac{1}{12}(\psi_B - \psi_A)(\psi^1 - \psi^2).$$

To determine ψ_A and ψ_B we consider

$$\log\left(\frac{\text{Bexp}\,(\psi,\psi_2)}{\text{Bexp}\,(\psi,\psi_1)}\right) = \int_{\psi_1}^{\psi_2} g'(x-\psi)dx = (\psi_2 - \psi_1)g'(\psi_m - \psi)$$

for some $\psi_m \in (\psi^1, \psi^2)$. Because $g'(\psi_m - \psi)$ is a monotonically increasing function of ψ we find $\psi_A = \max_k \psi_k$ and $\psi_B = \min_k \psi_k$ if $\psi_2 > \psi_1$, and if $\psi_2 < \psi_1$ we have $\psi_A = \min_k \psi_k$ and $\psi_B = \max_k \psi_k$. It follows that

$$\phi^1 - \phi^2 \leq \frac{1}{12}(\max_k \psi_k - \min_k \psi_k)|\psi^1 - \psi^2|.$$

Because the superscripts 1 and 2 may be interchanged without changing the meaning of the right-hand side, this proves (30) and hence the theorem. □

The proof of the theorem, valid for zero recombination and zero source term, clearly shows that convergence may be slower if a source term for the continuity equations takes values that make the right-hand side of (32) smaller. No solution exists for the local nonlinear problem, if the source term makes the right-hand side of (32) negative. This means that large source terms can cause the non-existence of a solution. Hence, we have to face the possibility that the correction equations in the multigrid process have no solution if the right-hand side of the equation gets too large.

5 The coarse grid correction

If, for the solution of the nonlinear discrete equation,

$$N_h(q_h) = f_h, \tag{35}$$

we consider the usual nonlinear coarse grid correction stage of a two-grid process,

$$N_{2h}(\tilde{q}_{2h}) = N_{2h}(q_{2h}) + \mu\, \overline{R}_{2h,h}(f_h - N_h(q_h^{\text{old}}))\,, \tag{36}$$

$$q_h^{\text{new}} = q_h^{\text{old}} + P_{h,2h}(\tilde{q}_{2h} - q_{2h})/\mu, \tag{37}$$

we recognise five important components that influence the effect of this stage. In the first place, there are the three operators N_{2h}, $\overline{R}_{2h,h}$ and $P_{h,2h}$, and further the starting approximation on the coarser grid, q_{2h}, and the parameter $\mu \in I\!R$. For a nonlinear problem, the operator N_{2h} is often constructed by the same method as is used for N_h; in our case it is described in Section 3. In principle, the choice for the operators $\overline{R}_{2h,h}$ and $P_{h,2h}$ is free (as long as they are accurate enough), but in the context of our MFE discretisation there exist a natural prolongation and restriction associated with the discretisation, viz. those induced by the relations $L^2_{2h}(\Omega) \subset L^2_h(\Omega)$, and $\mathbf{H}^{BC}_{2h}(\mathrm{div}\,,\Omega) \subset \mathbf{H}^{BC}_h(\mathrm{div}\,,\Omega)$. These relations imply that the prolongation corresponds for the potentials with piecewise constant interpolation, and for both components of the fluxes with piecewise linear interpolation in one direction and piecewise constant interpolation in the other. The corresponding prolongation stencils are $\begin{bmatrix} 1 & 1 \\ 1 & 1 \end{bmatrix}$, for the potentials (associated with a cell), and $\begin{bmatrix} 1/2 & 1/2 \\ 1 & 1 \\ 1/2 & 1/2 \end{bmatrix}$, $\begin{bmatrix} 1/2 & 1 & 1/2 \\ 1/2 & 1 & 1/2 \end{bmatrix}$, for the fluxes (associated with a horizontal and a vertical edge respectively). The natural restriction $\overline{R}_{2h,h}$ is the transpose of the natural prolongation $P_{h,2h}$, because the spaces of test and trial functions in (12) are the same.

For strong nonlinear problems, also the choice of the starting approximation q_{2h} and the parameter μ are of importance because they determine to a very large extent the coarse grid problem that is solved. If the distance between q_h^{old} and the solution of (35) is small, it is clear that mainly q_{2h} determines the coarse grid problem, and it is wise to select q_{2h} in such a way that the problem (36) is well conditioned. If (36) is not ill-conditioned, the parameter μ can be used to keep $\tilde{q}_{2h}$ in a sufficiently small neighbourhood of q_{2h}. This may guarantee the existence of a solution of the correction equation. However, the effect of a small μ can be that only a very small neighbourhood of q_{2h} is considered, so that nonlinear effects in N_{2h} are neglected. Moreover, the factor $1/\mu$ in (37) can amplify the errors made in the solution of (36).

For the semiconductor equations (6) without a row scaling, the residual for the continuity equations correspond with the rate-of-change in the carrier concentrations, cf. eq. (1). In this unscaled form, the natural restriction operator has a "physical meaning": the sum of the rate-of-change in four small sub-cells corresponds with the total rate-of-change in the father cell. We believe that this is an advantageous property of the equations in their unscaled form. However, without row-scaling, the size of the residuals (as well as the size of the diagonal elements of the Jacobian matrix) may vary largely in magnitude. This introduces the difficulty that for some parts of the domain Ω the large residual requires a very small μ, whereas a larger μ would be allowed in other parts. An even more awkward situation is encountered if the values for a proper row-scaling differ strongly for the equations related with a coarse grid cell and the corresponding equations on the finer level. In this case a large residual on the fine grid may yield an improper large correction on the coarse grid. This effect is seen in regions where the character of the solution changes rapidly (transition between N- and P-region, depletion layer). The same effect was observed by de Zeeuw in [4] in the 1D-case and it leads to the introduction of a residual damping operator D_{2h}. This D_{2h} is a diagonal operator, depending on the current coarse and fine grid solution, which has entries in $[0,1]$. Hence, for the coarse grid correction we use

$$N_{2h}(\tilde{q}_{2h}) = N_{2h}(q_{2h}) + D_{2h}(q_{2h}, q_h^{\mathrm{old}})\,\overline{R}_{2h,h}(f_h - N_h(q_h^{\mathrm{old}}))\,, \tag{38}$$

$$q_h^{\mathrm{new}} = q_h^{\mathrm{old}} + P_{h,2h}(\tilde{q}_{2h} - q_{2h})\,. \tag{39}$$

This means that the coarse grid correction (38), (39) is not able to reduce *all* components of the residual that can be represented on the coarse grid, but an amount $(I_{2h} - D_{2h})\overline{R}_{2h,h}(f_h -$

$N_h(q_h^{\text{old}}))$ remains unaffected (in a single CGC sweep). The elements of D_{2h} are different from 1 only in small regions (the transition regions in the semiconductor), and the effect of the damping is compensated in these regions by additional relaxation on the fine grid. The precise construction of the operator D_{2h} is found in [14].

The selection of a proper coarse grid approximation

For the selection of q_{2h} in (36) or (38) two approaches are in common use. Either $q_{2h} = R_{2h,h}q_h^{\text{old}}$ is used, where $R_{2h,h}$ is a restriction operator for the solution, or for q_{2h} one takes simply the last approximation that is available in the full multigrid process, i.e. one starts with the approximate solution on the coarse grid as obtained in the nested iteration, and later -at each stage of the multigrid process- the last approximate solution on a given level is used as an initial approximation in the next stage.

In practice, it appears that the latter technique performs rather well. However, we consider it unreliable because in all later stages of the process the approximate solutions on the coarser levels depend on the complete history of the computational process, and there is no mechanism that forces such a coarse grid approximation to stay in the neighbourhood of a solution. In fact, such an approximation q_{2h} may loose properties that are required for a proper approximate solution, e.g. symmetry.

The first approach, however, requires the selection of an $R_{2h,h}$ and for our problem there is no reason to assume that e.g. the simple use of L^2-projection of the Slotboom variables -as suggested by the discretisation- will yield a proper problem (38). It seems a better choice to take mean values for ψ and to construct ϕ_n and ϕ_p such that the total amount of electrons and holes in a coarse cell equals the sum of the amounts in the corresponding smaller cells.

A third, more simple technique was adopted because of its good results: compute a reasonably accurate discrete approximation on the coarse grid during the nested iteration, and keep this value as q_{2h} during all the later stages of the computation.

In our case this last technique can be understood as a favourable approach for the following reason. For the homogeneous continuity equations (20b,c) with $R = 0$, the Scharfetter-Gummel discretisation has the property that the row-sum of the discrete matrix $(\partial J_p/\partial \phi_p)$ at cell i is equal to the residual of the discrete equation for that cell: $\sum_l \frac{\partial}{\partial \phi_{p_l}}(\sum_k h_k d_{ik} J_{pk}) = \sum_k h_k d_{ik} J_{pk}$, and, analogously, for the other continuity equation $\sum_l \frac{\partial}{\partial \phi_{n_l}}(\sum_k h_k d_{ik} J_{nk}) = -\sum_k h_k d_{ik} J_{nk}$. This follows from, (cf. equation (20 b2,c2)), $J_{pk} = \frac{\partial J_{pk}}{\partial \phi_{p_k}} + \frac{\partial J_{pk}}{\partial \phi_{p_i}}$, $J_{nk} = \frac{\partial J_{nk}}{\partial \phi_{n_k}} + \frac{\partial J_{nk}}{\partial \phi_{n_i}}$. Hence the row-sums in the Jacobian matrix vanish when q_{2h} is in the neighbourhood of the discrete solution. This implies that in the neighbourhood of the discrete solution the linearised operators in the Gummel process are M-matrices. A positive recombination only improves the situation. This shows that the stability of the linearised operators is better in the neighbourhood of a solution than at some distance from the solution.

Other transfer operators

A priori there is also no reason to assume that the natural grid transfer operators $P_{h,2h}$ and $\overline{R}_{2h,h}$ are the best, or even that they are sufficiently accurate (smooth) in order not to disturb reduction of the the high frequency components in the solution.

Indeed, for the one-dimensional case, in combination with the Vanka relaxation we observe by Fourier analysis that these simple transfer operators are too inaccurate. The non-damped 5-point Vanka relaxation can be considered as eliminating the fluxes and applying a collective Gauss-Seidel procedure to the remaining potentials. After elimination of the fluxes, the differential equations for the potentials are second order, and hence the rule applies that the sum of the HF orders of the prolongation and restriction should at least be two [7]. The orders of the natural prolongation and restriction, however, are one. To obtain a proper convergence of the MG algorithm, one should take more accurate transfer operators. This is analyzed in detail by two-grid Fourier analysis in [13]. The simplest operator that satisfies a sufficient accuracy condition is piecewise linear interpolation and its adjoint as a weighted restriction. The LF and HF order of this restriction is 2.

However, the same effect is not seen in 2D [13]. The Vanka relaxation damps sufficiently the HF modes that are allowed by the transfer operators, and in practical 2D computations piecewise constant interpolation, together with its transpose for a restriction, gave satisfactory results. We did not observe an improvement when more accurate restrictions were used instead of the natural restriction.

Because of the asymmetric character of the convection operator, and in view of the successful use of an asymmetric prolongation in a multigrid method for the one-dimensional semiconductor problem in [4] [5] [6], it is interesting to consider the possibility of an asymmetric prolongation for the two-dimensional problem as well. In 1D such an interpolation was based on the form (cf. eq. (6 b,c))

$$\Phi(x) - \Phi(a) = \int_a^x e^{\pm\psi(\xi)} J d\xi, \tag{40}$$

with the assumption of a piecewise constant J and a piecewise linear ψ over the area of integration (the dual boxes). In our MFE context, the same exponential interpolation formula is found in Section 3 as equation (19). The principle behind the construction of that prolongation in the one-dimensional case is the equal flux over corresponding coarse and fine grid edges. In two dimensions, however, such an explicit prolongation cannot be constructed. This is because in two dimensions the assumption of a piecewise constant J and the existence of a unique function Φ leads to an inconsistency. Independence of $\Phi(x)$ on the integration path means grad $\Phi = \exp(\pm\psi)J$. This relation only holds for ψ and J satisfying

$$0 = \text{rot grad } \Phi = \text{rot } (e^{\pm\psi} J) = e^{\pm\psi}(\text{rot } J \pm J \times \text{grad } \psi). \tag{41}$$

With the assumption of a constant J, this implies that J should be parallel with grad ψ. However, for a two-dimensional case, this is generally too restrictive a condition. From equation (41) follows that J has the general form $J = \text{grad } u \mp u\text{grad } \psi$, for an arbitrary scalar function u.

Assuming that the dependence of the integration path has only a minor influence, we might overlook the non-uniqueness of Φ and select an path, e.g. select the shortest line segment from the coarse cell center (with the known potential) to the fine cell center (where the potential has to be computed). Then the fluxes over corresponding edges in the coarse and the fine mesh are not equal, and in our experiments this interpolation appears no better than the piecewise constant interpolation.

6 Example

As a test problem we used a bipolar NPN transistor from the CURRY example set [9]. The geometry of the transistor is shown in Figure 1. There is an N-type emitter region, a P-type base region and an N-type collector region. The length of the device is 20 micron and the width is 8. For the precise description of the doping profile we refer to [9]. The Shockley-Read-Hall model 3 is used for the recombination, with carrier lifetimes $\tau_p = \tau_n = 10^{-6}$. Dirichlet boundary conditions are given at the contacts. On the remaining boundaries homogeneous Neumann boundary conditions apply.

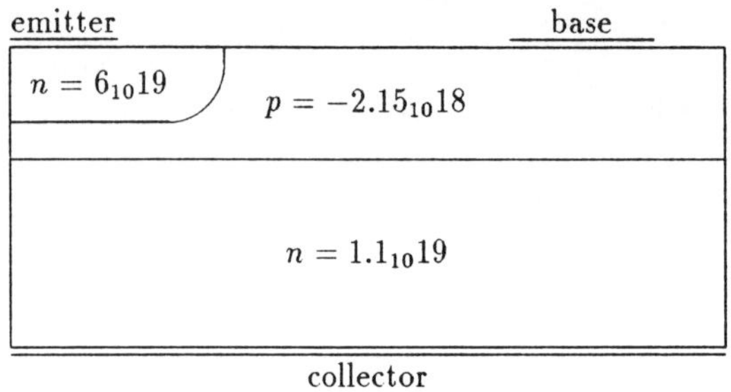

Figure 1: Geometry of the bipolar transistor

The voltages applied to the collector and base are kept constant at $V_{coll} = 1.0$V and $V_{base} = 0.0$V. The simulation is started with zero potential at the emitter. Then no currents are present because all P-N junctions are blocked. The simulation is continued, first by an increase of the emitter voltage to -0.5V and then in steps (-0.05V) to -0.8 Volts. Then currents are clearly present. In Table 1 we show the collector currents computed on a 16 × 40 , 32 × 80 and 64 × 160 mesh, together with a reference solution computed with the CURRY package on a non-uniform 56 × 62 grid. We see that the solution (i.c. the collector current) appears to converge for vanishing mesh-widths.

	MFEM uniform mesh			CURRY non-uniform
V_{emit}	16 × 40	32 × 80	64 × 160	56 × 62
0.0	5.3(-12)	5.1(-11)	5.2(-11)	7.2(-11)
-0.50	9.5(-5)	1.4(-5)	1.0(-5)	9.8(-6)
-0.55	5.8(-4)	9.5(-5)	7.0(-5)	6.7(-5)
-0.60	3.4(-3)	6.4(-4)	4.8(-4)	4.6(-4)
-0.65	1.8(-2)	4.3(-3)	3.3(-3)	3.1(-3)
-0.70	8.4(-2)	2.8(-2)	2.2(-2)	2.1(-2)
-0.75	3.2(-1)	1.7(-1)	1.4(-1)	1.3(-1)
-0.80	1.1(0)	7.9(-1)	7.1(-1)	6.9(-1)

Table 1: Collector currents (A/cm).

For the coarsest mesh, the device is divided into 4 × 10 (!) squares. We notice that this mesh is so coarse that the emitter boundary does not fit the edges of the cells. Therefore, for the discretisation, an obvious generalisation of the method described in Section 3 was used. In the discretisation, for each cell the current through such a boundary edge is determined by the boundary potential, the potential in that cell and the proportion of the edge that is covered by the contact. This treatment of the boundary prevents the obligation to use fine or irregular cells in the coarsest grids.

The initial estimates for the emitter voltages -0.5(-0.05)-0.8 were obtained from the solutions computed with the previous voltage. First the solution on the coarsest grid was accurately computed, and the solution on the finer grids was computed (approximately, by a few W-cycles) before an interpolation to the next finer grid was made. In the interpolation to the finer grid, the low frequencies in the solution were taken from the coarser grid, whereas the high frequency components were taken from the fine grid solution for the lower voltage. Thus mimicking a well known technique used for time dependent problems.

The MGM used, Convergence results

The multigrid method to solve the transistor problem applies the lumped MFE discretisation as described in Section 3. The natural prolongation and restriction operators were used, together with the residual damping as explained in Section 5. A single additional point-Vanka relaxation sweep was made over all fine grid cells for which the residual was damped in the coarse grid correction. As the initial estimate q_{2h} we kept the solution obtained initially on the coarse grid. Both in the pre- and in the post-relaxation stage a single sweep of the smoothing procedure was used.

Beside the symmetric lexicographic point-Vanka relaxation, also a (non-symmetric, but horizontal+vertical) line-Vanka relaxation was applied as smoothing procedure. In the results shown, only W-cycle results are given. As was shown earlier [6] [14] V-cycles are less robust for the semiconductor problem.

In Figure 2 convergence histories are shown for the multigrid solution process. On the horizontal axis the number of cycles is given, and on the vertical axis the scaled residual. The residual scaling was made pointwise, by means of the diagonal 3×3 blocks of the Jacobian matrix. Thus the residual corresponds with corrections that would occur if a pointwise collective Jacobi relaxation was used. Hence, the scaled residual can be associated with corrections for (ψ, ϕ_n, ϕ_p). For the resulting scaled residual the maximum was taken over the grid and over the three variables (ψ, ϕ_n, ϕ_p).

Convergence results are shown for the solution with 3, 4 and 5 grid levels, both for point-Vanka and for line-Vanka relaxation. It is observed that line-Vanka relaxation is the more efficient. It is seen that the convergence is not always stabilised to a constant factor after 10 iterations, but an almost grid independent convergence rate can be expected. In any case, convergence is fast and a limited number of iterations is sufficient to attain truncation error accuracy.

Anknowledgement

We would like to thank Dr. W.H.A. Schilders for reading the manuscript.

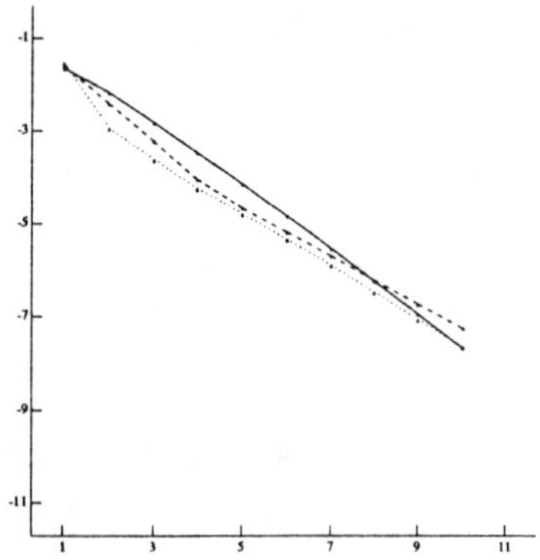

Case a: $V_e = 0.0$
Point-Vanka relaxation

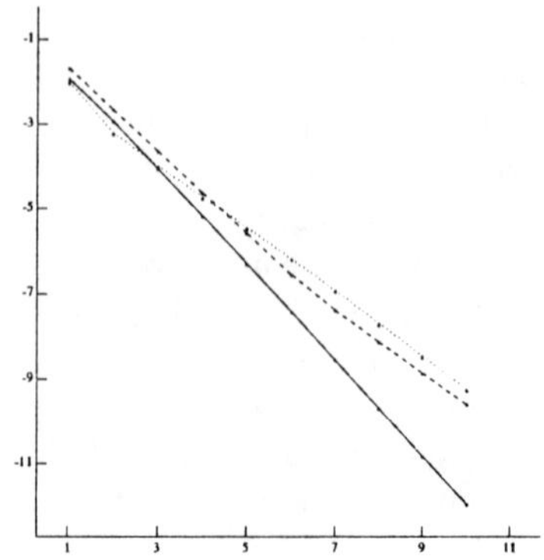

Case b: $V_e = 0.0$
Line-Vanka relaxation

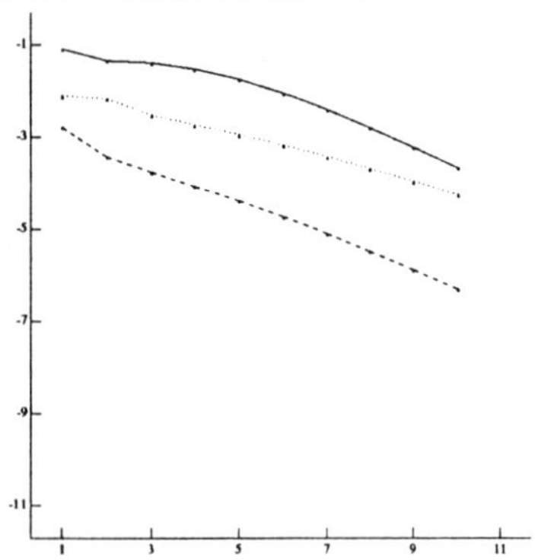

Case c: $V_e = -0.5$
Point-Vanka relaxation

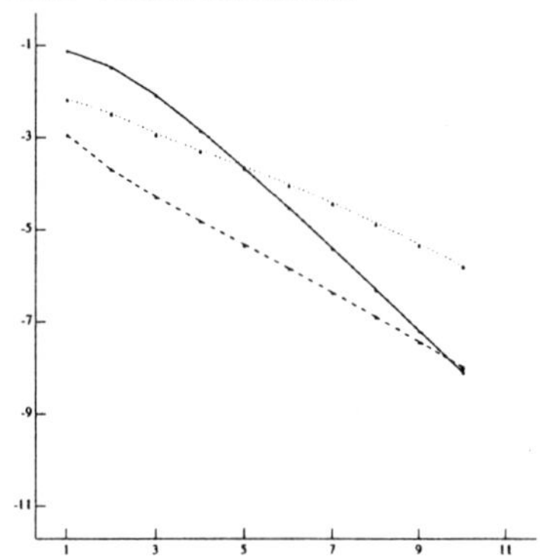

Case d: $V_e = -0.5$
Line-Vanka relaxation

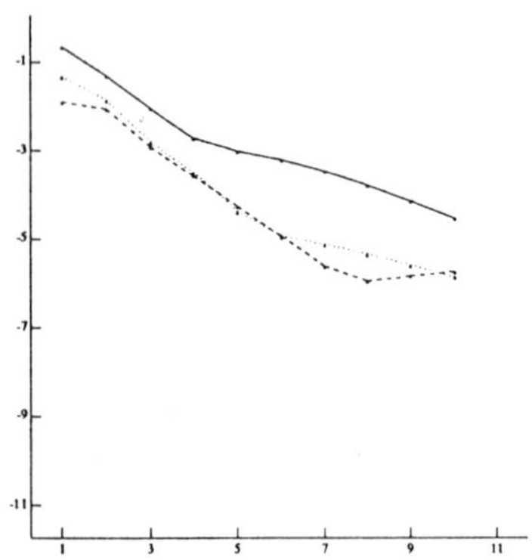

Case e: $V_e = -0.8$
Point-Vanka relaxation

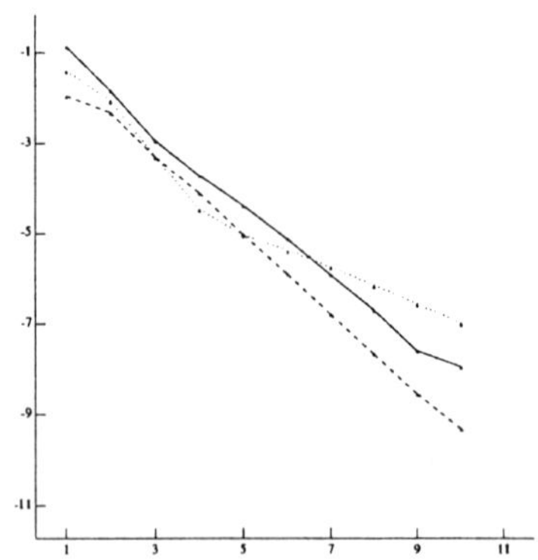

Case f: $V_e = -0.8$
Line-Vanka relaxation

Figure 2: Convergence of the MG method (W-cycle).
The 10-log of the scaled residual against the iteration number.
Solid line: 16×40 mesh; dotted line: 32×80 mesh; dashed line: 64×160 mesh.

References

[1] R.E. Bank, W. Fichtner, D.J. Rose, and R.K. Smith. Algorithms for semiconductor simulation. In P. Deuflhard and B. Engquist, editors, *Numerical Analysis of Singular Perturbation Problems*, pages 3–21. Birkhauser, 1988.

[2] A. Brandt. Multi-level adaptive techniques for singular perturbation problems. In P.W. Hemker and J.J.H. Miller, editors, *Numerical Analysis of Singular Perturbation Problems*, pages 53–142, Londen, New York, 1979. Academic Press.

[3] F. Brezzi, L.D. Marini, and P. Pietra. Two-dimensional exponential fitting and applications to semiconductor device equations. Technical Report 597, Inst. Anal. Numerica CNR, 1987.

[4] P.M. de Zeeuw. Nonlinear multigrid applied to a 1D stationary semiconductor model. Technical Report NM-R8905, CWI, Dept.NM, Amsterdam, 1989.

[5] P.W. Hemker. A nonlinear multigrid method for one-dimensional semiconductor device simulation. In Guo Ben-yu, J.J.H. Miller, and Shi Zhong-ci, editors, *BAIL V, Proceedings of the 5th International Conference on Boundary And Interior Layers - Computational and Asymptotic Methods*, pages 18–29, Dublin, 1988. Boole Press.

[6] P.W. Hemker. A nonlinear multigrid method for one-dimensional semiconductor device simulation: Results for the diode. *J. Comput. Appl. Math*, 30:117 126, 1990.

[7] P.W. Hemker. On the order of prolongations and restrictions in multigrid procedures. *J. Comput. Appl. Math*, 1990. to appear.

[8] T. Kerkhoven. On the effectiveness of Gummel's method. *SIAM J.S.S.C.*, 9:48–60, 1988.

[9] C. Lepoeter. CURRY example set. Technical Report No. 4322.271.6005, Philips, Corp. CAD Centre, Eindhoven, 1987.

[10] J.-F. Maitre, F. Musy, and P. Nigon. A fast solver for the Stokes equations using multigrid with a Uzawa smoother. In D. Braess, W. Hackbusch, and U. Trottenberg, editors, *Advances in Multi-Grid Metods*, volume 11 of *Notes on Numerical Fluid Dynamics*, pages 77–83, Braunschweig, 1985. Vieweg.

[11] P.A. Markowich. *The Stationary Semiconductor Device Equations.* Springer Verlag, Wien, New York, 1986.

[12] S.F. McCormick. *Multilevel Adaptive Methods for Partial Differential Equations*, volume 6 of *Frontiers in Applied Mathematics.* SIAM, 1989.

[13] J. Molenaar. A two-grid analysis of the combination of mixed finite elements and Vanka relaxation. Technical Report to appear, CWI, Dept.NM, Amsterdam, 1990.

[14] J. Molenaar and P.W. Hemker. A multigrid approach for the solution of the 2D semiconductor equations. *IMPACT*, page (to appear), 1990.

[15] S.J. Polak, C. den Heijer, W.H.A. Schilders, and P. Markowich. Semiconductor device modelling from the numerical point of view. *Int. J. Num. Meth. Engng.*, 24:763–838, 1987.

[16] S.J. Polak, W.H.A. Schilders, and H.D. Couperus. A finite element method with current conservation. In G. Baccarani and M. Rudan, editors, *Simulation of Semiconductor Devices an Processes*, volume 3, Bologna, Italy, 1988. Technoprint.

[17] G.H. Schmidt and F.J. Jacobs. Adaptive local grid refinement and multi-grid in numerical reservoir simulation. *J. Comput. Phys.*, 77:140–165, 1988.

[18] W. van Roosbroeck. Theory of flow of electrons and holes in germanium and other semiconductors. *Bell Syst. Techn. J.*, 29:560–607, 1950.

[19] S.P. Vanka. Block-implicit multigrid solution of Navier-Stokes equations in primitive variables. *J. Comp. Phys.*, 65:138–158, 1986.

Version: October 9, 1990

International Series of Numerical Mathematics, Vol. 98, © 1991 Birkhäuser Verlag Basel

Multiscale Monte Carlo Algorithms in Statistical Mechanics and Quantum Field Theory

P. G. Lauwers
Physikalisches Institut der Universität Bonn
Nußallee 12, 5300 Bonn 1
Federal Republic of Germany

Abstract

Conventional Monte Carlo simulation algorithms for models in statistical mechanics and quantum field theory are afflicted by problems caused by their locality. They become highly inefficient if investigations of critical or nearly-critical systems, i.e., systems with important large scale phenomena, are undertaken. We present two types of multiscale approaches that alleveate problems of this kind: stochastic cluster algorithms and multigrid Monte Carlo simulation algorithms. Another formidable computational problem in simulations of phenomenologically relevant field theories with fermions is the need for frequently inverting the Dirac operator. This inversion can be accelerated considerably by means of deterministic multigrid methods, very similar to the ones used for the numerical solution of differential equations.

1 Introduction

Multigrid methods have been very successful at speeding up the numerical solution of linear partial differential equations. By introducing several grids and exchanging information between them in a sophisticated way, different length scales that play an important role in the problem can be treated more or less independently; so procedures have been invented to generate the solution of the problem in a dramatically more efficient way as compared to the conventional relaxation algorithms on a single grid.

Monte Carlo simulations of models in statistical mechanics and of theories in elementary particle physics (quantum field theory) face similar problems. The conventional local Monte Carlo algorithms are very efficient at simulating the high-frequency aspects of the theory. Some of the most relevant aspects, however, are low-frequency phenomena, i.e., phenomena that extend over very long distances as compared to the elementary lattice distance. In this case, the conventional algorithms become forbiddingly slow. This slowing down occurs because these local algorithms can not treat large-scale phenomena in an efficient way. Lately multiscale algorithms of several types have been developed. Some of them have already brought about a real improvement in the quality of simulation data and have led to a better understanding of the relevant physical phenomena. Other multiscale algorithms are very promising and can reduce by a considerable factor the required CPU-time. In realistic theories of elementary particles one can not avoid introducing dynamical fermionic fields, describing quarks and leptons. The simulations of these theories require frequent inversions of the Dirac operator, a huge matrix. Due to the nature of this matrix, this can be accelerated considerably by deterministic multigrid methods, very similar to the ones used for solving partial differential equations.

The purpose of this review is to give a short overview of multiscale algorithms used in the simulation of the physical systems mentioned above. The emphasis is on the practical aspects of these algorithms and less on theoretical considerations. A complete review of this very active field of research would by now fill a book, and, therefore, this presentation can not avoid omitting some interesting developments.

Although many similarities exist between the numerical solution of partial differential equations and the Monte Carlo simulation of physical systems, there are some basic differences with drastic consequences for the algorithms. Therefore we start out in Section 2 by summarizing the basics of Monte Carlo simulations as they are used in statistical mechanics and quantum field theory. The problems that beset the conventional algorithms are explained in Section 3. In Section 4 we describe the stochastic cluster algorithms, a whole family of algorithms that reduce or even eliminate critical slowing down in the simulation of spin models and some related theories. Multigrid Monte Carlo simulation algorithms and their results are summarized in Section 5; although this approach is more difficult to implement in practice, it could be, in the long run, the most promising one for the physically important lattice gauge theories. In Section 6 we describe the application of deterministic multigrid methods to the inversion of the Dirac operator in a simple quantum field theory with fermions; a successful generalization to the phenomenological, realistic theories would bring about a real breakthrough in the numerical investigation of these models. Finally, Section 7 is devoted to a summary and an outlook.

2 Monte Carlo simulations in Statistical Mechanics and Quantum Field Theory

2.1 Statistical Mechanics

Statistical mechanics studies and describes the properties of macroscopic systems, generally consisting of a very large number of elementary microscopic building blocks. Everyday examples of such systems are the content of a bottle of wine, a balloon filled with a mixture of gases, and a small permanent magnet to keep the refrigerator door shut. A common situation and therefore of great interest for physics is the one where such macroscopic systems are in thermal equilibrium with their surroundings at absolute temperature T. In statistical mechanics such systems are described by the canonical ensemble [1].

Let us be concrete and introduce one of the oldest and best understood systems in statistical mechanics: the two-dimensional ferromagnetic Ising model [2, 3]. Consider a very large two-dimensional simple square lattice; its area is V and its vertices are labeled by a pair of integers $(i_1, i_2) \equiv i$. A spin variable S_i that takes the values $+1$ or -1 sits at each of the

(a)	(b)
+ + + + + +	+ + − + − +
+ + − + + +	+ − − + + −
+ + + + + +	− + + − − −
+ + + − − +	+ − + + − +
+ + + + + +	+ − + − + −
+ + + + + +	− − − − + +

Figure 1: Sections of Ising system configurations: (a) $T \ll T_c$, (b) $T \gg T_c$.

vertices i. If one takes the plane of the lattice to be horizontal, one can think of the spin variables S_i as little magnets that can point either up ($S_i = +1$) or down ($S_i = -1$). Two examples of possible spin configurations are given in Figure 1. In order to get an oversimplified but still remarkably good model of a ferromagnet, it is sufficient to assume that the energy $E(c)$, corresponding to a spin configuration c, is given by the real expression

$$E(c) = -\sum_{\langle k,l \rangle} S_k\, S_l - h \sum_{m} S_m \,, \tag{1}$$

where h with $h \geq 0$ denotes an external magnetic field; the first sum runs over all pairs $\langle k, l \rangle$ of nearest neighbor lattice points, while the second sum runs over all lattice points m. From this expression it follows that two nearest neighbor spins that are lined up, i.e., both are pointing up or both are pointing down, contribute -1 to the energy E of the whole system. On the other hand, if they are pointing in opposite directions, their contribution is positive. These considerations indicate that lining up spins tends to lower the energy of the system. If $h > 0$, analogous arguments show that turning the spins in the upward direction also lowers the energy. The most relevant macroscopic physical quantities corresponding to a spin configuration c are

- the average magnetization density m:

$$m = \frac{1}{V} \sum_i S_i \,, \tag{2}$$

- the average internal energy density ϵ:

$$\epsilon = \frac{1}{V} [- \sum_{\langle k,l \rangle} S_k\, S_l] \,, \tag{3}$$

- the average connected spin-spin correlation function $C(R)$ for spins at a distance R (periodic boundary conditions are assumed):

$$C(R) = \frac{1}{V}[\sum_{(i_1,i_2)} S_{(i_1,i_2)} S_{(i_1,i_2+R)}] - m^2 . \tag{4}$$

The physical question of interest is: if the Ising model is in thermal equilibrium with its surroundings at absolute temperature T, what are the average values of m, ϵ and $C(R)$, and how do they depend on T and h?

Starting from a few plausible assumptions, it is shown in statistical mechanics that all macroscopic properties of systems, described by a real energy functions $E(c)$ and in thermal equilibrium with their surroundings at absolute temperature T, can be derived in a simple way. An important role is played by the partition function

$$\mathcal{Z}(\beta) = \sum_c \exp[-\beta\, E(c)] , \tag{5}$$

where $\beta \equiv \frac{1}{kT}$ and k is called the Boltzmann constant. In the definition of the partition function $\mathcal{Z}$, the sum runs over all possible configurations c of the variables of the system. Expectation values $\langle A \rangle$ for physically interesting quantities such as defined in Equations 2 through 4 can then be obtained as

$$\langle A \rangle = \frac{\sum_c A(c) \exp[-\beta\, E(c)]}{\mathcal{Z}(\beta)} , \tag{6}$$

where A stands for the quantity under consideration; $A(c)$ is its value for the configuration c of the variables of the model. Equation 6 can be rewritten in a more suggestive way:

$$\langle A \rangle = \sum_c A(c) w(c) , \tag{7}$$

with

$$w(c) = \frac{\exp[-\beta\, E(c)]}{\mathcal{Z}(\beta)} ; \tag{8}$$

it is useful to think of $w(c)$ as the probability with which a particular configuration c can occur.

We now have a prescription to answer all relevant physical questions about this kind of systems. In practice, however, the summations over all configurations are impossible to carry out explicitly, except for a few relatively simple models. The first two-dimensional system for which these sums have been evaluated exactly is the ferromagnetic Ising model, defined by Equation 1 with $h = 0$ [4]; the more general case with nonzero magnetic field h has still not been solved analytically up to now. The Ising model with zero magnetic field is interesting because it exhibits a *phase transition*, i.e., there is a certain value of T, usually denoted by T_c, where the macroscopic properties of the system change dramatically. Such transitions are known to occur in real materials. Water, for example, boils under atmospheric pressure at $T = 373.15^0 K$ ($100.00^0 C$) and changes from a liquid to a gas. A permanent magnet heated up above the Curie temperature (for iron $T_c = 1043^0 K$) looses its spontaneous magnetization completely. The two-dimensional Ising model with $h = 0$ was the first microscopic model that predicted such a behavior and in which the properties of the phase transitions could be studied quantitively. In the high-temperature phase the system is very disordered and the expectation value for the average magnetization density $\langle m \rangle$ is zero. In the low-temperature phase, however, $\langle m \rangle \neq 0$ and the system becomes more and more ordered for decreasing values of T; for $T = 0$ it is completely ordered, i.e., we will find either $\langle m \rangle = +1$ or $\langle m \rangle = -1$. In Figure 1(a) we present a section of a typical low-temperature Ising system, and in Figure 1(b) a typical high-temperature Ising system. The high-temperature phase and the low-temperature phase are separated by a phase transition at $T = T_c = \frac{2}{k \log(\sqrt{2}+1)}$; at the phase transition the system is said to be critical and T_c is the critical temperature. Investigating the expectation value for the connected spin-spin correlation function $\langle C(R) \rangle$, we find that, both in the high-temperature and in the low-temperature phase, for fixed T, the function $\langle C(R) \rangle$ decreases for increasing values of the distance R: for very large values of R this decay is dominated by an exponential factor $\exp(-R/\xi)$. This way we have defined the correlation length ξ, which has an interesting behavior as a function of T: the correlation length ξ grows as one moves closer in T to the critical value T_c, both from above and from below. In the limit $T \to T_c$, the correlation length ξ actually diverges. The other physical quantities $\langle \epsilon \rangle$ and $\langle m \rangle$ do not diverge for $T = T_c$; they are continuous functions of T. The behavior of the two-dimensional Ising model for $T = T_c$ and $h = 0$ (diverging correlation length ξ, continuous $\langle \epsilon \rangle$ and $\langle m \rangle$) is an example of a *second-order phase transition.*

2.2 Principles of Monte Carlo simulations in statistical mechanics

For the large majority of physically relevant models, no analytic solution has been found, and one is forced to be less ambitious. Several alternative methods exist: mean-field theory, series expansions, etc.; these approximation methods work well in some cases, but completely fail for other cases. Monte Carlo simulations [8, 9], on the other hand, are a very general alternative: they work for almost all relevant statistical models. A drawback, however, is the statistical nature of this method and its results.

The ideas behind a Monte Carlo simulation are the following. Instead of summing $A(c)$ in expressions like Equation 7 over all possible configurations of the system with the weight $w(c)$, one generates a Markov chain of sample configurations $\{c_1, c_2, \ldots, c_i, \ldots, c_j, \ldots\}$ with $w(c)$ as probability distribution. The generation of such a chain is governed by the transition probability matrix $P = \{P_{cc'}\} = \{\text{probability}(c \to c')\}$. If the system is presently in configuration c, then the next element in the Markov chain must be picked stochastically between all possible configurations, each coming with a probability equal to the corresponding element in the transition probability matrix. If the transition probability matrix satisfies

- stationarity:

$$\sum_{c'} w(c')\, P_{c'c} = w(c) \; ; \tag{9}$$

- irreducibility: The probability to generate any configuration c' from any other configuration c in a finite number of steps should be greater than zero.

One can prove that the average of the $A(c_i)$, where the c_i are the configurations of the Markov chain, is a good estimator for the expectation value $\langle A \rangle$, at least for very long Markov chains [8, 9].

2.3 Quantum Field Theory

The interactions of the basic consituents of matter (electrons, photons, quarks, etc.), also called elementary particles, are described by quantum field theories. In the framework of these theories a large number of physical phenomena can be analyzed and predicted well by perturbative series expansions, especially in the area of the electromagnetic and weak interactions [5]. Quantum Chromodynamics, the field theory that describes the strong interactions, should also predict the mass of the proton and the neutron, the constituents of all nuclei. This is, however, just one example of many more very important questions that do not lend themselves to investigations by perturbative expansions. Such phenomena can be investigated by studying euclidean field theories in four dimensions. By discretizing time and space, the theory is transformed into a lattice model that, besides the unusual number of dimensions (four!), resembles statistical mechanics models of the type discussed before [6]. Consequently, one can use all methods of statistical mechanics, including Monte Carlo simulations, to study the properties of quantum field theories. I will present the simplest example: the bosonic field theory describing free massive scalar particles. In this case the partition function Z is expressed as

$$\mathcal{Z} = \int [d\phi] \exp[-\frac{1}{2} \sum_{i,j} \phi_i D_{ij} \phi_j \ - \ \frac{1}{2} \sum_i m^2 \phi_i^2] \tag{10}$$

and the expectation value for the two-point function is given by

$$\langle \phi_i \phi_j \rangle = \frac{\int [d\phi] \phi_i \phi_j \exp[-\frac{1}{2} \sum_{i,j} \phi_i D_{ij} \phi_j \ - \ \frac{1}{2} \sum_i m^2 \phi_i^2]}{\mathcal{Z}} \ . \tag{11}$$

The variables ϕ_i of this theory are real numbers and they are located at the vertices i of a simple cubic four-dimensional lattice; D_{ij} are elements of a sparse matrix coming from the discretization of derivatives. The sum over all configurations of the system in Equations 10 and 11 is now represented symbolically by $\int [d\phi]$, the integral sign replacing $\sum$ because the variables ϕ_i are continuous real variables and not discrete integers.

Many elementary particles, such as electrons, neutrinos and quarks, must be described by fermionic fields; simulating quantum field theories with fermions is still quite controversial. These fields are anticommuting numbers that are very hard to handle in a practical way on computers [6]. Luckily they can be integrated out analytically from the expressions for the partition function and the expectation values, but this procedure creates a computational problem of a different kind. After this integration the expression for the partition function looks schematically like

$$\mathcal{Z} = \int [d\phi] \det\{\Delta[\phi]\} \exp\{-S_0[\phi]\}, \tag{12}$$

where $\Delta[\phi]$ stands for a sparse but extremely large matrix (volume of lattice $\times$ volume of lattice); the elements of this matrix are functions of the bosonic fields. The expectation value for the two-point function of the fermionic fields ψ has become

$$\langle\bar{\psi}_i\psi_j\rangle = \frac{\int[d\phi]\{\Delta[\phi]^{-1}\}_{ij}\det\{\Delta[\phi]\}\exp\{-S_0[\phi]\},}{Z}. \tag{13}$$

A more careful and complete discussion of the problems relating to the introduction of fermions would require several chapters and can not be pursued here. From an algorithmic viewpoint the important thing to know is that the most promising method for simulations with dynamical fermions, the hybrid Monte Carlo method [7] , requires frequent inversions of the huge matrix $\Delta[\phi]$. Inverting this matrix with conventional methods, like the conjugate gradient algorithm, uses an extravagant part of the total CPU time required for the simulation. Because of the physical nature of the matrix that must be inverted, one can be hopeful that multigrid methods can reduce considerably the CPU time needed.

3 Local Monte Carlo algorithms and Critical Slowing Down

3.1 Autocorrelation times

A Monte Carlo algorithm based on a Markov chain that satisfies the criteria listed in Section 2.2 (stationarity, irreducibility) is exact, but not necessarily efficient, i.e., the simulation could take centuries on the most sophisticated computers. The reason is that the sample configurations $\{c_1, c_2, \ldots, c_i, \ldots, c_j, \ldots\}$ can be very correlated statistically. In that case, the chain must be extremely long before the average of the $A(c_i)$ becomes a good estimator of $\langle A\rangle$.

Let us consider the series of "measurements" $\{A(c_1), A(c_2), \ldots, A(c_i), \ldots, A(c_j), \ldots\}$ of the physical quantity $A(c)$ corresponding to the chain of sample configurations in order to quantify this effect. The unnormalized autocorrelation function $C_{AA}(t)$ can now be defined by

$$C_{AA}(t) = \Big[\frac{1}{N-t}\sum_{i=1}^{N-t} A(c_i)A(c_{i+t})\Big] \ - \ \Big[\frac{1}{N}\sum_{i=1}^{N} A(c_i)\Big]^2, \tag{14}$$

and the normalized autocorrelation function $\rho_{AA}(t)$ by

$$\rho_{AA}(t) = \frac{C_{AA}(t)}{C_{AA}(0)}. \tag{15}$$

In all of these considerations, we will assume that the total number of elements N in the chain is much larger than the values of t we are interested in. In general, $\rho_{AA}(t)$ decays exponentially ($\sim e^{\frac{-|t|}{\tau}}$) for large values of t. Then the *exponential autocorrelation time* corresponding to quantity A can be defined as

$$\tau_{exp,A} = \lim_{t\to\infty} \quad \sup \frac{t}{-\log|\rho_{AA}(t)|}. \tag{16}$$

Of course, $\tau_{exp,A}$ depends on the physical quantity A under consideration, and therefore we define the overall exponential autocorrelation time τ_{exp} as the exponential autocorrelation time corrsponding to the slowest mode:

$$\tau_{exp} = \sup_f \tau_{exp,f} \ . \tag{17}$$

Another relevant quantity is the *integrated autocorrelation* time $\tau_{int,A}$ corresponding to the quantity A defined by

$$\tau_{int,A} = \frac{1}{2} \sum_{t=-\infty}^{+\infty} \rho_{AA}(t) \ . \tag{18}$$

These autocorrelations are very significant for practical simulations. A Markov chain is started off with an initial configuration c_1. In order to avoid any bias in the results due to the particular choice of c_1, one must discard the first few configurations generated. Based on theoretical considerations and practical experience, no bias is expectied if at least 20 τ_{exp} configurations are dropped before collecting statistics is started. The integrated autocorrelation time $\tau_{int,A}$, on the other hand, is a good measure for how many consecutive measurements $A(c_i)$ in the chain are statistically correlated. Consequently, the number of statistically independent measurements that should be used to estimate the statistical error of the average of the $A(c_i)$ is $N/(2\tau_{int,A})$, where N stands for the total number of measurements.

3.2 Local update algorithms and critical slowing down

Until a few years ago almost all Monte Carlo simulations of physical models were performed using either the Metropolis [10] or the related heat-bath algorithm [11]. For the two-dimensional Ising model, I will describe how the heat-bath algorithm, starting from a given spin configuration c_i, generates the next configuration c_{i+1} in the Markov chain of sample configurations. Such a heat-bath update of the whole lattice is accomplished by updating all spins consecutively, preferably in a random order, and the procedure to update one spin S_i consists of two steps. First the sum of the four nearest neighbor spins is computed:

$$\Sigma_i = \sum_{j=1}^{4} S_{n_j} \ ; \tag{19}$$

then the spin S_i is assigned the "new" value $+1$ with probability $P(+1)$ or the value -1 with probability $P(-1)$, where

$$\begin{aligned} P(+1) &= \frac{\exp[-\beta(-\Sigma_i - h)]}{\exp[-\beta(-\Sigma_i - h)] + \exp[-\beta(+\Sigma_i + h)]} \\ P(-1) &= 1 - P(+1) \ . \end{aligned} \tag{20}$$

In practice a random number r is drawn, equally distributed between 0 and 1; if $r < P(+1)$ the spin S_i is assigned the new value $+1$, otherwise the value -1. A crucial property of this procedure is that the probabilities $P(+1)$ and $P(-1)$ only depend upon the nearest neighbor spins, i.e., the procedure is very local. A direct consequence of this locality is that many update sweeps of the whole system will be required before information is exchanged between remote parts of the system.

The spatial correlation in a system is described by the longest correlation length ξ as defined above in Section 2.1. It is experimentally found that for all models the autocorrelation times behave in the following way

$$\begin{aligned} \tau_{exp} &\sim \min(L,\xi)^{z_{exp}} \\ \tau_{int,A} &\sim \min(L,\xi)^{z_{int,A}} , \end{aligned} \tag{21}$$

where L stands for the linear dimension of the system; z_{exp} and $z_{int,A}$ are *dynamical critical exponents of critical slowing down.* Generally, one is interested in the behavior of infinite systems. In simulations, however, one is forced to work with finite volumes. In order to keep the systematic error due to the finiteness of the system small, it is necessary to keep the linear dimension of the system L much larger than the correlation length ξ. In that case, the previous equations simplify to the more familiar forms

$$\begin{aligned} \tau_{exp} &\sim \xi^{z_{exp}} \\ \tau_{int,A} &\sim \xi^{z_{int,A}} . \end{aligned} \tag{22}$$

Because of their locality the conventional Monte Carlo algorithms have dynamical critical exponents around 2 for all models. Heuristic arguments based on treating information propagation through the system as a random walk give exactly 2. This makes the local algorithms unsuitable for a thorough investigation of second-order phase transitions, i.e. systems where the correlation length diverges. From Equation 22, it follows that in this case the autocorrelation times diverge approximately quadratically in the correlation length ξ and this means that the number of Monte Carlo update sweeps of the system required to generate statistically independent configurations diverges in the same way. The phenomenon that the autocorrelation times diverge together with the correlation length ξ is called *critical slowing down* and the values of the dynamical critical exponents z are a measure for it.

Physicists are especially interested in systems with very large or diverging correlation lengths, hence the importance of developing algorithms without critical slowing down ($z = 0$) or at least strongly reduced critical slowing down. It is almost self-evident that the problem could be alleveated by using nonlocal algorithms. However, inventing nonlocal algorithms that also satisfy the criteria of irreducibility and stationarity, strict requirements for a valid Monte Carlo simulation, turned out to be a formidable task. Only during the last few years, real progress was made and in the following sections I will discuss two successful approaches: stochastic cluster algorithms and multigrid Monte Carlo methods.

4 Stochastic Cluster Algorithms

4.1 The Swendsen-Wang algorithm for the Potts model

The discovery of this cluster algorithm in 1987 by Swendsen and Wang [12] was an algorithmic breakthrough, and started off a real scramble for algorithms that eliminate critical slowing down in other interesting models. I will describe the algorithm for the ferromagnetic Ising model ($h = 0$) in two dimensions, which is identical to the 2-state Potts model. Just as for the heat-bath algorithm, I will describe how, starting from a spin configuration c_i, this cluster algorithm generates the next configuration c_{i+1} in the Markov chain of sample configurations. The procedure consists of three steps:

1. Visit all links connecting nearest neighbor spins S_k and S_l. If $S_k = S_l$ put a *bond* on that link with probability $p(\text{bond}) = 1 - \exp(-\beta)$ and no bond otherwise. If $S_k \neq S_l$, never put a bond on that link.

2. Construct the stochastic clusters of lattice points on the lattice. The criterion for two lattice points to belong to the same cluster is that there exists a path connecting them completely covered with the bonds that were put in step 1.

3. Visit all the stochastic clusters, including the ones that consist of one lattice point only. For each cluster pick at random the value +1 or −1 and set all spins of that cluster equal to the selected value.

The spin configuration obtained in this way is the next configuration in the Markov chain. It is not difficult to show that this procedure satisfies the criteria of irreducibility and stability of the probability distribution [12]. The authors of the method measured the dynamical critical exponent $z_{exp,E}$ (Equation 21), and found $z_{exp,E} = 0.35$ as compared to 2.15 for the conventional local algorithms. Since then, lower values have been quoted for the Swendsen-Wang algorithm, finally culminating in the claim by Heermann and Burkitt [13] that both the exponential and the integrated critical exponents measured for the magnetization would actually be compatible with zero. The situation is still fairly confusing but, in any case, critical slowing down has been strongly suppressed, if not completely eliminated.

In the context of a different model Wolff [14] developed an interesting *single-cluster modification* of this algorithm. The next configuration c_{i+1} in the Markov chain of sample configurations is now generated as follows. First a lattice point is picked at random, then the cluster to which this point belongs is constructed according to the rules given above, and finally all spins of this cluster are flipped. This modification was found to be superior to the conventional Swendsen-Wang method, even more so in three dimensions than in two dimensions[15].

4.2 Successful generalizations to other models

In the past two years stochastic cluster algorithms have been developed for a whole series of spin models and even two lattice gauge theories. Although all of these algorithms are based on the same principles as the original Swendsen-Wang algorithm for the Potts model, none of these extensions are really straightforward, and all of them require some fresh input. To show what is meant by this, I will give a somewhat unorthodox presentation of the stochastic cluster algorithm for the so-called $O(n)$ models [16] . To make it as simple as possible, let us consider the $O(2)$ model in two dimensions, better known as the XY-model in two dimensions. Consider a regular simple square lattice as for the Ising model, but with two-dimensional unit vectors $\mathbf{e}_i$ at the sites i taking the place of the Ising spins S_i. The energy function $E(c)$ for the XY-model has the form

$$E(c) = -\sum_{\langle k,l \rangle} \mathbf{e}_k . \mathbf{e}_l \,, \tag{23}$$

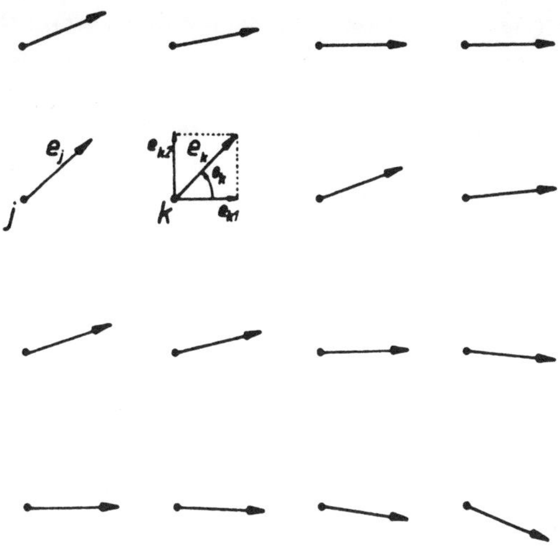

Figure 2: Section of an XY-system configuration with definition of variables.

where $\mathbf{e}_k.\mathbf{e}_l$ stands for the usual scalar product

$$\mathbf{e}_k.\mathbf{e}_l = e_{k1}e_{l1} + e_{k2}e_{l2} = \cos(\theta_k - \theta_l) \tag{24}$$

of two-dimensional vectors. In Figure 2 a section of a typical system is presented. Starting from a given configuration c_i, the procedure to generate the next configuration c_{i+1} in the Markov chain of sample configurations consists now of the following steps:

- Pick a random angle ϕ between 0 and 2π and rotate all vectors $\mathbf{e}_i$ of the system over this angle. Notice that this procedure does not change the energy of the system. From now on $\mathbf{e}_i$ denotes the vectors after they have been rotated over the angle ϕ.

- The energy function can be rewritten in the following way

$$\begin{aligned} E(c) &= -\sum_{\langle k,l\rangle}[e_{k1}e_{l1} + e_{k2}e_{l2}] \\ &= -\sum_{\langle k,l\rangle}[S_k S_l \; |e_{k1}| \; |e_{l1}| \; + e_{k2}e_{l2}] \, , \end{aligned} \tag{25}$$

  with $S_i \equiv \text{sgn}(e_{i1})$. The variables S_i defined in this way only take the values $+1$ or -1 and are therefore Ising-like variables.

- These newly defined S_i are updated with the conventional Swendsen-Wang algorithm as it was described in Section 4.1, except for the fact that the probability to put a bond on the link connecting the sites k and l should now be taken to be $p(\text{bond}) = 1 - \exp(-\beta|e_{k1}| \; |e_{l1}|)$. After finishing this procedure and remembering the definition $S_i \equiv \text{sgn}(e_{i1})$, we have constructed the new configuration c_{i+1}.

Wolff[16] actually used the single-cluster modification, which was discussed at the end of Section 4.1. The XY-model in two dimensions has a high-temperature phase (*vortex* phase) and a low-temperature phase (*spin-wave* phase). The spin-wave phase has the peculiar property that the system is critical everywhere in this phase, i.e., the correlation length $\xi = \infty$. In both phases the dynamical critical exponents of critical slowing are found to be in agreement with 0, i.e. critical slowing down has been eliminated completely. This made it possible to investigate the phase transition, which is not a second-order phase transition, much more thoroughly than was possible with the conventional algorithms, and some important observations about the nature of the transition could be made [16].

In this presentation it is impossible to give a complete overview of all the models for which successful stochastic cluster algorithms have been developed; more complete information can be found in a couple of excellent reviews, that have been written on the subject [9, 17]. I am forced to limit myself to a brief, incomplete, and necessarily subjective selection, with some emphasis on newer developments that have not been covered in these reviews.

Using an Ising embedding similar to the one for the XY-model, a stochastic cluster algorithm was developed for the ϕ^4 theory in four dimensions, a theory relevant for a better understanding of the weak and electromagnetic interactions in elementary particle physics [18]. A recent investigation indicates that, using the single-cluster modification, critical slowing down is probably eliminated completely [19].

In the area of lattice gauge theories the equivalent of the Swendsen-Wang algorithm was developed for the $Z(2)$ theory in three dimensions [20]. For the $SU(2)$ lattice gauge theory at finite temperatures (four dimensions with periodic boundary conditions, the size of the lattice in the fourth dimension is 1) a combination of a conventional algorithm with a stochastic cluster algorithm leads to a strong reduction of critical slowing down [21].

Statistical mechanics models that describe surface phenomena are also beset by critical slowing down. For the discrete Gaussian model (SOS) the situation was drastically improved by a recently developed cluster algorithm [22].

A generalization of the Swendsen-Wang algorithm had been developed to investigate the behavior of the Potts models with magnetic field near criticality [23]. The high-precision measurements, made possible by this algorithm, make it practically feasible to determine the central charge of the conformal field theory corresponding to critical two-dimensional lattice models by means of Monte Carlo simulations [24].

4.3 Prospects and limitations

Soon after the stochastic cluster algorithm had been developed for the Potts model [12], these ideas were combined with the multigrid philosophy [25]; this approach was successfully followed to investigate both the two-dimensional and three-dimensional Ising model [25, 26]. Instead of allowing the clusters to grow all over the lattice, according to the rules explained in Section 4.1, one restricts the growth in some way; the restricted clusters are then the spin variables for the next-coarser grid. This theoretically attractive procedure can of course be used recursively and one can introduce V-cycles and W-cycles. Because of high efficiency and great simplicity of the regular Swendsen-Wang algorithms, this more complicated approach has not been actively pursued by many researchers up to now for practical simulations. This stochastic multigrid method has, however, theoretical advantages, it shares with other multigrid methods. Recently it was shown that besides eliminating critical slowing down, stochastic multigrid Monte Carlo methods, used in the optimal way, drastically reduce the number of finest-grid configurations that are needed to reduce the statistical error [27]. It was shown explicitly for the Ising model in two dimensions, that it is possible to derive quantities like the critical temperature T_c in the thermodynamic limit, i.e. for an infinitely large system, to accuracy ϵ at a computational cost that does not grow faster than ϵ^{-2}; this is of course the best one can hope to reach with simulation methods.

The list of successes for the stochastic cluster algorithms is impressive. The euphoria has died down, however, since it has become clear that such algorithms can only be developed for a fairly small group of theories, which does include neither the relevant field theories for elementary particle phenomenology nor many interesting models in statistical mechanics as for example the $CP(n-1)$ models. A *no-go theorem* for many of these theories has actually been formulated recently [28]. Another practical drawback of the cluster-algorithms is the fact that they do not vectorize or parallelize well; recently some progress has been achieved, at least on vector processors [29, 30].

The theoretical derivation of the important aspects of these algorithms, e.g., the values for the dynamical critical exponents z, is still lacking, except for some trivial cases.

5 Multigrid Monte Carlo simulations in statistical mechanics

The origin of critical slowing down in simulation algorithms is very similar to the origin of the inefficiency of local relaxation algorithms (Gauss-Seidel, Jacobi, etc.) at handling low-frequency components of the error in the numerical solution of differential equations. The latter problem was overcome by the invention and careful application of multigrid techniques; the long list of topics covered at this conference tells the story. It is surprising that it took many years before these ideas were adapted to the simulations of statistical models [31, 32, 33]. For a detailed description of the background of this method I refer to the literature [36].

5.1 Multigrid Monte Carlo algorithm for the two-dimensional XY-model

I will return to the model that was already discussed in the framework of the stochastic cluster algorithms: the two-dimensional XY-model (Section 4.2). This model was studied by means of this method by two different groups with complete agreement as for the results [34, 35]; I will closely follow their method.

When multigrid methods are used for the numerical solution of partial differential equations, the finest grid is chosen directly as a function of how well we want the discretized solution to approximate the continuum solution. In statistical mechanics one wants in general information about infinite systems. The main criterion to fix the size of the lattice is then usually how large must the lattice be, in order to keep the finite-size effects, i.e., the systematic errors due to the finiteness of the system, below a certain chosen level. Let us assume that it has been decided to perform the simulation on a $2^M \times 2^M$ lattice, the finest grid Ω^M that we will consider; in this section upper indices will be grid labels. The coarser

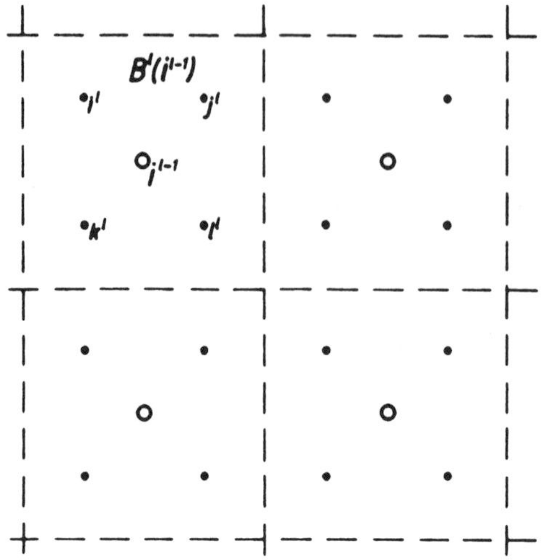

Figure 3: Coarsening of the grid for the XY-model: the points i^l, j^l, k^l and l^l of the Ω^l grid are represented by one grid point i^{l-1} on the Ω^{l-1} grid.

grids that we will use are denoted by Ω^l with $l = M-1,\ldots,1$. As energy function E^l on each of the grids Ω^l, we take a generalized version of Equation 23 :

$$E^l(c^l) = -\sum_{\langle k^l, l^l\rangle} [\alpha^l_{i^l j^l} \cos(\theta^l_{i^l} - \theta^l_{j^l}) + \beta^l_{i^l j^l} \sin(\theta^l_{i^l} - \theta^l_{j^l})] \,. \tag{26}$$

Keeping in mind the definition of the model on the finest grid Ω^M (Equation 23), we have

$$\begin{aligned} \beta^M_{i^l j^l} &= 0 \\ \alpha^M_{i^l j^l} &= \beta \,. \end{aligned} \tag{27}$$

An essential ingredient of any multigrid algorithm is the *interpolation* procedure I^l_{l-1}, that translates the corrections obtained on the coarser grid Ω^{l-1} back to the next-finer grid Ω^l. As usual we identify one coarse-grid point with a 2×2 cell on the next-finer grid. In Figure 3 we see how the coarse grid point i^{l-1} is identified with the 2×2 cell $B^l(i^{l-1})$ consisting of the fine-grid points i^l, j^l, k^l, and l^l. One out of many possible choices for the interpolation is piecewise-constant:

$$\theta^l_{i^l} = \theta^{l-1}_{i^{l-1}} \qquad \text{if} \qquad i^l \in B^l(i^{l-1}) \,. \tag{28}$$

In the multigrid Monte Carlo method, there is little freedom as for choosing how to coarsen the energy function E^l. We must generate the finest-grid configurations c^M with the correct probability given by Equation 8, otherwise the whole procedure is worthless. This condition is satisfied if the coefficients on the coarse grid $\alpha^{l-1}_{i^{l-1} j^{l-1}}$ and $\beta^{l-1}_{i^{l-1} j^{l-1}}$ are chosen as a function of $\alpha^l_{i^l j^l}$, $\beta^l_{i^l j^l}$ and $\theta^l_{i^l}$ on the fine grid in such a way that for arbitrary values of the coarse grid variables $\theta^{l-1}_{i^{l-1}}$ the following equation is satisfied:

$$E^{l-1}(c^{l-1}) = E^l(c^l + I^l_{l-1} c^{l-1}) \,; \tag{29}$$

The right hand side of this equation stands for E^l where the variables $\theta^l_{i^l}$ have been replaced by their original value $\theta^l_{i^l}$ plus the correction obtained from the coarse grid variables by means of the interpolation operation I^l_{l-1}.

Finally, the updating algorithm on the different grids is fundamentally different from the relaxation algorithms that are used when solving partial differential equations. In the latter case, a fully deterministic procedure is followed as, e.g., the Gauss-Seidel or Jacobi schemes. This is now replaced by a stochastic procedure $UP(\Omega^l, c^l)$; for the XY-model both a heat-bath algorithm [34] and some other procedures including the conventional Metropolis algorithms have been used [35].

Now a complete γ-cycle multigrid Monte Carlo scheme to generate the next configuration c_{i+1} in the Markov chain can be defined by means of the recursive definition of the operation $MGMC(\Omega^l, c^l)$ on the variables of the grid Ω^l:

1. perform ν_1 stochastic update sweeps $UP(\Omega^l, c^l)$ on the configuration of grid Ω^l,
2. **if** $l = 1$, then go to 4,
   else compute the parameters $\alpha^{l-1}_{i^{l-1}j^{l-1}}$ and $\beta^{l-1}_{i^{l-1}j^{l-1}}$ of E^{l-1} according to the coarsening procedure described above,
   $\theta^{l-1}_{i^{l-1}} \leftarrow 0$,
   perform γ times the procedure $MGMC(\Omega^{l-1}, c^{l-1})$,
   endif,
3. perform the correction $\theta^l_{i^l} \leftarrow \theta^l_{i^l} + I^l_{l-1}\theta^{l-1}_{i^{l-1}}$ with $i^l \epsilon B^l(i^{l-1})$,
4. perform ν_2 stochastic update sweeps $UP(\Omega^l, c^l)$ on the configuration on grid Ω^l .

The best results for the XY-model are obtained with full W-cycles ($\gamma = 2$). Measuring the integrated auto correlation times for several physical quantities, both groups agree that the z_{int} corresponding to the slowest mode is compatible with 0 in the low-temperature (spin-wave) phase. In the high-temperature (vortex) phase they find a value around 1.4, which should be compared with the value obtained for the heat-bath algorithm: 2.1 ± 0.3 [34]. Hence, the multigrid Monte Carlo algorithm eliminates critical slowing down completely for the spin-wave phase. In the vortex phase the algorithms suffers less from slowing down than the heat-bath algorithm, but it can not compete with the simpler stochastic cluster algorithm, which seems to eliminate critical slowing down completely in both phases.

5.2 Other applications and prospects

Because this method is more complicated to implement than the stochastic cluster algorithm and also because of the remarkable successes of the cluster method, the multigrid Monte Carlo method has not yet been tried out on many models.

The one-component ϕ^4 model was investigated by means of this method [42, 43]. The method does not eliminate critical slowing down at all in this case: the values for the dynamical critical exponents z are the same as the ones obtained for the conventional local algorithms [42]; the authors also provide arguments why this somewhat disappointing result should be expected. Nevertheless a considerable gain in computational speed is found because $\tau_{int}(\text{multigrid}) \approx \frac{1}{20}\tau_{int}(\text{heat-bath})$.

For the free bosonic field theory that we introduced as an example in Section 2.3, one can prove analytically that the autocorrelation times are bounded if the multigrid Monte Carlo method is used [32], i.e., no more critical slowing down.

This method is in principle generalizable to lattice gauge theories and many interesting models in statistical mechanics for which the stochastic clusters are known not to work [28]. It is also known that, in the physically relevant parameter range, assymptotically free lattices gauge theories as $SU(2)$ and $SU(3)$ have important properties in common with the free bosonic theory and the spin-wave phase of the XY-model, two models for which the multigrid Monte Carlo works well. Therefore it is not impossible that this method will become very useful for the simulation of lattice gauge theories on large lattices for high values of β were the dynamics in the system is quite similar to the spin-wave dynamics of the XY-model.

The efficiency of the method can be improved even more by measurements of physical quantities every time the coarse grids are updated [27]; this improvement is similar in spirit to the *improved estimators*, which were found to be useful in the framework of the stochastic cluster algorithms [44].

Finally, a practical advantage of this method is that it can be vectorized and parallelized efficiently in a straightforward way.

6 Deterministic multigrid methods for inverting the Dirac operator

The interactions of elementary particles are described by quantum field theories. Because fermions (electrons, quarks, etc.) play an important role, one can not hope to have a good description of nature, if the model does not contain fermions. Introducing them in the four-dimensional euclidean theories, which are studied on the lattice, is still a problem, which has not been solved in a completely satisfactory way. There exist a few competing prescriptions, but I will consider just one of them: the staggered fermions [37]. As was mentioned in Section 2.3, these fermionic fields are anticommuting; they must be integrated out analytically because there is no other practical way to treat them correctly in a simulation. This procedure, however, creates an additional computational problem. A huge (lattice volume × lattice volume) matrix, called the *Dirac operator*, must be inverted frequently, if the sytem is updated by means of the *hybrid Monte Carlo method* [7]; this method is at present the only known scheme that is at the same time rigorous and efficient enough to make realistic simulations possible on reasonably sized lattices (typical size is 24×16^3), albeit at a tremendous computational cost (many thousands of hours on one processor of a CRAY-Y-MP). There are strong arguments that any practical and rigorous algorithm with fermions will involve frequent inversions of the Dirac operator. Recently it was shown in the framework of a somewhat simpler theory, the *massive Schwinger model* ($U(1)$ gauge theory coupled to massive fermions) in two dimensions, that the inversion of the Dirac operator can be accelerated considerably by means of deterministic multigrid methods, well-known from the numerical solution of differential equations [39, 40, 41]. Although the details of this work are technically

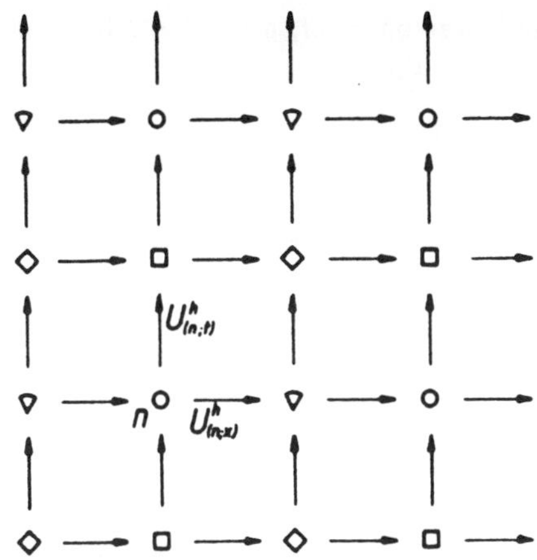

Figure 4: The massive Schwinger model: the four types of vertices indicate the interleaving sublattices (staggered fermions).

quite complicated and can not be presented here, I will point out in Section 6.1 where the similarities and the differences between this problem and the solution of differential equations lie.

6.1 Application to the massive Schwinger model

The massive Schwinger model is defined on a square two-dimensional lattice Ω^h with periodic boundary conditions. The lattice points are labeled by two dimensional vectors $\mathbf{n} \equiv (n_s, n_t)$ with n_s and n_t integers; the mesh size is h. In this section the upper indices denote the mesh size. The model describes massive fermions (the mass is denoted by m_q) coupled to a massless $U(1)$ gauge field. The gauge group elements (complex numbers of norm 1 in the $U(1)$ case) are located on the links of the lattice and they are denoted by $U^h_{(n_s,n_t;\mu)}$ for the link starting at the point (n_s, n_t) in the positive μ-direction, where μ is a direction vector and can only have the values $\mathbf{x} = (1,0)$ for the x-direction and $\mathbf{t} = (0,1)$ for the t-direction; for the opposite direction the complex conjugate value must be taken. The second kind of fields are the fermion fields ψ. In the staggered fermion formalism [37], the fermionic field ψ is actually spread out over a 2×2 cell, i.e., the fermionic field has four components, which are located on four interleaving lattices and can be treated as separate fields. In Figure 4, we show a section of the lattice: the arrows on the links indicate the direction for which the link variables U^h are defined and the four different kinds of symbols for the lattice points denote the four interleaving lattices, each of them corresponding to one component of the fermionic field.

Giving details about the hybrid Monte Carlo algorithm, or the quenched approximation [45], would lead us too far, but the computational problem that must be faced can be defined in a few lines. The part of the procedure that requires a very large share of the total CPU time is the frequent inversion of the Dirac operator, a (lattice volume × lattice volume) matrix $M^h[U^h]$. The elements $M^h[U^h]_{\mathbf{n},\mathbf{m}}$ of this matrix are defined by

$$\begin{aligned} M^h[U^h]_{\mathbf{n},\mathbf{m}} &= m_q\,\delta(\mathbf{n},\mathbf{m}) \;+\frac{1}{h}[\delta(\mathbf{m},\mathbf{n}-\mathbf{x})U^{h\,\dagger}_{(\mathbf{n}-\mathbf{x};\mathbf{x})} - \delta(\mathbf{m},\mathbf{n}+\mathbf{x})U^h_{(\mathbf{n};\mathbf{x})}] \\ &\quad +\frac{1}{h}(-1)^{n_x}[\delta(\mathbf{m},\mathbf{n}-\mathbf{t})U^{h\,\dagger}_{(\mathbf{n}-\mathbf{t};\mathbf{t})} - \delta(\mathbf{m},\mathbf{n}+\mathbf{t})U^h_{(\mathbf{n};\mathbf{t})}]\,, \end{aligned} \tag{30}$$

where †denotes the complex conjugate. The computation of the inverse of $M^h[U^h]$ is achieved by solving the linear system

$$\sum_{\mathbf{m}} M^h[U^h]_{\mathbf{n},\mathbf{m}}\Phi^h[U^h]_{\mathbf{m}} = \delta(\mathbf{n},\mathbf{0})\,. \tag{31}$$

It can readily be seen that the solution $\Phi^h[U^h]_{\mathbf{m}}$ is one of the columns of the the inverse of $M^h[U^h]$. All of this is extremely similar to the numerical problem that must be handled when solving numerically linear partial differential equations. There are, however, two essential differences. The first one is that the elements of the matrix $M^h[U^h]$ depend upon the value of the group elements U^h on the links, which have been generated by a stochastic process; this dependence is a consequence of the gauge invariance of the theory. The second difference is due to the four components of the fermionic field in the staggered formulation; this shows up as the factor $(-1)^{n_x}$ in the definition of the Dirac operator (Equation 30).

The overall procedure used in this case is identical to the one for differential equations. One uses a series of increasingly coarser grids Ω^h, Ω^{2h}, Ω^{4h}, etc. . A restriction operation I_h^{2h} is defined for moving residuals to the next-coarser grid as well as an interpolation operation I_{2h}^h for translating the computed corrections back to the next-finer grid. Finally one must define the Dirac operator on the coarser grids and choose a relaxation algorithm, e.g., in this case the Kaczmarz algorithm was selected [38]. V-cycles or W-cycles can then be used to accelerate the solution of the linear set in Equation 31. Building on the physical principle that the the form of the equations and of the Dirac operator should not change when the mesh size h is doubled, the expression for $M^{2h}[U^{2h}]$ is assumed to be the same as in Equation 30 with the obvious replacements $h \to 2h$ and $U^h \to U^{2h}$; this procedure implies of course that a restriction operation for the link elements U^h must be defined. The gauge invariance of the theory, which is the origin of the dependence of the Dirac operator $M^h[U^h]$ on the link elements U^h, must be taken into account in the definition of the restriction and interpolation operation as well as in the coarsening of $M^h[U^h]$. A simple example how this is done in the restriction operation is the following. If one computes the average of the fine-grid residuals $R^h_{(n_x,n_t)}$ and $R^h_{(n_x+2,n_t)}$ to put the result on a coarse-grid point that can be identified with the fine-grid point (n_x+1, n_t), one must *parallel transport* the contributions to the point (n_x+1, n_t), i.e., the value of the restricted residual on the coarser grid would be $\frac{1}{2}[U^h_{(n_x+1,n_t;\mathbf{x})}R^h_{(n_x+2,n_t)} + U^{h\,\dagger}_{(n_x,n_t;\mathbf{x})}R^h_{(n_x,n_t)}]$. Also in the interpolation operation and in the definition of the Dirac operator on the coarser grid, the link elements of the finer grid must be used as *parallel transporters*. For this reason, the authors of the method chose to call it the parallel-transported multigrid method. Keeping the form of the Dirac operator M unchanged on the different grids also means that, on the coarser grids, one also has four different components of the fermionic fields, i.e., four different interleaving lattices as in Figure 4. Consequently the restriction and interpolation operations should respect this property; for example, the residual on a coarser grid corresponding to a particular component of the fermionic field should only get contributions from residuals on the finer grids corresponding to the same fermionic component.

The efficiency of this method depends strongly on the two parameters of the massive Schwinger model: the mass of the fermions m_q, which appears explicitly in the expression of the Dirac operator, and β, which is an important parameter in the expression of the energy $E(c)$ for this model. For high values of β, the gauge link configurations are in general *smooth*, where smoothnes has a special meaning in the framework of lattice gauge theories. Two lattice points can be connected by many directed paths of links, and for each path, keeping track of its direction, one can compute the product of the U^h's situated on these links. If for each pair of points all these products have the same value, the configuration of the U^h's is said to be perfectly smooth. On the other hand, if for a pair of points close together and considering only relatively short paths, a wide variety of values for these products is found, the configuration is considered to be not smooth at all. Summarizing the results of the investigation, one can say that for very smooth configurations, i.e., for very high values of β, full W-cycles with 3 sweeps before and after every coarsening or interpolation work extremely well: there is a gain of several orders of magnitude in efficiency, if the multigrid algorithm is compared with one-level Kaczmarz relaxations. For lower values of β the efficiency is lower and from time to time configurations appear for which the above described method actually diverges. For these difficult configurations convergence is restored, if cut-off W-cycles are used; a W-cycle, cut off at level $\Omega^{2^n h}$, is the same as a full W-cycle, except that the corrections coming from grids coarser than $\Omega^{2^n h}$ are disregarded.

6.2 Prospects

The results for the massive Schwinger model obtained with the parallel-transported multigrid method are very encouraging [40, 41]. It should be possible to develop a robust multigrid solver for a large physically interesting range of parameters, if the routine is made adaptive in the following sense: it must choose by itself between several alternative procedures (e.g. full W-cycle, cut-off W-cycle, etc.), basing its choice on the properties of the configuration under investigation.

The method can be generalized to the phenomenologically important theories as, e.g., $SU(3)$ lattice gauge theory with fermions in four dimensions (QCD). In order to develop a solver that is useful for the large-scale simulations that are presently being performed for these theories, a careful investigation of the different possible coarsening and interpolation operations must be undertaken. To make the solver robust, one must almost certainly make it adaptive in the same way as for the massive Schwinger model; possibly more sophisticated alternatives will be needed for handling certain configurations, e.g., coarse-level polynomial acceleration.

7 Conclusions

Critical systems, i.e. systems with diverging correlations, are very important both in standard statistical mechanics and in the study of quantum field theories. In order to obtain very accurate results by Monte Carlo simulations, one must be able to approach criticality closely and collect good statistics. With the conventional local algorithms, this is practically impossible, even on supercomputers, because of severe critical slowing down. The development of algorithms with strongly reduced or no critical slowing down is therefore an important task.

Stochastic cluster algorithms and their stochastic multigrid extension are very efficient for a whole series of discrete models and simple continuous models where Ising variables can be embedded; in some cases they eliminate critical slowing down completely. It is highly improbable, however, that this approach can be generalized to lattice gauge theories and more complicated statistical models with continuous symmetries. Another practical drawback is that these algorithms can not be vectorized or parallelized in a straightforward way, although some progress has been made in this area.

Multigrid Monte Carlo methods are promising for models with continous symmetries, especially for the investigation of phases with properties resembling those of the spin-wave phase in the XY-model. In principle these methods can be generalized to lattice gauge theories, and they are expected to be effective in the physically relevant range of parameters. They can be vectorized and parallelized very well.

In order for quantum field theories to describe the elementary particles and their interactions in a realistic way, fermionic fields must be included. This makes frequent inversions of the Dirac operator necessary and this requires an extravagantly large share of the total CPU time for these simulations. Deterministic multigrid methods can reduce this share considerably.

Finally, multigrid methods can be developed to compute thermodynamic quantities (infinitely large systems) in the optimally efficient way by using information obtained from coarser grids efficiently.

References

[1] The different ensembles are described in every Statistical Mechanics textbook; a good pedagogical introduction to the field can be found in F. Reif, *Fundamentals of statistical and thermal physics* (McGraw-Hill Book Company,1965).

[2] E. Ising, Z. Physik 31 (1925) 253.

[3] B. M. McCoy and T. T. Wu, *The two-dimensional Ising Model* (Harvard University Press,1973).

[4] L. Onsager, Phys. Rev. 65 (1944) 117; for an clearer and shorter derivation see Reference [3].

[5] A large number of books stresses the continuum approach to quantum field theories; one of the most comprehensive textbooks of this kind: C. Itzykson and J.-B. Zuber, *Quantum Field Theory* (McGraw-Hill International Book Company, 1980).

[6] M. Creutz, *Quarks, gluons and lattices* (Cambridge University Press, 1983).

[7] S. Duane, A. D. Kennedy, B. J. Pendleton, and D. Roweth, Phys. Lett. B195 (1987) 216.

[8] J. M Hammersley and D. C. Handscomb, *Monte Carlo Methods* (Chapman and Hall, London, 1964) Chapter 9.

[9] A. D. Sokal, *Monte Carlo methods in Statistical Mechanics: Foundations and New Algorithms*, Cours de Troisième Cycle de la Physique en Suisse Romande (Lausanne, June 1989).

[10] N. Metropolis, A. W. Rosenbluth, M. N. Rosenbluth, A. H. Teller, and E. Teller, J. Chem. Phys 21 (1953) 1087.

[11] K. Binder, in *Phase Transitions and Critical Phenomena*, edited by C. Domb and M. S. Green (Academic Press, New York, 1976).

[12] R. H. Swendsen and J.-S. Wang, Phys. Rev. Lett. 58 (1987) 86.

[13] D. W. Heermann and A. N. Burkitt, Physica A 162 (1990) 210.

[14] U. Wolff, Phys. Rev. Lett. 62 (1989) 361.

[15] U. Wolff, Phys. Lett. B 228 (1989) 379.

[16] U. Wolff, Nucl. Phys. B322 (1989) 759.

[17] U. Wolff, *Critical Slowing Down*, in *LATTICE 89*, Proceedings of the 1989 Symposium on Lattice Field Theory, Capri, Italy, 18-21 September 1989, edited by N. Cabibbo et al., Nucl. Phys. B (Proc. Suppl.) 17 (1990) 93.

[18] C. Frick, K. Jansen, and P. Seuferling, Phys. Rev. Lett. 63 (1989) 2613.

[19] K. Jansen and C. B. Lang, *Fractal dimension of critical clusters in the Φ_4^4 model*, Graz University preprint UNIGRAZ UTP-06-11-90, University of California at San Diego Preprint UCSD/PTH 90-25.

[20] R. Ben-Av, D. Kandel, E. Katznelson, P. G. Lauwers, and S. Solomon, J. Stat. Phys. 58 (1990) 125.

[21] R. Ben-Av, H.-G. Evertz, M. Marcu, and S. Solomon, *Critical acceleration of finite temperature SU(2) gauge simulations*, Tel-Aviv University preprint TAUP 1840-90.

[22] H.-G. Evertz, M. Hasenbusch, M. Marcu, K. Pinn, and S. Solomon, *Stochastic cluster algorithms for discrete Gaussian (SOS) models*, DESY preprint DESY 90-126, Tel-Aviv University preprint TAUP 1836-90.

[23] P. G. Lauwers and V. Rittenberg, Phys. Lett. B233 (1989) 197.

[24] P. G. Lauwers and G. Schütz, *Estimation of the Central Charge by Monte Carlo Simulations*, Bonn University preprint BONN-HE-90-12.

[25] D. Kandel, E. Domany, D. Ron, A. Brandt, and E. Loh, Phys. Rev. Lett. 60 (1988) 1591.

[26] R. Ben-Av and S. Solomon, *3D-Ising Model with no Critical Slowing Down*, Weizmann Institute preprint WIS/89-81 JAN-PH.

[27] A. Brandt, M. Galun, and D. Ron, *Optimal Multigrid Algorithms for Calculating Critical Temperature and Thermodynamic Quantities*, Weizmann Institute preprint (1990).

[28] S. Caracciolo, R. G. Edwards, A. Pelissetto, A. D. Sokal, *Generalized Wolff-type embedding algorithms for nonlinear σ-models*, to be published in *LATTICE 90*, Proceedings of the 1990 Symposium on Lattice Field Theory, Tallahasse, Florida, U.S.A..

[29] H. Mino, *A vectorized algorithm for cluster formation in the Swendsen-Wang Dynamics*, YAMANASHI-90-02.

[30] R. G. Edwards, X.-J. Li, and A. D. Sokal, *Sequential and Vectorized Algorithms for Computing the Connected Components of an Undirected Graph*, in preparation.

[31] A. Brandt, D. Ron and D. J. Amit, *Multilevel approaches to discrete state and stochastic problems*, in *Multigrid Methods II*, W. Hackbusch and U. Trottenberg editors (Springer, 1986).

[32] J. Goodman and A. D. Sokal, Phys. Rev. Lett. 56 (1986) 1015.

[33] G. Mack, *Multigrid Methods in Quantum Field Theory*, in *Non perturbative Quantum Field Theory*, Cargèse Lectures July 1987, G. 't Hooft et al. editors (Plenum Press, 1988).

[34] R. G. Edwards, J. Goodman and A. D. Sokal, *Multigrid Monte-Carlo: II. Two-dimensional XY-model*, Florida State University preprint FSU-SCRI-90-99.

[35] A. Hulsebos, J. Smit and J. C. Vink, *Multigrid simulation of the XY-model*, Amsterdam University preprint ITFA-90-17.

[36] J. Goodman and A. D. Sokal, Phys. Rev. D 40 (1989) 2035.

[37] L. Susskind, Phys. Rev. D16 (1977) 3031.

[38] S. Kaczmarz, *Angenäherte Auflösung von Systemen Linearer Gleichungen*, Bull. Acad. Polon. Sci. Lett. A., 35 (1937) 355.

[39] R. Ben-Av, A. Brandt, and S. Solomon, Nucl. Phys. B329 (1990) 193.

[40] R. Ben-Av, A. Brandt, M. Harmatz, E. Katznelson, P. G. Lauwers, S. Solomon, K. Wolowesky, *Fermion Simulations Using Parallel Transported Multigrid*, Bonn University preprint BONN-HE-90-11, to be published in Phys. Lett. B.

[41] M. Harmatz, P. G. Lauwers, R. Ben-Av, A. Brandt, E. Katznelson, S. Solomon, and K. Wolowesky, *Parallel-transported multigrid and its application to the Schwinger model*, Bonn University preprint BONN-HE-90-15, to be published in *LATTICE 90*, Proceedings of the 1990 Symposium on Lattice Field Theory, Tallahasse, Florida, U.S.A..

[42] J. Goodman, A. D. Sokal, Phys. Rev. D40 (1989) 2035.

[43] G. Mack and S. Meyer, *The effective action from multigrid Monte Carlo*, DESY preprint DESY 89-009.

[44] F. Niedermayer, Phys. Rev. Lett. 61 (1988) 2026.

[45] F. Fucito, E. Marinari, G. Parisi, and C. Rebbi, Nucl. Phys. B180 [FS2] (1981) 369.

TWO FAST SOLVERS BASED ON THE MULTI-LEVEL SPLITTING OF FINITE ELEMENT SPACES

PETER LEINEN, HARRY YSERENTANT*

Abstract. The hierarchical basis preconditioner and the recent method of Bramble, Pasciak and Xu are described and their structure is compared. Some algorithmic aspects are discussed and a numerical example for an adaptively generated, nonuniform grid is given.

1. Introduction. The numerical solution of the linear systems arising from finite element discretizations on adaptively generated, strongly nonuniform grids requires specialized techniques. For example, a local refinement should not require a complete reconstruction of the data structures, or the theory behind the solver should not be based on regularity properties of the continuous problem or on quasiuniformity assumptions for the involved sequence of grids because such assumptions are often not fulfilled in practice.

There are two solvers which are especially attractive for such applications, the hierarchical basis method [14] together with its Gauß-Seidel variant, the hierarchical basis multigrid method [2], and the recent method of Bramble, Pasciak and Xu [13], [5]. For a unified treatment with a special emphasis on nonuniformely refined general grids see [15].

Here in this paper we give a short description of the hierarchical basis- and the Bramble–Pasciak–Xu preconditioner, compare their algorithmic realization and give a numerical example.

2. The Preconditioners. Both solvers are closely related and have a very similar structure.

Assume that a sequence of triangulations $\mathcal{T}_0, \mathcal{T}_1, \ldots, \mathcal{T}_j$ of a domain $\Omega_0 \subseteq \mathbb{R}^2$ is given, beginning with a coarse initial triangulation $\mathcal{T}_0$. For the simple case of a uniform refinement, one obtains the triangles forming $\mathcal{T}_{k+1}$ by subdividing each triangle of $\mathcal{T}_k$ into four congruent subtriangles. More general schemes allowing a nonuniform refinement are presented in [3], [12], [9], or [11]; see also [1], [2] or [6]. The three-dimensional case is more complicated; possible approaches are described in [16] or [4].

With every triangulation $\mathcal{T}_k$ we associate a finite element space $\mathcal{S}_k$ consisting of functions which vanish on a boundary piece Γ of Ω (depending on the boundary value problem under consideration). For simplicity here we restrict our attention to piecewise linear functions for the solution of simple scalar second order problems. Clearly, $\mathcal{S}_k$ is a subspace of $\mathcal{S}_l$ for $l \geq k$.

Let $\mathcal{N}_k = \{x_1, \ldots, x_{n_k}\}$ be the set of vertices of the triangles in $\mathcal{T}_k$ not lying on the boundary piece Γ. Then $\mathcal{S}_k$ is spanned by the nodal basis functions $\psi_i^{(k)}$, $i = 1, \ldots, n_k$, which are defined by

$$\psi_i^{(k)}(x_l) = \delta_{il}\ , \ x_l \in \mathcal{N}_k\ . \tag{1}$$

The hierarchical basis functions are

$$\widehat{\psi}_i = \psi_i^{(0)}\ , \ x_i \in \mathcal{N}_0\ , \tag{2}$$

* Mathematisches Institut, Universität Tübingen, D-7400 Tübingen, Germany.

and

$$\widehat{\psi}_i = \psi_i^{(k)} \ , \ x_i \in \mathcal{N}_k \setminus \mathcal{N}_{k-1} \ . \tag{3}$$

$\widehat{\psi}_i$, $i = 1, \ldots, n_k$, is the hierarchical basis of $\mathcal{S}_k$.

We fix a final level j and consider the discrete boundary value problem to find a function $u \in \mathcal{S} = \mathcal{S}_j$ satisfying

$$a(u,v) =< f, u >, \quad v \in \mathcal{S}. \tag{4}$$

Here $a(u,v)$ denotes the bilinear form in the weak formulation of a second order elliptic boundary value problem. This bilinear form is assumed to be selfadjoint and coercive. f is a linear functional on $\mathcal{S}$ representing the right hand side of the differential equation.

If $\mathcal{S}^*$ denotes the space of linear functionals on $\mathcal{S}$, the boundary value problem (4) can be written as

$$Au = f \tag{5}$$

with a linear operator

$$A : \mathcal{S} \to \mathcal{S}^*. \tag{6}$$

The hierarchical basis- and the Bramble–Pasciak–Xu preconditioner are approximate inverses

$$C : \mathcal{S}^* \to \mathcal{S} \tag{7}$$

of this operator A.

The hierarchical basis preconditioner $C_H : \mathcal{S}^* \to \mathcal{S}$ is given by

$$C_H r = u_0 + \sum_{k=1}^{j} \sum_{x_i \in \mathcal{N}_k \setminus \mathcal{N}_{k-1}} \frac{1}{d_i} < r, \widehat{\psi}_i > \widehat{\psi}_i \tag{8}$$

where $u_o \in \mathcal{S}_0$ is the solution of

$$a(u_0, v) =< r, v >, \quad v \in \mathcal{S}_0. \tag{9}$$

The d_i are scaling factors. For the case of the hierarchical basis preconditioner one can choose

$$d_i = a(\widehat{\psi}_i, \widehat{\psi}_i). \tag{10}$$

The formulation (8), (9) of the hierarchical basis preconditioner has been given by Xu [13]. Reformulated in terms of the nodal basis, the hierarchical basis preconditioner is

$$C_H r = u_0 + \sum_{k=1}^{j} \sum_{i=n_{k-1}+1}^{n_k} \frac{1}{d_i} < r, \psi_i^{(k)} > \psi_i^{(k)} \tag{11}$$

with $u_0 \in \mathcal{S}_0$ defined by (9). Formally, the Bramble–Pasciak–Xu preconditioner $C_X : \mathcal{S}^* \to \mathcal{S}$ is quite similar to this operator. It is given by

$$C_X r = u_0 + \sum_{k=1}^{j} \sum_{i=1}^{n_k} \frac{1}{d_i^{(k)}} < r, \psi_i^{(k)} > \psi_i^{(k)} \tag{12}$$

with certain scaling factors $d_i^{(k)}$ and $u_0 \in \mathcal{S}_0$ given by (9). The proper choice of the scaling factors $d_i^{(k)}$ for nonuniformely refined grid is discussed in [15].

For the mathematical properties of the preconditioners we refer to [14] and [2] and to [13] and [5], respectively, and to [15], especially as it concerns the application of the Bramble–Pasciak–Xu preconditioner to nonuniformely refined grids.

Here we mention only that, for two-dimensional applications, the quotient of the maximum and the minimum eigenvalue of $C_H^{-1}A$ and $C_X^{-1}A$, respectively, grows like $O(j^2)$ with the number j of refinement levels. For uniformely refined grids of gridsize h this means a growth like $O(|\log h|^2)$ and is nearly optimal. This behavior is independent of regularity properties of the continuous problem or the quasiuniformity of the grids.

Unfortunately, for three-dimensional problems the hierarchical basis preconditioner deteriorates. In contrast, the condition number estimate for the Bramble–Pasciak–Xu preconditioner is dimension independent. In addition, the Bramble–Pasciak–Xu preconditioner can take advantage of the regularity of the boundary value problem. In the extreme case, the behavior improves from $O(j^2)$ as above to $O(j)$.

Of course, for three-dimensional applications the Bramble–Pasciak–Xu preconditioner is superior. For two-dimensional problems this is less clear because, for adaptively refined grids, the algorithms and data structures become more complicated and expensive.

3. Some remarks on the Algorithmic Realization. The computation of $C_H r$, as defined by (8) and (9), is cheap. Beginning with the values

$$< r, \psi_i^{(j)} >, \qquad i = 1, \ldots, n_j, \tag{13}$$

which are the components of the right hand side in a matrix formulation with respect to the nodal basis fo $\mathcal{S}_j$, first one computes

$$< r, \widehat{\psi}_i >, \qquad i = 1, \ldots, n_j, \tag{14}$$

recursively advancing from one level to the next coarser level.

As $\widehat{\psi}_i$, $i = 1, \ldots, n_0$, is the nodal basis of the inital space $\mathcal{S}_0$, the computation of $u_0 \in \mathcal{S}_0$ additionally requires only the solution of a low-dimensional level 0 matrix equation.

Finally, all terms must be summed up. This is done recursively beginning with the values of $C_H r$ at the nodes x_i, $i = 1, \ldots, n_0$, of the initial level and ending with

$$(C_H r)(x_i), \qquad i = 1, \ldots, n_j.$$

As exactly one hierarchical basis function $\widehat{\psi}_i$ is associated with every node x_i, the overall work and storage for evaluating $C_H r$ remains strictly proportional to the number of unknowns, independently of the distribution of the unknowns among the levels. The $< r, \psi_i^{(j)} >$ can be overwritten with the values $< r, \widehat{\psi}_i >$.

The Bramble–Pasciak–Xu preconditioner can be realized along the same lines as the hierarchical basis preconditioner.

Note, however, that the double sum in (8) and (11), respectively, consists of

$$\sum_{k=1}^{j}(n_k - n_{k-1}) = n_j - n_0 \tag{15}$$

terms, whereas in (12)

$$n_j \leq \sum_{k=1}^{j} n_k \leq j n_j \tag{16}$$

terms occur. The reason is that a changing number of basis functions $\psi_i^{(k)}$ is associated with every node x_i.

In the general framework of adaptively refined grids, this leads to considerable complications. For a naive realization both the work and storage can grow like $O(jn_j)$. In our current preliminary code, the Bramble–Pasciak–Xu preconditioner has the same storage requirements as the hierarchical basis preconditioner, but the work per iteration step can be bounded only by $O(jn_j)$.

4. A numerical example. As a standard test example [1], [2] we consider the solution of the Laplace equation

$$-\Delta u = 0 \tag{17}$$

on the unit circle with a slit as in Fig. 1.

The boundary conditions on the circle, expressed in polar coordinates, are

$$u = r^{\frac{1}{4}} \sin \tfrac{\varphi}{4}. \tag{18}$$

On the two sides of the slit, the boundary conditions are $u = 0$ and $\partial u / \partial n = 0$, respectively. The solution of the boundary value problem on the whole domain is given by (18) and exhibits a typical corner- or crack singularity.

This problem has been solved using a modified version of the adaptive code KASKADE [8], [6], once with the hierarchical basis method as preconditioner, and then with the Bramble–Pasciak–Xu preconditioner.

For both cases, the initial triangulation consisted of 8 triangles with 10 nodes; one node counts twice. The initial grid has been refined several times as discussed in [6]. After every refinement step, the error in the solution of the discrete problem has been reduced to the level of the discretization error, giving the starting values for the next grid. This process has been stopped with a grid of depth $j = 22$ consisting of 15244 triangles with 7768 nodes for the hierarchical basis case and a grid of depth $j = 16$ consisting of 9049 triangles with 4862 nodes for the Bramble–Pasciak–Xu case. These grids are shown in Fig. 2 and 3, respectively.

The convergence histories for these examples are shown in Fig. 4 and Fig. 5. The vertical lines indicate new grids obtained by the adaptive refinement processes. DISTL is the distance, measured in the energy norm, of the actual approximation to the exact piecewise linear finite element solution on the final grid, and DISTQ denotes the distance to the piecewise quadratic solution on the final grid. This line shows that it is senseless to perform more than a few iterations after every refinement.

Thus, in the given example both algorithms lead to comparable, very satisfying results. Our general, at the moment still limited experience is that in terms of iteration steps, the Bramble–Pasciak–Xu preconditioner is usually a little bit faster, approximately compensating the higher cost of the single step.

REFERENCES

[1] Bank, R.E.: *Software Package for Solving Elliptic Partial Differential Equations.* Philadelphia: SIAM (1990).

[2] Bank, R. E., Dupont, T., Yserentant, H.: *The Hierarchical Basis Multigrid Method.* Numer. Math. **52**, pp. 427–458 (1988).

[3] Bank, R. E., Sherman, A. H., Weiser, A.: *Refinement Algorithms and Data Structures for Regular Local Mesh Refinement.* In: Scientific Computing (eds.: R. Stepleman et al.), Amsterdam: IMACS/North Holland, pp. 3–17 (1983).

[4] Bänsch, E.: *Local mesh refinement in 2 and 3 dimensions.* Report No. 6, SFB 256, Universität Bonn (1989).

[5] Bramble, J. H, Pasciak, J. E., Xu, J.: *Parallel Multilevel Preconditioners.* Math. Comp. **55**, pp. 1-22 (1990).

[6] Deuflhard, P., Leinen, P., Yserentant, H.: *Concepts of an Adaptive Hierarchical Finite Element Code.* IMPACT of Computing in Science and Engineering **1** , pp. 3–35 (1989).

[7] Hackbusch, W.: *Multigrid Methods and Applications.* Berlin, Heidelberg, New York: Springer (1985).

[8] Leinen, P.: *Ein schneller adaptiver Löser für elliptische Randwertprobleme auf Seriell und Parallelrechnern.* Doctoral thesis, FB Mathematik, Universität Dortmund (1990).

[9] Mitchell, W.F.: *Unified Multilevel Adaptive Finite Element Methods for Elliptic Problems.* Report No. UIUCDCS-R-88-1436, Department of Computer Science, University of Illinois (1988).

[10] Ong, M. E. G.: *Hierarchical Basis Preconditioners for Second Order Elliptic Problems in Three Dimensions.* Technical Report No. 89–3, Department of Applied Mathematics, University of Washington, Seattle (1989).

[11] Rivara, M.C.: *Algorithms for refining triangular grids suitable for adaptive and multigrid techniques.* Int.J.Numer.Meth.Engrg. **20**, pp. 745-756 (1984).

[12] Sewell, E.G.: *Automatic generation of triangulations for piecewise polynomial approximations.* Ph.D.Thesis, Purdue University, West Lafayette (1972).

[13] Xu, J.: *Theory of Multilevel Methods.* Report No. AM48, Department of Mathematics, Pennsylvania State University (1989).

[14] Yserentant, H.: *On the Multi-Level Splitting of Finite Element Spaces.* Numer. Math **49**, pp. 379–412 (1986).

[15] Yserentant, H.: *Two Preconditioners Based on the Multi-Level Splitting of Finite Element Spaces.* Numer. Math. **58**, 163-184 (1990).

[16] Zhang, S.: *Multi-Level Iterative Techniques.* Ph.D.Thesis, Department of Mathematics, Pennsylvania State University (1988).

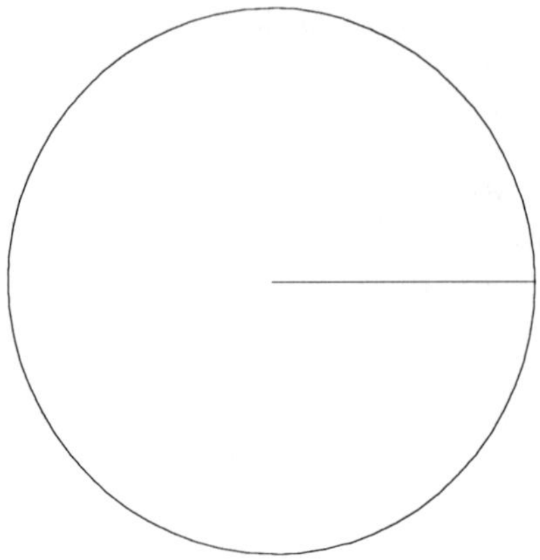

FIG. 1. *The domain* Ω

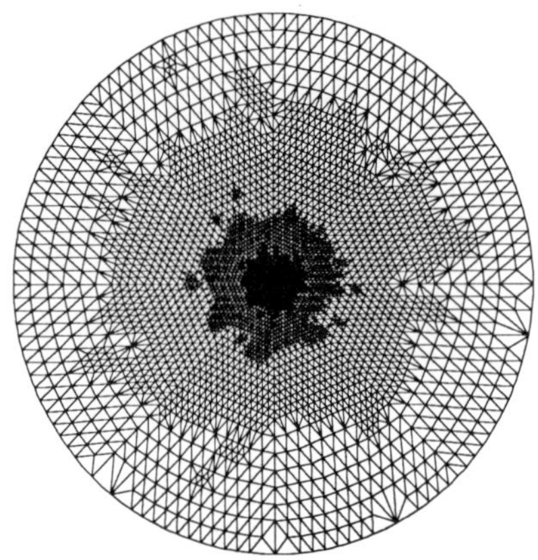

FIG. 2. *The final triangulation for the hierarchical basis case (7768 nodes)*

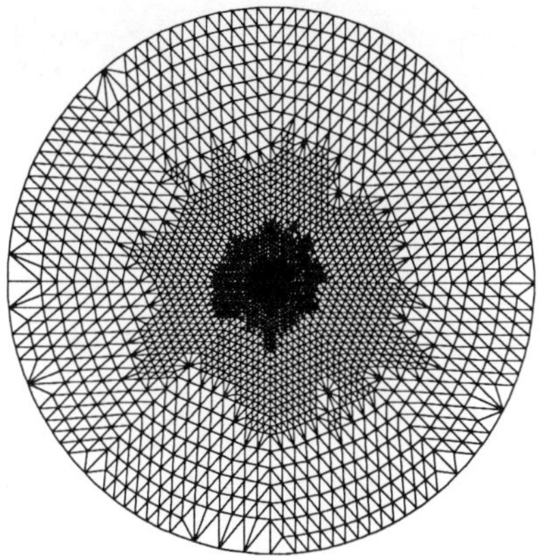

FIG. 3. *The final triangulation for the Bramble–Pasciak–Xu case (4862 nodes)*

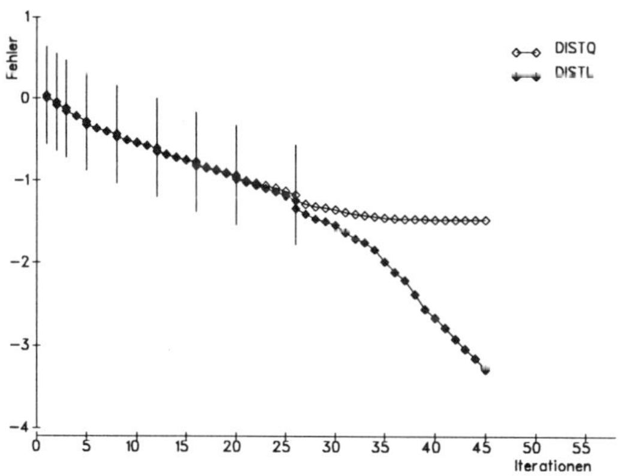

FIG. 4. *Convergence history for the hierarchical basis case*

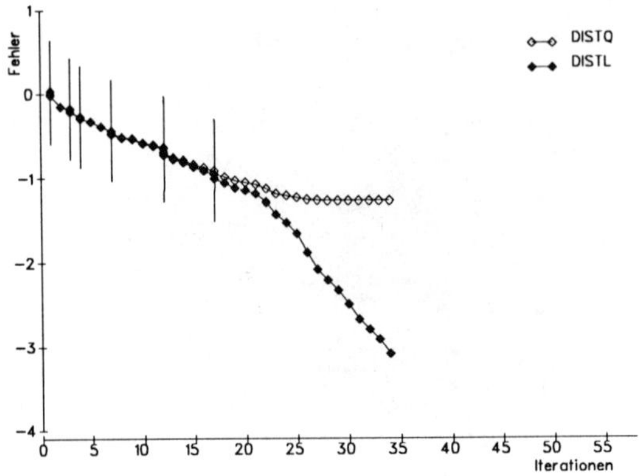

FIG. 5. *Convergence history for the Bramble–Pasciak–Xu case*

International Series of Numerical Mathematics, Vol. 98, © 1991 Birkhäuser Verlag Basel

Multigrid Methods for Turbulent Flow Problems

John Ruge
University of Colorado at Denver, Denver, Colorado

Achi Brandt
Weizmann Institute of Science, Rehovot, Israel

Jim McWilliams
Ralph Milliff
National Center for Atmospheric Research, Boulder, Colorado

Abstract: This paper covers work in progress in applying multigrid ideas to turbulent flow problems in oceanography. As a model problem, we use the incompressible 2-D time-dependent Navier-Stokes equations. The preferred current approach is to use predictor-corrector methods to get second-order in time, and solve the spatial problem using spectral methods or finite differences combined with FFT solvers. There is a great opportunity for improving existing methods with multigrid algorithms, ranging from the introduction of Poisson solvers to fully adaptive space-time multigrid approaches. Here, we report on our experiences with multigrid in several different directions, including Lagrangian methods within the predictor-corrector framework and semi-Lagrangian discretizations with implicit time-stepping.

Introduction:

The accurate solution of turbulent flow problems in oceanography requires the modelling of large bodies of water, while the solution features can range from small eddies to global currents. However, the domain resolution is limited by practical considerations of available computer memory and solution times with current solution methods. Multigrid methods have proven to be very useful for the fast solution of large problems, and although some work has been done (c.f., [1], [3]), it is not yet

---

This work was supported in part by the AFOSR under Contract number 89-0126 and by the National Center for Atmospheric Research .

used extensively in problems in oceanography. Ultimately, we would like to develop multigrid algorithms fully adaptive in space and time for such problems, and toward this goal have performed some preliminary work in several different directions. These include a straightforward implementation of multigrid as a Poisson solver within existing time-stepping methods, Lagrangian approaches, and multigrid for the fully implicit equations, while testing new ideas in discretizations.

Our model problem is the 2-d, incompressible, time-dependent Navier-Stokes equations in streamfunction-vorticity form. We consider the case of small or no viscosity, which is consistent with the application we have in mind. This allows for for interesting nonlinear turbulent effects.

Currently, the preferred approach to such problems is to separate the space and time discretizations, using predictor-corrector methods to get second-order in time. The spatial problem is then solved using spectral methods or finite differences combined with FFT solvers. An obvious improvement here is to apply multigrid methods to the solution of the finite difference problem. This is not only faster (at least for large enough problems), but extends easily to irregular domains.

In addition, we are experimenting with characteristic and near-characteristic methods in order to allow for larger time steps without decreasing accuracy and to minimize the introduction of artificial viscosity when the flow is not aligned with the grids. One approach is to follow the flow on a Lagrangian grid, using a predictor-corrector method to update the trajectories. This requires solution of a Poisson equation on the Lagrangian grid, which we solve by transferring the problem to the original fixed grid where multigrid cycling is cheaper.

We also considered a fully implicit discretization, using "narrow upstream differences" (NUD) which bracket the particle paths as closely as possible in space-time. This combines the space and time discretizations in order to approximate the material derivative of the vorticity. This results in nonlinear equations which can be solved by multigrid. At the same time, we tested other, more standard discretizations for comparison.

We discuss our experiences (both positive and negative) with these approaches and present some results.

The Model Problem

The model problem we consider is the 2-D incompressible time-dependent Navier-Stokes equations:

(1a) $$u_t+uu_x+vu_y+p_x=\nu\Delta u$$

(1b) $$v_t+uv_x+vv_y+p_y=\nu\Delta\zeta$$

(1c) $$u_x+v_y=0$$

We take our domain Ω as the unit square and use consider doubly periodic boundary conditions. We consider the equations in streamfunction-vorticity form. (1c) implies that there exists a streamfunction ψ with

(2) $$u=-\psi_y \quad v=\psi_x$$

Vorticity, defined as the curl of the velocity field, satisfies

(3) $$\zeta=v_x-u_y=\Delta\psi$$

By taking the curl of the (1a) and (1b), pressure is eliminated, and the equations (along with the definition of vorticity) become

(4a) $$\zeta_t+\psi_x\zeta_y-\psi_y\zeta_x=\nu\Delta\zeta$$

(4b) $$\Delta\psi=\zeta$$

The spatial derivative operator on the left side of (4a) is called the Jacobian and is denoted by

(5) $$J(\psi,\zeta)=\psi_x\zeta_y-\psi_y\zeta_x$$

For the applications we are considering, there are several concerns. First, since we are focusing on ocean modelling, diffusion effects are very short range relative to domain size, we take ν to be small. In fact, in some tests we use a fourth order, rather than second order, diffusion term. Second, in this preliminary work we are not interested in boundary effects, and thus consider only doubly periodic boundary conditions.

In judging approximations obtained with different solution methods we pay attention to certain aspects of the solution. Some of these are mentioned when specific test problems are discussed. However, some general criteria used are conservation relations derived from (4). The first is the rate of change of energy over time:

(6) $$\frac{d}{dt}\int_\Omega\nabla\psi\cdot\nabla\psi=-2\nu\int_\Omega\zeta^2$$

Others are integrals of powers of ζ. The first two are:

$$\frac{d}{dt}\int_{\Omega}\zeta = 0 \qquad \frac{d}{dt}\int_{\Omega}\zeta^2 = -2\nu\int_{\Omega}\nabla\zeta\cdot\nabla\zeta \tag{7}$$

There are an infinite number of such conservation relations, but these two are those used most. The second gives the rate of change of the quantity termed enstrophy over time.

Test Problems

We take two test problems. The first, which we call the "turbulence model", uses only fourth order diffusion with $\nu=10^{-6}$. An initial distribution of Fourier modes is specified for the vorticity, and we solve for the corresponding streamfunction. This initial state has features on all grid scales. The quality of the solutions obtained by multigrid methods are judged against those produced by FFT methods, mainly considering the conservation relations, the resulting distributions of Fourier modes, and the "look" of the developing vortices. It should be noted that, in this case, the solution at large times depends very sensitively on initial conditions, so that a pointwise comparison of solutions obtained by different methods is not considered useful.

The second test problem is the modon (c.f., [2]). This is an exact solution to the inviscid problem (i.e. $\nu=0$) and is basically a self-propagating vorticity "dipole", with paired vortices rotating in opposite directions moving at a fixed speed through the fluid. Thus the solution at any time is simply a translation of the initial condition. Approximations to this solution are judged by the conservation relations, the accuracy of the modon speed compared to the phase speed of the continuous solution, the maintainence of the shape and integrity of the modon and the absence of "noise" in the wake.

Current Solution Methods

One usual approach to this problem is to separate the space and time discretization questions, and to use predictor corrector methods to get second order accuracy in time. The spatial problem is then solved using spectral methods or finite differences combined with FFT solvers. When the solution has features on all scales, the work necessary for given accuracy is about the same in both cases. Here, we concentrate on the finite difference approach.

Uniform grids are used, with mesh size denoted by h. The value of a function Ψ at a point (ih,jh) is denoted by $\Psi_{i,j}$. The usual 5-point stencils are used for the Laplace operators. A preferred discretization of $J(\Psi,\zeta)$ is the Arakawa Jacobian, given by

$$
\begin{aligned}
J(\psi,\zeta)=\frac{1}{12h^2}[&(\psi_{i,j-1}+\psi_{i+1,j-1}-\psi_{i,j+1}-\psi_{i+1,j+1})(\zeta_{i+1,j}-\zeta_{i,j})\\
&+(\psi_{i-1,j-1}+\psi_{i,j-1}-\psi_{i-1,j+1}-\psi_{i,j+1})(\zeta_{i,j}-\zeta_{i-1,j})\\
&+(\psi_{i+1,j}+\psi_{i+1,j+1}-\psi_{i-1,j}-\psi_{i-1,j+1})(\zeta_{i,j+1}-\zeta_{i,j})\\
&+(\psi_{i+1,j-1}+\psi_{i+1,j}-\psi_{i-1,j-1}-\psi_{i-1,j})(\zeta_{i,j}-\zeta_{i,j-1})\\
&+(\psi_{i+1,j}-\psi_{i,j+1})(\zeta_{i+1,j+1}-\zeta_{i,j})+(\psi_{i,j-1}-\psi_{i-1,j})(\zeta_{i,j}-\zeta_{i-1,j-1})\\
&+(\psi_{i,j+1}-\psi_{i-1,j})(\zeta_{i-1,j+1}-\zeta_{i,j})+(\psi_{i+1,j}-\psi_{i,j-1})(\zeta_{i,j}-\zeta_{i+1,j-1})]
\end{aligned}
\tag{8}
$$

Initially, we thought that this discretization may be needlessly expensive. This approximation has the property that it is spatially conservative locally, so that the quantities mentioned above are conserved to the level of time discretization error.

The predictor corrector method used is described as follows, with time step Δt and superscript k in parentheses denoting a function at time $k\Delta t$. We assume that $\psi^{(k)}$ and $\psi_t^{(k)}$ are known for $k<n+1$.

Predictor: $\psi^{(n+1)}=\psi^{(n-1)}+2\,\Delta t\,\psi_t^{(n)}$

Corrector: ***Evaluate***: $f=-J(\psi^{(n+1)},\Delta\psi^{(n+1)})+\nu\,\Delta\psi^{(n+1)}$

Solve: $\Delta^h\psi_t^{(n+1)}=f$

Correct: $\psi^{(n+1)}=\psi^{(n)}+\frac{\Delta t}{2}(\psi_t^{(n+1)}+\psi_t^{(n)})$

The spatial operators are discretized as described above, and the solution step in the corrector is generally done using an FFT solver. The corrector can be repeated as many times as necessary, and converges to the Crank-Nicolson solution. In practice, two corrector steps are generally sufficient for convergence.

Multigrid Approaches

One obvious approach we used was to replace the FFT solver with a multigrid Poisson solver. For large enough problems, multigrid should be more efficient since it can solve the problem in O(n) operations, where n is the number of grid points. Another advantage of multigrid is the ability to deal with irregular domains with the same efficiency, which the FFT solver can't do.

The next area we explored was trajectory methods, with the idea that we could trace the flow using Lagrangian methods, while using the fixed, or Eulerian, grid for multigrid processing since the equations and solution should be cheaper there. The primary advantages anticipated were the possibility of larger time steps, and limited introduction of artificial viscosity.

Finally, we considered implicit time-stepping methods where we combined the space and time discretizations using "narrow upstream differences" (NUD) in which we constructed stencils in space-time which followed the trajectories as closely as possible. The goals here were also to increase the time step and to limit artificial viscosity. Our experiences and conclusions in these three approaches are given below.

Multigrid as a Poisson Solver

The first step we took was a straightforward implementation of multigrid as a replacement for the FFT Poisson solver in the correction step. Since this algorithm is well known, the details will not be given here, but can be found in a number of references (e.g., [2], and [4]). Briefly, we used a full multigrid (FMG) algorithm with (1,1) V-cycles and an initial approximation $\psi_t^{(n+1)} = \psi_t^{(n)}$ so that we were solving for the time increment. We used a code written by John Adams at NCAR optimized for the CRAY XMP.

Results were not surprising. Multigrid did well, surpassing the FFT solver at h=1/256. One FMG cycle was sufficient for the modon case, although the quality of the solution for the turbulence model was judged not as good as the FFT solution. Additional multigrid work was able to improve the solution quality.

We also experimented with different discretizations for the Jacobian, hoping to reduce the overall work since the Arakawa Jacobian was somewhat expensive to compute. Instead, we tried a central approximation to the Jacobian. Results for the turbulence model were not good. Although the modon fared better, there was a loss of conservation of both energy and enstrophy and noise was left in its wake. We then tried a global renormalization of the solution to maintain the energy balance. Results did not improve. The turbulence results were still bad, and enstrophy in the modon case tended to grow. We considered adding some viscosity proportional to the increased enstrophy, which would tend to damp high frequencies in the solution, thus reducing the resulting enstrophy. However, we came to realize that our real problem was the presence of "red/black" modes in the solution caused by the central approximation to the Jacobian. We attempted to cure this problem with a "double discretization" in which we used the central Jacobian in computing a right-hand side for the Poisson equation for use in residual transfer, but used an upstream approximation tp compute the right-hand side for use in relaxation. However, results were not immediately promising and we decided our time would be better spent exploring other multigrid questions. We concluded that, at least for the present, the Arakawa Jacobian was better than the alternatives.

Lagrangian (Trajectory) methods

The second approach we took was to have the grid follow particle paths, or trajectories, as in a Lagrangian method, but to retain the fixed, or Eulerian grid for multigrid processing. A predictor corrector method is used as before. The advantages we saw in such an approach were large time steps, small artificial viscosity (proportional to the actual viscosity) and fast multigrid solution (because of the uniform grid). In addition, no discretization of the Jacobian is necessary, allowing possible additional savings. Some disadvantages we foresaw were the need to restart the grid occasionally and the difficulty of detecting steady state.

The equations in the Lagrangian case simplify to

(9)
$$\frac{D\zeta}{Dt}-\nu\Delta\zeta$$
$$\Delta\psi-\zeta$$

where the time derivative is along trajectories. To describe the method we developed, some definitions and notation are needed. The fixed grid is the usual uniform grid $\Omega_F - \{(ih,jh):0\le i,j<1/h\}$. We assume that the original grid is advected by the flow, with trajectories defined by the paths of the grid points. The advected grid at time $n\Delta t$ is denoted by $\Omega_A^{(n)} - \{(\xi_{ij}^{(n)},\eta_{ij}^{(n)})\}$. We can assume that $(\xi_{ij}^{(0)},\eta_{ij}^{(0)})-(ih,jh)$. Functions defined on the advected grid are denoted by a subscript A, while those defined on the fixed grid are denoted by a subscript F. For transferring functions between grids, we have the two operators, to be defined later. They are denoted as follows, with the time level omitted.

I_F^A *transfer from* Ω_F *to* Ω_A
I_A^F *transfer from* Ω_A *to* Ω_F

With this approach, we also found it useful to maintain approximations to the velocity, $u^{(n)}$ and $v^{(n)}$.

Now, the predictor-corrector method we chose is as follows, assuming we know $\Omega_A^{(n)}$, $\zeta_A^{(n)}$, $\psi_A^{(n)}$, $u_A^{(n)}$ and $v_A^{(n)}$

Predictor*: *Compute $\Omega_A^{(n+1)}$

$$\xi_{i,j}^{(n+1)} - \xi_{i,j}^{(n)} + \Delta t\, u_{A\,i,j}^{(n)}$$

$$\eta_{i,j}^{(n+1)} - \eta_{i,j}^{(n)} + \Delta t\, v_{A\,i,j}^{(n)}$$

Advect ζ_A

$$\zeta_{A\,ij}^{(n+1)} - \zeta_{A\,i,j}^{(n)} + \Delta t\, \nu (\Delta^A \zeta_A^{(n)})_{i,j}$$

Corrector*: *Solve $\Delta^A \psi_A^{(n+1)} - \zeta_A^{(n+1)}$

Compute $u_A^{(n+1)}$ ***and*** $v_A^{(n+1)}$

Correct $\Omega_A^{(n+1)}$

$$\xi_{i,j}^{(n+1)} - \xi_{i,j}^{(n)} + \frac{\Delta t}{2}(u_{A\,i,j}^{(n)} + u_{A\,i,j}^{(n+1)})$$

$$\eta_{i,j}^{(n+1)} - \eta_{i,j}^{(n)} + \frac{\Delta t}{2}(v_{A\,i,j}^{(n)} + v_{A\,i,j}^{(n+1)})$$

Readvect ζ_A

$$\zeta_{A\,i,j}^{(n+1)} - \zeta_{A\,i,j}^{(n)} + \frac{\Delta t\, \nu}{2}(\Delta^A \zeta_A^{(n)} + \Delta^A \zeta_A^{(n+1)})$$

The corrector step can be repeated, and converges to the Crank-Nicolson solution.

There are several points here which are not yet well defined. First, the grid transfer operators must be specified. We used bilinear interpolation for transferring functions from the fixed to the advected grid, and the transpose of this interpolation for the transfer from the advected to the fixed grid. There were several reasons for this choice. One was that a true interpolation from the advected grid to the fixed grid would be difficult and expensive, while going the other way was relatively easy. On the other hand, using the transpose of interpolation guaranteed conservation of the sum of ζ. In addition, this choice made sense if we viewed the fixed grid as providing a correction to a problem on the advected grid.

Second, note that the operator Δ^A, which denotes the Laplacian on the advected grid, appears several times. Here, we plan to transfer the problem to the fixed grid, perform the required operation and transfer the result back. That is

$$\Delta^A \zeta_A - I_F^A \Delta_F I_A^F \zeta_A$$
$$(\Delta^A)^{-1} \zeta_A - I_F^A (\Delta_F)^{-1} I_A^F \zeta_A \tag{10}$$

Finally, note is that the method for computing velocities on the advected grid is not specified. In practice, we tried several options including computing velocities directly on the advected grid and computing velocities on the fixed grid (since Ψ is available on the fixed grid) and transferring them to the advected grid.

The above choices, particularly with respect to the grid transfer operators, were mainly motivated by the fact that we wanted to do our multigrid processing on the fixed grid for efficiency reasons. Thus, the fixed grid can be viewed as providing a correction to the advected grid problem. In this context, using the transpose of interpolation to transfer functions to the fixed grid makes sense, since this is a common practice in standard multigrid algorithms. An alternative to try would have been a true interpolation from the advected grid to the fixed grid, but this would be difficult and expensive, while going the other way was relatively easy. On the other hand, using the transpose of interpolation guaranteed preservation of the sum of ζ, which seemed attractive from a conservation standpoint.

Our results showed a fast loss of conservation. We soon realized that there were several problems with the approach as we first envisioned it. One of these was that, even for incompressible flow, the transpose of interpolation does not preserve constants, so smooth errors may not be well approximated by the fixed grid. Another problem was that the transpose of interpolation can easily have a nontrivial null space, so that certain solution components may not be found using the fixed grid alone. We concluded that some form of relaxation on the advected grid would be necessary. We did make some efforts to overcome these problems, but could not escape the conclusion that our original idea of performing all multigrid processing on the fixed grid was not practical. We again decided our time could be better spent exploring other possibilities.

Narrow Upstream Differences and Other Implicit Approaches

The last approach we considered was to use implicit time-steppin with "narrow" stencils which bracket the traject ry in space-time. The advantages we saw in this approach wer large time steps (due to the fact that the stencil approximates a trajectory method), small artificial viscosity (because of the small cross-stream cross-section of the stencil) and fast multigrid solution.

Assume we have a uniform grid defined as before and that we have solution approximations to $\psi^{(k)}$ and $\zeta^{(k)}$ for $k \leq n$. To define the solution at time $(n+1)\Delta t$ using fully implicit time stepping, we must discretize the equations (4) at that time. We discretize (4b) using the usual 5-point stencil for the Laplacian, denoted by Δ^h and the x and y derivatives of ψ in (4a) using central differences, which we denote by δ_x^h and δ_y^h, respectively. According to (2), this gives us the velocity. Now, let (x,y,t) be a point in space-time with $(x,y) \in \Omega^h$ and $t=(n+1)\Delta t$ Then, for the moment assume that the velocity (u,v) at that point is known. The goal is to approximate the operator

$$\zeta_t + u\zeta_x + v\zeta_y \tag{11}$$

The trajectory in space-time (in the backward time direction) is in the direction $-(u,v,1)$. We want to choose three "directions" of the following form which bracket the trajectory vector as closely as possible.

$$(i_l, j_l, k_l)\ ,\ |i_l|,\ |j_l|,\ |k_l| \leq 1\ , l=1,2,3 \tag{12}$$

This is sometimes not well defined, but an arbitrary choice can be made. We can choose these directions so that $|i_1| = |j_2| = |k_3| = 1$. Once these directions are chosen, each can be used to construct a second order difference stencil (actually we consider two difference stencils)

$$\begin{aligned} \delta_l^u \zeta(x,y,t) &= \frac{3}{2}\zeta(x,y,t) - 2\zeta(x+i_l h, y+j_l h, t+k_l\Delta t) + \frac{1}{2}\zeta(x+2i_l h, y+2j_l h, t+2k_l\Delta t) \\ \delta_l^c \zeta(x,y,t) &= \frac{1}{2}[\zeta(x-i_l h, y-j_l h, t-k_l\Delta t) - \zeta(x+i_l h, y+j_l h, t+k_l\Delta t)] \end{aligned} \tag{13}$$

The first is an upstream difference, while the second is a central difference. Whichever one we decide to use, we denote in a generic way as δ_l^h. In either case, a Taylor series expansion shows that

$$\delta_l \zeta(x,y,t) = -i_l h\zeta_x - j_l h\zeta_y - k_l \Delta t\zeta_t + \tau \tag{14}$$

where τ is a second order error term. Now, we wish to approximate (11) with a stencil of the form

$$\sum_l \alpha_l \delta_l^h \zeta \sim \zeta_t + u\zeta_x + v\zeta_y \tag{15}$$

Using (14), we can ignore the error term and set up a linear system to solve for the α's. We obtain a solution of the form

$$\alpha_l = a_l u + b_l v + c_l, \quad l=1,2,3. \tag{16}$$

where the a's, b's and c's are dependent on the directions chosen, but independent of velocity once the direction is chosen. Thus, the directions are picked based on an initial approximation to the velocity at each point, and can be held fixed. As long as the general direction of the trajectory does not change, an upstream stencil remains an upstream stencil. The discretization of (4) then becomes

$$\sum_l (-a_l \delta_y^h \psi + b_l \delta_x \psi + c_l) \delta_l^h \zeta - \nu \Delta^h \zeta \tag{17}$$
$$\Delta^h \psi = \zeta$$

The necessary coefficients are computed at the beginning of the time step. Similar stencils can be formed on the coarser grids, using the same time step, Δt, and the coarse grid coefficients are easily computed from those on the fine grid. We will not present the details here. The resulting nonlinear system is easy to solve using a full approximation storage scheme (FAS), at least for the range of time steps we have worked with.

Simple variations in the algorithm allow us to form a number of different stencils for testing and comparison. The one described, when the upstream differences are used, is referred to as the narrow upstream difference (NUD). Other variations are listed below:

NSUD - Narrow space upstream difference. The same as NUD, but we set $i_3 = j_3 = k_1 = k_2 = 0$. This gives the usual second order backward Euler discretization in time, while giving a narrow spatial upstream difference.

NSCD - Narrow space central difference. This is the same as NSUD, but central rather than upstream spatial differences are used.

UD - Upstream difference. This is the standard second order upstream difference, and is the same as NSUD, but we force $i_2 = j_1 = 0$.

CD - Central difference. This is the standard second order central difference, and is the same as UD, but central differences are used.

Finally, we also include the following discretization:

Arakawa - Arakawa discretization. This is the implicit version. The time derivative term is discretized by second order backward Euler, while the Arakawa Jacobian is used at time n+1 for the spatial derivatives.

The following table shows the percent of energy remaining after 10, 20 and 30 time steps with each of the discretization methods on the modon problem with h=1/64.

Discretization	**10 steps**	**20 steps**	**30 steps**
Arakawa	99.98	99.96	99.97
CD	99.11	98.2	99.6
UD	103.7	108.2	112.4
NSCD	98.	91.	84.
NSUD	80.	67.	58.
NUD	72.88	54.	43.

Table I. Percent of energy loss for the modon problem.

Although the time scales involved are quite short, energy loss (or gain) is quite large in some cases. Generally, the more narrow or upstream the discretization, the worse the energy conservation. This could be partly due to the form of the truncation error. Although the leading term of the truncation error is smaller in the case of narrow or upstream differences, the third order term, which is absent in central differences, is quite large and may dominate when the solution varies on the scale of the grid. We did find improvement (compatible with second order truncation error) as we refined the grid, but energy conservation was still better for central differences. Still more work is needed to determine the real source of problems here.

Conclusions:

Multigrid implementation does not always follow predictable paths, as many people know. More work is needed in order to determine which of the ideas tried are really promising, but it appears that the best approach developed uses familiar stencils in an implicit time-stepping context. Generally, except for the trajectory methods, the discretization was more of a problem than the multigrid solvers used, which for the most part worked well. We will continue to explore the implementation of new multigrid ideas.

References:

[1] J. R. Bates, F. H. M. Semazzi, R. W. Higgins and S. R. M. Barros, Integration of the Shallow Water Equations on the Sphere Using a Vector Semi-Lagrangian Scheme with a Multigrid Solver, Monthly Weather Review, August 1990.

[2] A. Brandt, Multilevel Techniques: 1984 Guide with Applications to Fluid Dynamics, GMD Studien Nr 85. Gesellschaft fuer Mathematik und Datenverarbeitung, St. Augustin, 1984.

[3] S. R. Fulton, P. E. Ciesielski and W. H. Schubert, Multigrid Methods for Elliptic Problems, Monthly Weather Review, 1985.

[4] K. Stueben and U. Trottenberg, Multigrid Methods: Fundamental Algorithms, Model Problem Analysis and Applications, in Multigrid Methods, Proceedings of conference held at Koln-Porz, Nov. 23-27, 1981, W. Hackbusch & U. Trottenberg, eds., Lecture Notes in Mathematics 960, Springer-Verlag, Berlin, 1982, pp. 1-176.

A survey of Fourier smoothing analysis results

P. Wesseling
Delft University of Technology,
Department of Technical Mathematics and Informatics,
P.O. Box 356,
2600 AJ Delft, The Netherlands

Abstract

Collecting and completing results scattered in the literature, an extensive survey of Fourier smoothing analysis results is presented. Recent developments concerning the influence of damping and modifying incomplete factorizations, and concerning a heuristic method by which the influence of boundary conditions can be successfully accounted for, will be discussed. A new Gauss-Seidel variant, called Gauss-Seidel-Jacobi, is presented. The set of test problems consists of the rotated anisotropic diffusion equation and the convection diffusion equation. Only a few smoothing methods work for all test problems. Unfortunately, these methods have certain drawbacks. Either they do not lend themselves well for vector and parallel computing, or their generalization to general systems of equations is problematic.

1. Introduction

To assess the robustness of smoothing methods for multigrid algorithms two test problem are important, namely the rotated anisotropic diffusion equation (3.1) and the convection-diffusion equation (3.2). These equations model effects that often occur simultaneously in applications, namely a mixed derivative, high mesh aspect rations due to numerical grid generation techniques, elliptic behaviour in parts of the domain, and (almost) hyperbolic behaviour in other parts of the domain. For the development of efficient multigrid algorithms it is desirable to have smoothing methods that are efficient for both test problems (robustness). Because results are scattered through the literature and incomplete, Fourier smoothing analysis results are presented for most, if not all, well-known smoothing methods for both test problems. Relatively recent insights concerning the way in which the type of boundary condition can be accounted for in a heuristic fashion ([57], [5], [23]) are incorporated, as are recently proposed incomplete factorization variants ([28], [23], [54], [57]). It is found that the robustness of "zebra" methods is enhanced considerably by damping. Furthermore, a new Gauss-Seidel variant, called Gauss-Seidel-Jacobi, is presented. First, a brief introduction to Fourier smoothing analysis is given.

2. Fourier smoothing analysis

For an introduction to Fourier smoothing analysis, see [34]. Here we give a brief outline and describe a way to approximately take into account the influence of boundary conditions, following [57], [5] and [23].

Let there be given a uniform grid G:

$$G = \{x \in \mathbb{R}^2 : \; x = j \cdot h, \; j = (j_1, j_2), \; h = (h_1, h_2), \; j_\alpha = 0, 1, 2, ..., n_\alpha, \; h = 1/n_\alpha\} \tag{2.1}$$

Let the algebraic system $A\mathbf{u} = \mathbf{f}$ to be solved on this grid be denoted in stencil notation by

$$\sum_k A(j,k) u_{j+k} = f_j \tag{2.2}$$

Usually, smoothing methods are basic iterative methods. That is, they are based on a splitting $A = M - N$ and defined by

$$M\mathbf{u}^{n+1} = N\mathbf{u}^{(n)} + \mathbf{f} \tag{2.3}$$

The error amplification matrix is $S = I - M^{-1}A$. If the coefficients of the partial differential equation are constant and if the boundary conditions are periodic then (usually) the stencils $[A], [M], [N]$ do not depend on the first argument j (which will henceforth be deleted), and the eigenfunctions of S are

$$\begin{aligned} \psi_j(\theta) &= \exp(ij\theta), \theta \in \Theta \,, \\ \Theta = \{\theta &= (\theta_1, \theta_2) : \; \theta_\alpha = 2\pi k_\alpha / n_\alpha, \; k_\alpha = -m_\alpha, -m_\alpha + 1, ..., \\ & \quad m_\alpha + 1, \; m_\alpha = n_\alpha/2 - 1\} \end{aligned} \tag{2.4}$$

assuming n_α to be even, and the corresponding eigenvalues are

$$\lambda(\theta) = \sum_k N(k) \exp(ik\theta) / \sum_k M(k) \exp(ik\theta), \quad \theta \in \Theta \,. \tag{2.5}$$

This is also called the amplification factor (of Fourier mode $\boldsymbol{\psi}(\theta)$). If under- or overrelaxation with factor ω is applied then $\lambda(\theta)$ becomes

$$\lambda(\theta) := \omega\lambda(\theta) + 1 - \omega. \tag{2.6}$$

Unless mentioned otherwise, $\omega = 1$. For "pattern" smoothing methods, such as white-black Gauss-Seidel, $\psi_j(\theta)$ are not eigenvectors, but certain higher-dimensional subspaces spanned by linear combinations of $\psi_j(\theta)$ are invariant. All methods to be considered here leave following four-dimensional subspaces invariant:

$$\boldsymbol{\Psi}(\theta) = (\boldsymbol{\psi}(\theta^1), \boldsymbol{\psi}(\theta^2), \boldsymbol{\psi}(\theta^3), \boldsymbol{\psi}(\theta^4)) \tag{2.7}$$

with

$$\begin{aligned} \theta^1 \in \Theta_{\tilde{s}} &= \Theta \bigcap [-\pi/2, \pi/2)^2, \; \theta^2 = (\theta_1^1 - sign(\theta_1^1)\pi, \theta_2^1 - sign(\theta_2^1)\pi), \\ \theta^3 &= \theta^1 - (0, sign(\theta_2^1)\pi), \; \theta^4 = \theta^1 - (sign(\theta_1^1)\pi, 0) \end{aligned} \tag{2.8}$$

An arbitrary periodic grid function can be written as

$$u_j = \sum_{\theta \in \Theta_{\tilde{s}}} c_\theta^T \boldsymbol{\Psi} j_j(\theta) \tag{2.9}$$

For the pattern smoothing methods considered here we have that if the error before smoothing is $c_\theta^T \boldsymbol{\Psi}(\theta)$, then after smoothing it is given by $(\Lambda(\theta) c_\theta)^T \boldsymbol{\Psi}(\theta)$, with $\Lambda(\theta)$ a 4x4 matrix, called the amplification matrix. With under- or overrelaxation we have

$$\Lambda(\theta) := \omega\Lambda(\theta) + (1-\omega)I \tag{2.10}$$

The sets of smooth and rough wavennumbers are defined by

$$\Theta_s = \Theta \bigcap (-\pi/2, \pi/2)^2, \; \Theta_r = \Theta \backslash \Theta_s \tag{2.11}$$

and the Fourier smoothing factor is defined by

$$\rho = \max(|\lambda(\theta)| : \theta \in \Theta_r) \tag{2.12}$$

In the case of pattern smoothing ρ is defined as follows. Note that $\psi(\theta^s)$, $s = 2, 3, 4$ are rough Fourier modes, whereas $\psi(\theta^1)$ is smooth, except when $\theta_1^1 = -\pi/2$ or $\theta_2^1 = -\pi/2$. These values have to be included in Θ_r, because this is found to make a large difference for certain cases. The projection operator of $\psi(\theta)$ on the rough Fourier modes is the diagonal matrix

$$Q(\theta) = \begin{pmatrix} \delta(\theta) & & & \\ & 1 & & \\ & & 1 & \\ & & & 1 \end{pmatrix} \tag{2.13}$$

with $\delta(\theta) = 1$ if $\theta_1 = -\pi/2$ and/or $\theta_2 = -\pi/2$, otherwise $\delta(\theta) = 0$. A suitable definition of the Fourier smoothing factor is

$$\rho = \max\{k(Q(\theta)\Lambda(\theta)) : \theta \in \Theta_{\bar{s}}\} \tag{2.14}$$

with k the spectral radius.

With Dirichlet boundary conditions, the eigenstructure is no longer of harmonic type. But the influence of Dirichlet boundary conditions can be taken into account heuristically as follows. The error at the boundary is always zero, and therefore we exclude wavenumbers with $\theta_1 = 0$ and/or $\theta_2 = 0$, so that the set of rough wavenumbers to be taken into account becomes

$$\Theta_r^D = \Theta_r \bigcap \{\theta \in \Theta \;: \theta_1 \neq 0 \text{ and/or } \theta_2 \neq 0\} \tag{2.15}$$

The corresponding smoothing factor is denoted by ρ_D:

$$\rho_D = \max\{|\lambda(\theta)| : \theta \in \Theta_r^D\} \tag{2.16}$$

Similarly, in the case of pattern smoothing, wavenumbers of the type $(0, \theta_2^s)$ and $(\theta_1^s, 0)$, $s = 1, 3, 4$ are to be excluded ($\theta_\alpha^2 = 0$ cannot occur). That is, the corresponding elements of c_θ are to be replaced by zero. This can be implemented by replacing $Q\Lambda$ by $PQ\Lambda$, with P a diagonal matrix given by

$$P(\theta) = \begin{pmatrix} p_1(\theta) & & & \\ & 1 & & \\ & & p_3(\theta) & \\ & & & p_4(\theta) \end{pmatrix} \tag{2.17}$$

with $p_1(\theta) = 0$ if $\theta_1 = 0$ and/or $\theta_2 = 0$; $p_3(\theta) = 0$ if $\theta_1 = 0$ (hence $\theta_1^3 = 0$), and $p_4(\theta) = 0$ if $\theta_2 = 0$ (hence $\theta_2^4 = 0$); otherwise, $p_s(\theta) = 1, s = 1, 3, 4$. The definition of the smoothing factor becomes

$$\rho_D = \max\{k(P(\theta)Q(\theta)\Lambda(\theta)) : \theta \in \Theta_{\bar{s}}\} \tag{2.18}$$

The above definitions of the smoothing factor depend on n_α, and hence are mesh-size dependent. Mesh-independent definitions are obtained if the discrete set Θ is replaced by $\{\theta = (\theta_1, \theta_2) : \theta_\alpha \in [-\pi, \pi]\}$. This makes the smoothing factor more difficult to compute numerically, and gives less realistic results in cases where the type of boundary condition has much influence. The definitions given above are especially appropiate if n_α is chosen equal to the values used when running multigrid in practice. Of course one wants $\rho < 1$ uniformly in n_α, at least when the influence of the boundary conditions is not significant. The behaviour of ρ when $n_\alpha \to \infty$ can be investigated numerically. In practice $n_\alpha > 10^3$ is very rare.

3. Test problems

In order to investigate and compare efficiency and robustness of smoothing methods the following test problems, representative of a wide class of problems, are useful:

$$-(\varepsilon c^2 + s^2)\frac{\partial^2 u}{\partial x^2} - 2(\varepsilon - 1)cs\frac{\partial^2 u}{\partial x \partial y} - (\varepsilon s^2 + c^2)\frac{\partial^2 u}{\partial y^2} = 0 \qquad (3.1)$$

$$-\varepsilon(\frac{\partial^2 u}{\partial x^2} + \frac{\partial^2 u}{\partial y^2}) + c\frac{\partial u}{\partial x} + s\frac{\partial u}{\partial y} = 0 \qquad (3.2)$$

where $c = \cos\beta, s = \sin\beta$. There are two parameters to be varied: $\varepsilon > 0$ and β. Equation (3.1) is called the rotated anisotropic diffusion equation, because it results from $-\varepsilon\partial^2 u/\partial x^2 - \partial^2 u/\partial y^2 = 0$ by rotating the axes over an angle β. Equation (3.2) is the convection-diffusion equation. In is desirable that a smoothing method is robust, i.e. is efficient for both (3.1) and (3.2), for all ε and β.

Finite difference discretization leads to the following stencils for (3.1):

$$[A] = (\varepsilon c^2 + s^2)[-1 \;\; 2 \;\; -1] + (\varepsilon - 1)cs\begin{bmatrix} 1 & -1 & 0 \\ -1 & 2 & -1 \\ 0 & -1 & 1 \end{bmatrix} + (\varepsilon s^2 + c^2)\begin{bmatrix} -1 \\ 2 \\ -1 \end{bmatrix} \qquad (3.3)$$

or

$$[A] = (\varepsilon c^2 + s^2)[-1 \;\; 2 \;\; -1] + \frac{1}{2}(\varepsilon - 1)cs\begin{bmatrix} 1 & 0 & -1 \\ 0 & 0 & 0 \\ -1 & 0 & 1 \end{bmatrix} + (\varepsilon s^2 + c^2)\begin{bmatrix} -1 \\ 2 \\ -1 \end{bmatrix} \qquad (3.4)$$

Upwind discretization of (3.2) leads to

$$[A] = \varepsilon\begin{bmatrix} & -1 & \\ -1 & 4 & -1 \\ & -1 & \end{bmatrix} + \frac{h}{2}[-c - |c| \;\; 2|c| \;\; c - |c|] + \frac{h}{2}\begin{bmatrix} s - |s| \\ 2|s| \\ -s - |s| \end{bmatrix} \qquad (3.5)$$

These test problems are thought to be adequate to mimick the effect of general curvilinear coordinates (causing a mixed derivative) and large mesh aspect rations occurring in applications, and to model physical conservation laws in which diffusion and transport play a role, and which are of elliptic or hyperbolic nature.

In the tests described hereafter the parameter β will be sampled according to $\beta = k\pi/12, k = 0, 1, 2, ..., 23$, unless stated otherwise. The worst case found will be included in the results.

4. Fourier smoothing analysis results

In the literature many smoothing analysis results may be found. Some sources are: [34], [21], [35], [22], [46], [47]. But often, tests have been carried out only for a subset of the test problems (3.1), (3.2), the possibility of damping has not been sufficiently exploited, and the influence of the boundary conditions has been insufficiently recognized. Here we will make an attempt at unification and completion of the material. For brevity, we will leave out many details, such as explicit expressions for amplification factors and matrices, when these are easy to obtain or well-known. Also, well-known smoothing methods will not be defined in detail. We take $n_1 = n_2 = n$.

4.1 Line Jacobi

Point Jacobi will not be considered, because it is not robust. With vertical line Jacobi for (3.1) with $\beta = 0$ and damping parameter $\omega = 1$ the amplification factor is given by

$$\lambda(\theta) = \varepsilon \cos \theta_1 / (1 + \varepsilon - \cos \theta_1) \tag{4.1}$$

We note immediately that $|\lambda(0, \pi)| = 1$, so that this seems not to be a good smoother. This is surprising, because as $\varepsilon \downarrow 0$ the method becomes an exact solver. This apparent contradiction is resolved by taking boundary conditions into account. It may be shown that (details not given here)

$$\rho_D = |\lambda(\pi, \varphi)| = \varepsilon / (1 + \varepsilon - \cos \varphi), \quad \varphi = 2\pi/n \tag{4.2}$$

so that indeed $\lim_{\varepsilon \downarrow 0} \rho_D = 0$. In cases where the boundary conditions are influential ρ_D is generally found to predict actual smoothing behaviour better than ρ.

An efficient and robust smoother is found to be *alternating damped Jacobi*. This consists of vertical line Jacobi followed by horizontal line Jacobi. Each step is damped separately with a fixed damping parameter $\omega = 0.7$. The value $\omega = 0.7$ was chosen after some experimentation. This smoothing method seems to have been overlooked until now. Table 4.1 presents results for (3.1) with discretizations (3.3) and (3.4). We have $\rho_D = \rho$ in the cases

	(3.3)		(3.4)	
ε	ρ, ρ_D	β	ρ, ρ_D	β
1	0.28	any	0.28	any
10^{-2}	0.95	45^0	0.44	45^0
10^{-5}	1.00	45^0	0.45	45^0
10^{-8}	1.00	45^0	0.45	45^0

Table 4.1: Fourier smoothing factors for the rotated anisotropic diffusion equation with alternating Jacobi; $\omega = 0.7; n = 64$.

listed in table 4.1. Table 4.2 gives results for (3.2). Clearly, this is an efficient smoother for both test problems, provided the mixed derivative is discretized according to (3.4). Finer sampling of β and increasing n does not change this conclusion. It is of practical importance that the method lends itself well for vectorization and parallelization.

ε	ρ	β	ρ_D	β
1	0.28	0^0	0.28	0^0
10^{-2}	0.29	0^0	0.29	0^0
10^{-5}	0.40	0^0	0.30	0^0
10^{-8}	0.39	0^0	0.30	0^0

Table 4.2: Fourier smoothing factors for the convection-diffusion equation with alternating Jacobi; $\omega = 0.7; n = 64$.

4.2 Point Gauss-Seidel

For $\varepsilon = 10^{-5}, n = 64, \beta = 105^o$ we find $\rho = 0.77, \rho_D = 0.71$, for symmetric PGS (forward followed by backward). Hence, contrary to popular belief, symmetric PGS is not a very good smoother for (3.2). Defining $P_1 = ch/\varepsilon$, $P_2 = sh/\varepsilon$ and choosing $P_1 = -\alpha P_2$, $P_2 \gg 1$, $\alpha P_2 \gg 1$ one finds

$$|\lambda(\pi/2, 0)| \cong (1+\alpha)^{-2} \tag{4.3}$$

so that ρ approaches 1 if α is small. This happens only for a small set of values of β, so in practice good smoothing may be observed in many cases. A robust version is four-direction PGS, making four sweeps, each starting in a different corner. Results are given in table 4.3.

ε	ρ	ρ_D	β
1.0	0.040	0.040	0°
10^{-1}	0.043	0.042	0°
10^{-3}	0.16	0.12	0°
10^{-5}	0.20	0.0015	15°

Table 4.3: Fourier smoothing factors for (3.2) discretized according to (3.5); four-direction PGS, $n = 64$.

4.3 Line Gauss-Seidel

Alternating line Gauss-Seidel (ALGS) consists of forward vertical line followed by forward horizontal line Gauss-Seidel. No value of ω could be found for which this smoother is robust for (3.3). Table 4.4 gives results for (3.4)

For the convection-diffusion equation (3.2) ALGS is not robust. If $P_2 < 0$, $P_1 = \alpha P_2$, $\alpha > 0$, $|P_2| \gg 1$ and $|\alpha P_2| \gg$ then

$$|\lambda(0, \pi/2)| \cong \alpha\{(1+\alpha)^2 + 1\}^{-1/2} \tag{4.4}$$

which tends to 1 if $\alpha \gg 1$. Symmetric (forward followed by backward) horizontal GS (SHGS) and vertical GS (SVGS) are robust for this test problem, as illustrated by table 4.5 for SVGS.

Numerically we find for $\beta = 0$ and $\varepsilon \ll 1$ that $\rho = |\lambda(0, \pi/2)|$, so that

$$\rho = (1 + P_1)/(9 + 3P_1) \cong 1/3 \tag{4.5}$$

ε	ρ $\beta = 0^o, 90^o$	ρ, β	ρ_D $\beta = 0^o, 90^o$	ρ_D, β
1	0.15	0.15, any	0.15	0.15, any
10^{-1}	0.37	$0.37, 15^o$	0.36	$0.37, 15^o$
10^{-3}	0.45	$0.58, 15^o$	0.085	$0.58, 15^o$
10^{-5}	0.45	$0.59, 15^o$	0.001	$0.59, 15^o$

Table 4.4: Fourier smoothing factors for (3.1) discretized according to (3.4); ALGS smoothing; $\omega = 1; n = 64$.

ε	ρ	β	ρ_D	β
1	0.20	90°	0.20	0°
10^{-1}	0.20	90°	0.20	90°
10^{-3}	0.30	0°	0.26	0°
10^{-5}	0.33	0°	0.002	75°

Table 4.5: Fourier smoothing factors ρ, ρ_D for (3.2) discretized according to (3.5); SVGS smoothing; $n = 64$.

We may conclude that alternating symmetric line Gauss-Seidel is robust for both test problems, provided that (3.1) is discretized according to (3.4). A disadvantage of this smoother is that it does not lend itself to vector or parallel computing. Jacobi and Gauss-Seidel with pattern orderings are more favorable in this respect.

4.4 Pattern Gauss-Seidel

With pattern Gauss-Seidel the grid points are colored according to various patterns, and points of the same color are updated simultaneously. With white-black Gauss-Seidel the points are colored chess board fashion. This smoother is far from robust for our test problems. But for equations close to Poisson's equation without a mixed derivative it is a superior smoother ($\rho = 1/4$ for Poisson) with excellent vectorization potential.

With zebra Gauss-Seidel horizontal or vertical grid lines are alternately colored white and black. Alternating zebra Gauss-Seidel (AZGS) uses both horizontal and vertical lines. Following the suggestion of [34] AZGS is arranged as follows in four steps: horizontal white, horizontal black, vertical black, vertical white. This gives slightly better smoothing factors than other arrangements, and gives identical results for $\beta = 0^o$ and $\beta = 90^o$ for (3.1).

Table 4.6 gives some results for the rotated anisotropic diffusion equation. Regardless of the value of ω, AZGS does not work for discretization (3.3). We see that AZGS is robust for this test problem. For table 4.6 β has been sampled with an interval of 2^0. Results with damping ($\omega = 0.7$) have been included. Clearly, damping is not needed in this case and is even somewhat disadvantageous. However, as shown by table 4.7, AZGS may need damping for the convection-diffusion equation. Although ρ_D looks good with $\omega = 1, \rho \to 1$ as $\varepsilon \downarrow 0$, and when n is increased ρ_D approaches ρ of course. Numerical experiments show

	$\omega = 1$		$\omega = 0.7$	
ε	ρ, ρ_D	β	ρ, ρ_D	β
1	0.05	any	0.32	any
10^{-3}	0.50	8^0	0.65	8^0
10^{-5}	0.54	4^0	0.67	8^0
10^{-8}	0.54	4^0	0.67	8^0

Table 4.6: Fourier smoothing factors for (3.1) discretized according to (3.4); AZGS smoothing; $n = 64$.

that $\omega = 0.7$ is a suitable fixed value to make ρ small. For table 4.7 β has been sampled with an interval of 2^0.

	$\omega = 1$				$\omega = 0.7$	
ε	ρ	β	ρ_D	β	ρ, ρ_D	β
1	0.05	0^0	0.05	0^0	0.32	0^0
10^{-3}	0.41	24^0	0.37	28^0	0.37	44^0
10^{-5}	0.95	4^0	0.58	22^0	0.44	4^0
10^{-8}	1.00	2^0	0.59	22^0	0.45	4^0

Table 4.7: Fourier smoothing factors for the convection-diffusion equation; AZGS smoothing; $n = 64$.

The only robust and easily vectorizable/parallelizable smoothers found until now are damped alternating Jacobi and damped alternating zebra. These methods involve simultaneous solution along grid lines. For more difficult problems involving systems of nonlinear differential equations this may be a bit awkward. Therefore we present the following variant, which is robust for (3.2) only, but which involves pointwise updates only, and easily allows vector and parallel processing. The method will be called alternating white-black Gauss-Seidel (AWBGS). This smoothing method consists of four steps. First the horizontal grid lines are visited in forward order, then in backward order. Next, the vertical lines are visited in forward order, and then in backward order. However, the lines are not solved for exactly, which would give us alternating symmetric line Gauss-Seidel, because this impedes vectorization and parallelization. Instead the grid points on the line under consideration are divided in white (j_α is even, $\alpha = 1$ for horizontal lines) and black points, and the points of equal colour are solved for simultaneously. A smoothing method using this ordering for the incompressible Navier-Stokes equations has been proposed in [36]. Because the lines are not solved exactly this smoother does not work for the anisotropic diffusion equation. We will present results for the convection-diffusion equation only, in table 4.8, for which β has been sampled with an interval of 2^0; the worst cases are presented. For $\omega = 1$, $\rho \to 1$ as $\varepsilon \downarrow 0$, but ρ_D remains reasonably small. When n increases, $\rho_D \to \rho$. To keep ρ bounded away from 1 as $\varepsilon \downarrow 0$ damping may be applied. Numerical experiments show that $\omega = 0.75$ is a suitable fixed value. We see that this smoother is efficient and robust for the convection-diffusion equation.

	$\omega = 1$				$\omega = 0.75$			
ε	ρ	β	ρ_D	β	ρ	β	ρ_D	β
1.0	0.02	0^0	0.02	0^0	0.26	0^0	0.26	0^0
10^{-1}	0.02	0^0	0.02	0^0	0.27	0^0	0.27	0^0
10^{-2}	0.05	0^0	0.04	0^0	0.28	0^0	0.28	0^0
10^{-3}	0.20	0^0	0.17	0^0	0.40	0^0	0.35	0^0
10^{-5}	0.87	2^0	0.52	10^0	0.50	0^0	0.42	4^0
10^{-8}	0.98	2^0	0.53	10^0	0.50	0^0	0.43	6^0

Table 4.8: Fourier smoothing factors for the convection-diffusion equation; AWBGS smoothing; $n = 64$.

4.5 Incomplete factorization smoothing

For details on incomplete factorization (IF) iterative methods and their application to multigrid, see [26], [25], [38], [12], [13], [53], [21], [51], [52], [50], [14], [8], [49], [48], [45], [47], [46], [33], [32], [22], [28], [54], [57], [23]. IF methods are of the splitting type described in section 2, and their amplification factor is given by (2.5). A difficulty in the Fourier smoothing analysis of IF methods is that the stencils $[M]$ and $[N]$ are not independent of the location in the grid, even if $[A]$ is. But usually (though not always) $[M]$ and $[N]$ rapidly tend to constant stencils away from the grid boundaries. These constant stencils are substituted in (2.5).

Seven-point ILU (incomplete LU) factorization is defined as follows. We write

$$A = LD^{-1}U - N \tag{4.6}$$

with L, D, U as follows. Let

$$[A] = \begin{bmatrix} f & g & \\ c & d & q \\ & a & b \end{bmatrix} \tag{4.7}$$

and

$$[L] = \begin{bmatrix} 0 & 0 & \\ \gamma & \delta & 0 \\ & \alpha & \beta \end{bmatrix}, \; [D] = \begin{bmatrix} 0 & 0 & \\ 0 & \delta & 0 \\ & 0 & \end{bmatrix}, \; [U] = \begin{bmatrix} \xi & \eta & \\ 0 & \delta & \mu \\ & 0 & 0 \end{bmatrix} \tag{4.8}$$

We require $M = LD^{-1}U = A + N$, with

$$[N] = \begin{bmatrix} p_2 & 0 & 0 & 0 & \\ & 0 & p_3 & 0 & \\ & 0 & 0 & 0 & p_1 \end{bmatrix} \tag{4.9}$$

where $p_3 = \sigma(|p_1| + |p_2|)$. Here p_1, p_2 are (necessarily) left free, and σ is a parameter. With $\sigma = 0$ we have standard ILU, with $\sigma \neq 0$ modified ILU. A nonlinear algebraic system for $\alpha, \beta, ..., \xi, \eta$ is obtained, which is easily solved numerically in the same way as will be described for 9-point ILU. One finds $p_1 = \beta\mu/\delta, p_2 = \gamma\xi/\delta$. It has been found in [28], [54], [57], [23] that choosing $\sigma \neq 0$ may improve smoothing efficiency and robustness. As remarked by Wittum [54], [57] modification is better than damping, because if N is small with $\sigma = 0$, it is still small with $\sigma \neq 0$. The optimum value of σ depends on the problem. We prefer to use a fixed value of σ for all problems. From the analysis and experiments of Wittum [54], [57] and our own experiments it follows that $\sigma = 0.5$ is a good choice. We will present results only for $\sigma = 0$ and $\sigma = 0.5$.

For the anisotropic diffusion equation discretized according to (3.3) we have symmetry: $\mu = \gamma$, $\xi = \beta$, $g = a$, $f = b$, $q = c$, $p_1 = p_2 = p$, and (2.5) results in

$$\begin{aligned} \lambda(\theta) &= \{\sigma p + p\cos(2\theta_1 - \theta_2)\}/\{a\cos\theta_2 + b\cos(\theta_1 - \theta_2) \\ &+ c\cos\theta_1 + d/2 + \sigma p + p\cos(2\theta_1 - \theta_2)\} \end{aligned} \tag{4.10}$$

With rotation angle $\beta = 90^o$ and $\varepsilon \ll 1$ one finds

$$\begin{aligned} &0 \le \sigma < 1/2: \ \rho = |\lambda(0,\pi)| \cong \frac{(1-\sigma)p}{2\varepsilon + \sigma p - p} \\ &\rho_D = |\lambda(\varphi,\pi)| \cong \frac{\sigma - 1 + 2\varphi^2}{\delta^2(2 + \varphi^2/2\varepsilon) + \sigma - 1} \\ &1/2 \le \sigma \le 1: \ \rho = |\lambda(0,\pi/2)| \cong \sigma p/(\varepsilon + \sigma p) \\ &\rho_D = |\lambda(\varphi,\pi/2)| \cong \frac{\sigma + 2\varphi}{\delta^2(1 + \varphi^2/2\varepsilon) + \sigma - 2\varphi} \end{aligned} \tag{4.11}$$

where $\varphi = 2\pi/n$. We see that the boundary conditions make much difference. The parameter σ is also influential. Table 4.9 gives results. With $\sigma = 0.5$ the smoother is robust; finer sampling and increasing n gives results indicating that ρ an ρ_D are bounded away from 1. But for some values of β this smoother is not very effective.

ε	σ	ρ $\beta = 0^o$	ρ $\beta = 90^o$	ρ, β	ρ_D $\beta = 0^o$	ρ_D $\beta = 90^o$	ρ_D, β
1	0	0.13	0.13	0.13, any	0.12	0.12	0.12, any
10^{-1}	0	0.17	0.27	$0.45, 75^o$	0.16	0.27	$0.44, 75^o$
10^{-3}	0	0.17	0.84	$1.69, 75^o$	0.02	0.16	$1.55, 75^o$
10^{-5}	0	0.17	0.98	$1.74, 75^o$	10^{-4}	2.10^{-3}	$1.59, 75^o$
1	0.5	0.11	0.11	0.11, any	0.11	0.11	0,11, any
10^{-1}	0.5	0.089	0.23	$0.50, 60^o$	0.087	0.23	$0.50, 60^o$
10^{-3}	0.5	0.091	0.31	$0.82, 60^o$	0.029	0.097	$0.82, 60^o$
10^{-5}	0.5	0.086	0.31	$0.83, 60^o$	4.10^{-4}	10^{-3}	$0.82, 60^o$

Table 4.9: Fourier smoothing factors ρ, ρ_D for (3.1) discretized according to (3.3); 7-point ILU smoothing; $n = 64$.

One might try other values of σ to diminish ρ_D, but we did not find a fixed, problem-independent value that would do. A more efficient and robust smoother of ILU type will be introduced shortly. Oertel and Stüben ([28]) give multigrid results with this smoother, which are in agreement with our smoothing analysis results.
Table 4.10 gives results for (3.2). It is found numerically that $\rho \ll 1$ and $\rho_D \ll 1$ for $\varepsilon \ll 1$, except for β close to 0° or 180°. Numerically we find that for $\varepsilon \ll 1$ and $|\sin\beta| \ll 1$ we have $\rho \cong |\lambda(0,\pi/2)|$, both for $\sigma = 0$ and $\sigma = 1/2$. After some manipulation one obtains

	$\sigma=0$			σ	
ε	ρ	ρ_D	β	ρ, ρ_D	β
1	0.13	0.12	90°	0.11	0°
10^{-1}	0.13	0.13	90°	0.12	0°
10^{-3}	0.44	0.43	165°	0.37	165°
10^{-5}	0.58	0.54	165°	0.47	165°

Table 4.10: Fourier smoothing factors ρ, ρ_D for (3.2) discretized according to (3.5); 7-point ILU smoothing; $n = 64$.

$$|\lambda(0,\pi/2)|^2 \cong \frac{(P_2+1)^2(\sigma^2+1)}{\{(P_2+2)(1-2\tan\beta)+\sigma(1+P_2)^2\}^2+(1-2P_2\tan\beta)^2} \tag{4.12}$$

Hence

$$\rho^2 \lesssim (\sigma^2+1)/(\sigma+1)^2 \tag{4.13}$$

Choosing $\sigma = 1/2$, (4.12) gives $\rho \lesssim 1/3\sqrt{5} \cong 0.75$, so that the smoother is robust. With $\sigma = 0$ equation (4.12) gives

$$\lim_{P_2\to 0} = 1/\sqrt{5},\ \lim_{P_2\to\infty} = (1-4\tan\beta+8\tan^2\beta)^{-1/2} \tag{4.14}$$

so there is a lack of robustness with $\sigma = 0$, for a small range of β.

When A has a nine-point stencil

$$[A] = \begin{bmatrix} f & g & p \\ c & d & q \\ z & a & b \end{bmatrix} \tag{4.15}$$

7-point ILU is not effective, and 9-point ILU is more appropriate:

$$[L] = \begin{bmatrix} 0 & 0 & 0 \\ \gamma & \delta & 0 \\ z & \alpha & \beta \end{bmatrix},\ D = \begin{bmatrix} 0 & 0 & 0 \\ 0 & \delta & 0 \\ 0 & 0 & 0 \end{bmatrix},\ U = \begin{bmatrix} \xi & \eta & p \\ 0 & \delta & \mu \\ 0 & 0 & 0 \end{bmatrix} \tag{4.16}$$

One now finds $M = A + N$ with

$$[N] = \frac{1}{\delta}\begin{bmatrix} \gamma\xi & 0 & 0 & 0 & \\ z\xi & 0 & \sigma n & 0 & \beta p \\ & 0 & 0 & 0 & \beta\mu \end{bmatrix} \tag{4.17}$$

with $n = |\gamma\xi| + |z\xi| + |\beta p| + |\beta\mu|$ and σ a parameter. The elements $\alpha, \beta, ..., \xi, \mu$ may be found iteratively as follows:

$$\begin{aligned}
&\alpha_0 = a,\ \beta_0 = b,\ \gamma_0 = c,\ \delta_0 = d,\ \mu_0 = q,\ \xi_0 = f, \\
&\eta_0 = g,\ j = 0,1,2,... : \\
&\alpha_{j+1} = a - z\mu_j/\delta_j,\ \beta_{j+1} = b - \alpha_{j+1}\mu_j/\delta_j, \\
&\gamma_{j+1} = c - (z\eta_j + \alpha_{j+1}\xi_j)/\delta_j, \\
&n_{j+1} = (|\beta_{j+1}\mu_j| + |z\xi_j| + |\beta_{j+1}p| + |\gamma_{j+1}\xi_j|)/\delta_j, \\
&\delta_{j+1} = d - (zp + \alpha_{j+1}\eta_j + \beta_{j+1}\xi_j + \gamma_{j+1}\mu_j)/\delta_j + \sigma n_{j+1}, \\
&\mu_{j+1} = q - (\alpha_{j+1}p + \beta_{j+1}\eta_j)/\delta_{j+1},
\end{aligned} \tag{4.18}$$

$$\xi_{j+1} = f - \gamma_{j+1}\eta_j/\delta_j,\ \eta_{j+1} = g - \gamma_{j+1}p/\delta_{j+1}$$

When $z = p = 0$ 9-point ILU is identical to 7-point ILU, so we discuss only test problem (3.1) discretized according to (3.4). Table 4.11 gives results; for $\varepsilon = 1$ see table 4.9. Clearly, this smoother is not robust for $\sigma = 0$. But also for $\sigma = 1/2$ there are values of β for which

	$\sigma = 0$		$\sigma = 0.5$
ε	ρ $\beta = 75^o$	ρ_D $\beta = 75^o$	ρ, ρ_D $\beta = 75^o$
10^{-1}	0.52	0.50	0.42
10^{-3}	1.87	1.62	0.68
10^{-5}	1.92	1.66	0.68

Table 4.11: Fourier-smoothing factors ρ, ρ_D for (3.1) discretized according to (3.4); 9-point ILU smoothing; $n = 64$.

smoothing efficiency is not great. For example, with finer sampling of β around 75° one finds a local maximum of about $\rho_D = 0.73$ for $\beta = 85^o$.

A more robust and efficient ILU-type smoother is alternating ILU (AILU), proposed by Oertel and Stüben [28]. AILU consists of two steps. The first step consists of ILU smoothing as just discussed. For the second step the following factorization is used, restricting ourselves to the 7 point case for brevity, assuming that $[A]$ is given by (4.6):

$$[\bar{L}] = \begin{bmatrix} 0 & \bar{\gamma} & \\ 0 & \bar{\delta} & \bar{\alpha} \\ & 0 & \bar{\beta} \end{bmatrix},\ [\bar{D}] = \begin{bmatrix} 0 & 0 & \\ 0 & \bar{\delta} & 0 \\ & 0 & 0 \end{bmatrix},\ [\bar{U}] = \begin{bmatrix} \bar{\xi} & 0 & \\ \bar{\nu} & \bar{\delta} & \\ & \bar{\mu} & 0 \end{bmatrix} \tag{4.19}$$

This ILU factorization corresponds to a backward ordering of the grid points. We now have $A = \bar{M} - \bar{N}$ with

$$[\bar{N}] = \begin{bmatrix} \bar{p}_2 & & \\ 0 & 0 & 0 \\ 0 & \bar{p}_3 & 0 \\ 0 & 0 & 0 \\ & & \bar{p}_1 \end{bmatrix} \tag{4.20}$$

with $\bar{p}_1 = \bar{\beta}\bar{\mu}/\bar{\delta},\ \bar{p}_2 = \bar{\gamma}\bar{\xi}/\bar{\delta},\ \bar{p}_3 = \sigma(|\bar{p}_1| + |\bar{p}_2|)$. The element $\bar{\alpha},\ \bar{\beta}, ..., \bar{\xi}$ can be determined in a similar way as $\alpha, \beta, ..., \xi$. Table 4.12 gives results for test case (3.1). We see that with $\sigma = 0.5$ we have an efficient and robust smoother for this test case. These results are in harmony with the multigrid results obtained in [28]. Similar results are obtained for discretization (3.4) with 9-point AILU smoothing. Table 4.13 gives results for test problem (3.2). Clearly, AILU with $\sigma = 0.5$ is a very efficient smoother for all test problems. Computing cost is twice as large as for ILU, but this is more than compensated by smaller smoothing factors. Additional storage equivalent to the size of A is required.

Another type of incomplete factorization is incomplete block factorization or incomplete block LU decomposition (IBLU). For descriptions and theory of IBLU see [3], [6], [33], [1], [2], [4], [30]. We will assume that $[A]$ is given by (4.15). Then the matrix A has the following structure on a $n_1 x n_2$ grid:

ε	σ	ρ $\beta = 0^o, 90^o$	ρ_D $\beta = 0^o, 90^o$	ρ, ρ_D	β
1	0	9.10^{-3}	9.10^{-3}	9.10^{-3}	any
10^{-1}	0	0.021	0.21	0.061	30°
10^{-3}	0	0.057	3.10^{-3}	0.61	45°
10^{-5}	0	0.064	10^{-6}	0.94	45°
1	0.5	4.10^{-3}	4.10^{-3}	4.10^{-3}	any
10^{-1}	0.5	0.014	0.014	0.028	15°
10^{-3}	0.5	0.026	2.10^{-3}	0.090	45°
10^{-5}	0.5	0.028	0	0.11	45°

Table 4.12: Fourier smoothing factors ρ, ρ_D for (3.1) discretized according to (3.3); 7-point AILU smoothing; $n = 64$.

	$\sigma = 0$		$\sigma = 0.5$	
ε	ρ, ρ_D	β	ρ, ρ_D	β
1	9.10^{-3}	0°	4.10^{-3}	0°
10^{-1}	9.10^{-3}	105°	4.10^{-3}	0°
10^{-3}	0.063	105°	0.027	120°
10^{-5}	0.086	105°	0.036	105°

Table 4.13: Fourier smoothing factors ρ, ρ_D for (3.2) discretized according to (3.5); 7-point AILU smoothing; $n = 64$.

$$A = \begin{pmatrix} B_1 & U_1 & & & \\ L_2 & B_2 & U_2 & & \\ & \ddots & \ddots & \ddots & \\ & & \ddots & \ddots & U_{n_2-1} \\ & & & L_{n_2} & B_{n_2} \end{pmatrix} \tag{4.21}$$

with L_j, B_j, U_j $n_1 \mathrm{x} n_2$ matrices. There is a matrix D such that

$$A = (L+D)D^{-1}(D+U) \tag{4.22}$$

with D a block diagonal matrix consisting of $n_1 \mathrm{x} n_1$ blocks D_j satisfying

$$D_1 = B_1, \quad D_j = B_j - L_j D_{j-1}^{-1} U_j \tag{4.23}$$

Most of the D_j are full. In IBLU D_j is replaced by a tridiagonal approximation, for example by replacing $L_j D_{j-1}^{-1} U_j$ by its tridiagonal part:

$$\tilde{D}_1 = B_1, \quad \tilde{D}_j = B_j - \text{ tridiag } (L_j \tilde{D}_{j-1}^{-1} U_j) \tag{4.24}$$

The IBLU factorization is defined by $A = M - N$,

$$M = (L + \tilde{D})\tilde{D}^{-1}(\tilde{D} + U) \tag{4.25}$$

For algorithms to compute $\tilde{D}$ and $\tilde{D}^{-1}$, see the publications listed above. Far enough from the grid boundaries, $[M]$ becomes constant, and this stencil must be determined for the application of smoothing analysis, as before. This can be done by the following method, which is essentially based on the algorithm for $\tilde{D}_j$ given in [33]. Let $[\tilde{D}] = [\tilde{b}\ \tilde{a}\ \tilde{c}]$. Let $\tilde{D} = (\tilde{E}+I)\tilde{F}^{-1}(I+\tilde{G})$ be a triangular factorization of $\tilde{D}$, with the non-zero elements of $\tilde{E}, \tilde{F}, \tilde{G}$ given by $\tilde{e}_{i,i-1} = \tilde{e}$, $\tilde{f}_{ii} = \tilde{f}$, $\tilde{g}_{i,i+1} = \tilde{g}$. The elements $\tilde{a}, ..., \tilde{g}$ can be computed with the following iteration method:

Algorithm 1:

$\tilde{b} = c;\ \tilde{a} = d;\ \tilde{c} = q;\ \tilde{f} = 1/\tilde{a};\ \tilde{g} = \tilde{c}f;$

<u>do</u> <u>until</u> convergence

$\tilde{e} = \tilde{b}\tilde{f};\ \tilde{f} = 1/(\tilde{a} - \tilde{e}\tilde{g}/\tilde{f});\ \tilde{g} = \tilde{c}\tilde{f};$

$\tilde{s}_0 = \tilde{f}/(1-\tilde{g}\tilde{e});\ \tilde{s}_{-1} = -\tilde{e}\tilde{s}_0;\ \tilde{s}_{-2} = -\tilde{e}\tilde{s}_{-1};$

$\tilde{s}_{-3} = -\tilde{e}\tilde{s}_{-2};\ \tilde{s}_1 = -g\tilde{s}_0;\ \tilde{s}_2 = -\tilde{g}\tilde{s}_1;\ \tilde{s}_3 = -\tilde{g}\tilde{s}_2;$

$k = -2, -1, ..., 2:\ \sigma_k = z\tilde{s}_{k+1} + a\tilde{s}_k + b\tilde{s}_{k-1};$

$k = -1, 0, 1:\ \tilde{t}_k = f\sigma_{k+1} + \sigma_k g + p\sigma_{k-1};$

$\tilde{b} = c - \tilde{t}_{-1};\ \tilde{a} = d - \tilde{t}_0;\ \tilde{c} = q - \tilde{t}_1$

<u>od</u>

Once $[\tilde{D}]$ has been determined, Fourier smoothing analysis proceeds as follows. The amplification factor $\lambda(\theta)$ is given by (2.5). The constant coefficient (Toeplitz) operators $L, \tilde{D}, U$ and A share the same eigenvectors $\psi_j(\theta)$ defined by (2.3). Therefore we have

$$\sum_k M(k)\psi_{j+k}(\theta) = \{\lambda_1(\theta)\lambda_3(\theta)/\lambda_2(\theta)\}\psi_j(\theta) \tag{4.26}$$

with

$$\begin{aligned} \lambda_1(\theta) &= \sum_k \{L(k) + \tilde{D}(k)\}\psi_k(\theta) \\ \lambda_2(\theta) &= \sum_k \tilde{D}(k)\psi_k(\theta) \\ \lambda_3(\theta) &= \sum_k \{\tilde{D}(k) + U(k)\}\psi_k(\theta) \end{aligned} \tag{4.27}$$

Furthermore,

$$\begin{aligned} &\sum_k N(k)\psi_{j+k}(\theta) = \sum_k \{M(k) - A(k)\}\psi_{j+k}(\theta) \\ &= \lambda_1(\theta)\lambda_3(\theta)/\lambda_2(\theta) - \lambda_A(\theta) \end{aligned} \tag{4.28}$$

where $\lambda_A(\theta) = \sum_k A(k)\psi_k(\theta)$. Hence, $\lambda(\theta)$ is given by

$$\lambda(\theta) = 1 - \lambda_2(\theta)\lambda_A(\theta)/\lambda_1(\theta)\lambda_3(\theta) \tag{4.29}$$

Table 4.14 gives results for the rotated anisotropic diffusion equation (3.1) discretized according to (3.4). In cases where algorithm 1 does not converge rapidly, in practical applications the elements of $\tilde{D}$ do not settle down quickly to values independent of location as one moves away from the domain boundary, so that in these cases Fourier smoothing analysis

ε	ρ $\beta = 0^o$	ρ $\beta = 90^o$	ρ_D $\beta = 0^o$	ρ_D $\beta = 90^o$
1	0.058	0.058	0.056	0.056
10^{-1}	0.108	0.133	0.102	0.116
10^{-3}	0.164*	0.194	0.025*	5.10^{-3}
10^{-5}	0.141*	0.200	0.000*	0.000

Table 4.14: Fourier smoothing factors ρ, ρ^D for (3.1) discretized according to (3.4); IBLU smoothing; $n = 64$. The symbol * indicates that algorithm 1 did not converge within 6 decimals in 100 iterations.

is not realistic. Table 4.15 gives results for the convection-diffusion equation. It is clear

ε	ρ	β	ρ_D	β
1.0	0.058	0°	0.056	0°
10^{-1}	0.061	0°	0.058	0°
10^{-3}	0.173	0_o	0.121	0°
10^{-5}	0.200	0°	10^{-3}	15°

Table 4.15: Fourier smoothing factors ρ, ρ_D for (3.2) discretized according to (3.5); IBLU smoothing; $n = 64$.

that IBLU is an efficient smoother for all cases. This is confirmed by the multigrid results obtained in [33].

For aspects of vectorization see [1], [4].

4.6 Multistage smoothing methods

Multistage smoothing methods are also a type of iterative method (2.3) (of the semi-iterative kind, as we will see), but in the multigrid literature they have been developed as techniques to solve systems of ordinary differential equations, arising from the spatial discretization of systems of hyperbolic or almost hyperbolic partial differential equations. Accordingly, we will apply multistage smoothing only to test problem (3.2). Multistage smoothing has been introduced in [18].

A time-derivative is added to the equation to be solved:

$$\frac{\partial u}{\partial t} - \varepsilon(\frac{\partial^2 u}{\partial x^2} + \frac{\partial^2 u}{\partial y^2}) + c\frac{\partial u}{\partial x} + s\frac{\partial u}{\partial y} = 0 \tag{4.30}$$

Spatial discretization according to (3.5) gives a system of ordinary differential equations denoted by

$$\frac{d\mathbf{u}}{dt} = -h^{-2}A\mathbf{u} \tag{4.31}$$

where A is the operator defined in (3.5). A p-stage (Runge-Kutta) time-stepping method for (4.31) is defined by

$$\begin{aligned} \mathbf{u}^{(0)} &= \mathbf{u}^n \\ \mathbf{u}^{(k)} &= \mathbf{u}^{(0)} - c_k \nu h^{-1} A \mathbf{u}^{(k-1)}, \ k = 1, 2, ..., p \\ \mathbf{u}^{n+1} &= \mathbf{u}^{(p)} \end{aligned} \tag{4.32}$$

Here superscript n denotes the time-level, superscript (k) denotes stage number k, and $\nu \equiv \Delta t/h$ is the CFL (Courant-Friedrichs-Lewy) number. One always has $c_P = 1$ for consistency. Since the time-derivative in (4.31) is an artefact, c_k is chosen not too optimize accuracy but stability and smoothing behaviour.

Eliminating $\mathbf{u}^{(k)}$ equation (4.33) can be rewritten as

$$\mathbf{u}^{n+1} = P_p(-\nu h^{-1} A)\mathbf{u}^n \tag{4.33}$$

with the *amplification polynominal* P_p defined by

$$P_p(z) = 1 + z(1 + c_{p-1}z(1 + c_{p-2}z(...(1 + c_1 z)...))) \tag{4.34}$$

Obviously, equation (4.33) can be interpreted as an iterative method for solving $h^{-2}A\mathbf{u} = 0$ with iteration matrix

$$S = P_p(\nu h^{-1} A) \tag{4.35}$$

Such methods for which S is a polynomial in the matrix of the system to be solved are called *semi-iterative methods*; see [37] for the theory of such methods. For $p = 1$ we obtain the damped Jacobi method with diagonal scaling (diag $(A) = I$), also known as the one-stage Richardson method; as an initial value problem solver this is known as the forward Euler method. Following the trend in the multigrid literature, we will analyze (4.33) as a multistage method for differential equations.

The CFL number ν is restricted by stability. In order to assess stability and the smoothing behaviour of (4.33), the Fourier modes $\psi(\theta)$ defined by (2.5) are substituted for $\mathbf{u}$. We have $\nu h^{-1} A\psi(\theta) = \nu h^{-1}\mu(\theta)\psi(\theta)$. With A defined by (3.5) we have

$$\begin{aligned} \mu(\theta) &= 4\varepsilon + h(|c| + |s|) - (2\varepsilon + h|c|)\cos\theta_1 \\ &- (2\varepsilon + h|s|)\cos\theta_2 + ih(c\sin\theta_1 + s\sin\theta_2) \end{aligned} \tag{4.36}$$

whereas with central differences we have

$$\mu(\theta) = 4\varepsilon - 2\varepsilon(\cos\theta_1 + \cos\theta_2) + ih(c\sin\theta_1 + s\sin\theta_2) \tag{4.37}$$

Equation (4.33) gives $\mathbf{u}^{n+1} = \lambda(\theta)\mathbf{u}^n$ with the amplification factor $\lambda(\theta)$ given by

$$\lambda(\theta) = P_p(-\nu\mu(\theta)/h) \tag{4.38}$$

The stability restriction on ν is that one should have

$$|\lambda(\theta)| \le 1, \quad \theta \in \Theta \tag{4.39}$$

The smoothing factors ρ and ρ_D are defined by (2.12) and (2.16), respectively, as before.

When the coefficients c and s in (4.30) are replaced by general variable coefficients v_1 and v_2 the appropriate definition of the CFL number ν is

$$\nu = (|v_1| + |v_2|)\Delta t/h \tag{4.40}$$

Choosing Δt the same in every grid point this results in a variable ν. For smoothing purposes it is better to fix ν at some favorable value, resulting in a Δt which varies over the grid points. This is called *local time stepping*; this is allowed because temporal accuracy is irrelevant. For the purpose of Fourier analysis the coefficients c, s and ε must be assumed constant, of course.

The optimum values of ν and c_k are problem-dependent. Some analysis of the optimization problem involved may be found in [24]. In general, this optimization problem can only be solved numerically.

Let us assume that the flow is aligned with the grid ($\beta = 0^o$ or $\beta = 90^o$); take $\beta = 0^o$. Furthermore, let $\varepsilon = 0$. Then $\mu(0, \theta_2) = 0$ and $\lambda(0, \theta_2) = 1, \forall\theta_2$, so that we have no smoother. This is typical for multistage smoothing: when the flow is aligned with the grid, waves perpendicular to the flow are not damped, if there is no cross-flow diffusion term. The boundary conditions take care of such modes; we have $\rho_D < 1$.

We proceed with two examples.

A four-stage method Based upon an analysis of Catalano and Deconinck (private communication), in which optimal values for c_k and ν are determined for the upwind discretization (3.5), we choose

$$c_1 = 0.07, \quad c_2 = 0.19, \quad c_3 = 0.42, \quad \nu = 2.0 \tag{4.41}$$

Table 4.16 gives some results. It is found that ρ_D differs very little from ρ.

ε/β	0°	15°	30°	45°
0	1.00	0.593	0.477	0.581
10^{-5}	0.997	0.591	0.482	0.587

Table 4.16: Smoothing factor ρ for (3.2) discretized according to (3.5); four-stage method; $n = 64$.

It is not necessary to choose β outside $[0^o, 45^o]$, since results are symmetric in β. For $\varepsilon \gtrsim 10^{-3}$ the method becomes unstable for certain values of β. Hence, for problems in which the mesh-Péclet number varies widely in the domain it would seem necessary to adapt c_k and ν to the local stencil.

A five-stage method The following method has been proposed in [17] for a central discretization of the Euler equations of gasdynamics:

$$c_1 = 1/4, \quad c_2 = 1/6, \quad c_3 = 3/8, \quad c_4 = 1/2 \tag{4.42}$$

The method has also been applied to the compressible Navier-Stokes equations in [19]. We will apply this method to (4.30) with central discretization. With $\mu(\theta)$ given by (4.37) and $\varepsilon = 0$ we have $|\lambda(0, \pi)| = 1$, so that we have no smoother. Therefore an artificial dissipation term is added to (4.31), which becomes

$$\frac{d\mathbf{u}}{dt} = -h^{-2}A\mathbf{u} - h^{-1}B\mathbf{u} \tag{4.43}$$

with

$$B = \kappa \begin{bmatrix} & & 1 & & \\ & & -4 & & \\ 1 & -4 & 12 & -4 & 1 \\ & & -4 & & \\ & & 1 & & \end{bmatrix} \tag{4.44}$$

where κ is a parameter. We have $B\psi(\theta) = \eta(\theta)\psi(\theta)$ with

$$\eta(\theta) = 4\kappa\{(1-\cos\theta_1)^2 + (1-\cos\theta_2)^2\} \tag{4.45}$$

For reasons of efficiency Jameson and Baker ([17]) update the artificial dissipation term only in the first two stages. This gives a five-stage method with amplification polynomial $P_5(z_1, z_2)$ defined by

$$\begin{aligned} &P_1 = 1 - c_1(z_1+z_2), \quad P_2 = 1 - c_2(z_1+z_2)P_1, \\ &P_3 = 1 - c_3 z_1 P_2 - c_3 z_2 P_1, \quad P_4 = 1 - c_4 z_1 P_3 - c_4 z_2 P_1, \\ &P_5(z_1,z_2) = 1 - z_1 P_4 - z_2 P_1 \end{aligned} \tag{4.46}$$

where $z_1 = \nu h^{-1}\mu(\theta), \ z_2 = \nu\eta(\theta)$.

In one dimension Jameson and Baker ([17]) advocate $\kappa = 0.04$ and $\nu = 3$; for stability ν should not be much larger that 3. In two dimensions $\max\{\nu h^{-1}|\mu(\theta)|\} = \nu(c+s) \le \nu\sqrt{2}$. Therefore we choose $\nu = 3/\sqrt{2} \cong 2.1$. With $\nu = 2.1$ and $\kappa = 0.04$ the results of Table 4.17 are obtained, both for $\varepsilon = 0$ and $\varepsilon = 10^{-5}$.

β	0°	15°	30°	45°
ρ	0.70	0.77	0.82	0.82

Table 4.17: Smoothing factor ρ for (3.2) with central discretization and artificial dissipation; five-stage method; $n = 64$.

Again, $\rho_D \cong \rho$. This method allows only $\varepsilon \ll 1$; for example, for $\varepsilon = 10^{-3}$ and $\beta = 45^o$ one finds $\rho = 0.96$. According to table 4.17 we have a smoother; varying n does not influence ρ very much. But the smoothing efficiency is not very inpressive. This is the only smoother known to the author for a central discretization of (3.2). The application of the other smoothers discussed before to the centrally discretized convection-diffusion equation with higher order artificial dissipation added has not been investigated.

5. Conclusions

Fourier smoothing analysis of most if not all well-known smoothing methods has been carried out for the two-dimensional test problems (3.1) and (3.2), discretized according to (3.3) or (3.4) and (3.5). The following methods work for both problems, without having to adapt parameters:

- Alternating damped Jacobi (provided (3.1) is discretized with (3.4));
- Alternating symmetric line Gauss-Seidel (provided (3.1) is discretized with (3.4));

- Damped alternating zebra Gauss-Seidel (provided (3.1) is descretizd with (3.4));
- Alternating incomplete LU-factorization with damping (AILU);
- Incomplete block LU-factorization (IBLU).

It is important to take the type of boundary condition into account. The heuristic way in which this has been done within the framework of Fourier smoothing analysis correlates well with multigrid convergence results obtained in practice.

Alternating symmetric line Gauss-Seidel is rather unsuitable for vector and parallel processing. There are techniques to implement incomplete factorizations efficiently on vector and parallel computers; see [16], [15], [43], [42], [41], [39], [31], [40], [44], [1], [4]. A simple vectorizing smoother that works for (3.2) but not for (3.1) is alternating white-black Gauss-Seidel.

Incomplete factorizations are difficult to generalize to systems, except in cases where they effectively operate on single equations, see for example [55], [56] for the case of the incompressible Navier Stokes equations.

Symmetric point Gauss-Seidel is not such an excellent smoother for the convection-diffusion equation as is generally believed, but four-direction Gauss-Seidel is. A better vectorizable variant is alternating white-black Gauss-Seidel.

Of course, in three dimensions robust and efficient smoothers are more elusive than in two. IBLU, the most powerful smoother in two dimensions, is not robust in 3D [22]. Robust 3D smoothers can be found among methods that solve exactly in planes (plane Gauss-Seidel) ([35]). For a successful multigrid approach to a complicated three-dimensional problem using ILU-type smoothing, see [48], [45], [47], [46].

Multistage smoothing methods work only for a limited class of problems, and require tuning to the problem at hand. They are very suitable for vectorized and parallel computing, and easily extendable to central discretizations of hyperbolic systems, which probably explains their current popularity in computational gasdynamics.

References

1. O. Axelsson. Analysis of incomplete matrix factorizations as multigrid smoothers for vector and parallel computers. *Appl. Math. Comp. 19*, 3–22, 1986.

2. O. Axelsson. A general incomplete block-matrix factorization method. *Lin. Alg. Appl. 74*, 179–190, 1986.

3. O. Axelsson, S. Brinkkemper, and V.P. Il'in. On some versions of incomplete block matrix factorization iterative methods. *Lin. Algebra Appl. 59*, 3–15, 1984.

4. O. Axelsson and B. Polman. On approximate factorization methods for block matrices suitable for vector and parallel processors. *Lin. Alg. Appl. 77*, 3–26, 1986.

5. T.F. Chan and H.C. Elman. Fourier analysis of iterative methods for elliptic boundary value problems. *SIAM Rev. 31*, 20–49, 1989.

6. P. Concus, G.H. Golub, and G. Meurant. Block preconditioning for the conjugate gradient method. *SIAM J. Sci. Stat. Comp. 6*, 220–252, 1985.

7. W. Hackbusch, editor. *Efficient solutions of elliptic systems*, Vieweg, Braunschweig/Wiesbaden, 1984. Notes on Numerical Fluid Mechanics 10.

8. W. Hackbusch. *Multi-grid methods and applications*. Springer-Verlag, Berlin, 1985.

9. W. Hackbusch, editor. *Robust multi-grid methods*, Vieweg, Braunschweig/Wiesbaden, 1988. Proc. 4th GAMM-Seminar, Kiel, 1988.

10. W. Hackbusch and U. Trottenberg, editors. *Multigrid methods*, Springer-Verlag, Berlin, 1982. Lecture Notes in Mathematics (960).

11. W. Hackbusch and U. Trottenberg, editors. *Multigrid Methods II*, Springer-Verlag, Berlin, 1986. Lecture Notes in Mathematics 1228.

12. P.W. Hemker. The incomplete LU-decomposition as a relaxation method in multi-grid algorithms. In J.H. Miller, editor, *Boundary and Interior Layers - Computational and Asymptotic Methods*, pages 306–311, Boole Press, Dublin, 1980.

13. P.W. Hemker. On the comparison of Line-Gauss-Seidel and ILU relaxation in multigrid algorithms. In J.J.H. Miller, editor, *Computational and Asymptotic Methods for Boundary and Interior Layers*, pages 269–277, Boole Press, Dublin, 1982.

14. P.W. Hemker, R. Kettler, P. Wesseling, and P.M. De Zeeuw. Multigrid methods: development of fast solvers. *Appl. Math. Comp.*, 13:311–326, 1983.

15. P.W. Hemker and P.M. De Zeeuw. Some implementations of multigrid linear systems solvers. In D.J. Paddon and H. Holstein, editors, *Multigrid Methods for Integral and Differential Equations*, pages 85–116, Clarendon Press, Oxford, 1985.

16. P.W. Hemker, P. Wesseling, and P.M. de Zeeuw. A portable vector-code for autonomous multigrid modules. In B. Enquist and T. Smedsaas, editors, *PDE Software: Modules, Interfaces and Systems*, pages 29–40, 1984.

17. A. Jameson and T.J. Baker. *Multigrid Solution of the Euler Equations for Aircraft Configurations*. AIAA-Paper 84-0093, 1984.

18. A. Jameson and T.J. Baker. *Solution of the Euler Equations for Complex Configurations*. AIAA-Paper 83-1929, 1983.

19. M. Jayaram and A. Jameson. *Multigrid Solution of the Navier-Stokes Equations for Flow over Wings*. AIAA-Paper 88-0705, 1988.

20. J.Mandel, S.F. McCormick, J.E. Dendy, Jr., C. Farhat, G. Lonsdale, S.V. Parter, J.W. Ruge, and K. Stüben, editors. *Proceedings of the Fourth Copper Mountain Conference on Multigrid Methods*, SIAM, Philadelphia, 1989.

21. R. Kettler. Analysis and comparison of relaxation schemes in robust multigrid and conjugate gradient methods. In *Hackbusch and Trottenberg*, pages 502–534, 1982.

22. R. Kettler and P. Wesseling. Aspects of multigrid methods for problems in three dimensions. *Appl. Math. Comp.*, 19:159–168, 1986.

23. M. Khalil. *Analysis of Linear Multigrid Methods for Elliptic Differential Equations with Discontinuous and Anisotropic Coefficients.* PhD thesis, Delft University of Trechnology, Delft, The Netherlands, 1989.

24. B. Van Leer, C.-H. Tai, and K.G. Powell. *Design of Optimally Smoothing Multistage Schemes for the Euler Equations.* AIAA Paper 89-1933-CP, 1989.

25. J.A. Meijerink and H.A. Van der Vorst. Guidelines for the usage of incomplete decompositions in solving sets of linear equations as they occur in practical problems. *J. Comput. Phys.*, 44:134–155, 1981.

26. J.A. Meijerink and H.A. Van der Vorst. An iterative solution method for linear systems of which the coefficient matrix is a symmetric M-matrix. *Math. Comp.*, 31:148–162, 1977.

27. K.W. Morton and M.J. Baines, editors. *Numerical Methods for Fluid Dynamics II*, Clarendon Press, Oxford, 1986.

28. K.-D. Oertel and K. Stüben. Multigrid with ILU-smoothing: systematic tests and improvements. In , editor, *Hackbusch(1988)*, pages 188–199.

29. D.J. Paddon and H. Holstein, editors. *Multigrid Methods for Integral and Differential Equations*, Clarendon Press, Oxford, 1985.

30. B. Polman. Incomplete blockwise factorizations of (block) H-matrices. *Lin. Alg. Appl.*, 90:119–132, 1987.

31. J. J. F. M. Schlichting and H.A. Van der Vorst. Solving 3D block bidiagonal linear systems on vector computers. *J. of Comp. and Appl. Math.*, 27:323–330, 1989.

32. P. Sonneveld, P. Wesseling, and P.M. de Zeeuw. Multigrid and conjugate gradient acceleration of basic iterative methods. In *Morton and Baines (1986)*, pages 347–368, 1986.

33. P. Sonneveld, P. Wesseling, and P.M. de Zeeuw. Multigrid and conjugate gradient methods as convergence acceleration techniques. In *Paddon and Holstein (1985)*, pages 117–168, 1985.

34. K. Stüben and U. Trottenberg. Multigrid methods: fundamental algorithms, model problem analysis and applications. In *Hackbusch and Trottenberg (1982)*, pages 1–176, 1982.

35. C.-A. Thole and U. Trottenberg. Basic smoothing procedures for the multigrid treatment of elliptic 3D-operators. *Appl. Math. Comp.*, 19:333–345, 1986.

36. S.P. Vanka and K. Misegades. *Vectorized Multigrid Fluid Flow Calculations on a CRAY X-MP48*. AIAA Paper 86-0059, 1986.

37. R.S. Varga. *Matrix Iterative Analysis*. Prentice-Hall, Englewood Cliffs, New Jersey, 1962.

38. R.S. Varga, E.B. Saff, and V. Mehrmann. Incomplete factorizations of matrices and connections with H-matrices. *SIAM J. Numer. Anal.*, 17:787–793, 1980.

39. H.A. Van der Vorst. Analysis of a parallel solution method for tridiagonal linear systems. *Parallel Computing*, 5:303–311, 1987.

40. H.A. Van der Vorst. ICCG and related methods for 3D problems on vector computers. *Computer Physics Comm.*, 53:223–235, 1989.

41. H.A. Van der Vorst. Large tridiagonal and block tridiagonal linear systems on vector and parallel computers. *Parallel Computing*, 5:45–54, 1987.

42. H.A. Van der Vorst. The performance of FORTRAN implementations for preconditioned conjugate gradients on vector computers. *Parallel Computing*, 3:49–58, 1986.

43. H.A. Van der Vorst. A vectorizable variant of some ICCG methods. *SIAM J. Sci. Stat. Comp.*, 3:350–356, 1982.

44. H.A. Van der Vorst and K. Dekker. Vectorization of linear recurrence relations. *SIAM J. Sci. Stat. Comp.*, 10:27–35, 1989.

45. A.J. Van der Wees. FAS multigrid employing ILU/SIP smoothing: a robust fast solver for 3D transonic potential flow. In *Hackbusch and Trottenberg (1986)*, pages 315–331.

46. A.J. Van der Wees. Impact of multigrid smoothing analysis on three-dimensional potential flow calculations. In *Mandel et al. (1989)*, pages 399–416.

47. A.J. Van der Wees. *A nonlinear multigrid method for three-dimensional transonic potential flow*. PhD thesis, Delft University of Technology, 1988.

48. A.J. Van der Wees. Robust calculation of 3d potential flow based on the nonlinear FAS multi-grid method and a mixed ILU/SIP algorithm. In J.G. Verwer, editor, *Colloquium Topics in Applied Numerical Analysis*, pages 419–459, Centre for Mathematics and Computer Science, Amsterdam, 1984. CWI Syllabus.

49. A.J. Van der Wees, J. Van der Vooren, and J.H. Meelker. *Robust Calculation of 2D Transonic Potential Flow Based on the Nonlinear FAS Multi-Grid Method and Incomplete LU Decompositions*. AIAA Paper 83-1950, 1983.

50. P. Wesseling. Multigrid solution of the Navier-Stokes equations in the vorticity-streamfunction formulation. In *Hackbusch (1984)*, pages 145–154, 1984.

51. P. Wesseling. A robust and efficient multigrid method. In *Hackbusch and Trottenberg (1982)*, pages 614–630.

52. P. Wesseling. Theoretical and practical aspects of a multigrid method. *SIAM J. Sci. Stat. Comp. 3*, 387–407, 1982.

53. P. Wesseling and P. Sonneveld. Numerical experiments with a multiple grid and a preconditioned Lanczos type method. In R. Rautmann, editor, *Approximation methods for Navier-Stokes problems*, pages 543–562, Springer-Verlag, Berlin, 1980. Lecture Notes in Math., 771.

54. G. Wittum. Linear iterations as smoothers in multigrid methods: theory with applications to incomplete decompositions. *Impact of Computing in Science and Engineering*, 1:180–215, 1989.

55. G. Wittum. Multi-grid methods for Stokes and Navier-Stokes equations with transforming smoothers: algorithms and numerical results. *Numer. Math.*, 54:543–563, 1989.

56. G. Wittum. On the convergence of multi-grid methods with transforming smoothers. *Num. Math.*, 57:15–38, 1990.

57. G. Wittum. On the robustness of ILU smoothing. *SIAM J. Sci. Stat. Comp.*, 10:699–717, 1989.

Part II
Contributed Papers

International Series of Numerical Mathematics, Vol. 98, © 1991 Birkhäuser Verlag Basel

Multilevel Methods for Fast Solution of N-Body and Hybrid Systems

by

Dinshaw S. Balsara and Achi Brandt

I) Introduction

Many physical problems entail the task of simultaneously solving Liouville's / Boltzmann's equation along with the fluid / magnetofluid equations. The representation of Liouville's equation is usually by way of superparticles, Hockney and Eastwood (1981), which map out the trajectory of a distribution in phase space. The representation of the fluid phase is usually best done on a grid. Such composite systems where a grid-based representation and a particle-based representation are simultaneously required are called *hybrid* systems. The time-evolution of such systems always entails making a consistent evaluation of the potential and forces at all particle positions and grid positions. Thus, for each point "i", whether particle or grid position, we have to evaluate :

$$\psi_i^h = \sum_j K_{i\,j}^{h\,h} \rho_j^h \quad \text{or in matrix notation :-} \quad \psi^h = K^{h\,h} \rho^h \tag{1}$$

where "j" extends over all other particle and grid positions. Taken as it stands, eq (1) entails making O (n^2) evaluations. This evaluation has to be made for *both* phases *self-consistently*. We call the task of evaluating (1) the problem of *multi-integration*. Schemes can be devised for accurately evaluating (1) for one phase or the other, eg. Greengard and Rokhlin (1987) which works only for special kernels and only when simple analytic multipolar expansions are available. We wish to evaluate (1) for both phases with *optimal* efficiency, i.e. in O (s n) operations, where "n" is the number of particles and grid points at which we need to have the potential and force evaluation and "s" is a small number that depends only on the desired accuracy ε of the evaluation (typically $s = \log(1/\varepsilon)^q$ for the methods described here). Moreover, we wish to have *adaptibility* in both space and time. This seems to rule out the use of fast Fourier techniques since they try to represent the problem in spectral rather than real space. Fast Fourier techniques also suffer from the problem that sometimes the kernel may not be of convolution type and then fast Fourier techniques would not work. Moreover, when boundaries are irregular, Fourier techniques can't be used.

Techniques developed here are likely to have applicability to astrophysics, plasma physics and in the simulation of semiconductor devices. Generalization of the techniques developed here to oscillatory kernels is described in Brandt (1990). In Sect. II we review the basic Brandt and Lubrecht (1990) (BL hence forth) paper along with a more concise derivation of their basic result. In Sect. III we introduce the idea of softening of kernels that is essential to this work and discuss the method's efficiency. In Sect. IV we give an accuracy analysis for a system of two particles. In Sect. V we give a timing analysis. In Sect. VI we discuss future directions.

II) Multi-Integration on Uniform Grids

BL essentially concentrate on solving the multi-integration problem, eq. (1), when the density ρ_i^h is defined on a grid of mesh spacing "h". We need to evaluate ψ_i^h in eq (1). The question is how to do it in O (s n) steps? BL show that this can be done in the following four steps. The method is recursive. Thus assume there exists a coarse grid of size H = 2 h and that a p order interpolation matrix I_H^h is being used to transfer solutions from grid "H" to "h".

Step I :- Adjoint Interpolation to Coarse Grid

Evaluate

$$\rho^H = \frac{1}{2^d} \left(I_H^h \right)^T \rho^h \tag{2}$$

on the coarse grid "H". "d" here is the dimensionality of the space we are dealing with. Superscript "T" stands for adjoint.

Step II :- Solve on Coarse Grid

On the coarse grid evaluate

$$\psi^H = 2^d K^{H\,H} \rho^H \tag{3}$$

Here $K^{H\,H}$ is the injection of $K^{h\,h}$ onto the coarse grid, i.e. $K_{I\,J}^{HH} = K_{2i\,2j}^{h\,h}$.

Step III :- Interpolate to Fine Grid

Interpolate the coarse grid solution to the fine grid as follows

$$\tilde{\psi}^h = I_H^h \psi^H \tag{4}$$

Step IV :- Make Local Corrections on the Fine Grid

Combining (1), (2), (3) and (4) above we get

$$\psi^h - \tilde{\psi}^h = (K^{hh} - [I_H^h K^{HH} (I_H^h)^T]) \rho^h \tag{5}$$

As written, (5) has the sum extending over all points in the domain. No gain in efficiency is obtained from that. However, most kernels of interest in physics such as those displayed in eqs (8) and (9) below have the property of asymptotic smoothness. We say that a kernel K is asymptotically smooth (on grid "h") if it has the property that

$$\| \partial_x^p K (x, y) \| < C_p \xi^{q - p} \tag{6}$$

for all $\xi \geq O (h)$. Here $\xi = \| x - y\|$, ∂_x^p is any p order derivative with respect to x, q is independent of p and C_p depends only on p. If the kernel satisfies (6) unconditionally on the grid "h" then the kernel is truly smooth on the grid "h" and the local corrections on that grid are not even needed as shown by BL. For asymptotically-smooth kernels and p order interpolation,

$$K_{ij}^{hh} - [I_H^h K^{HH} (I_H^h)^T]_{ij} \approx O (h^p \xi^{q - p}) \tag{7}$$

and $| h^p \xi^{q - p} |$ is a rapidly decreasing function of $\xi = \|j - i\|$. Thus for suitable p (depending on the accuracy ε) $\exists$ m such that for $\|j - i\| >$ mh, $| K_{ij}^{hh} - [I_H^h K^{HH} (I_H^h)^T]_{ij} |$ is smaller than some suitably small fraction ε of $|K_{ij}^{hh}|$ and is neglected.

This completes the description of the two level scheme. Now realize that interpolation is an O (s n) process and so are the local corrections. Thus if Step II on the coarse grid is made O (s n) then the whole scheme becomes O (s n). This is easy to do since the recursion can be continued down to a level which has O ($\sqrt{n}$) grid points, then the multi-integration can be done by direct evaluation on that grid in O (s n) operations. Thus the whole process of evaluating (1) is done in O (s n) steps.

III) Softened Kernels

In most of this work we will test kernels that are singular-smooth and of the form

$$K(\vec{x}_i, \vec{x}_j) = \frac{1}{\|\vec{x}_i - \vec{x}_j\|} \tag{8}$$

and

$$K(\vec{x}_i, \vec{x}_j) = \frac{\vec{x}_i - \vec{x}_j}{\|\vec{x}_i - \vec{x}_j\|^3} \tag{9}$$

though nowhere are we specifically committed to this form of the kernel.

The BL work applies most efficiently to ordered grids. A system of particles can be thought of as a disordered grid, i.e. one where the grid points occupy arbitrary positions in space. The most "natural" extension of the BL work to disordered grids is complicated as we see below.

Consider a set of particles with positions given by $\{\vec{x}_i \mid i = 1,\dots, n\}$ and masses $\{m_i \mid i = 1,\dots, n\}$ We need to evaluate

$$\psi(\vec{x}_j) = \sum_i K(\vec{x}_j; \vec{x}_i)\, m_i \quad \forall\ i, j = 1,\dots, n \tag{10}$$

Obviously, we need an ordered grid on which the task of multi-integration can be reduced in complexity. So let us assume we have a fine grid $\{\vec{X}_I \mid I = 1,\dots, N\}$ with grid spacing "h" available to us. Let us also assume that we have a continuous assignment function $w_{i\,I}$ function available which assigns the particle's mass, concentrated at $\vec{x}_i$ as a density distribution to the grid points $\vec{X}_I$ around $\vec{x}_i$. We assume, as is usual, that $w_{i\,I}$ is of finite support. It usually proves most advantageous, especially for hybrid systems, to take $w_{i\,I}$ to be the polynomial version of the discrete interpolation I_H^h used in the previous section. That way, the grid based density does not even need to go through an assignment step which further improves the efficiency. Our purpose will be to show, by a careful set of stages, how the transition of the problem, i.e the density and potential, from the disordered to the ordered grid can be effected. This will then give us a clear idea of what errors are made in the process and, therefore, what local corrections are needed. First we transfer the masses defined at particle positions to a density defined on the grid through the following approximation :-

$$\psi(\vec{x}_j) \approx \sum_{iI} K(\vec{x}_j; \vec{X}_I)\, w_{iI}\, m_i \tag{11a}$$

or

$$\psi(\vec{x}_j) \approx \sum_{I} K(\vec{x}_j; \vec{X}_I)\, \rho_I \tag{11b}$$

where

$$\rho_I \equiv \sum_{i} w_{iI}\, m_i \tag{12}$$

To really arrive at a grid based representation we need to make an interpolation of the potential on the grid to the particle using an interpolation function $\overline{w}_{jJ}$ as follows

$$\psi(\vec{x}_j) \approx \sum_{I} K(\vec{x}_j; \vec{X}_I)\, \rho_I \tag{13a}$$

$$\psi(\vec{x}_j) \approx \sum_{IJ} \overline{w}_{jJ}\, K(\vec{X}_J; \vec{X}_I)\, \rho_I \tag{13b}$$

so that

$$\psi(\vec{x}_j) \approx \sum_{J} \overline{w}_{jJ}\, \psi_J \tag{13c}$$

where

$$\psi_J \equiv \sum_{I} K(\vec{X}_J; \vec{X}_I)\, \rho_I \tag{14}$$

ψ_J now can be evaluated on a grid using efficient methods for multi-integration. To really get the exact solution, we need to correct for the above assignment and interpolation steps in the following way

$$\psi(\vec{x}_j) = \sum_{J} \overline{w}_{jJ}\, \psi_J + \sum_{i} [\, K(\vec{x}_j; \vec{x}_i) - \sum_{JI} \{ \overline{w}_{jJ}\, K(\vec{X}_J; \vec{X}_I)\, w_{iI} \}]\, m_i \tag{15}$$

Again as in BL we need to make local corrections. For an ordered grid the term in square brackets is tabulated. This will not work for singular-smooth kernels and arbitrarily distributed particles. To see why, realize that the second term in the square bracket cannot be tabulated and its direct evaluation is computationally prohibitive. This pathology is intimately related to the singularity of the kernel. Notice that the second term in the square bracket in eq (15) is essentially an interpolation of the kernel onto itself. If the kernel were suitably smooth on scale "h" then the kernel, on interpolation, would produce a fair representation of itself. Then the kernel would have to have a scale "h" built into it. However, the singular-smooth kernels, like (8) and (9) here, are essentially scaleless while the grid has a scale viz. size "h" and phenomena that vary faster than scale "h" cannot be represented on the grid. Thus, a singular-smooth kernel (say $k(r) = 1/r$) and the interpolation of itself on a grid of size "h" (denoted by $\widetilde{k}(r)$) will differ substantially over a distance "mh" in the neighborhood of the singularity. See Fig 1. "m" here is a small number the choice of which will become evident from eq (16) later.

Now, consider a kernel $k_p(r)$ which is a softened form of the singular-smooth kernel, i.e. $k_p(r) = k(r)$ for r > mh and all derivatives of $k_p(r)$ upto order "p" match up with the "p" derivatives of $k(r)$ at r = mh while for r < mh $k_p(r)$ is smooth and differs substantially from $k(r)$. Thus $k_p(r)$ is smooth everywhere. Then, for "m" chosen large enough, $k_p(r)$ matches up substantially with $\widetilde{k_p}(r)$ the interpolation of itself on a grid of size "h". See Fig 2. "p" is the order of interpolation used. The choice of both "m" and "p" determines the accuracy of the evaluation. Graphical demonstration of this dependence on "m" and "p" is deferred to Sec. IV. Thus we now have

$$\sum_{JI} \{ \overline{w}_{jJ} K_p (\vec{X}_J; \vec{X}_I) w_{iI} \} = K_p (\vec{x}_j; \vec{x}_i) + O (h^p K^{(p)}(mh)) \quad (16)$$

Thus choosing "m" large enough but independent of "h", the correction term on the right hand side of eq (16) can be made negligible. In particular, we can choose it so that $|h^p K^{(p)}(mh)|$ is smaller than the specified error accuracy times $|K(mh)|$. Thus by using $K_p (\vec{X}_J; \vec{X}_I)$ for the grid based evaluation we have

$$\psi_J \equiv \sum_I K_p (\vec{X}_J; \vec{X}_I) \rho_I \quad (17)$$

and the particle's force, with interpolation and local correction becomes

$$\psi(\vec{x}_j) = \sum_J \overline{w}_{jJ}\,\psi_J + \sum_i [K(\vec{x}_j; \vec{x}_i) - K_p(\vec{x}_j; \vec{x}_i)]\, m_i + O(h^{p+1} K^{(p+1)}) \quad (18)$$

Thus by extending the correction (and softening) to a sufficiently large distance "mh" we can indeed bound the error and still arrive at an O (n) process. There is still one caveat. The caveat is that the number of grid points should be proportional to the number of particles that they underlie. Only then can the number of local corrections be held down to a small number.

The local adaptivity of the method has still to be demonstrated. One possible resolution that we have explored is to have a softening of "m" grid points on each level so that the picture of the softened kernel on two grid levels is as shown in Fig 3. Notice that on grid "h" we wish to have the multi-integration evaluated with a kernel softened out to "m" grid points, i.e. out to a distance "mh". On the coarser grid "2h" the softening is done out to "m" grid points *on that coarser grid*, i.e. out to a distance "2mh". Thus the potential (or forces) evaluated on the grid "2h" can be transferred by interpolation to the grid "h". Because of the softening on grid "2h", this transferred potential is an accurate representation of the multi-integration evaluated with a kernel softened out to a distance "2mh". But on the grid "h" we want to have the multi-integration with the kernel's softening done only out to "m" grid points, i.e. a distance "mh". Thus the corrections on the grid "h" have to be carried out to a distance "2mh" after transferring the solution from grid "2h" to grid "h". The ability to do local refinements is now evident and we can *locally* extend the process out to any desired level of refinement. Note too that this adaptibility springs essentially from the fact that we keep the solution in physical space and do not, at any point in the evaluation process, transform to Fourier space. This has two advantages 1) We may take the idea of superparticles further and cluster particles on one level to form yet a superparticle on a coarser level. 2) We may use the same subroutines to do the local refinement as we use for obtaining the solution on the global grid.

The above paragraph, where softening is applied on each level (global or local) out to a distance of "m" times of the level's zone size is one way of doing the multi-integration in O (s n) steps. Notice too that it relies on a coarsening of the number of grid zones by a factor of 2^d at each level. It naturally guarantees that the force evaluated on two particles is equal and opposite up to machine accuracy. This (combined with linearity) ensures strict momentum conservation which is a very desirable property in many physical systems. An alternative approach exists which is to use the BL formulation (with directional coarsening) for solving the multi-integration on the global grid. Thus on the finest uniform grid we want to evaluate the potential softened out to a distance of "m" grid zones. Then transfers to local regions of refinement can either be accomplished using the two grid trick described above or through further directional coarsening (alternated directionally, of course,) on the locally refined grids with successive reduction of the distance of softening at each level where the grid becomes isotropic.

Either way, the algorithm has been formulated by us so that, with a slight extension, it can even be made adaptive in time.

IV) Accuracy Analysis

Due to the superposition (linearity) inherent in the method, the best test is to do two particle tests. Two particles are put on the grid and are moved apart in very small steps in a specified direction. The force between them evaluated by our method at each position is then compared to the real force evaluated analytically. The accuracy of the result depends on the order of interpolation and improves with higher order interpolation. The accuracy also depends on the choice of "m", the softening distance, and improves with larger choice of "m". Fig 4 shows both these trends.

V) Timing Analysis

We carried out the timing analysis on a Cray 2 supercomputer. All loops were fully vectorized. Because the optimal operation can be obtained when there is about one particle per grid point, a good guess about the timing is obtained by considering the speed with which the evaluation is done on a grid. We have used $m = 6$ and $p = 6$ here. For an O(n) process we expect that the number of particles updated per second should remain a constant as the problem size is increased. This is what we got.

fine grid	levels	time (in sec)	zones/sec
80X80	4	.581	11015
120X120	5	.910	15806
160X160	5	1.387	18451
200X200	6	1.91	20931
250X250	6	2.68	23315

Thus the speed goes up as the grid becomes larger. This obviously reflects the better vectorization, especially at coarser levels as the problem size is increased.

VI) Future Directions

The timings given in the previous section represent evaluating forces in each direction as convolution sums. Efforts are underway to obtain the forces directly from differentiation of the potential. That should result in substantial speedups. Also, the second method described here, with directional coarsenings, may prove to be substantially faster than the first. In the interest of providing a rather extensive discussion of the evaluation step the time-integration has not been fully described and will be done in a later publication. The possibility of using superparticles on higher levels is yet an unexploited possibility as is the possibility of making two-body relaxation time considerations to decide on an optimal softening for different physical systems and/or on different length scales. All of these we defer to future works.

References

Brandt, A., and Lubrecht, A. A., (1990), J.C.P. (vol. 90, 349-370)

Brandt, A., (1990) IMACS 1st Int. Conf. on Comp. Phys.

Greengard, L., and Rokhlin, V., (1987), J.C.P., 73, 325

Hockney, R. W. and Eastwood, J. W. (1981), Computer Simulation Using Particles, (McGraw Hill)

Figure Captions

Fig 1 The kernel $k(r)$ and $\tilde{k}(r)$, the interpolation of $k(r)$ (as defined in the text) on a grid of size "h".

Fig 2 The softened kernel $k_p(r)$ and $\tilde{k}_p(r)$, the interpolation of $k_p(r)$ on to itself (as defined in the text) on a grid of size "h".

Fig 3 The softened kernels, shown hypothetically with m=4, for a two level system.

Fig 4 The analytic two particle force is compared with the force evaluated by our method. The percentage error is plotted for (a) p=2, m=6; (b) p=4, m=6; (c) p=4, m=8; (d) p=6, m=6; (e) p=6, m=8; (f) p=6, m=9.5. An 80X80 grid with 4 levels is used.

Dinshaw S. Balsara
Physics and Astronomy Dept.
Johns Hopkins University
Baltimore, Maryland, U.S.A.

Achi Brandt
Dept. of Applied Mathematics
Weizmann Institute
Rehovot, Israel

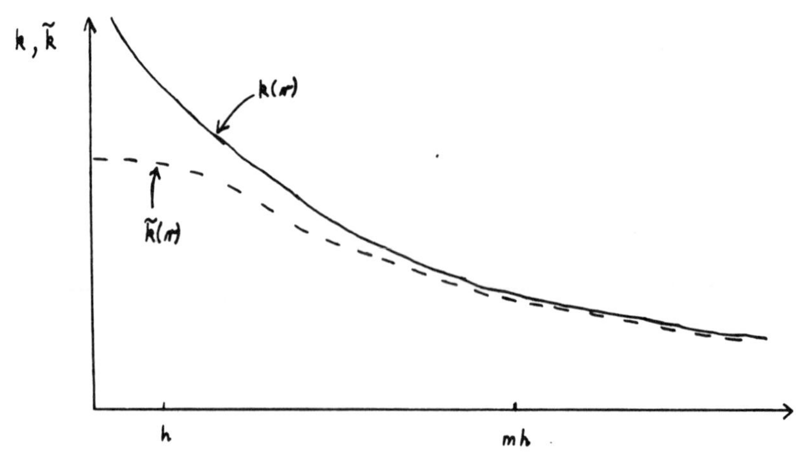

Fig. 1

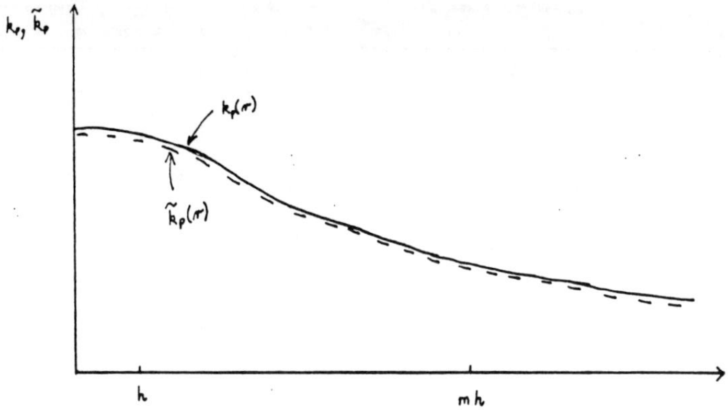

Fig 2

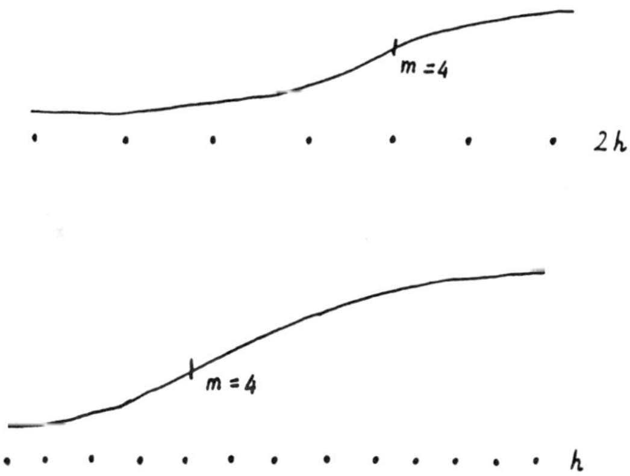

Fig 3

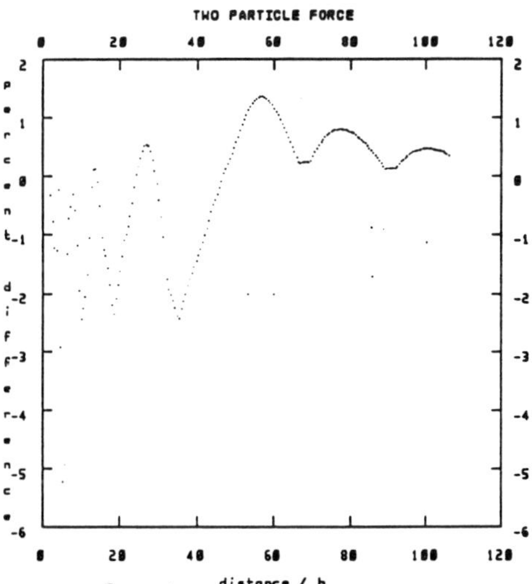

Fig 4(a)

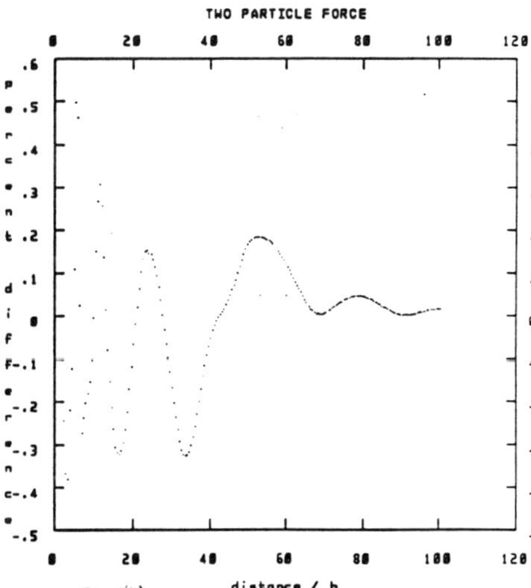

Fig 4(b)

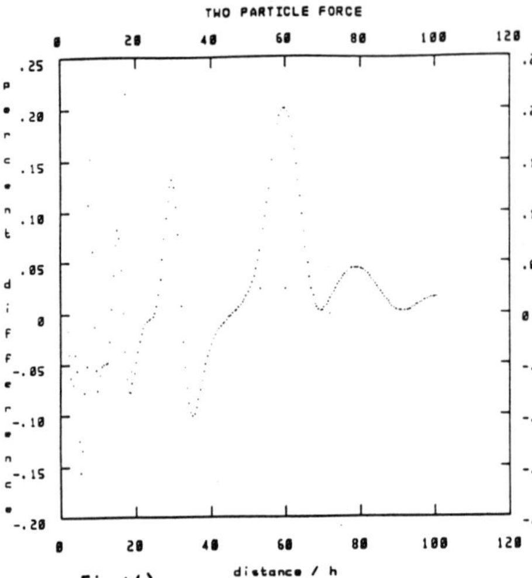

Fig 4(c)

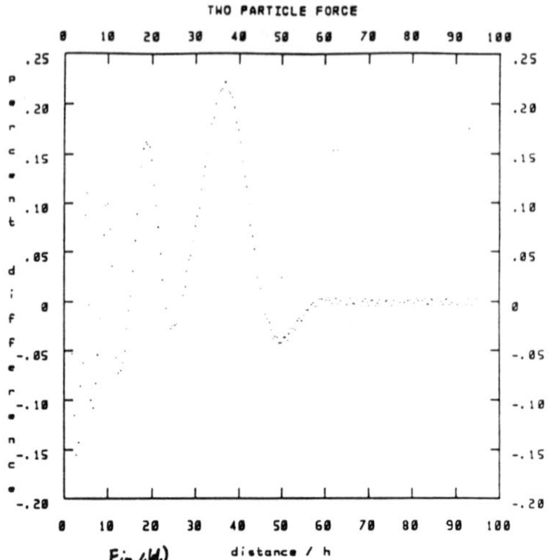

Fig 4(d)

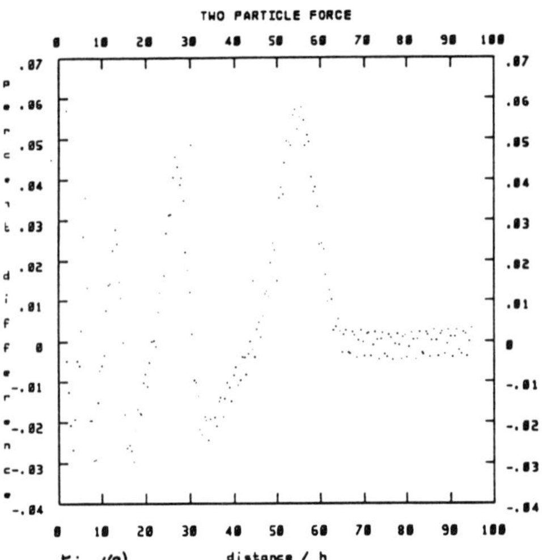

Fig 4(e)

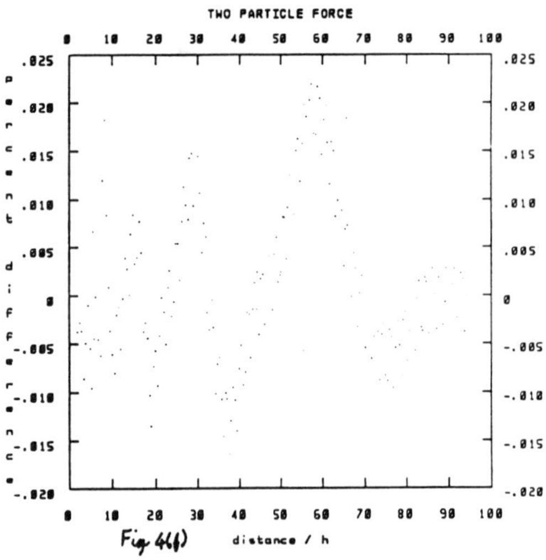

Fig 4(f)

Parabolic Multigrid Revisited

Achi Brandt* and Joseph Greenwald*
Applied Mathematics and Computer Science
Weizmann Institute of Science
Rehovot, Israel, 76100

Abstract

We review multigrid algorithms for solving the equations arising from the discretization of parabolic problems. Several open points regarding the construction of the most efficient algorithm are resolved. In addition, we demonstrate that Red/Black relaxation ordering may ruin the large time behavior of the algorithm, unless properly incorporated into the solver.

Contents

*Research supported by the Air-Force Office of Scientific Research, United States Air Force under grant numbers AFOSR-86-0126 and AFOSR-86-0127 and by the National Science Foundation under grant number NSF DMS-8704169.

1 Introduction

In developing multigrid solvers for sequences of boundary-value problems such as those arising in solving time-dependent equations, a few basic guidelines should be observed. First, since changes in the solution between problem steps are usually dominated by smooth components, one should solve for these smooth components before cycling on the finest grid. These smooth components should be eliminated by an analog of the FMG algorithm designed for problem sequences: the algorithm's structure is a modified F-cycle. A second, related point is that the FMG step solves for the *increment function,* i.e., the difference between two consecutive solutions. In particular, only the coarse grid (CG) approximation to the increment, not the CG approximation to the solution, is interpolated to the fine grid. Finally, the amount of computational work the algorithm needs should be proportional to the increment function. For example at steady state when the increment is zero, the algorithm should not need to do work. Likewise, wherever the increment is *smooth* (as in most regions of most parabolic problems), the algorithm should only rarely (e.g., once in many time steps) activate finer grids.

An additional guideline for parabolic time-dependent problems is to avoid feeding high-frequency (HF) errors into their aliasing low-frequency (LF) ones at the end of a time step. Otherwise one may ruin the asymptotic time behavior of the algorithm, since LF error components decay much more slowly than HF ones. As discussed in §4.5, one often overlooked source for such destructive feeding is the red/black ordering for local relaxation schemes.

We begin our discussion by reviewing the equations for the FMG-type algorithm in §3. This algorithm had first been outlined in [4, §3.9]. A somewhat different one has been proposed in [1], and the difference between the two is discussed. In particular it is pointed out that the algorithm in [1] is non-stationary, i.e., it drives the solution away from any approached steady state, and hence it cannot decrease work to be proportional to the size of the increment function (or even just the size of that function's HF components). In §4.2 we show that FMG-based algorithms are more efficient than simple multigrid cycling algorithms. Then in §4.3 we discuss results regarding the order of the FMG interpolation operator, showing that second order (i.e., linear interpolation) is sufficient. Higher orders are needed only at those time steps where accuracy is desired in terms of some solution *derivatives.*

The result of this work is an efficient algorithm for solving discretizations of parabolic problems. Additional multigrid techniques for parabolic problems are discussed elsewhere [10], [11], [12]. They include the *frozen-τ* technique for adaptively activating fine grids; a parallelization algorithm in space *and time*; and a *double-discretization* technique which uses two time discretizations (Crank-Nicholson and fully implicit) to generate more accurate approximations than either discretization alone would.

2 The Problem and Its Discretizations

We consider parabolic initial value problems of the form

$$U_t(x,t) - LU(x,t) = F(x,t) \qquad x \in D \subset R^2, \quad t \in [0,T] \tag{1}$$

where L is a second-order elliptic operator (e.g., the Laplacian) and appropriate initial conditions are supplied. Let $A^h \underset{def}{=} \frac{1}{k} - \beta L^h$, where $0 \le \beta \le 1$ and L^h denotes a grid-h approximation of L (e.g., $L^h = \Delta^h =$ the five point Laplacian), and let u_n^h and F_n^h be the grid-h approximations to $U(x, nk)$ and $F(x,(n-1+\beta)k)$ respectively. The one-level (in time) discretization of (1) with time step k and uniform spatial mesh size h can be written at $t = nk$ as:

$$A^h u_n^h = F_n^h + (A^h + L^h)u_{n-1}^h \underset{def}{=} f_n^h. \tag{2}$$

2.1 Choice of β

Typical values for β are either 0, .5 or 1. For $\beta = 0$ the scheme is *explicit;* the convenience it affords for generating approximations for d-dimensional problems is often outweighed by the constraint $k/h^2 < 2^{-d}$ imposed by the stability requirement. At the other end of the range, $\beta = 1$, the fully implicit (FI) scheme places no restriction on the time-step size, but as with the explicit scheme, it is only $O(k)$ accurate. The scheme for $\beta = .5$ is the Crank-Nicholson (CN) scheme; it has the attractive properties of $O(k^2)$ accuracy without any time-stepping restriction.

For smooth, i.e., low-frequency (LF) solutions, then, the CN scheme would be the discretization of choice. It is well known, however, that for the CN scheme, high-frequency (HF) error modes exhibit oscillatory behavior instead of the differential decay. For these HF components the FI scheme approximates the differential decay much better. Instead of compromising upon either one of these schemes it is possible within the multigrid (MG) algorithm to use both schemes so as to yield an approximation exhibiting the HF behavior of the FI scheme and the LF behavior of the CN scheme, [10] .

2.2 Relationship to Elliptic Multigrid

From (2) one sees that we must solve a sequence of related elliptic problems. Since the operator A^h is more strongly elliptic than its steady-state counterpart L^h, these problems were natural candidates for extending multigrid methods to time dependent problems. The main subtlety from a multigrid perspective introduced by these problems is that the right-hand side is a function of the solution at the previous time: errors propagate in time. Since smooth error components decay much more slowly than HF ones, good algorithms should try to inhibit the feeding of HF data into smooth errors (cf. §4.5 below).

2.3 Increment Equation

It is often instructive to analyze the algorithms in terms of the *time-increment function* $\delta_n^h = u_n^h - u_{n-1}^h$. For linear L^h the grid-h *increment equation* is

$$A^h \delta_n^h = F_n^h + L^h u_{n-1}^h \underset{def}{=} s_n^h. \tag{3}$$

If $F(x,t) = F(x)$, or more generally as $t \to \infty$ if $F(x,t)$ varies slowly, then a steady state, $u(x,\infty)$, or a quasi-steady-state solution exists. When $u(x,\infty)$ exists, (3) takes the particularly simple form $A^h\delta_n^h = L^h(u_{n-1}^h - u_\infty^h)$, where u_∞^h is the grid-h approximation to $u(x,\infty)$, satisfying $-L^h u_\infty^h = F_\infty^h$. The increment is simply dependent upon the deviation from the steady-state solution.

Notice that the truncation errors of (2) and (3) may be quite different. For example, as steady state is approached the latter decay (usually very rapidly at first) while the former tends to $u(\cdot,\infty) - u^h_\infty$. To avoid error accumulation with time, the goal of *any* algorithm for either (2) or (3) should be to solve them with errors smaller than the *incremental truncation error,* i.e., the discretization error of (3), not of (2). The same is true even when L^h is nonlinear: (2) should be solved, but to the accuracy of the truncation error in δ^h_n.

3 Solution Algorithms

Normally for sequences of problems, the increment δ^h_n is a smooth function. For parabolic problems its smoothness usually increases with time except near forcing terms which change nonsmoothly with time (including near nonsmooth boundaries or boundary conditions). Assuming a smooth δ^h_n, the naive approach of iterating with a multigrid correction cycle would be inefficient: to solve the problem to the level of its incremental-truncation error, the number of multigrid iterations needed is h-dependent; see §4. Based upon this consideration and those of §4, an FMG-type alorithm was proposed in [4] (elaborated in [7]). The FMG-based algorithm has two phases at each time step: that for generating the first approximation for the increment, and that of the subsequent correction cycles. To avoid ambiguity when discussing the first-approximation step we need to differentiate between the *extrapolated approximation* (derived from the solution at previous times), which is the *input* to that step, and the *incremented first approximation,* (derived from solving on the coarser girds), which is its output.

We assume previous knowledge of elliptic multigrid; for relevant background, as well as the notation which we follow, see for example [6]. The coarse grid (CG) equations for two-grid (meshsizes h and H=2h) algorithms will be reviewed first; then for programming convenience a complete multigrid algorithm will be presented in §3.2. Regarding notation, I^H_h and $\hat{I}^H_h$ denote two, possibly distinct, fine-to-coarse grid transfer operators, the first transfers residuals (or RHS), the second transfers (approximate) solutions. I^h_H is a coarse-to-fine grid interpolation operator used in the CG correction, while Π^h_H is the coarse-to-fine FMG interpolation operator (see §3.1.1). L^h and L^H denote respectively the grid h and H discretizations of L. Hence, the grid-H representation of A^h is $A^H = \frac{1}{k} - \beta L^H$.

3.1 Full Approximation Scheme (FAS)

For nonlinear L, the multigrid equations should be written in FAS form, (cf. [6, §8.1]). In fact, FAS is desirable even for *linear* time-dependent problems, since it affords a convenient environment for the advanced techniques of [11], [10].

To easily understand the reasoning behind the algorithm, the reader can assume L^h to be linear and replace the FAS with a "Correction Scheme"; i.e., in §3.1.1 replace the variable u^h_n by $u^h_{n-1}+\delta^h_n$ and u^H_n by $\hat{I}^H_h u^h_{n-1}+\delta^H_n$, and regard δ^h_n as the unknown function on grid-h and δ^H_n as its grid-H approximation. Some terms will then cancel out and the equation's meaning will become more transparent (and perhaps also somewhat easier to program). Similarly, in §3.1.3 replace u^h_n by $\tilde{u}^h_n + v^h_n$ and u^H_n by $\hat{I}^H_h \tilde{u}^h_n + v^H_n$, where $\tilde{u}^h_n$ is the current grid-h approximation so that v^h_n is the current grid-h error and v^H_n is its grid-H approximation.

3.1.1 First-Approximation Step

As the extrapolated approximation, we take the previous time solution u^h_{n-1}. The FAS grid-H equations are then

$$A^H u^H_n = I^H_h f^h_n + \tau^H_h(u^h_{n-1}), \tag{4}$$

where $\tau^H_h(u^h_{n-1}) = A^H \hat{I}^H_h u^h_{n-1} - I^H_h A^h u^h_{n-1}$ is the "fine-to-coarse defect correction" (cf. [6, §8.2]) measured at the previous time.

Remark: Another FMG-type algorithm, introduced in [1], uses instead of (4)

$$A^H u^H_n = I^H_h F^h_n + (A^H + L^H)\hat{I}^H_h u^h_{n-1}. \tag{5}$$

Note that at steady state $f^h_n - A^h u^h_{n-1} = 0$, therefore for (4) $u^H_n = \hat{I}^H_h u^h_{n-1}$ as it should be. On the other hand, (5) introduces a grid-H error that may be larger than the current grid-h increment. To emphasize this difference we will refer to (4) and (5) as the *stationary* and the *nonstationary* equations, respectively. While (5) introduces its own unnecessary error forcing calculations to proceed on grid h to retain grid h accuracy, (4) can be used in conjunction with the frozen-τ algorithm of [11] to save work by *adaptively* determining when one can time step on grid H while retaining the accuracy of the grid-h. In a multilevel algorithm, this difference would be even more significant, see §4.4.

Having solved (4), the incremented first approximation on grid h is

$$\tilde{u}^h_n = u^h_{n-1} + \Pi^h_H(u^H_n - \hat{I}^H_h u^h_{n-1}). \tag{6}$$

In FMG algorithms for second-order elliptic operators, Π^h_H is usually chosen to be of order four (cubic or bicubic interpolation; cf. [6, §7.1]). In §4.3 we will discuss the fact that for parabolic problems, using linear interpolation for Π^h_H is sufficient in terms of the evolution of the solution. Only if one likes to monitor norms based upon higher-order derivatives, (e.g. residual norms) is the higher-order interpolation necessary. Numerical results are presented in §4.

3.1.2 Other Starting Approximations

We have taken u^h_{n-1} to be our starting "extrapolated approximation". For several reasons it is indeed unnecessary to extrapolate from previous times a better approximation. First, any other extrapolation is costly in that it is a grid-h task. In a multigrid environment, such grid-h extrapolation makes sense only to better approximate the HF components of δ^h_n, since the smooth components are inexpensively obtained from the CG. Usually these HF components are already of small amplitude, but even when they are not, since they usually decay exponentially and substantially at each time step, a higher-order algebraic extrapolation is often a worse approximation than the zero order one.

Obtaining an extrapolated approximation by an explicit time scheme, as in some predictor-corrector schemes, also makes no sense here. In fact, if the time step k is larger than those at which the explicit scheme is stable, then this first approximation will give *larger* HF errors than our simple first approximation $\tilde{\delta}^h_n = 0$.

If any other extrapolated approximation is nonetheless used, it should of course replace $\hat{I}^H_h u^h_{n-1}$ in equations (4) and (6).

3.1.3 Correction Cycle

Assume a fine-grid approximation $\tilde{u}_n^h$. It is then updated by ν_1 relaxation sweeps (pre-smoothing step). The exact solution of (2) can now be written as $u_n^h = \tilde{u}_n^h + v^h$ where v^h, the current error, is now a smooth function. The coarse grid variable is then $u_n^H = \hat{I}_h^H \tilde{u}_n^h + v^H$, where $I_H^h v^H$ is designed to approximate v^h. The coarse grid correction (CGC) equations are:

$$\begin{aligned} A^H u_n^H &= I_h^H(f_n^h - A^h \tilde{u}_n^h) + A^H \hat{I}_h^H \tilde{u}_n^h \\ &= I_h^H f_n^h + \tau_h^H(\tilde{u}_n^h). \end{aligned} \tag{7}$$

Having solved (7), the coarse grid correction is

$$\tilde{u}_n^h = u_{n-1}^h + I_H^h(u_n^H - \hat{I}_h^H u_{n-1}^h). \tag{8}$$

To smooth the HF errors associated with the interpolation, the correction cycle ends with post-smoothing by ν_2 relaxation sweeps; often $\nu_2 = 0$, as explained in §4.5.

3.2 FAS-FMG Algorithm

For this section, grid levels are simply denoted by a natural ordering, in which grid $j+1$ is the direct refinement of grid j. On any grid j the approximation to L is denoted by L^j, and at any stage the problem we are solving has the form

$$A^j u_n^j = R^j, \tag{9}$$

where $A^j = \frac{1}{k} - \beta L^j$ and for each stage R^j is defined below. I_{j-1}^j denotes an interpolation operator from grid $j-1$ to grid j, and I_{j+1}^j is the residual fine-to-coarse (grid $j+1$ to grid j) transfer (or restriction) operator. The operator Π_{j-1}^j is the FMG interpolation from grid $j-1$ to j and $\hat{I}_{j+1}^j$ is the fine-to-coarse FAS *solution* transfer operator. The smoothing (relaxation with (9)) operator acting on a grid function u^j is denoted $S_j(R^j)u^j$, so that $u^j \leftarrow S_j(R^j)^\nu u^j$ represents the replacement of u^j by the result of ν relaxation sweeps starting with u^j. Similarly updating the approximation u^j by the application of γ multigrid cycles for grid j is written as $u^j \leftarrow MG_j(R^j)^\gamma u^j$. The following algorithm skeleton, TFMG (time-step FMG), is the algorithm used at each time step to solve (2), where u_n^h, L^h and F_n^h are now denoted u_n^m, L^m and F_n^m, respectively. We present only the *FAS* algorithm with *stationary* CG equation, (4).

Begin TFMG:

Global Parameters: m = finest defined grid; k= time step; $\beta = \begin{cases} 1 & \text{FI Scheme} \\ 0.5 & \text{CN Scheme} \end{cases}$

CALL $RHSDEF(m, F_n^m, u, u_*, R, k, \beta)$; (Define RHS).

CALL $FMG(m, R, u, u_*)$; (FMG algorithm with u_* as initial approximation).

End TFMG.

Procedure: $RHSDEF(m, F_n^m, u, u_*, R, k, \beta)$: (Define the R^j for the FMG step).

c u, R are large arrays which store the grid functions u^j and R^j for $j = 1, \ldots, m$.

c u_* is an optional large array which stores the grid functions u_*^j for $j = 1, \ldots, m$.

c Storing it saves some calculations in the subroutines FMG and MG.

Begin

$u_*^m = u_{n-1}^m$; (need not store u_{n-1}^m separately)
c ($L^m u_{n-1}^m$ can be calculated *once* for the ensuing 2 lines without additional storage).
$s^m = F_n^m + L^m u_{n-1}^m$;
$R^m = F_n^m + \frac{1}{k} u_{n-1}^m + (1-\beta) L^m u_{n-1}^m$;
Do $j = m-1, 1, -1$
$u_*^j = \hat{I}_{j+1}^j u_*^{j+1}$;
$s^j = I_{j+1}^j s^{j+1}$;
$R^j = s^j + A^j u_*^j$;
End Do
End Procedure *RHSDEF*.

Remark: In case of another extrapolated approximation ($u_*^m \neq u_{n-1}^m$) the definition of s^m should be replaced by $s^m = F_n^m + L^m \left((1-\beta)u_{n-1}^m + \beta u_*^m\right) + \frac{1}{k}\left(u_{n-1}^m - u_*^m\right)$.

Procedure: $FMG(m, R, u, u_*)$:
c (FMG Step for $A^m u_n^m = R^m$ with extrapolated approximation u_*^m).
Solve $A^1 u_n^1 = R^1$, yielding the approximate solution u^1;
Do $l = 2, m, 1$
$u^l \leftarrow u_*^l + \Pi_{l-1}^l (u^{l-1} - u_*^{l-1})$;
$u^l \leftarrow MG_l(R^l)^{\eta(l)} u^l$; (perform $\eta(l)$ times the multigrid cycle defined below).
End Do
$u_n^m = u^m$ End Procedure *FMG*.

Procedure: $MG_j(R^j, u^j)$: (Multigrid Cycle: $u^j \leftarrow MG_j(R^j)u^j$)
Integer j; Array u^j, R^j;
Step 1: *(pre-smoothing by ν_1 relaxation sweeps)*: $u^j \leftarrow S_j(R^j)^{\nu_1} u^j$;
Step 2: *(CG solution)*: If $j = 1$ go to Step 4; Else,
$u_*^{j-1} = \hat{I}_j^{j-1} u^j$;
$R^{j-1} = I_j^{j-1}(R^j - A^j u^j) + A^{j-1} u_*^{j-1}$;
$u^{j-1} \leftarrow MG_{j-1}(R^{j-1})^{\gamma(j-1)} u_*^{j-1}$;
Step 3: *(CG correction)*: $u^j \leftarrow u^j + I_{j-1}^j (u^{j-1} - u_*^{j-1})$;
Step 4: *(post-smoothing by ν_2 relaxation sweeps)*: $u^j \leftarrow S_j(R^j)^{\nu_2} u^j$;
End Procedure MG_j.

When $\eta(l) = \eta$ (independent of l) the algorithm is called η-FMG. For fixed $\gamma(j) = \gamma$, γ is called the cycle *index*. A cycle with the index 1 is called a V(ν_1, ν_2) cycle; with index 2, a W(ν_1, ν_2) cycle. Usually 1-FMG is used with V(1,0) or V(2,0) cycles; for judging the quality of results, comparisons with results for larger γ and/or η are sometimes made.

The overall algorithm TFMG can also be viewed as a variant of the $F(\nu_1, \nu_2)$ cycle defined in [6, §6.2]. A few modifications had to be made. The first modification to the standard F-cycle is that the first CG correction is performed before any grid-h relaxation. This is based upon the assumption that the increment is a smooth function so that grid-h relaxation will be most effective only after the smoothness of the increment is accounted for. The second modification is that in this first CG correction the possibly higher-order FMG interpolation is used. One further modification, discussed in more detail in §4.5, is that often the final smoothing (i.e. any smoothing not followed by a CG correction in the same time step) should

Fig. 1: Schematic Diagram of TFMG

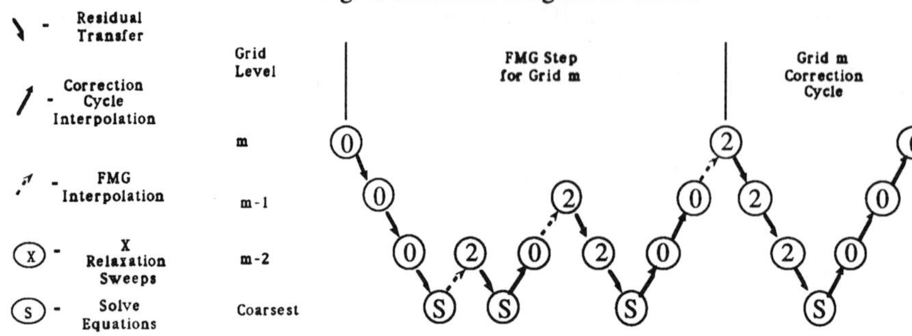

be skipped. A schematic figure of this modified F-cycle appears in Fig. 1. The case shown is $\nu_1 = 2, \nu_2 = 0$. On the coarsest grid the equations are solved by any appropriate way; e.g., directly, or by $\nu_1 + \nu_2$ relaxation sweeps (as in the procedure MG).

4 Two-Level Analysis and Numerical Results

Having presented the algorithm of (§3.1), we will now discuss its efficiency. Several topics concerning the algorithm's performance will be analyzed and tested numerically.

4.1 Model Problem and General Results

The model problem for the numerical tests reported herein is the heat equation with periodic boundary conditions:

$$\begin{aligned} U_t(x_1,x_2,t) - \Delta U(x_1,x_2,t) = F(x_1,x_2); \quad t \in [0,T]; \quad (x_1,x_2) \in [0,2\pi) \times [0,2\pi) \\ U(x_1,x_2,0) = G(x_1,x_2); \quad U(x_1,x_2,t) = U(x_1+2\pi,x_2,t) = U(x_1,x_2+2\pi,t). \end{aligned} \tag{10}$$

The use of periodic boundary conditions allows us to focus upon the behavior of the algorithm in the interior of more general domains. For the figures presented I_h^H and $\hat{I}_h^H$ are "full weighting" transfer operators (cf. [6, §4.4]). The smoothing rates are the same as that of the Helmholtz operator, $-\Delta + c$ with $c = 1/k$ (cf. [9, pg. 113]). From the smoothing rates reported therein and the considerations of §4.2 below, TFMG using Gauss-Seidel red/black (RB) relaxation with one V cycle ($\eta = \gamma = \nu_1 = \nu_2 = 1$) should be sufficient to solve below truncation error at each time step. For smooth δ_n^h experiments showed that instead of the W(1,1) cycle a V(1,0) cycle was sufficient. Using the frozen τ techniques of [12], [11] even cheaper cycles were possible as δ_n^h became more smooth with time by skipping relaxation completely on the finer grids for a few time steps. On the other hand, when δ_n^h was not smooth using RB relaxation with even a W(2,1) cycle was not sufficient for the long time evolution behavior (see §4.5) . Alternatively for such instances, one could sacrifice some smoothing efficiency and use lexicographic ordering with a W(2,1) cycle. Double discretization algorithms [10] allow the safe use of RB relaxation with one V(1,1) cycle even for nonsmooth δ_n^h.

4.2 MG Cycles Versus FMG-Type Algorithms

Since usually $||\delta_n^h|| \ll ||u_n^h||$ and the asymptotic two-grid convergence factors are very good, one might believe that a simple multigrid cycle for δ_n^h is sufficient. However, this need not be true even if one desires only $O(k)$ global truncation error. Indeed, if solution to $O(k^p + h^q)$ global accuracy is desired, the *incremental* truncation error should be $O(k^{p+1} + kh^q)$. To reduce the starting $O(k)$ algebraic error to this accuracy, a reduction by a factor $O(k^p + h^q)$ is needed, which in general would require $O\left(p\log(\frac{1}{h}) + q\log(\frac{1}{k})\right)$ multigrid cycles.

For the algorithm TFMG, on the other hand, after the incremented-first-approximation the error is already $O(k^{p+1} + kH^q)$, hence it only need be reduced by a factor just somewhat smaller than 2^{-q}. This is well within the limits of a MG cycle.

Also, as mentioned in §3, TFMG eliminates the smooth components of the error before any grid-h relaxation is performed. For very smooth increments, as can be expected in the large time behavior of parabolic problems, starting with relaxation on the finest grid would not significantly decrease the HF error components; in fact, relaxation may even *increase* them.

4.3 FMG Interpolation Order

Our main result regarding the order of Π_H^h is that linear interpolation is as good as any higher order interpolation for the long time evolution of the solution. Only near a time step at which accuracy is desired in *derivatives* of the solution, higher order Π_H^h should be used (cf. the rules in [6, §7.1]). This is true even for the CN scheme for which solution derivatives appear in the RHS, and even when δ_n^h is not small compared with u_n^h.

The role of cubic interpolation for Π_H^h is to keep the HF interpolation errors small. For parabolic problems, however, the HF errors committed at one time step should decay rapidly in future time steps. Thus, the HF errors introduced by linear interpolation do not have a chance to accumulate.

This explanation does not hold for the CN scheme for which HF errors oscillate instead of decaying. Let the HF induced by interpolation at time step n have amplitude A; in the next time step this error would have amplitude near $-A$. However, there is a new interpolation error of amplitude A; the net effect is that the errors almost cancel each other so that again no accumulation of HF errors occurs with time. Fig. 2 exemplifies this: the norm of the difference between the exact CN solution and the solution of TFMG with linear interpolation (TFMG(Linear)) practically coincides with that of TFMG(Cubic), both being well below the norm of the truncation error (U-CN). Similar results were obtained for any choice of smooth or nonsmooth $F(x_1, x_2)$ and $G(x_1, x_2)$.

4.4 Stationary and Non-stationary Equations Compared

Denote the incremented-first-approximation for the stationary equations (4) by $\tilde{u}_n^h$ and for the nonstationary equations (5) by $\bar{u}_n^h$, and let the correspondiing errors be $\tilde{v}_n^h \underset{def}{=} \tilde{u}_n^h - u_n^h$

Fig. 2: Typical Time Behavior of Algebraic and Truncation Errors

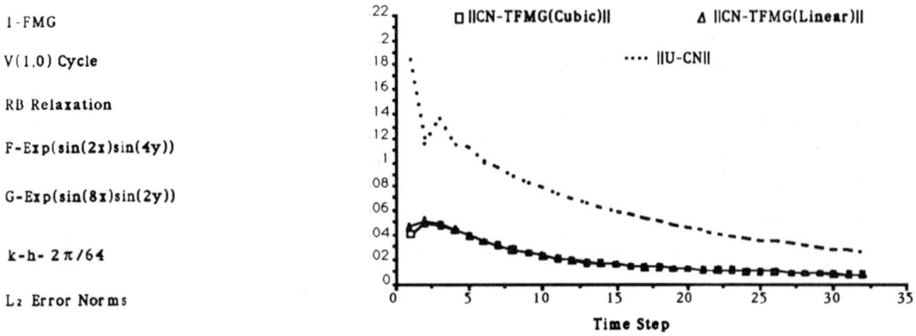

and $\bar{v}_n^h \underset{def}{=} \bar{u}_n^h - u_n^h$. Then as a steady state u_∞^h is approached

$$\lim_{n\to\infty} \tilde{v}_n^h = 0 \qquad \text{while} \qquad \lim_{n\to\infty} \bar{v}_n^h = \lim_{n\to\infty} -\Pi_H^h (A^H)^{-1} \tau_{n-1}^H \approx -\Pi_H^h (A^H)^{-1} \tau_\infty^H$$

where $\tau_{n-1}^H \underset{def}{=} L^H \hat{I}_h^H u_{n-1}^h - I_h^H L^h u_{n-1}^h$ and $\tau_\infty^H \underset{def}{=} L^H \hat{I}_h^H u_\infty^h - I_h^H L^h u_\infty^h$.

A few comments about $\bar{v}_n^h$ are important. First because of its dependence upon τ_∞^H, its magnitude increases substantially when HF components are present in u_∞^h. Since the HF modes converge to steady state rapidly, $\bar{v}_n^h$ quickly achieves its maximal magnitude. At each time step, the algorithm must remove this algorithm-induced error. Finally, and perhaps most significantly, as reported in [11], (5) forces us to return at each time step to grid-h to get an approximation to τ_{n-1}^H even though it may have changed very little from the last time step. On the other hand (4) fits naturally into the frozen-τ algorithms (cf.[4], [11]) which allow for substantial computational savings when changes in grid-h HF modes are small. Fig. 3 shows an experiment with the FI ($\beta = 1$) discretization, demonstrating that whereas the stationary FMG algorithm converges to steady state with very little work, the nonstationary FMG algorithm forces the algorithm away from steady state. Notice that this result was obtained even with cubic Π_H^h and with I_h^H and $\hat{I}_h^H$ set equal to full weighting to minimize the HF contribution to τ_{n-1}^H (injection, as suggested in [1], may give even worse results).

4.5 Relaxation and Mode Coupling

Normally, red-black (RB) Gauss-Seidel relaxation is recommended for elliptic problems because of its good smoothing rate, low cost and potential for parallelization [6, §3.6]. It is usually insignificant that the excellent smoothing rate of this and other "pattern relaxation" schemes actually results from the conversion of HF error components into their aliasing LF ones. Not so in parabolic problems: smooth errors in solving the algebraic system at each time step are much more destructive than HF errors, since they affect the solution for many more time steps. The RB conversion of HF into LF components at the end of a time step is thus more harmful than helpful. This is especially true for the CN discretization, in which HF deviations from steady state linger for many time steps and can thus feed many times

Fig. 3: Asymptotic Time Behavior For Errors Using Stationary and Nonstationary Equations

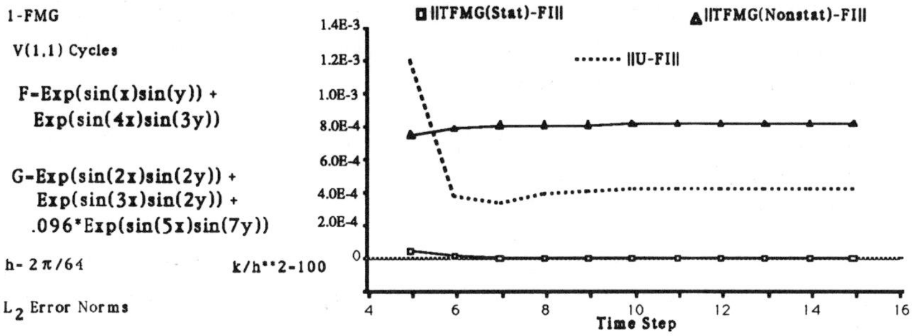

Fig. 4: Temporal Effect on Errors of RB Relaxation

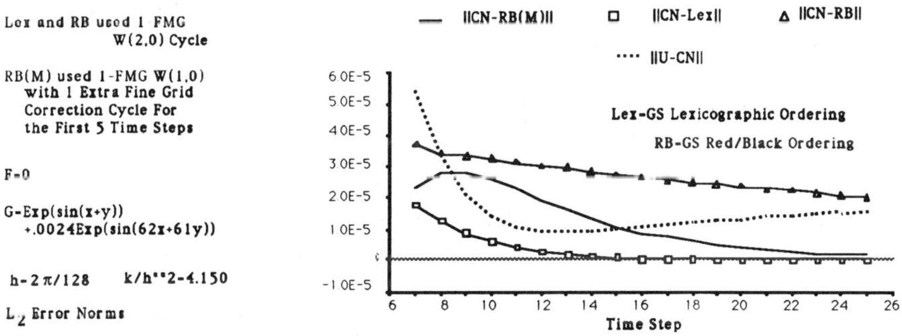

into LF components. Numerical experiments with (10) show these effects very clearly (see Fig. 4 and more in [12] and [10]).

The good smoothing rates and parallelizability of RB relaxation nonetheless argue in favor for its use. Usually this can be safely done if just one of the following two precautions is employed. (I) Use W, not V, cycles. (II) Avoid final relaxation, that is, any relaxation not followed (within the same time step) by CG correction. As Fig. 4 shows for very significant HF, the cheapest solution is to use an extra finest grid correction cycle for a few time steps, (note that by using the extra cycle for 5 time steps, we could use the W(1,0) cycle instead of the W(2,0) one).

The most efficient treatment of HF increments is by double discretization, where RB relaxation can again safely be used (see [10] and [12]).

References

[1] Hackbush,W. "Parabolic Multi-Grid Methods. " Computing Methods in Applied Sciences and Engineering, VI. North Holland, Amsterdam, 20-45, (1984).

[2] Hackbush,W. Multigrid Methods and Applications, Springer Series in Comp. Math. 4, Springer-Verlag, Berlin, (1985).

[3] Brandt, A. Numerical stability and fast solutions to boundary value problems. Boundary and interior layers-computational and asymptotic methods. Proceedings, Dublin, June 1982. Boole Press, Dublin, 29-49, (1982).

[4] Brandt, A. " Multi-level Adaptive Finite-Element Methods–I: Variational Problems". Special Topics of Applied Mathematics. J. Frehse, D. Pallaschke, U. Trottenberg (eds.). North-Holland Publishing Company, 91-128, (1980).

[5] Brandt, A. Multigrid Solvers for non-elliptic and singular-perturbation steady-state problems, Research Report, Dept. of Applied Mathematics, Weizmann Institute of Science, Rehovot, Israel, (1981).

[6] Brandt, A. Multigrid Techniques: 1984 Guide with Applications to Fluid Dynamics. GMD-Studien Nr 85, Bonn, (1984).

[7] Gendler, Efim. Multigrid solution of parabolic problems. Master's Thesis. Weizmann Institute of Science, Rehovot, Israel, (1986).

[8] Kroll, Thomas. Multigrid solution of parabolic problems. Master's Thesis, Institut fur Angewandte Mathematik, Universitat Bonn, (1981).

[9] Trottenberg, U. and K. Stuben. Multigrid Methods: Fundamental Algorithms, Model Problem Analysis and Applications. Multigrid Methods (Proceedings) Lecture Notes in Mathematics. Ed. W. Hackbusch and U. Trottenberg. Springer-Verlag Berlin Heidelberg New York, (1982)

[10] Brandt, A. and J. Greenwald. Multigrid Time-Dependent Double Discretization Methods. In preparation. (1991)

[11] Brandt, A. and J. Greenwald. Frozen-τ techniques for Parabolic Problems. In preparation. (1991).

[12] J. Greenwald. Multigrid Methods For Sequences of Problems. Thesis Final Report. Weizmann Institute of Science. (Feb. 1991).

International Series of Numerical Mathematics, Vol. 98, © 1991 Birkhäuser Verlag Basel

Time-Parallel Multigrid Solution of the Navier-Stokes Equations

by

Jens Burmeister and Graham Horton

Abstract. *We consider the two-dimensional unsteady incompressible Navier-Stokes equations. Conventional methods for the solution of this system when discretized by an implicit method such as backward Euler proceed by solving a sequence of problems iteratively. It is shown that despite the sequential nature of this process, several processors may be employed to solve at several time steps simultaneously. Details of the basic multi-grid method and of the smoothing technique are presented and results are given for a simple unsteady model problem.*

1. Introduction

The solution of large non-linear systems of unsteady partial differential equations is one of the major classes of computationally intensive problems encountered in scientific and technical fields. For this reason, they have been studied intensively with a view to solution on parallel processors. Most approaches derive their parallelism from a subdivision of the computational domain in the space directions, yielding for example grid partitioning or domain decomposition methods. However, many unsteady problems possess a degree of complexity in the time direction which can greatly exceed that of the space directions. Thus it seems natural to ask whether parallelism can additionally be achieved along the time axis.

A new approach for parabolic equations based on the multigrid technique was introduced by Hackbusch in 1984 [4], where it was shown how the multigrid idea can be extended to apply to a sequence of time-steps simultaneously, as opposed to the standard solution procedure, whereby successive time-steps are treated in a sequential manner. In the following year, one of the authors analysed the one-dimensional unsteady heat equation, obtaining convergence results similar to those for elliptic equations [2]. Additionally it could be shown that the convergence is independent of the number of time-steps to be solved simultaneously, which encouraged further investigation of the problem, including a parallel implementation on the memory-coupled nearest-neighbour multiprocessor DIRMU [1].

In this paper, the method is extended to a system of non-linear equations, the two-dimensional incompressible Navier-Stokes equations :

$$
\begin{aligned}
\frac{\partial(\rho u)}{\partial t} + \frac{\partial}{\partial x}(\rho u^2 - \mu\frac{\partial u}{\partial x}) + \frac{\partial}{\partial y}(\rho uv - \mu\frac{\partial u}{\partial y}) &= -\frac{\partial p}{\partial x} + f^x \\
\frac{\partial(\rho v)}{\partial t} + \frac{\partial}{\partial x}(\rho uv - \mu\frac{\partial v}{\partial x}) + \frac{\partial}{\partial y}(\rho v^2 - \mu\frac{\partial v}{\partial y}) &= -\frac{\partial p}{\partial y} + f^y \\
\frac{\partial(\rho u)}{\partial x} + \frac{\partial(\rho v)}{\partial y} &= 0
\end{aligned}
$$

Note the particular form of the continuity equation due to the assumption of incompressibility $\rho = const.$ We further assume constant viscosity $\mu = const.$

In the following section, the non-linear time-parallel multigrid method is presented in a general form. The restriction and prolongation operators are seen to be similar to standard procedures, whereas the smoothing part has to be studied in more detail. The smoothing procedure is based on the well-known SIMPLE procedure of Patankar and Spalding [6], and is coupled with a pointwise ILU decomposition method due to Stone [7] for the solution of the resulting systems.

In section 3 the special time-parallel SIMPLE-like smoothing procedure is considered in detail, showing how the standard method can be extended to solve a set of non-linear problems at successive time-steps. Particular attention will be paid to the identification of those algorithmic aspects significant to the parallel implementation. It will be seen that from the parallel point of view, the problem to be solved reduces to the parallel solution of scalar linear parabolic equations. This can be achieved with a pipelined parallel ILU procedure described in section 4.

Methods such as the time-parallel method presented here have been called "parallelization of the method", as opposed to the standard " parallelization of the data" type grid partitioning algorithms. One characteristic of such a method is that its numeric behaviour is dependent on the number of processors used. This motivates a more detailed analysis of the efficiency obtained by the method, which is outlined in section 5.

A test case based on the driven-cavity model problem is briefly described and results are presented for a variety of parameters. The results are discussed based on the concept of splitting the total efficiency obtained into numerically dependent and implementation-dependent components.

2. Time-Parallel Multi-Grid Method

Let a nonlinear partial differential equation for an unknown time and space dependent function $u = u(t, x, y)$ be given

$$\frac{\partial u}{\partial t} + \mathcal{L}(u) = q(t, x, y), \quad (x, y) \in \Omega \subset R^2, \quad 0 < t \leq T .$$

The problem is assumed to have suitable boundary conditions and initial values for $t = 0$. Following the method of lines, the problem is discretized first in space by a discretization method like finite elements, finite differences or finite volumes. For this purpose a hierarchy of grids is used, indicated by a level $l \in \{0, \ldots, lmax\}$. The time integration is performed using the backward Euler formula. For ease of presentation, the time step Δt is fixed.

The resulting discrete counterpart of the nonlinear problem for gridlevel l on a fixed time interval $[t_k, t_{k+m}]$ reads as follows. Using a given grid-function $u_{l,k}$, solve the set of discrete nonlinear problems

$$\tfrac{1}{\Delta t} u_{l,k+j} + \mathcal{L}_l \left(u_{l,k+j} \right) = q_{l,k+j} + \tfrac{1}{\Delta t} u_{l,k+j-1} \ , \ j = 1, \ldots, m \ .$$

Here $u_{l,k+j}$ denotes a grid function which is defined on a grid with mesh size h_l at time $t_{k+j} = (k + j)\Delta t$.

The usual solution procedure is to calculate the unknown grid functions $u_{l,k+1}, u_{l,k+2}, \ldots$ in a sequential manner. Multigrid methods can be applied to the discrete nonlinear problem at each individual time step [5].

Following the ideas proposed in a paper by Hackbusch [4] and the implementation strategy described in [1], the set of nonlinear problems will be solved iteratively by means of multigrid techniques. In the following the nonlinear two-grid version will be discussed.

The i-th iterate $\left\{ u_{l,k+j}^{(i)} \right\}_{j=1,\ldots,m}$ may be given or already be calculated. The algorithm contains of a smoothing part and a coarse grid correction.

Smoothing: Two different smoothing strategies are proposed. Both are constructed using an iteration scheme $\mathcal{S}_l$ as a module. The iteration $\mathcal{S}_l$ should be well suited as a smoother for nonlinear problems like

$$\tfrac{1}{\Delta t} u_{l,*} + \mathcal{L}_l \left(u_{l,*} \right) = \textit{right-hand-side} \quad .$$

The application of $\mathcal{S}_l$ $\nu-$times will be denoted by $\mathcal{S}_l{}^{\nu}$.

- Gauss-Seidel-like smoothing:

$$u_{l,k+j}^{(i+\frac{1}{2})} = \mathcal{S}_l{}^{\nu} \left(u_{l,k+j}^{(i)}, \mathcal{L}_l \left(\cdot \right), u_{l,k+j-1}^{(i+\frac{1}{2})} \right) \ , \ j = 1, \ldots, m \ .$$

- Jacobi-like smoothing:

$$u_{l,k+j}^{(i+\frac{1}{2})} = \mathcal{S}_l{}^{\nu} \left(u_{l,k+j}^{(i)}, \mathcal{L}_l \left(\cdot \right), u_{l,k+j-1}^{(i)} \right) \ , \ j = 1, \ldots, m \ .$$

Coarse grid correction:

- calculate the defects

$$d_{l,k+j} = \frac{1}{\Delta t} u_{l,k+j}^{(i+\frac{1}{2})} + \mathcal{L}_l \left(u_{l,k+j}^{(i+\frac{1}{2})} \right) - q_{l,k+j} - \frac{1}{\Delta t} u_{l,k+j-1}^{(i+\frac{1}{2})} \ , \ j = 1, \ldots, m \ .$$

- restrict the defects and the current iterates to the coarse grids (with a given restriction operator r):

$$\tilde{d}_{k+j} := r \ d_{l,k+j} \quad \text{and} \quad \tilde{u}_{k+j} := r \ u_{l,k+j}^{(i+\frac{1}{2})} \ , \ j = 1, \ldots, m \ .$$

- solve the coarse grid equation ($v_{l-1,k} := r \ u_{l,k}$):

$$\frac{1}{\Delta t} v_{l-1,k+j} + \mathcal{L}_{l-1} (v_{l-1,k+j}) = \frac{1}{\Delta t} \tilde{u}_{k+j} - \frac{1}{\Delta t} \tilde{u}_{k+j-1} + \mathcal{L}_{l-1} (\tilde{u}_{k+j}) - \tilde{d}_{k+j}$$
$$+ \frac{1}{\Delta t} v_{l-1,k+j-1} \ , \ j = 1, \ldots, m \ .$$

- Prolongation (with a given prolongation operator p) and correction:

$$u_{l,k+j}^{(i+1)} = u_{l,k+j}^{(i+\frac{1}{2})} + p \ (v_{l-1,k+j} - \tilde{u}_{k+j}) \ , \ j = 1, \ldots, m \ .$$

The algorithm can easily be extended to a multi-grid version by solving the coarse grid equation (a discrete nonlinear problem at level $l - 1$) iteratively with starting guess $v_{l-1,k+j} := \tilde{u}_{k+j}$, $j = 1, \ldots, m$; in addition post-smoothing steps may be defined. For the choice $m = 1$ (one timestep to solve) the above algorithm reduces to the *full approximation storage* algorithm (FAS) . Main parts of the algorithm like defect calculation, restriction, prolongation and correction may be chosen in a standard way and moreover can be computed independently, and therefore in parallel in different timesteps. The two smoothing variants differ in the use of values from preceding timesteps; the new values in the Gauß-Seidel version using values from the same iteration (index $i + \frac{1}{2}$), whilst the Jacobi version uses values from the preceding iteration (index i). Thus the Jacobi-like smoother is trivially parallelizable across the timesteps.

The generalization of the above time-parallel multigrid method to nonlinear systems of equations is straightforward and can be done for the discrete counterpart of the incompressible Navier-Stokes equations

$$\begin{aligned} D_l^x u_{l,k+j} + Q_l^x(u_{l,k+j}, v_{l,k+j}) + G_l^x p_{l,k+j} &= f_{l,k+j}^x + D_l^x u_{l,k+j-1} \\ D_l^y v_{l,k+j} + Q_l^y(u_{l,k+j}, v_{l,k+j}) + G_l^y p_{l,k+j} &= f_{l,k+j}^y + D_l^y v_{l,k+j-1} \\ -K_l^x u_{l,k+j} - K_l^y v_{l,k+j} &= 0 \end{aligned}$$
$$, \ j = 1, \ldots, m \quad .$$

The nonlinear operators $Q_l^x(\cdot,\cdot)$ and $Q_l^y(\cdot,\cdot)$ include the nonlinearity of the Navier-Stokes equations, whilst the linear operators (matrices) G_l^x, G_l^y and K_l^x, K_l^y are discrete analogues of the gradient and the divergence operators. When finite differences are used, the diagonal matrices D_l^x, D_l^y become $\frac{1}{\Delta t} I_l$. In the case of finite volumes, the diagonal matrices may have varying coefficients. The discretization technique is the same as that described in [8],[9].

The particular smoothing procedure $\mathcal{S}_l$ applicable to the discrete Navier-Stokes-problem will be described in the following section, identifying in particular the parallelizability of the various components. In section 4 it will be shown that the remaining components may also be parallelized, even in the Gauß-Seidel sense.

3. Time-Parallel SIMPLE Algorithm

The equations in this section are considered for a specific grid level l, which will be omitted henceforth for clarity. For ease of presentation let the set of unknowns denoted by $(u_j, v_j, p_j)^T,\ j = 1, \ldots, m$.

The SIMPLE-method (Semi-Implicit Method for Pressure-Linked Equations) was introduced by Patankar and Spalding [6] and is widely used as an iteration scheme for the solution of Navier-Stokes problems, both directly and as a smoother in multigrid solvers. The arising nonlinear problem for a fixed time-step t_j

$$
\begin{aligned}
D^x u_j + Q^x(u_j, v_j) + G^x p_j &= f_j^x + D^x u_{j-1} \\
D^y v_j + Q^y(u_j, v_j) + G^y p_j &= f_j^y + D^y v_{j-1} \\
-K^x u_j - K^y v_j &= 0
\end{aligned}
$$

can be interpreted as a stationary Navier-Stokes problem enhanced with diagonal entries reflecting the time discretization.

One iteration step of the SIMPLE-method consists of a linearization of the nonlinear problem, a factorization of the system matrix and the sparse approximation of dense matrix inverses. The linearization is performed by a fixpoint iteration, where values from the previous iteration are used to determine coefficients (this is equivalent to a Newton-iteration with a simplified Jacobian)

$$
\begin{bmatrix} D^x + Q^x(u_j^{old}, v_j^{old}) & 0 & G^x \\ 0 & D^y + Q^y(u_j^{old}, v_j^{old}) & G^y \\ -K^x & -K^y & 0 \end{bmatrix}
\begin{bmatrix} u_j^{new} \\ v_j^{new} \\ p_j \end{bmatrix}
=
\begin{bmatrix} f_j^x + D^x u_{j-1} \\ f_j^y + D^y v_{j-1} \\ 0 \end{bmatrix} .
$$

By grouping together the velocity grid-functions, the resulting system matrix is algebraically factorized

$$
\left[\begin{array}{cc|c} Q_j^x & 0 & G^x \\ 0 & Q_j^y & G^y \\ \hline -K^x & -K^y & 0 \end{array}\right]
=: \begin{bmatrix} Q_j & G \\ -K & 0 \end{bmatrix}
= \begin{bmatrix} Q_j & 0 \\ -K & K Q_j^{-1} G \end{bmatrix}
\begin{bmatrix} I & Q_j^{-1} G \\ 0 & I \end{bmatrix} ;
$$

whereby the abbreviations $Q_j^x := D^x + Q^x(u_j^{old}, v_j^{old})$ and $Q_j^y := D^y + Q^y(u_j^{old}, v_j^{old})$ are used. Because Q_j^{-1} is a dense matrix, a sparse approximation R_j is used instead. The original SIMPLE-method is defined by using the diagonal part of Q_j

$$R_j^{-1} := \mathrm{diag}(Q_j)$$

yielding the approximate sparse factorization

$$\begin{bmatrix} Q_j & G \\ -K & 0 \end{bmatrix} \approx \begin{bmatrix} Q_j & 0 \\ -K & KR_jG \end{bmatrix} \begin{bmatrix} I & R_jG \\ 0 & I \end{bmatrix} .$$

To obtain the time-parallel SIMPLE method, consider the extended system obtained by writing the discrete Navier-Stokes problem for m time-steps together:

$$\begin{bmatrix} Q_1^x & 0 & G^x & & & & & & \\ 0 & Q_1^y & G^y & & & & & & \\ -K^x & -K^y & 0 & & & & & & \\ -D^x & 0 & 0 & Q_2^x & 0 & G^x & & & \\ 0 & -D^y & 0 & 0 & Q_2^y & G^y & & & \\ 0 & 0 & 0 & -K^x & -K^y & 0 & & & \\ & & & \ddots & & \ddots & & & \\ & & & -D^x & 0 & 0 & Q_m^x & 0 & G^x \\ & & & 0 & -D^y & 0 & 0 & Q_m^y & G^y \\ & & & 0 & 0 & 0 & -K^x & -K^y & 0 \end{bmatrix} \begin{bmatrix} u_1 \\ v_1 \\ p_1 \\ u_2 \\ v_2 \\ p_2 \\ \vdots \\ u_m \\ v_m \\ p_m \end{bmatrix} = \begin{bmatrix} f_1^x + D^x u_0 \\ f_1^y + D^y v_0 \\ 0 \\ f_2^x \\ f_2^y \\ 0 \\ \vdots \\ f_m^x \\ f_m^y \\ 0 \end{bmatrix}$$

Analog to the standard SIMPLE method, an approximate factorization, which ignores some fill-in, is performed on the system to obtain the block matrices L and U:

$$L = \begin{bmatrix} Q_1 & 0 & & & & \\ -K & KR_1G & & & & \\ -D & 0 & Q_2 & 0 & & \\ 0 & 0 & -K & KR_2G & & \\ & & \ddots & & \ddots & \\ & & -D & 0 & Q_m & 0 \\ & & 0 & 0 & -K & KR_mG \end{bmatrix}$$

$$U = \begin{bmatrix} I & R_1G & & & & \\ 0 & I & & & & \\ & & I & R_2G & & \\ & & 0 & I & & \\ & & & & \ddots & \\ & & & & I & R_mG \\ & & & & 0 & I \end{bmatrix}$$

Note that the systems at different time-steps in the block U equation are independent and may therefore be solved in parallel at each time-step. The block L equation however contains dependencies via the diagonal matrices D which must be considered more carefully with a view to parallelization.

The block L system can rearranged, grouping u, v and p equations together:

$$\begin{bmatrix} Q_1^x & & & \\ -D^x & Q_2^x & & \\ & \ddots & \ddots & \\ & & -D^x & Q_m^x \end{bmatrix} \begin{bmatrix} u_1^{temp} \\ u_2^{temp} \\ \vdots \\ u_m^{temp} \end{bmatrix} = \begin{bmatrix} f_1^x + D^x u_0 \\ f_2^x \\ \vdots \\ f_m^x \end{bmatrix}$$

$$\begin{bmatrix} Q_1^y & & & \\ -D^y & Q_2^y & & \\ & \ddots & \ddots & \\ & & -D^y & Q_m^y \end{bmatrix} \begin{bmatrix} v_1^{temp} \\ v_2^{temp} \\ \vdots \\ v_m^{temp} \end{bmatrix} = \begin{bmatrix} f_1^y + D^y v_0 \\ f_2^y \\ \vdots \\ f_m^y \end{bmatrix}$$

$$\begin{bmatrix} KR_1G & & & \\ & KR_2G & & \\ & & \ddots & \\ & & & KR_mG \end{bmatrix} \begin{bmatrix} p_1 \\ p_2 \\ \vdots \\ p_m \end{bmatrix} = \text{right-hand-side}$$

The p equations at different time-steps are independent of each other and may thus be processed in parallel. However they depend on new values of both u^{temp} and v^{temp} at the same time-step, and must therefore be processed after these two equations.

Note that the u^{temp} and v^{temp} equations are independent of each other and moreover that they have the same structure, both corresponding to unsteady linear scalar convection-diffusion type equations.

The problem of time-parallelization of the Navier-Stokes problem thus reduces to the parallel solution of unsteady linear scalar convection-diffusion type equations. This is done with the pointwise incomplete decomposition method described in the next section.

4. Pipelined Time-Parallel ILU Method

As mentioned in [1], the processor network should be a ring configuration. All variables of one time-step will be assigned to one processor. Thus the number of processors used is equal to the number of time-steps to be computed simultaneously.

Assuming standard 5-point discretization for the space derivatives, backward Euler for the time derivative and lexicographic ordering of the grid points within each time-step, the matrices $Q_j^x, Q_j^y, j = 1, \ldots, m$ are pentadiagonal and matrices D^x, D^y are diagonal. Let the resulting bidiagonal blockstructured matrix be denoted by A and the system of linear equations to be solved denoted by $Ax = b$.

To obtain the Gauß-Seidel-like smoothing, we perform the pointwise ILU decomposition on matrix A which is defined by requiring that the approximate factors $\bar{L}$ and $\bar{U}$ may only have non-zero diagonals in places corresponding to those of the matrix A. The Jacobi-like smoothing is obtained by ignoring the matrices D^x or D^y. In addition, the decomposition may be improved using the technique of Stone [7]. One step $x^{(i)} \to x^{(i+1)}$ of the corresponding ILU-iteration is defined by the following sequence of operations:

$$\begin{array}{ll} \text{calculate} & d = b - A\,x^{(i)} \quad , \\ \text{solve} & \bar{L}\,z = d \quad , \\ \text{solve} & \bar{U}\,y = z \quad , \\ \text{correct} & x^{(i+1)} = x^{(i)} + y \quad . \end{array}$$

The matrix $\bar{U}$ contains only upper diagonals, and is therefore trivially parallelizable across the timesteps, as can the defect computation and the correction; $\bar{L}$ is a lower triangular matrix, which in the Gauß- Seidel case contains coefficients from matrices D^x or D^y. The system $\bar{L}\,z = d$ may be solved in parallel using the following pipeline algorithm:

Algorithm 1 *Pipelined Solution of* $\bar{L}z = d$

```
On each processor k=1,...,m do
define gridfunction z[1..n][1..n]
FOR j := 1 TO n DO
    FOR i := 1 TO n DO
        receive_from_previous_processor(z_{i,j,k-1})
        compute(z_{i,j,k})
        send_to_next_processor(z_{i,j,k})
    END FOR
```

To obtain the Jacobi-like procedure, the diagonal D^x or D^y is not present, and the lines beginning `receive` and `send` may be deleted from the above algorithm, thus removing communication from the body of the smoothing procedure.

5. Measurement of Efficiency

The time-parallel method presented in this paper is an example of an algorithm with computational redundancy. This means that the number of operations performed by the parallel algorithm exceeds that of the serial method. Thus one may expect, in addition to the losses in efficiency incurred by synchronization and communication, a further loss owing to the additional work carried out. In order to identify these losses in efficiency more precisely, the total efficiency of the algorithm E_{tot} may be considered as a product of the *numeric* efficiency E_{num} and the *parallel* efficiency E_{par} defined as follows:

$$\begin{aligned} E_{par} &= \frac{T_{Iter}(1)}{T_{Iter}(p)} \\ E_{num} &= \frac{\#Iter(1)}{p * \#Iter(p)} \\ E_{tot} &= E_{num} * E_{par} \end{aligned}$$

where $T_{Iter}(p)$ represents the computation time for one iteration on p processors, and $\#Iter(p)$ the total number of iterations required by p processors to perform the integration.

Numeric efficiency describes the increase in the number of iterations needed by the parallel method to integrate a certain number of time steps to a specified accuracy compared to the serial time-step-by-time-step approach. This increase is caused largely by the deterioration in the rate of convergence observed with increasing number of processors.

Parallel Efficiency is defined as the ratio of computation time on one and on p processors. It describes losses due to communication overhead and synchronization. The classically defined efficiency

$$E = \frac{T(1)}{p * T(p)}$$

where $T(p)$ represents the total computation time on p processors, was found to be almost equal to E_{tot}, differing only as a result of the small overhead incurred in the parallel algorithm between iterations. The results presented in the following section include both parallel and numeric efficiency, and a discussion on the factors that influence them.

6. Model problem and Results

The model problem used to test the algorithm is a variation on the standard Driven Cavity problem, where the lid of the cavity has velocity $u = \sin(t)$, instead of the constant velocity assumed for the steady case. The problem was discretized with a 32×32 finest grid and solved using V-cycles with $\nu_1 = \nu_2 = 3$. Tests were carried out for Reynolds numbers 1 and 100 based on cavity width and for time step sizes $\frac{\pi}{32}$ and $\frac{\pi}{3200}$. Table 1 gives the results obtained with the Gauß-Seidel smoothing variant on up to 8 processors. T_{tot} is the total computation time in seconds, and T_{Iter} the computation time per iteration in seconds.

Δt	Re	p	$\#Iter$	T_{Tot}	T_{Iter}	$E_{num}(\%)$	$E_{par}(\%)$	$E_{tot}(\%)$
$\frac{\pi}{32}$	1	1	536	7911	14.7	100	100	100
		2	272	4264	15.6	99	94	93
		4	141	2289	16.2	95	91	86
		8	76	1293	16.9	88	87	76
	100	1	345	5095	14.7	100	100	100
		2	176	2753	15.6	98	94	93
		4	91	1481	16.2	95	91	86
		8	48	816	16.9	90	87	78
$\frac{\pi}{3200}$	1	1	320	4727	14.7	100	100	100
		2	160	2504	15.6	100	94	94
		4	81	1318	16.2	99	91	90
		8	43	730	16.9	93	87	81
	100	1	256	3796	14.7	100	100	100
		2	128	2013	15.6	100	94	94
		4	65	1067	16.2	98	91	89
		8	37	636	16.9	86	87	75

Table 1: Efficiency results for the time-parallel method

The parallel efficiency for a given number of processors is, of course, identical for each problem, since the iteration performed is the same in every case. It is seen to vary between 94% and 87% on 2 and 8 processors respectively. These efficiencies, which one may have expected to be considerably lower, given the number of communications to be carried out in each iteration (at least $2048 = 2 \times 32 \times 32$). The high parallel efficiency is due both to the good ratio of communication setup to floating point operation times (about 9 for the T800 Transputer) and to the large amount of arithmetic to be performed between transfer operations.

The numeric efficiency is only weakly dependent on the values of the parameters Δt and Re that were chosen and is in all cases better than 85%.

7. Conclusions

In this paper a non-linear Multigrid method was used to solve the two- dimensional incompressible unsteady Navier-Stokes equations which allows several time-steps to be processed simultaneously. Attention was focussed on the smoothing procedure, which is based on the well-known SIMPLE method. It was shown how the question of parallelism could be reduced to the time- parallel solution of an unsteady scalar convection-diffusion type equation for which a pipelined strategy was given. Experimental results show the behaviour of the method for various parameters and indicate high efficiencies ($\geq 75\%$) for up to 8 processors.

The total efficiency was considered as a product of the parallel efficiency and the numeric efficiency, enabling a better understanding of losses incurred by communication and by a slight increase in the number of iterations needed in the parallel case.

The communication requirement of the time-parallel method is much higher than that of a standard grid partitioning approach. In the former, values of all grid points must be transferred to a neighbouring processor during an iteration, whilst in the latter case, only the values of edge points must be communicated. Thus the communicated data volume is considerably higher. Within the time-parallel method, the two smoothing variants Gauß-Seidel-like and Jacobi-like differ strongly in their communication profile. Whilst the Jacobi method only requires the grid data between iterations, which can be achieved with one transfer operation, the Gauß-Seidel smoother requires the values to sent pointwise, so that the number of messages equals the number of grid points at one time-step. This will obviously cause problems on parallel machines with a high communication setup time.

Although the time-parallel method presented here is shown to have good overall efficiency, we do not consider it primarily as an alternative to the standard grid partitioning approach, but more as an additional source of parallelism. Thus in the case where the efficiency of a standard method drops unacceptably for a increasing number of prcessors because of unequal load or short vector lengths, an approach with combined space and time parallelism may prove to be more efficient.

Acknowledgement

This work was supported in part by the Stiftung Volkswagenwerk within its program "Entwicklung von Berechnungsverfahren für Probleme der Strömungstechnik".

References

[1] Bastian, P., Burmeister, J., Horton, G.: *Implementation of a parallel multigrid method for parabolic partial differential equations.*, in [10]..

[2] Burmeister, J.: *Paralleles Lösen diskreter parabolischer Probleme mit Mehrgittertechniken.* Diplom thesis, Kiel, 1985.

[3] Glowinski, R., Lions, J.-R.(eds.): *Computing Methods in Applied Sciences and Engineering, VI.* Proceedings of the 6th International Symposium on Computing Methods in Applied Sciences and Engineering. Versailles, France, December 12-16, 1983. North Holland 1984.

[4] Hackbusch, W.: *Parabolic Multi-grid Methods.* In [3].

[5] Hackbusch, W.: *Multi-grid Methods and Applications.* Heidelberg: Springer-Verlag 1985.

Time-Parallel Multigrid for Navier-Stokes Equations

[6] Patankar S., Spalding D.B., *A Calculation Procedure for Heat, Mass and Momentum Transfer in Three-Dimensional Parabolic Flows*, Int. J. Heat Mass Transfer, 15, 1972.

[7] Stone, H. L., *Iterative Solution of Implicit Approximations of Multi-Dimensional Partial Differential Equations*, SIAM J. Num. Anal. 5, 1968.

[8] Hortmann, M., Peric, M., Scheuerer, G., *Finite Volume Multigrid Prediction of Laminar Natural Convection: Bench-Mark Solutions.*, Int. J. Num. Meth. Fluids, Vol. 11, pp. 189-207 (1990).

[9] Demirdzic, I., Peric, M., *Finite Volume Method for Prediction of Fluid Flow in Arbitrarily Shaped Domains with Moving Boundary.* Int. J. Num. Meth. Fluids, Vol. 10,pp. 771-790 (1990).

[10] Hackbusch, W.(ed.): *Parallel algorithms for pdes*, Proceedings of the 6th GAMM-Seminar Kiel, January 19-21, 1990, Vieweg-Verlag, Wiesbaden, to appear.

Jens Burmeister
Institut für Praktische Mathematik
Universität Kiel
Olshausenstr. 40
D-2300 Kiel 1

Graham Horton
Lehrstuhl für Rechnerstrukturen
Universität Erlangen-Nürnberg
Martensstr. 3
D-8520 Erlangen

SOLUTION OF 3-D PROBLEMS USING OVERLAPPING GRIDS AND MULTI-GRID METHODS

Laszlo Fuchs
Department of Gasdynamics
Royal Institute of Technology
S-100 44 Stockholm, SWEDEN
and
Scientific & Technical Computing,
IBM Svenska AB, Stockholm, SWEDEN

ABSTRACT: *Recent extensions of a Multi-grid method for calculating flows in stationary and moving geometries is described. The governing equations are discretized on a system of overlapping grids. Local grid-refinements can be included interactively. New local grids are topologically derived from the given grids. The grid refinement (mesh halving) may be done isotropically (all directions) or non-isotropically (refinement only along one coordinate line). The later scheme is mostly useful in adaptively localizing thin layers such as boundary layers. The discrete equations are solved by a Multi-grid method. The overlapping grid system (and also the local grids) require information exchange among the different sub-grids. This exchange can be made rather efficient if done as integral part of the Multi-grid process. This Multi-Grid method on overlapping systems of grid have been applied for the solution of some viscous flow problems.*

1. INTRODUCTION

A major difficulty in calculating flows is due to complex geometries and/or resolution difficulties. Usually one treats complex domain by using body-fitted coordinates. Grid generation problems become than an important issue in problem solution. The grid generation problem becomes even tougher when the geometry itself is time-dependent. Further complexity in CFD arises due to the generation of large gradients with apriori unknown thickness and location. The later problem is treated by different types of adaptive grid generation. Finite elements can be used to treat both the generation of complex grids and also for adaptive refinements (e.g. [1]). There are however two main draw backs with finite-element (or rather any unstructured mesh, independent of the discretization method of the PDEs): Unstructured grids require larger overhead both in terms of computer memory and additional computational operations. Typically, the unstructured-grid (using finite-volume approximations for transonic flow) solver requires 3-5 times more computational effort compared to a similar solver on a structured (finite-volume) grid. A second difficulty with single grid systems, occurs when the geometry itself is time-dependent. All single grid methods require a complete regeneration of the mesh in each time step. The grid generation process becomes a time consuming and complex process (e.g. how "good" is the grid?). Furthermore, calculated values at a given time must be redistributed to the new grid before integration in time can take place. This "rezoning" process introduces stability and accuracy problems.

An alternative approach which has been developed in recent years [2-6], is to use systems of overlapping grids. We have been using such methods only together with Multi-Grid (MG) CFD solvers [7-14]. Initially, the basic concepts of (adaptive) local-mesh refinement technique (in 2-D) was studied [7]. The basic overlapping grid/MG technique was implemented for flows in systems of straight channels [8,9]. Later, the method was extended to include general (stationary) 2-D object (using streamfunction and vorticity formulation [10]). Further extension included moving objects, undergoing linear or rotational acceleration [11-13]. For 3-D cases we have introduced local grid refinement [13] and here we report the application of the more general concept of overlapping 3-D grids. Naturally, local grid refinements are always possible. Local mesh-refinements may be isotropic or uni-directional. In the former case, the mesh spacing is halved in each direction, in the sub-region that is being refined. In the later case the mesh spacing is halved only in one direction and is unchanged in the other two. This type of local refinement is of interest for boundary-layer like flows, where the local scales become unisotropic. The usage of MG solvers is most important for keeping the competitive edge of the overlapping grid method. Here, we demonstrate the scheme for steady and unsteady incompressible flows in different geometries. In another paper, we report the progress in solving model equations in completely arbitrary 3-D domains [14].

In the following we describe the overlapping grid system and the basic Navier-Stokes/MG solver. The scheme is applied for a 2-D case of (cold) flows past a flame-holder in a wind tunnel. Further calculations are for flows in an isothermal combustion chamber of gasturbine and an Otto engine model.

2. NUMERICAL METHODS

Governing equations

The governing equations for incompressible viscous flows is given by

$$\frac{D\mathbf{u}}{Dt} + \mathbf{u} \text{ grad } \mathbf{u} - \text{grad } p - \frac{1}{Re} \text{ div grad } \mathbf{u} = 0$$

$$\text{div } \mathbf{u} = 0 \tag{1}$$

with appropriate boundary conditions at inflow/outflow and no-slip conditions on solid boundaries.

In pure 2-D cases one may replace eq.s (1) by the streamfunction and vorticity formulation.

Spatial discretization

The governing equations are discretized on systems of overlapping grids. These grids are generated by imposing the following constraints:

a. Each subgrid is defined independently of the others, using its own local system of coordinates. Each object has also its own coordinate system (and thus a corresponding grid). For moving objects, the grid topologies remain unchanged.

b. The union of all subgrids "cover" the whole computational domain.

c. Section of a subgrid with the others is non-empty.

d. Locally refined grids are obtained by halving the mesh spacing (in the computational space) in at least in one coordinate direction. The extent of this new grid is normally less than that of the full subgrid.

These basic rules are complemented with certain priorities:

e. We use structured grids only (at least for the time being),

f. The grid system chosen to discretized a certain domain is cartesian, orthogonal or of general body-fitted type. We prefer to use the former grids as much as possible since the number of terms and hence the number of computer operations needed to express the discrete approximations is minimal. Good mesh distributions is obtained by widely using local grid refinements.

The overlapping and the locally refined grid systems, fit well into the MG structure of grids. Some implementation details are given in [11]. Basically, the used pointer system allows the addition and cancellation of subgrids. When general body fitted grids are used, metrics data is stored only for the involved grid points. The pointer system allows simple addressing among the grids. Such a system is a crucial part of the scheme since one has to manage both parent/child relation for the MG process, and overlapping "boundary" surfaces for interzonal information exchange. The total "overhead" required for the locally refined and overlapping grid systems is small (requiring storage of information with a dimension of at least one lower than the dimension of the problem). Interzonal "boundaries" are marked logically and their addresses are stored in special arrays, together with additional data about the cell in which they are contained in one of the overlapping subgrids. This information is enough to dynamically define the interpolation "molecule", used for exchanging data among subgrids. In most cases, with the exception of the presence of discontinuities, we use isotropic interpolation formulae. The order of interpolation is problem dependent (order 3 or 5 for the primitive variable and the streamfunction and vorticity formulations, respectively). In some cases discontinuities (e.g. shocks) cross subgrid boundaries. In such cases, isotropic interpolations may lead to local oscillations in the solution (Gibb's phenomena). To eliminates such oscillation we have employed [15], adaptively generated, biased interpolation molecule. This molecule is found by selecting a sided interpolation formula, where the function is smoothest possible. In practice, the molecule is chosen so that the interpolation becomes one sided, and it excludes the large gradient region.

Discrete approximations

The governing equations are approximated by finite-differences on the given grids. The basic approximation use central finite-differences for all terms except the convective terms which are discretized by an upwind (first order) approximation. To enhance accuracy, we use, when so desired, higher order approximations in a few-step defect correction manner [16]. This type of method has been applied in some cases to attain accuracy of 4-th and 6-th orders. When the cell Reynolds number is small (less than 2) one may use second order central difference approximations for all terms.

An additional complexity arises when the streamfunction and vorticity formulation is used and when the body-surface values of the streamfunction can not be determined apriori. In such cases, additional equations (equal in number to the number of such surfaces) has to be added. These equations are found by imposing a unique value on the pressure as one integrates the momentum equations around the objects (for details see ref. [10]).

At outflow sections we assume, when appropriate, that the flow may be "parabolized": it is assumed that there is a main flow direction and that the diffusion along this direction is small compared to the transverse terms. The resulting equations give rise to a non-local boundary condition which was found to be most useful in internal flow calculations [17].

For unsteady flows, we use a two step time-integration technique:

a. Predictor: an explicit Euler or multi-stage Runge-Kutta steps.

b. Corrector: Implicit in the viscous terms.

This type of time stepping is required when the physical changes are relatively slow and the limitation in time stepping is set by the stability due to the viscous terms of the governing equations.

Solution procedure

To enable fast convergence we have used the MG method that we employed in the past. This solver uses, for steady flows, a lexicographic point relation scheme for the momentum equation and a distributive scheme, updating all dependent variables simultaneously, when the continuity equation is updated.

The restriction operators are cell area or cell volume based, resulting in mass conservation in local basis (i.e. up to round-off). Prolongation of the corrections is done linearly and isotropically.

Interzonal exchange is done as an integral part of the relaxation scheme, allowing exchange of function values on the finest levels, or the corrections on the coarse grids. The information exchange is done by Lagrange interpolation formula, up to order five, is used. Previously reported tests show that the presence of interzonal "boundaries" do not cause reduction in the convergence rate of the basic MG solver.

For unsteady flows, we use the Stokes (or NS) solver in the implicit corrector step. The amount of work required in each implicit step is usually equal to a single MG cycle, leading to a reduction in the residual between one and two orders of magnitudes.

In the case of relative motion of the different subgrid, one has to add linear and rotational acceleration terms to the momentum equations, and also adjust the interzonal interpolation to account for the differences in the relative speed of the different frames of references. As mentioned above, no other terms are affected due to the unsteadiness of the geometry.

3. COMPUTED EXAMPLES

First, consider a 2-D case of cold low past a flame-holder in a wind tunnel. The geometry of this case is similar to the one used in experiments at Volvo Flygmotor AB. The geometry is first considered as 2-D, even though we have made also some 3-D calculations. The flow in the wind tunnel is relatively slow (Re=50 000, based on the height of the triangular flame-holder) and is incompressible. The shape of the cross section of the flame-holder is triangular. For this problem we generated a cartesian grid attached to the walls of the tunnel, and a body-fitted (orthogonal) grid. The triangle has in this model rounded edges (so that a smooth O-mesh can be used). The two grids overlap, as shown in Figure 1. As seen, the triangle itself and a region in the cartesian grid, in its immediate neighborhood ar "shut-down".

These points are eliminated only logically, thus keeping the structure of the grid system. Each sub-grid, contains also a sequence of coarse grids that are used in the MG process. Some of the finer levels may be only local, derived by isotropic refinement (covering only the neighborhood) of the triangle, or by non-isotropic refinement (in the direction perpendicular to the walls of the tunnel). The extent to which two sub-grids should overlap, is determined by the method of implementing the information exchange procedure: It is unwise to have too few cells in the common region, since errors are than fed back and forth through the interzonal exchange procedure. We usually use at least one cell more than the number of relaxation sweeps that are used on grids that participate in the information exchange.

In the calculation done on this types of grids, we use a time-dependent solver. The number of grid points used in different runs, was in the range from 50 000 to over 200 000 nodes, including those belonging to locally refined subgrids. Both first order and second order accurate solutions have been calculated.

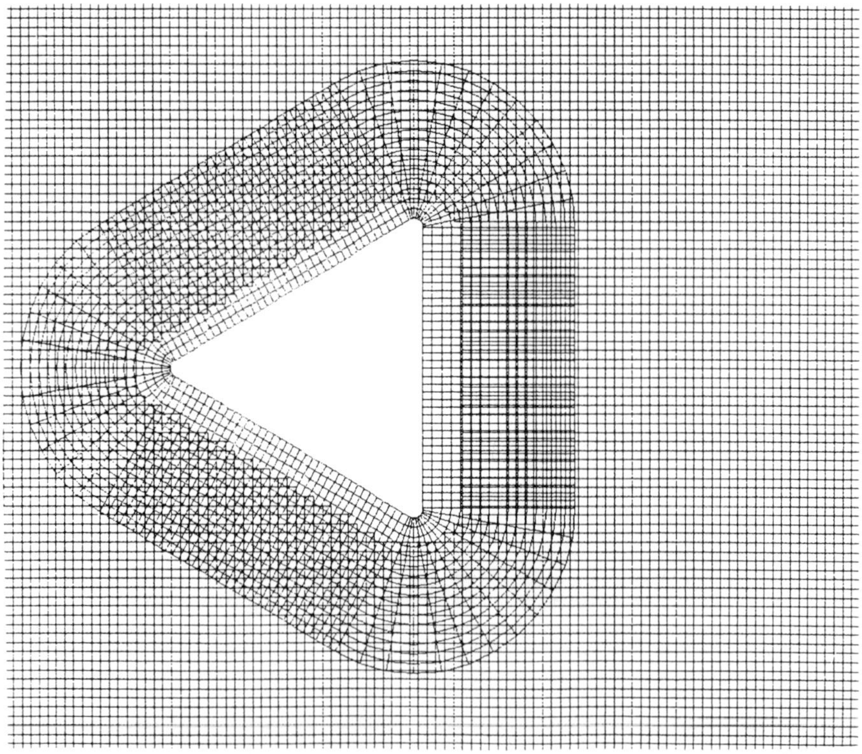

Figure 1: An overlapping grid system for the flame-holder geometry. The cartesian grid is refined locally near the triangular object. Note that the grids spacings in both system are roughly similar.

The later solutions were obtained by adding a defect term to the momentum equations. The calculations showed that for Re = 50 000 there is a vortex shedding behind the flame-holder. For the given geometry (flame-holder of side size of 40 mm, tunnel height of 120 mm) the period of the oscillatory motion was about 30 ms (Figure 2). The mean length of the separation bubble behind the flame-holder is longer than those found in wind-tunnel measurements. In parallel calculations, using a k-epsilon turbulence model, we have found that the length of the separation bubble is rather sensitive in the boundary conditions imposed on the variables of the turbulence models. The assumption of 2-D flow is also a major source for the discrepancy between calculated and experimental results. Further calculations, using a Large Eddy Simulation (LES) model, are to be conducted with the overlapping grid code.

Steady 3-D flow in a gasturbine combustion chamber model is considered in Fig. 3. The grid is composed of a cartesian part (at the neighborhood of the axis of the cylinder, and a cylindrical part (the one shown on the figure). The two grids overlap each other. Film cooling is simulated by allowing air to injected from and almost parallel to the outer walls.

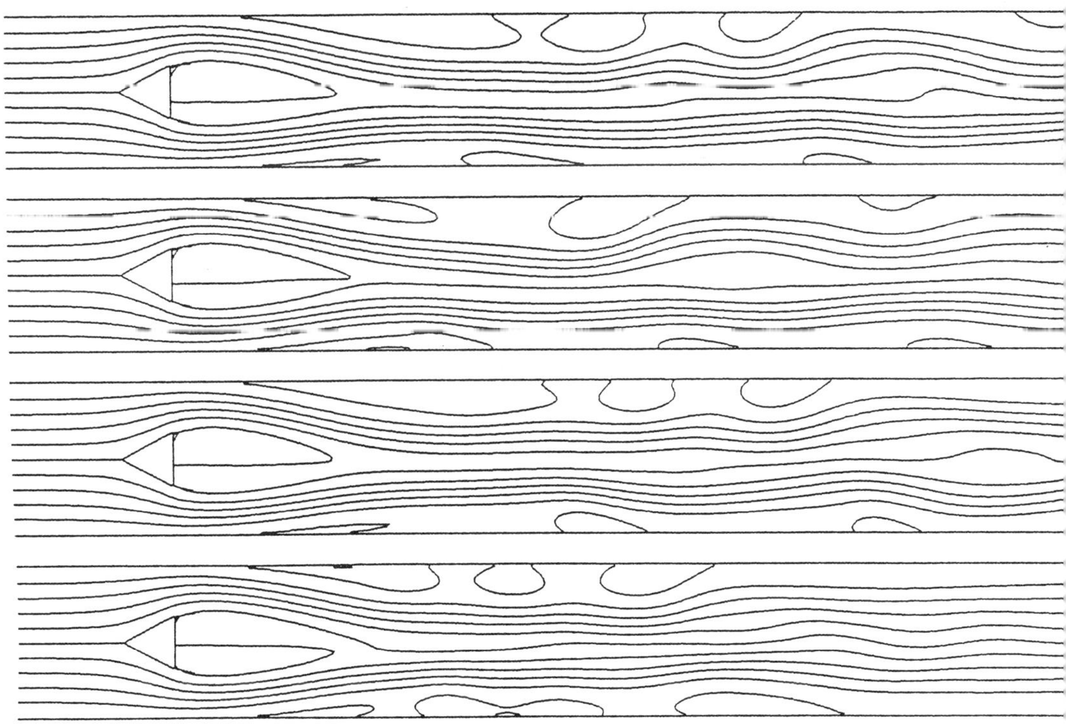

Figure 2: Instantaneous streamlines at 3 different times.

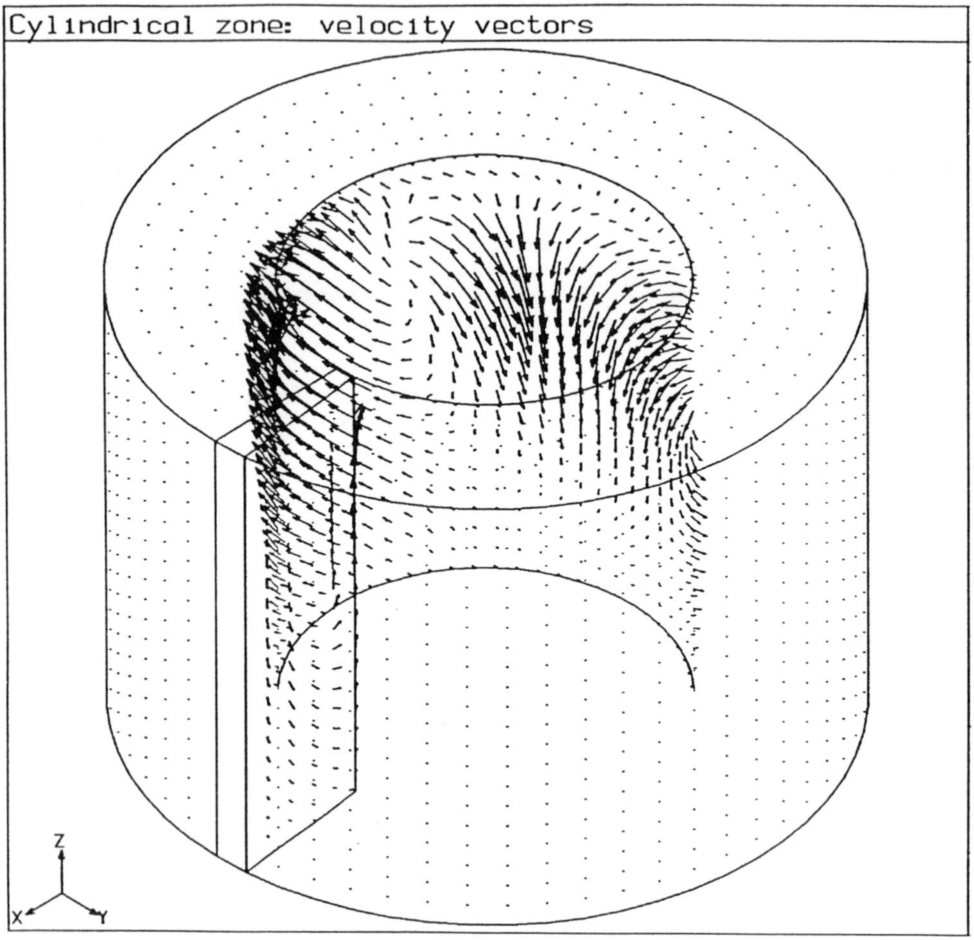

Figure 3: Velocity vectors in a gasturbine combustion chamber model. The computational grid is composed of a cylindrical grid overlapping a cartesian grid. Only the cylindrical part is shown.

Next, consider the flow in an Otto-engine model. The geometry is essentially similar to the one in the previous case: We assume that the geometry is cylindrical, with a piston moving back and forth in the cylinder. The compression ratio is 10. In the top of the cylinder there are 4 circular openings. At both suction and ejection stages, only two of the openings are open. At the inflow ports, there is a strong swirl imposed on the incoming jets, as seen in Figure 4. Figures 5-7 show the flow-field (as seen on a rather coarse level of the cartesian grid system) at 3 different stages of the engine-cycle: Largest cylinder volume, as the inlet valves open (Fig. 5). Figure 6. shows the stage when the piston reached the mid-point at the ejection stage. Figure 7. shows the

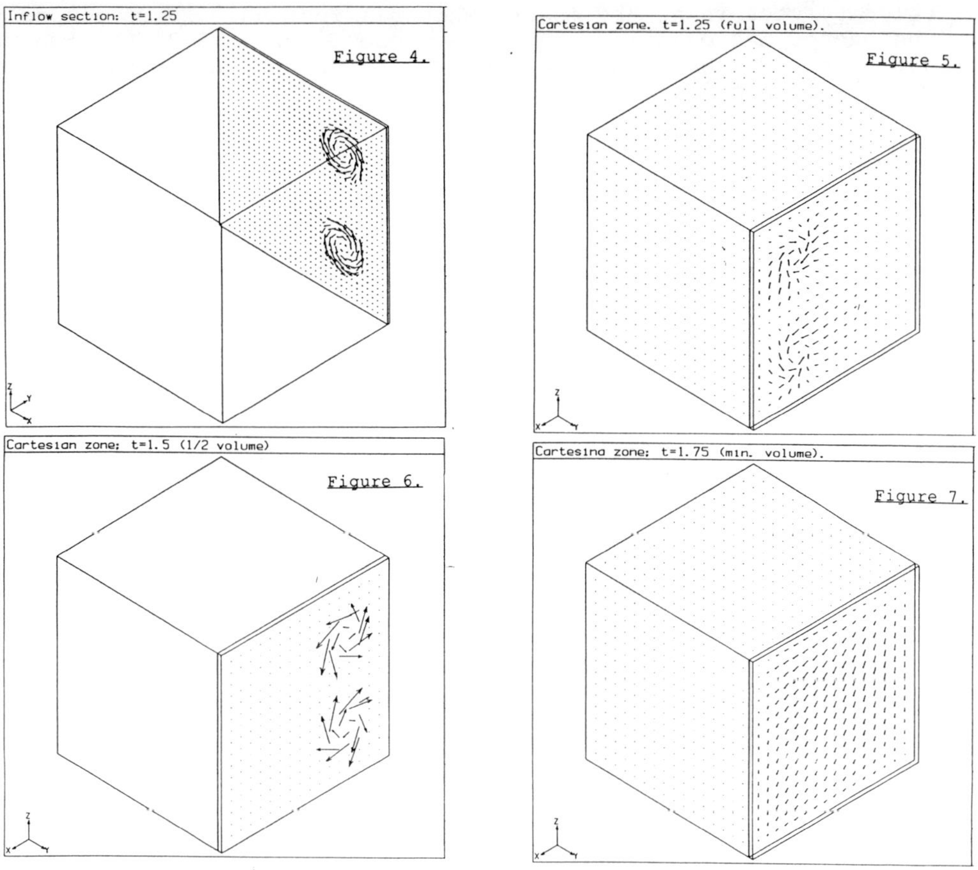

Figure 4-7: Flow in an Otto engine model. Two open valves are shown for a crank angle of 180°, fine grid view (fig. 4), coarse grid view (fig. 5). Crank angle 270° (fig. 6), and crank angle of 360° with all four valves closed (fig. 7).

velocity vectors at the moment when all four valves are closed at the end of the ejection stage. In these figures (5-7) the results are plotted as seen on a coarse grid, even though the calculations were made on a finer grid.

4. CONCLUDING REMARKS

The application of a 3-D overlapping grid technique, coupled with a multi-grid solver has been presented for the solution of some viscous flow problems. The main features of the schemes are the possibility of using simple (cartesian or orthogonal) grids in many parts of the computational geometry. Complex geometries are treated by using body-fitted coordinates locally and by using local (isotropic or uni-directional) mesh refinements. The combination of overlapping grids with MG solvers allows as efficient solvers as on non-overlapping grids, the gain is, however, in increased accuracy for given number of mesh points, or alternatively, reduced computational effort for a required accuracy.

REFERENCES

1. Mavripilis, D.J. and A. Jameson, Multigrid Solution of the Navier-Stokes equations on triangular meshes. AIAA J. **28** (1990), 1415-1425.
2. Atta, E.H. and J. Vadyak, A grid overlapping scheme for flowfield computations about multicomponent configurations. AIAA J. 21 (1983), 1271-1277.
3. Atta, E.H. and J. Vadyak, A grid interfacing zonal algorithm for three dimensional transonic flows about aircraft configurations. AIAA Paper 82-10117, 1982.
4. Hanshaw W.D. and G. Chesshire, multigrid on composite grids. SIAM J. Sci. Stat. Comp. **8** (1987), 914-923.
5. Berger, M.J. SIAM J. Sci. Stat. Comp. **7** (1986), 905.
6. Rai, M.M. A conservative treatment of zonal boundaries for Euler equation calculations. J. Comp. Phys. **62** (1986), 472-503.
7. Fuchs, L. A local mesh-refinement technique for incompressible flows. Computers & Fluids **14** (1985), 69-81.
8. Fuchs, L. Numerical flow simulation using zonal grids. AIAA Paper 85-1518, 1985.
9. Fuchs, L. Numerical computation od viscous incompressible flows in systems of channels. AIAA Paper. 87-367, 1987.
10. Fuchs, L. Calculation of viscous flows in multiply-connected domains. Notes on Numerical Fluid Mechanics **20** (ed. W. Hackbusch et al), Vieweg Verlag, (1988), 96-103.
11. Fuchs, L. Calculation of viscous incompressible flows in time-dependent domains. Notes on Numerical Fluid Mechanics, **30**, (ed W. Hackbusch), Vieweg Verlag (1990), 62-71.

12. Fuchs, L. Calculation of flow fields using overlapping grids. Notes on Numerical Fluid Mechanics, 29, (Ed. P. Wesseling), Vieweg Verlag, 138-147.
13. Fuchs, L. Calculation of flows in complex geometries using overlapping grids. AIAA Paper 90-1563, 1990
14. Tu, J.Y. and L. Fuchs Finite-Volume and multigrid methods on 3-D zonal overlapping grids. (to appear in Proc. 3rd European MG conference, 1990)
15. Gu C-Y. and L. Fuchs, A non-isotropic interpolation scheme applied to zonal-grid calculation of transonic flows. Numerical Methods in Laminar and Turbulent Flows 5 (Eds. C. Taylor, et. al.), Pineridge Press (1987), 975-989.
16. Fuchs, L. Defect corrections and higher numerical accuracy. Notes on Numerical Methods in Fluid Mechanics, 10 (Ed. W. Hackbusch), Vieweg Verlag, 52-66.
17. Fuchs, L. and H-S. Zhao, Solution of 3-D viscous incompressible flows by a multi-grid method. Int. J. Numer. Meth. in Fluids. **5** (1985) 311-329.

A Multigrid Method for the Solution of a Convection - Diffusion Equation with Rapidly Varying Coefficients

Jürgen Fuhrmann and Klaus Gärtner
Karl-Weierstraß-Institut für Mathematik
Mohrenstraße 39, Berlin D-O-1086

Abstract

Consider the following equations

$$-\operatorname{div}(\operatorname{grad} n - n \operatorname{grad} \psi) = f$$

and (for $u = e^{-\psi} n$)

$$-\operatorname{div}(e^{\psi} \operatorname{grad} u) = f$$

in a two- or threedimensional rectangular domain with the usual boundary conditions. Based on the idea to use mixed finite element methods to get a reformulation of the exponential fitting scheme and its higher dimensional box derivates [2], [12] a multigrid method for the solution of the two problems in rectangular domains has been developed and coded. The coarse grid matrix condensation uses harmonical averaging of the fine grid matrix coefficients fulfilling a Galerkin condition in the mixed formulation. The intergrid transfer operators are fitted to the upwind character of the discretization and are derived using the mixed formulation, too. The method shows good convergence for problems with ψ having unresolved jumps on the finest grid. The method is used in combination with conjugated gradients acceleration to tackle problems arising in semiconductor device simulation. It's extension to the 3D case is under consideration now.

1 Problem Formulation

Consider the following boundary value problem:

$$\begin{aligned} -\operatorname{div}(\operatorname{grad} n - n \operatorname{grad} \psi) &= f \text{ in } \Omega \subset R^d, \\ n &= 0 \text{ on } \partial\Omega. \end{aligned} \tag{1}$$

This is a special case of a convection-diffusion problem admitting a convection potential. Problems of this kind occur in semiconductor device simulation where n denotes the carrier density and ψ the electrostatical potential, which will be assumed here to be a given function. Introduce the variable $u = e^{-\psi}$ and transform (1) to

$$\begin{aligned} -\operatorname{div}(e^{\psi}\operatorname{grad} u) &= f \text{ in } \Omega, \\ u &= 0 \text{ on } \partial\Omega. \end{aligned} \tag{2}$$

This is a selfadjoint boundary value problem with strongly varying coefficients. It will be considered together with the nonselfadjoint one, which is of greater practical interest because of the range of u exceeding the usual floating point number representations. So, the theoretical considerations will be made for the selfadjoint problem and, after that, the back transform will be given.

The classical variational formulation of the problem (2) will be the following: Let $U = H_0^1(\Omega)$ and define for $u, v \in U$

$$l(u,v) := \int_\Omega e^{\psi}\,(\operatorname{grad} u, \operatorname{grad} v)_{R^d}\; d\omega$$

and , for $u \in U$,

$$f(u) := \int_\Omega fu \;\; d\omega.$$

Find $u \in U$ with

$$l(u,v) = f(v) \quad \forall v \in U. \tag{3}$$

There is the well known

Theorem 1 *For Ω with Lipschitz continuous boundary and $\psi \in L^\infty(\Omega)$ this problem has an unique solution $\hat{u} \in U$.*

By introducing the variable $\phi = e^{\psi}\operatorname{grad} u$ split (2) into

$$\left\{ \begin{aligned} \phi e^{-\psi} - \operatorname{grad} u &= 0 \\ -\operatorname{div}\phi &= f. \end{aligned} \right. \tag{4}$$

From this, letting $\Phi = (L^2(\Omega))^d$ and by defining for $\phi, \eta \in \Phi$

$$a(\phi,\eta) := \int_\Omega e^{-\psi}(\phi\,,\eta)_{R^d}\, d\omega$$

and for $\phi \in \Phi, u \in U$

$$b(u,\phi) = -\int_\Omega (\phi\,, \operatorname{grad} u)_{R^d}\;\; d\omega$$

one comes after using the Gauss theorem in the second equation to the following variational formulation of (4) : Find $(u,\phi) \in (U \times \Phi)$ with

$$\left\{ \begin{aligned} a(\phi,\eta) + b(u,\eta) &= 0 \quad \forall \eta \in \Phi \\ b(v,\phi) &= f(v) \quad \forall v \in U. \end{aligned} \right. \tag{5}$$

The following holds [12] :

Theorem 2 *Problem (5) has an unique solution $(\hat{u},\hat{\phi}) \in (U \times \Phi)$ with $\hat{u}$ being the unique solution of (3) and $\hat{\phi} \overset{L^2}{=} e^{\psi}\operatorname{grad}\hat{u}$.*

2 Discretization and Multigrid Components

2.1 The Idea

Using the ideas of [2],[3],[10], based on a mixed finite element interpretation of the exponential fitting scheme used to discretize (1), discrete operators, correspondig coarse grid and intergrid transfer operators will be constructed.

Let $U_h \subset U$ and $\Phi_h \subset \Phi$ be some finite element subspaces, assume $\psi_h \in U_h$. Define discrete operators using the operators defined in (5):

$$\begin{aligned} A_h : \Phi_h \to \Phi_h^\star \quad & \text{by} \quad (A_h\phi_h, \eta_h) = a(\phi_h, \eta_h) \quad \forall \eta_h \in \Phi_h, \\ B_h : U_h \to \Phi_h^\star \quad & \text{by} \quad (B_h u_h, \eta_h) = b(u_h, \eta_h) \quad \forall \eta_h \in \Phi_h, \\ E_h : U_h \to U_h \quad & \text{by} \quad E_h u_h = e^{-\psi_h} u_h. \end{aligned}$$

Then the discretized version of the selfadjoint problem is

$$\begin{cases} A_h\phi_h + B_h u_h = 0 \\ B_h^T u_h \qquad\quad = f_h. \end{cases} \tag{6}$$

Assume,the pair (A_h, B_h) fulfill the discrete LBB condition and A_h is diagonal or otherwise easily invertible by static condensation. This makes it possible to construct invertible discrete operators

$$L_h :- B_h^T \circ A_h^{-1} \circ B_h \quad : \quad U_h \to U_h^\star \quad \text{and} \tag{7}$$

$$M_h := L_h \circ E_h \quad : \quad U_h \to U_h^\star. \tag{8}$$

for the selfadjoint and the nonselfadjoint problem, respectively. Further, let (U_H, Φ_H) be a pair of coarse grid spaces with corresponding prolongation operators

$$p_u : U_H \to U_h \quad \text{and} \quad p_\phi : \Phi_H \to \Phi_h.$$

Then the following operators can be constructed

$$\begin{aligned} r_u &:= p_u^T, \\ r_\phi &:= p_\phi^T, \\ p_n &:= E_h^{-1} \circ p_u \circ E_H, \\ A_H &:= r_\phi \circ A_h \circ p_\phi, \\ B_H &:= r_\phi \circ B_h \circ p_u, \\ L_H &:= B_H^T \circ A_H^{-1} \circ B_H \quad \text{and} \\ M_H &:= L_H \circ E_A \end{aligned}$$

supposing (A_H, B_H) fulfil the same conditions as (A_h, B_h). The whole situation can be illustrated by the following commutative diagram:

$$\begin{array}{ccccccccc} U_h & \xrightarrow{E_h} & U_h & \xrightarrow{B_h} & \Phi_h^\star & \xrightarrow{A_h^{-1}} & \Phi_h & \xrightarrow{B_h^T} & U_h^\star \\ p_n \uparrow & & p_u \uparrow & & r_\phi \downarrow & & \uparrow p_\phi & & \downarrow r_u \\ U_H & \xrightarrow{E_H} & U_H & \xrightarrow{B_H} & \Phi_H^\star & \xrightarrow{A_H^{-1}} & \Phi_H & \xrightarrow{B_H^T} & U_H^\star. \end{array} \tag{9}$$

Here, A_H and B_H are constructed by a Galerkin condition, while L_H and M_H fulfill a Galerkin condition only in a weaker sense. Further, one has to remark that there is no special restriction r_n which could be constructed from p_n.

The idea to incorporate together multigrid and mixed finite elements is used by Reusken [8], Hemker [5], Molenaar [7], Schmidt and Jacobs [11] too, but they use other approaches to derive the multigrid components.

In the follwing two subsections there will be constructed the multigrid operators for the given equations explicitely for the onedimensional and the higher dimensional case, respectively.

2.2 The Onedimensional Case

To make the situation described in the previous section transparent, the 1D-case is considered first.

2.2.1 The Selfadjoint Problem

Let $\Omega = (x_0, x_{N_f}) \subset R^1$ be subdivided by the grid $x_0 < x_1 < \ldots < x_{N_f}$ into N_f elements $[x_i, x_{i+1}]$ Take

$$U_h := \{u \in U| \quad u|_{[x_i,x_{i+1}]} \in P_1[x_i, x_{i+1}], i = 0 \ldots N_f - 1\}.$$

For given i, let $\xi_i \in U_h$ be the unique function with $\xi_i(x_j) = \delta_{ij}, j = 1 \ldots N_f - 1$. Then $\{\xi_i\}_{i=1\ldots N_f-1}$ is the FEM basis of U_h. Let $\pi_h : U_h \to R^{N_f}$ be the canonical FEM isomorphism associated to this basis. In the sequel,the finite element spaces and their canonical images will identified by this isomorphism. Introduce the discrete flux space

$$\Phi_h := \{\phi \in \Phi| \quad \phi|_{[x_i,x_{i+1}]} \in P_0[x_i, x_{i+1}], i = 1 \ldots N_f - 1\}$$

with the basis $\{\chi_i\}_{i=0\ldots N_f-1}$ defined by $\chi_i \equiv \delta_{ij}$ on $[x_j, x_{j+1}]$. Construct the matrices A_h, B_h associated to Φ_h, U_h and the bilinear forms $a(\cdot,\cdot)$, $b(\cdot,\cdot)$ (see (6)). Assume $\psi = \sum \psi_i \xi_i$, $u_h = \sum u_i \xi_i \in U_h$. Then

$$\begin{aligned}(A_h\phi_h)_i &= a\left(\sum \phi_m \chi_m, \chi_i\right) = \phi_i a(\chi_i, \chi_i) \\ &= d_i^0 \phi_i\end{aligned}$$

with

$$d_i^0 = \int_{x_i}^{x_{i+1}} e^{-\psi} dx = \frac{e^{-\psi_{i+1}} - e^{-\psi_i}}{\psi_i - \psi_{i+1}}(x_{i+1} - x_i).$$

using the fact that ψ is linear on $[x_i, x_{i+1}]$. Further,

$$\begin{aligned}(B_h u_h)_i &= b\left(\sum u_m \phi_m, \chi_i\right) = u_i b(\phi_i, \chi_i) + u_{i+1} b(\phi_{i+1}, \chi_i) \\ &= u_{i+1} - u_i \\ (B_h^T \phi_h)_i &= \phi_i - \phi_{i+1}.\end{aligned}$$

Obviously, the matrix A_h is a diagonal one with diagonal elements bounded away from zero, so static condensation can be used to exclude the variable ϕ from the discrete system (6). One gets

$$\phi_i = \frac{u_{i+1} - u_i}{d_i^0} \quad \text{with} \quad (L_h u_h)_i = d_{i-1}(u_i - u_{i-1}) + d_i(u_i - u_{i+1})$$

where

$$d_i = (d_i^0)^{-1} = \frac{1}{\int_{x_i}^{x_{i+1}} e^{-\psi} dx} = \frac{\psi_i - \psi_{i+1}}{e^{-\psi_{i+1}} - e^{-\psi_i}} \cdot \frac{1}{x_{i+1} - x_i} \tag{10}$$

is the harmonical mean over the i-th element of the coefficient function e^{ψ} divided by the element size. Now, the coarse grid operator associated to L_h is constructed conforming to the mixed approach. Choose another partition $X_0 < X_1 < \ldots < X_{N_c}$ of Ω with $N_f = 2N_c - 1$, $x_{2i} = X_i$ and $x_{2i+1} = \frac{X_i + X_i \mathbf{t}_H}{2}$. Like above, one can define the space

$$U_H := \{u \in U| \quad u|_{[X_i, X_{i+1}]} \in P_1[=_i, X_{i+1}], i = 1 \ldots N_c - 1\}$$

and it's basis $\{\Xi_i\}_{i=1 \ldots N_c - 1}$. Assume there is some linear interpolation operator $p : U_H \to U_h$ given by the formula

$$(pu)_i = \begin{cases} u_{i/2} & , i \equiv 0(2) \\ \alpha_{\frac{i-1}{2}} u_{\frac{i-1}{2}} + \beta_{\frac{i-1}{2}} u_{\frac{i+1}{2}} & , i \equiv 1(2) \end{cases} , i = 1 \ldots N_f - 1 \tag{11}$$

with $\alpha_j + \beta_j = 1, j = 0 \ldots N_c - 1$. Define $r : U_h \to U_H$ by $r = p^T$. Introduce the coarse grid flux space

$$\Phi_H := \{\phi \in \Phi| \quad \phi|_{[X_i, X_{i+1}]} \in P_0[X_i, X_{i+1}], i = 1 \ldots N_c - 1\}$$

with the basis $\bar{\chi}_i$ defined by $\bar{\chi}_i \equiv \delta_{ij}$ on $[X_j, X_{j+1}]$. Further, define

$$(p_\phi)_i = \begin{cases} \phi_{i/2} & , i \equiv 0(2) \\ \phi_{\frac{i+1}{2}} & , i \equiv 1(2). \end{cases} , i = 1 \ldots N_f - 1 \tag{12}$$

Using (9) for $(u, \psi) \in (U_H \times \Phi_H)$ and $i - 2m$

$$\begin{aligned} (A_H)_m &= (r_\phi A_h p_\phi \phi)_m = (A_h p_\phi \phi)_i + (A_h p_\phi \phi)_{i+1} = d_i^0 (p_\phi \phi)_i + d_{i+1}^0 (p_\phi \phi)_{i+1} \\ &= D_m^0 \phi_m \end{aligned}$$

holds with

$$D_m^0 = d_{2m}^0 + d_{2m+1}^0.$$

Further,

$$\begin{aligned} (B_H u)_m &= (r_\phi B_h p_u u)_m = (B_h p_u u)_i + (B_h p_u u)_{i+1} \\ &= (p_u u)_{i+1} - (p_u u)_i + (p_u u)_{i+2} - (p_u u)_{i+1} \\ &= u_{m+1} - u_m. \end{aligned}$$

$B_H^T = (r_u B_h^T p_\phi)$ results in

$$(B_H^T \phi)_m = \phi_m - \phi_{m-1}.$$

So one gets

$$(L_H u)_m = D_{m-1}(u_m - u_{m-1}) + D_m(u_m - u_{m+1})$$

with

$$D_m = \frac{1}{\frac{1}{d_{2m}} + \frac{1}{d_{2m+1}}} \tag{13}$$

independent of the interpolation coefficients from (11) and fitting to the ideology of using harmonical means as matrix coefficients. So the interpolation p_u can be designed in this context without of any influence on L_H. Consider the coarse grid element $[X_m, X_{m+1}]$ and assume to be given an $u_H = \sum \Xi_M U_m \in U_H$. For $i = 2m$ and $u_h = p_u u_H = \sum \phi_i u_i$, $u_i = U_m$ and $u_{i+2} = U_{m+1}$ holds. Use implicitely $\phi_m = D_m(U_{m+1} - U_m)$. One has to look for a value of u_{i+1} which guarantees flux conservation, i.e.

$$(p_\phi \phi)_i = \phi_i = d_i(u_{i+1} - u_i) = d_{i+1}(u_{i+2} - u_{i+1}) = \phi_{i+1} = (p_\phi \phi)_i.$$

Using this, get

$$u_{i+1} = \frac{d_i u_i + d_{i+1} u_{i+2}}{d_i + d_{i+1}} = u_i \frac{1}{1 + \frac{d_{i+1}}{d_i}} + u_{i+1} \frac{1}{1 + \frac{d_i}{d_{i+1}}}.$$

So the prolongation coefficients can be proposed:

$$\alpha_i = u_i \frac{1}{1 + \frac{d_{i+1}}{d_i}} \quad \text{and} \quad \beta_i = 1 - \alpha_i. \tag{14}$$

Remark, that the same u_{i+1} results from solving the onedimensional boundary value problem

$$(e^{\psi_h} u')' = 0, \quad u(X_m) = u_m, \quad U(X_{m+1}) = U_{m+1}$$

on (X_m, X_{m+1}) with piecewise linear ψ_h and taking $u_{i+1} = u(x_{i+1})$. This is a variant of a so called "interpolation using the grid equation" which is recommended by many authors and which one can find in AMG procedures, too [9], [1]. Experiments showed that this interpolation together with the coarse grid matrix (9) yields convincing convergence results.

2.2.2 The Nonselfadjoint Problem

The transformation $n = E_h u_h = e^{-\psi_h} u$ allows to construct the matrix M_h as recommended in (9):

$$(M_h^n n_h)_i = -d_{i-1}^l n_{i-1} + (d_{i-1}^u + d_i^l) n_i - d_i^u n_{i+1}$$

where

$$d_i^l = d_i e^{-\psi_i}, \qquad d_i^u = d_i e^{-\psi_{i+1}}$$

are the lower and upper diagonal coefficients, respectively. This is the well known and practicable exponential fitting difference scheme (Scharfetter-Gummel, Il'in, Allen-Southwell scheme) used for partial differential equations with rapidly varying coefficients.

Now, the coarse grid matrix using (9) can be calculated. Write

$$(M_H^n n_H)_j = -D_{j-1}^l N_{j-1} + (D_{j-1}^u + D_j^l) N_j - d_j^u N_{j+1}$$

Then for $i = 2j$,

$$D_j^l = e^{-\psi_i} D_j = \frac{e^{-\psi_i}}{\frac{1}{d_i} + \frac{1}{d_{i+1}}} = \frac{1}{\frac{1}{e^{-\psi_i} d_i} + \frac{1}{e^{-\psi_i} d_{i+1}}}$$

$$= \frac{1}{\frac{1}{d_i^l} + \frac{1}{e^{\psi_{i+1}-\psi_i} d_{i+1}^l}} = \frac{1}{\frac{1}{d_i^l} + \frac{d_i^u}{d_i^l d_{i+1}^l}}$$

$$= \frac{d_i^l}{1 + \frac{d_i^u}{d_{i+1}^l}}$$

and

$$D_j^u = e^{-\psi_{i+1}} D_j = \frac{e^{-\psi_{i+1}}}{\frac{1}{d_i} + \frac{1}{d_{i+1}}} = \frac{d_{i+1}^u}{1 + \frac{d_{i+1}^l}{d_i^u}}.$$

The prolongation coefficients are calculated as follows:

$$n_{i+1} = e^{\psi_{i+1}} u_{i+1} = \frac{d_i e^{-\psi_i} n_i + d_{i+1} e^{-\psi_{i+2}} n_{i+2}}{e^{-\psi_{i+1}}(d_i + d_{i+1})} = \frac{d_i^l n_i + d_{i+1}^u n_{i+2}}{d_i^u + d_{i+1}^l}.$$

Remark that the weights of p_n do not fit into the scheme (11). The restriction r_n is the same as r_u, but it's coefficients in terms depending only on the coefficients of M_h are needed according to practical reasons.

$$(r_n v_h)_j = (1 - \alpha_{j-1}) v_{i-1} + v_i + \alpha_j v_{i+1}$$

with

$$\alpha_j = \frac{d_i}{d_i + d_{i+1}} = \frac{e^{-\psi_{i+1}} d_i}{e^{-\psi_{i+1}}(d_i + d_{i+1})} = \frac{d_i^u}{d_i^u + d_{i+1}^l}.$$

2.3 Extension to Higher Dimensions

For the discretization in higher dimensions, use the finite element spaces introduced in [3] which allow to carry over the 1D results. For simplicity, consider the two dimensional case, the generalization to three dimensions will be obvious. Let

$$\Omega = (x_0, x_{N_f^x}) \times (y_0, y_{N_f^y}) \subset R^2$$

be subdivided by the grid

$$(x_0, x_1, \ldots, x_{N_f^x}) \times (y_0, y_1, \ldots, y_{N_f^y})$$

into $N_f = N_f^x \cdot N_f^y$ elements $q_{ij} = [x_i, x_{i+1}] \times [y_j, y_{j+1}]$. Define

$$U_h := \{u \in U| \quad u|_{q_{ij}} \in Q_1(q_{ij}), i = 0 \ldots N_f^x - 1, j = 0 \ldots N_f^y - 1\}.$$

as the usual space of piecewise bilinear functions with the basis $\{\xi_{ij}\}$ defined by $\xi_{ij}(x_i, y_j) = \delta_{ij}$ In this space, u and the potential ψ will be approximated. Introduce the flux domains

$$F_{ij}^x = [x_i, x_{i+1}] \times \left\{ \begin{array}{ll} \left[y_0, \frac{y_0+y_1}{2}\right] & , j = 0 \\ \left[\frac{y_{N_f^y-1}+y_{N_f^y}}{2}, y_{N_f^y}\right] & , j = N_f^y \\ \left[\frac{y_{j-1}+y_j}{2}, \frac{y_j+y_{j+1}}{2}\right] & , \text{else} \end{array} \right\},$$

$$i = 0 \ldots N_f^x - 1, j = 0 \ldots N_f^y$$

$$F_{ij}^y = \left\{ \begin{array}{ll} \left[x_0, \frac{x_0+x_1}{2}\right] & , i = 0 \\ \left[\frac{x_{N_f^x-1}+x_{N_f^x}}{2}, x_{N_f^x}\right] & , i = N_f^x \\ \left[\frac{x_{i-1}+x_i}{2}, \frac{x_i+x_{i+1}}{2}\right] & , \text{else} \end{array} \right\} \times [y_j, y_{j+1}],$$

$$i = 0 \ldots N_f^x, j = 0 \ldots N_f^y - 1.$$

Define the flux space

$$\Phi_h := \{\phi = (\phi_x, \phi_y) \in \Phi | \quad \phi_x|_{F_{ij}^x} = \text{const and } \phi_y|_{F_{ij}^y} = \text{const } \forall i, j\}$$

as a space of piecewise constant vector functions. As a basis take the functions

$$\chi_{ij}^x = \left\{ \begin{array}{ll} (1,0) & \text{on } F_{ij}^x \\ (0,0) & \text{else} \end{array} \right. \quad \text{and} \quad \chi_{ij}^y = \left\{ \begin{array}{ll} (0,1) & \text{on } F_{ij}^y \\ (0,0) & \text{else} \end{array} \right.$$

To get the discrete operators, a special quadrature rule to calculate all occuring integrals of functions $f(x, y)$ over the domains F_{ij}^x, F_{ij}^y is introduced. Let

$$w_i^x = \left\{ \begin{array}{ll} \frac{x_1-x_0}{2} & , i = 0 \\ \frac{x_{N_f^x}-x_{N_f^x-1}}{2} & , i = N_f^x \\ \frac{x_{i+1}-x_{i-1}}{2} & , \text{else} \end{array} \right. \quad \text{and} \quad w_j^y = \left\{ \begin{array}{ll} \frac{y_1-y_0}{2} & , j = 0 \\ \frac{y_{N_f^y}-y_{N_f^y-1}}{2} & , j = N_f^y \\ \frac{y_{j+1}-y_{j-1}}{2} & , \text{else} \end{array} \right.$$

be the widths of the flux domains. Use

$$\iint_{F_{ij}^x} f(x, y)\, dxdy \approx \int_{x_i}^{x_{i+1}} f(x, y_j)\, dx \cdot w_j^y =: I_{ij}^x(f) w_j^y,$$

$$\iint_{F_{ij}^y} f(x, y)\, dxdy \approx \int_{y_j}^{y_{j+1}} f(x_i, y)\, dy \cdot w_i^x =: I_{ij}^y(f) w_i^x.$$

Now, one can calculate the discrete operators : Take $\phi_h = \sum_{ij} \phi_{ij}^x + \sum_{ij} \phi_{ij}^y$. Remark that ψ_h is linear at the edges with the node values ψ_{ij}.

$$\begin{aligned} a(\phi_h, \chi_{ij}^x) &= \phi_{ij}^x \cdot \iint_{F_{ij}^x} e^{-\psi_h(x,y)}\, dxdy \\ &\approx I_{ij}^x(e^{-\psi_h}) w_j^y = \frac{e^{-\psi_{i+1,j}} - e^{-\psi_{ij}}}{\psi_{ij} - \psi_{i+1,j}} (x_{i+1} - x_i) w_j^y, \\ a(\phi_h, \chi_{ij}^y) &= \phi_{ij}^y \cdot \iint_{F_{ij}^y} e^{-\psi_h(x,y)}\, dxdy \\ &\approx I_{ij}^y(e^{-\psi_h}) w_i^x = \frac{e^{-\psi_{i,j+1}} - e^{-\psi_{ij}}}{\psi_{ij} - \psi_{i,j+1}} (y_{j+1} - y_j) w_i^x. \end{aligned}$$

Further,

$$b(u_h, \chi_{ij}^x) = \iint_{F_{ij}^x} \frac{\partial u_h(x,y)}{\partial x}\, dxdy \approx I_{ij}^x \left(\frac{\partial u_h}{\partial x}\right) w_j^y = (u_{i+1,j} - u_{ij}) w_j^y,$$

$$b(u_h, \chi_{ij}^y) = \iint_{F_{ij}^y} \frac{\partial u_h(x,y)}{\partial y}\, dxdy \approx I_{ij}^y \left(\frac{\partial u_h}{\partial y}\right) w_j^y = (u_{i,j+1} - u_{ij}) w_i^y.$$

The resulting discrete operator A_h is invertible by static condensation. So the coefficients d_{ij}^x and d_{ij}^y can be defined:

$$\begin{aligned}\phi_{ij}^x &= (A_h^{-1} B_h u_h)_{ij}^x = d_{ij}^x (u_{i+1,j} - u_{ij}) = \frac{u_{i+1,j} - u_{ij}}{x_{i+1} - x_i} \cdot \frac{\psi_{ij} - \psi_{i+1,j}}{e^{-\psi_{i+1,j}} - e^{-\psi_{ij}}},\\ \phi_{ij}^y &= (A_h^{-1} B_h u_h)_{ij}^y = d_{ij}^y (u_{i,j+1} - u_{ij}) = \frac{u_{i,j+1} - u_{ij}}{y_{j+1} - y_j} \cdot \frac{\psi_{ij} - \psi_{i,j+1}}{e^{-\psi_{i,j+1}} - e^{-\psi_{ij}}}.\end{aligned}$$

Transposing B_h, get

$$\begin{aligned}L_h u_h &= B_h^T \phi_h\\ &= w_j^y(\phi_{i,j}^x - \phi_{i-1,j}^x) + w_i^x(\phi_{i,j}^y - \phi_{i,j-1}^y)\\ &= w_j^y d_{ij}^x (u_{i+1,j} - u_{ij}) + w_j^y d_{i-1,j}^x (u_{i-1,j} - u_{ij})\\ &\quad + w_i^x d_{ij}^y (u_{i,j+1} - u_{ij}) + w_i^x d_{i,j-1}^y (u_{i,j-1} - u_{ij}).\end{aligned}$$

This is exactly the Scharfetter-Gummel box scheme used to discretize (1) in higher dimensions. Now, define the restriction and prolongation operators. Assume the fine grid is a regularly refined coarse grid with function spaces U_H and Φ_H defined in the same manner as above with basis functions $\{\Xi_{kl}\}, \{\bar{\chi}_{kl}^*\}$ associated to the later. Remark that there is no canonical embedding $\Phi_H \hookrightarrow \Phi_h$. So define p_ϕ as an injection: For $i = 2l, k = 2j$ let

$$p_\phi(\bar{\chi}_{kl}^x) = \chi_{ij}^x + \chi_{i+1,j}^x \quad \text{and} \quad p_\phi(\bar{\chi}_{kl}^y) = \chi_{ij}^y + \chi_{i,j+1}^y.$$

Further, let

$$\begin{aligned}p_u(\Xi_{kl}) &= \phi_{ij}\\ &\quad + \alpha_{kl}^x \xi_{i+1,j} + \beta_{k-1,l}^x \xi_{i-1,j}\\ &\quad + \alpha_{kl}^y \xi_{i,j+1} + \beta_{k,l-1}^y \xi_{i,j-1}\\ &\quad + \gamma_{k-1,l-1} \xi_{i-1,j-1} + \gamma_{k,l-1} \xi_{i+1,j-1}\\ &\quad + \gamma_{k-1,l} \xi_{i-1,j+1} + \gamma_{k,l} \xi_{i+1,j+1}.\end{aligned}$$

with $\alpha_{kl}^* + \beta_{kl}^* = 1, \gamma$ defined by some other conditions. Using this, build $A_H = p_\phi^T A_h p_\phi$ and $B_H = p_\phi^T B_h p_u$

$$\begin{aligned}A_H \bar{\chi}_{kl}^x &= p_\phi^T A_h (\chi_{ij}^x + \chi_{i+1,j}^x)\\ &= p_\phi^T w_j^y \left(\frac{\chi_{ij}^x}{d_{ij}^x} + \frac{\chi_{i+1,j}^x}{d_{i+1,j}^x}\right)\\ &= w_j^y \left(\frac{1}{d_{ij}^x} + \frac{1}{d_{i+1,j}^x}\right) \bar{\chi}_{kl}^x\end{aligned}$$

and similarly,

$$A_H \bar{\chi}^y_{kl} = w^x_i \left(\frac{1}{d^y_{ij}} + \frac{1}{d^y_{i,j+1}}\right)\bar{\chi}^y_{kl}.$$

Further,

$$(p^T_\phi B_h p_u u_H)^x_{kl} = w^y_j (u_{k+1,l} - u_{kl}) \quad \text{and} \quad (p^T_j B_h p_u u_H)^y_{kl} = w^x_j (u_{k,l+1} - u_{kl})$$

holds, and the coefficients α, β are vanishing like in the onedimensional case, γ is not used at all. This yields a coarse grid operator

$$\begin{aligned}(L_H u_H)_{kl} &= w^y_j D^x_{kl}(u_{k+1,l} - u_{kl}) + w^y_j D^x_{k-1,l}(u_{k-1,l} - u_{kl}) \\ &\quad + w^x_i D^y_{kl}(u_{k,l+1} - u_{kl}) + w^x_i D^y_{k,l-1}(u_{k,l-1} - u_{kl}).\end{aligned}$$

with

$$D^x_{kl} = \frac{1}{\frac{1}{d^x_{i+1,j}} + \frac{1}{d^x_{ij}}} \quad \text{and} \quad D^y_{kl} = \frac{1}{\frac{1}{d^y_{i,j+1}} + \frac{1}{d^y_{ij}}}$$

corresponding to the onedimensional case. The coefficients α, β can be defined at the coarse grid flux elements:

$$\alpha^x_{kl} = u_i \frac{1}{1 + \frac{d^x_{i+1,j}}{d^x_{ij}}}, \quad \beta^x_{kl} = 1 - \alpha^x_{kl},$$

$$\alpha^y_{kl} = u_i \frac{1}{1 + \frac{d^y_{i,j+1}}{d^y_{ij}}}, \quad \beta^y_{kl} = 1 - \alpha^y_{kl}$$

Remark that there is no way to calculate the coarse grid cell midpoint values. In practice,the knowledge of them is not necessary because one can start smoothing with a red-black halfstep over the odd points. Correspondingly, before restricting one has to use this kind of smoothing to make the residuals in those points zero.

This will not work in three dimensions, and here is one of the outstanding questions - what is the real nature of this prolongation operator? However, from the algebraic form of the edge interpolation coefficients one gets some suggestions for an interpolation on the coarse grid cell faces. This seems to be something similar to [1].

For the n-variable, calculations as in the onedimensional case give the recipe to proceed in higher dimensions, too.

3 A Numerical Experiment

To explore the recipes described above there has been written a Multigrid Test System - a mixed C and FORTRAN code which can be used on VMS and UNIX machines. ILU smoothing and conjugated gradients acceleration routines working for both the selfadjoint and the nonselfadjoint problem using the possibility to symmetrize the nonselfadjoint problem by the diagonal transformation E_h is used. For details of this, see [4].

To test the algorithms, a problem (see Fig.1) has been considered which comes near to real world problems arising in semiconductor device simulation. In has been discretized by an

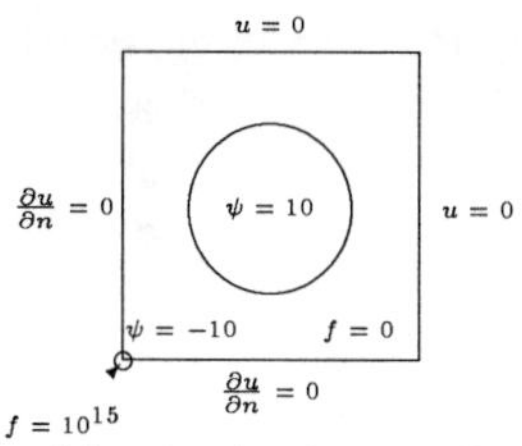

Fig.1: The 2D test problem in the domain $\Omega = [0, 10^{-3}] \times [0, 10^{-3}]$.

isotropic 65×65 grid, which isn't able to resolve the coefficient jumps. In Fig.2 one can see the convergence history for the solution of V(1,1) cycles and a 3×3 coarsest grid have been used. A 3D problem similar to the problem described above (with a high potential "ball") has

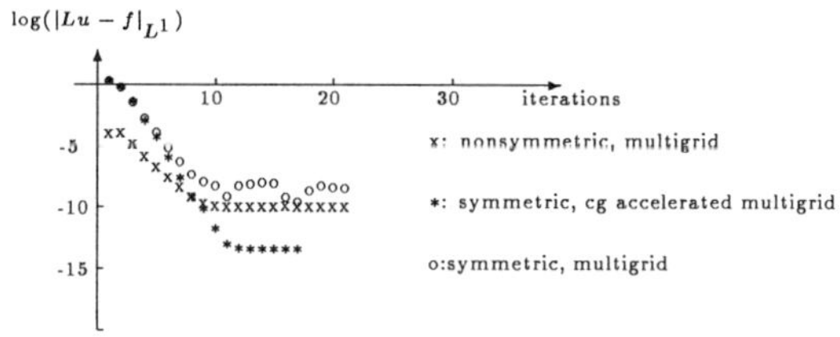

Fig.2: Convergence history for the 2D test problem on a 65×65 grid.

been considered using a coarse grid cell face interpolation by exploiting the edge interpolation idea. The convergence is shown in Fig.3.

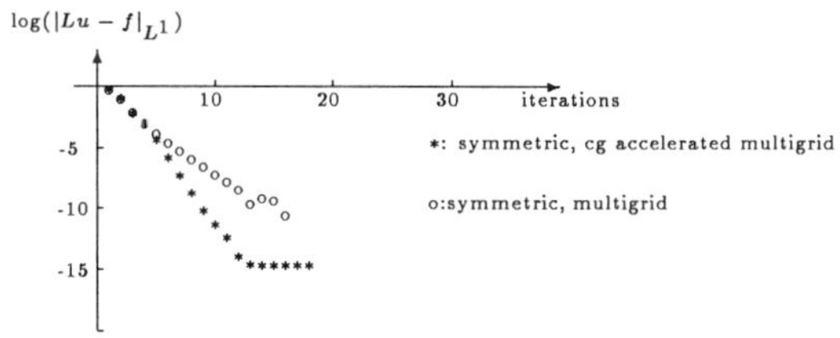

Fig.3: Convergence history for the 3D test problem on a $33 \times 33 \times 33$ grid.

References

[1] Alcouffe R.E., Brandt A., Dendy Jr J.E., Painter J.W.: The multi grid method for the diffusion equation with strongly discontinuous coefficients
Siam J. Sci. Stat. Comput. 2(1981)4,430-454

[2] Brezzi F., Marini L.D., Pietra P.: Two dimensional exponential fitting and applications to semiconductor device simulation, preprint No.579, IANCR, Pavia 1987

[3] Fuhrmann J. : An interpretation of the Scharfetter-Gummel scheme as a mixed finite element discretization in : Fourth Multigrid Seminar (ed. G.Telschow), Berlin 1990 Karl-Weierstraß-Institut für Mathematik: Report R-MATH-03/90

[4] Gärtner,K. : Iterative Algorithmen für die 3D-Bauelementesimulation, Tagungsbericht zur 4.Tagung Schaltkreisentwurf Dresden 1990

[5] Hemker P. W., Molenaar J.: An adaptive multigrid approach for the solution of the 2D semiconductor equations Proccedings of the Third European Multigrid Conference October 1 - 4,1990, Bonn, West Germany W.Hackbusch and U.Trottenberg eds, Birkhäuser Verlag

[6] Jung M., Langer U., Meyer A., Queck W., Schneider M.: Multigrid preconditioners and their applications in : Third Multigrid Seminar (ed. G.Telschow), Berlin 1989 Karl-Weierstraß-Institut für Mathematik: Report R-MATH-03/89

[7] Molenaar J.: A two grid analysis of the combination of mixed finite elements and Vanka relaxation Technical Report, CWI, Dept. NM, Amsterdam ,1990

[8] Reusken A: A multigrid method for mixed finite element discretizations of current continuity equations Eindhoven Univ. of Technology, Dept.of Math. and Computer Science RANA 90-09

[9] Ruge J.W., Stüben K.: Algebraic multigrid in: Multigrid Methods (ed. S. McCormick) Frontiers in Applied Mathematics no. 3, SIAM Philadelphia 1987

[10] Šaidurov V.V. : Mnogosetočnye metody konečnych elementov, Moskva, Nauka 1989

[11] Schmidt G.H., Jacobs F.J.: Adaptive local grid refinement and multgrid in numerical reservoir simulation. J. Comput. Phys. 77(1988),140-165

[12] Wang Song: Ph.D. Thesis, University of Dublin 1989

Parallel Multigrid Solution of 2D and 3D Anisotropic Elliptic Equations: Standard and Nonstandard Smoothing

U. Gärtel, A. Krechel, A. Niestegge, H.-J. Plum
Gesellschaft für Mathematik und Datenverarbeitung (GMD)
5205 Sankt Augustin, F.R. Germany

Abstract:
For the efficient multigrid solution of 2D and 3D elliptic partial differential equations on multiprocessor machines, parallel block relaxation methods such as line relaxation, block ILU smoothing and plane relaxation, have been implemented and applied successfully for problems with arbitrary anisotropies. High user program portability has been achieved by use of the ANL/GMD macros. Numerical results, particularly concerning multiprocessor efficiencies, are presented and discussed.

Keywords: parallel multigrid methods, 2D and 3D elliptic problems, anisotropies, block relaxation, ILU smoothing, portability, multiprocessor efficiency

1 Introduction

In recent years much progress has been made in developing multigrid methods for multiprocessor systems, especially in the German supercomputer project SUPRENUM (see for example

[14, 6, 11, 13, 16, 17]). We report here on parallel multigrid methods (developed as part of this project), which solve linear selfadjoint elliptic equations of the form

$$\begin{aligned} Lu &= f \qquad (\Omega), \\ u &= g \qquad (\partial\Omega). \end{aligned}$$

Ω denotes a 2D or 3D rectangular domain and L is a 2nd order differential operator with variable coefficients, allowing arbitrary anisotropies.

Parallel relaxation methods have been developed which have both good smoothing properties as well as a high multiprocessor performance. Whereas for isotropic problems checkered point relaxation has reasonable properties, in anisotropic cases more sophisticated methods such as line relaxation, (block) ILU smoothing or plane relaxation (in the 3D case) are worthwile. The resulting algorithms show the typical 'multigrid optimality'.

For the treatment of the tridiagonal systems arising in line relaxation a solver with high multiprocessor speedup, based on cyclic reduction, has been implemented. Plane relaxation is carried out by suitable parallel 2D multigrid methods. As to ILU relaxations, there are essentially two approaches of parallelization. The first is to parallelize the standard algorithms using e.g. 'frontal' methods, see [12]. However, these methods usually have a poor parallel efficiency due to a high amount of communication and significant idle times of particular processors. This motivates the second approach which mainly consists of a modification of the standard ILU pattern in order to get higher parallelism. We have chosen this way, although the numerical properties of the resulting smoother depend on the number of processes participating, see Section 3.3. Moreover, we have decided for the *block* ILU approach ('IBLU').

In Section 2 we describe the components of the sequential multigrid algorithms, which underly the parallel programs, and report on some convergence results. The type of (block) relaxation used depends on the kind of anisotropy of the given coefficients.

In Section 3 the parallelization techniques are explained. Following the principle of *grid partitioning*, the discrete grid is divided into subgrids, each of which (together with some overlapping region) is assigned to a particular process. Local grid operations (point relaxation, intergrid operations) are easily parallelized in this way, whereas the use of block (i.e. line, plane or IBLU) relaxations requires additional parallelization strategies.

The best possible portability was one major goal in the program design. This is achieved by the ANL/GMD macros ([2]), which are used directly in the linear algebra packages involved (i.e. in the tridiagonal system solver and in the block ILU package) and indirectly by calling routines of the SUPRENUM communications library ([5, 6]) (which is based on these macros). All mapping and communication tasks are performed by library subroutines or macros, so that machine dependent constructs are completely hidden behind them. This way, the application program can be used on every parallel system on which the ANL/GMD macros have been installed.

The numerical results presented here refer to experiments with the parallel codes on the INTEL iPSC/2-d5-VX. Although all programs were designed so that *vectorization* is possible, we did not use the VX vector facility. 3D Problems of reasonable size, making vectorization worthwile, cannot be loaded on the VX boards.

In Section 4 we outline future generalizations.

2 The Sequential Multigrid Algorithms

We investigate the 2D model problems

$$Lu = -(au_x)_x - (bu_y)_y, \tag{1}$$
$$Lu = -u_{xx} - u_{yy} + \tau u_{xy} \quad (|\tau| < 2), \tag{2}$$

and the 3D model problem [1]

$$Lu = -(au_x)_x - (bu_y)_y - (cu_z)_z. \tag{3}$$

The given problem is discretized on a grid Ω_h of uniform mesh size h by use of finite differences. This leads to grid equations

$$L_h u_h = f_h \qquad (\Omega_h), \tag{4}$$
$$u_h = g_h \qquad (\partial\Omega_h) \tag{5}$$

with a discrete operator L_h, which approximates L with consistency order h^2. (For (1) ((3)) the usual 5-point (7-point) stencil, for (2) a 7-point difference scheme is used.

We choose standard coarsening and, as to the intergrid transfer operators, linear interpolation and full weighting for equations (1) and (3), whereas 7-point operators (corresponding to the difference stencil) are used for (2).

Our emphasis lies on a careful design of the relaxation method. Error smoothing is carried out by *Gauss-Seidel relaxation* steps, performed point- or blockwise, or by *IBLU (=incomplete block LU) smoothing*, depending on the kind of anisotropy of the given coefficients. In order to achieve a good balance of efficiency and robustness, for a concrete problem the 'cheapest' among the suggested methods with sufficient smoothing properties should be chosen.

In the following we explain our choice of the smoothing method by distinguishing several typical cases for (1) – (3). (The relations '$\sim$' and '$\gg$' between the coefficients of L have the meaning 'of same order of magnitude as' and 'essentially much larger than', respectively.)

Smoothing method for (1):

1. $a \sim b$: *point relaxation*;
2. $a \gg b$: *x-line relaxation*;

[1] We did not yet integrate the treatment of mixed derivatives in our 3D codes.

3. $b \gg a$: *y-line relaxation*;

4. *general case: alternating line relaxation or IBLU smoothing*

Smoothing method for (2):

1. $|\tau| \ll 2$: *point relaxation*;

2. $|\tau| \sim 2$: *IBLU relaxation*;

Smoothing method for (3):

1. $a \sim b \sim c$: *point relaxation*;

2. $a \gg b \sim c$: *x-line relaxation*;

3. $a \sim b \gg c$: *(x, y)-plane relaxation*, performed by a 2D multigrid method, which uses point relaxation for error smoothing;

4. $a \gg b \gg c$: *(x, y)-plane relaxation*, performed by a 2D multigrid method, which uses x-line relaxation for error smoothing;

5. $a \gg b$, $a \gg c$: *(x, y)-plane relaxation*, performed by a 2D multigrid method, which uses alternating line relaxation for error smoothing;

6. *general case: alternating plane relaxation*, performed by 2D multigrid methods, which use alternating line relaxation for error smoothing.

When *plane relaxation* is used, it is sufficient to carry out each relaxation step by *one* step of a 'cheap' 2D multigrid method, described by the components: V(1,1)-cycle (i.e.: V-cycle, one relaxation step before and one after coarse grid correction), full weighting, linear interpolation, Gauss-Seidel relaxation (point- or linewise, depending on the anisotropy of the 2D problems in the planes).

The underlying idea for block Gauss-Seidel relaxations is to solve simultaneously for strongly coupled unknowns (i.e. for lines or for planes) in cases when the coefficients show special anisotropies all over the given domain Ω (see also [4], [15]). In cases when the (variable) coefficients are allowed to have changing anisotropies in different parts of Ω, alternating block methods or IBLU smoothing are used.[2]

Concrete 2D results demonstrating the high numerical efficiency of point- and (alternating) line relaxation in certain examples are well known. We want to show the success of IBLU smoothing in case of the τ problem (2); for sufficiently large $|\tau| < 2$, IBLU achieves the best multigrid efficiency when compared with point- or alternating line relaxation. In Table 1 the

Smoother	τ=1		τ=1.5		τ=1.9		τ=1.99	
	ρ	CPU time	ρ	CPU time	ρ	CPU time	ρ	CPU time
point, V(2,1)	.05	159	.1	201	.56	746	.93	7830
alt. line, V(1,1) (*)	.08	230	.15	305	.66	1245	.95	12666
IBLU, V(1,1)	.01	387	.01	354	.01	370	.01	331
(*) With alt. line relaxation, this problem could not be loaded on one process; CPU times extrapolated from the four process results, see Table 6.								

Table 1:
Different smoothers for problem (2).

CPU times (in sec.) for the solution up to some accuracy are shown, besides the convergence rates ρ of a complete cycle. The underlying grid has 192×192 points.

Table 2 contains convergence factors ρ that have been measured for concrete problems of form (3), covering the 6 cases cited above, using in each case the mentioned 'cheapest, safe' relaxation type.

Here, Ω is the 3D unit-cube and $\tilde{a}, \tilde{b}, \tilde{c}$ are polynomial coefficients of highly anisotropic behavior:

$$
\begin{aligned}
\tilde{a} &= a_1(x)a_2(y)a_3(z) \quad \text{with} \quad a_i(x) = 1 + 4(\gamma_i - 1)(x - x^2), \\
\tilde{b} &= b_1(x)b_2(y)b_3(z) \quad \text{with} \quad b_i(x) = 1 + (\gamma_i - 1)x, \\
\tilde{c} &= c_1(x)c_2(y)c_3(z) \quad \text{with} \quad c_i(x) = \gamma_i + (1 - \gamma_i)x, \\
&\quad \gamma_1 = 10, \gamma_2 = 2, \gamma_3 = 5.
\end{aligned}
$$

case	a	b	c	ρ^*	ρ
1.	1	1	1	.20	.19
2.	100	1	1	.074	.045
3.	100	100	1	.052	.030
4.	10000	100	1	.052	.0039
5.	10000	$\tilde{b}$	$\tilde{c}$		.036
6.	$\tilde{a}$	$\tilde{b}$	$\tilde{c}$		.10

Table 2:
Convergence factors for problem (3)

The numbers ρ^* are 3D two-grid-convergence factors, calculated by model problem analysis (see [15]). The numbers ρ have been computed with the described multigrid code, performing W(1,1)-cycles and using a finest grid Ω_h with $32 \times 32 \times 32$ intervals.

[2]We did not yet implement a 3D IBLU method. For certain 3D problems, it could be beneficial to use 2D IBLU within (alternating) plane relaxation.

3 The Parallel Algorithms

In the following, we use the quantities

$$S(P) := \frac{T(1)}{T(P)}, \qquad E(P) := \frac{S(P)}{P} \tag{6}$$

in order to judge the multiprocessor efficiency of a parallel algorithm. Here, $T(P)$ is the CPU time needed to solve the given problem on P processes.

$T(1)$ refers to the time needed by the parallel program (designed for general P) in the special case $P = 1$. Note, that the algorithm coming out for $P = 1$ may be quite different from the one for $P > 1$ (cf. the remark at the end of Section 3.2 and the 'Comments' in Sections 3.3 and 3.4). In such cases, S and E have to be interpreted with care.

3.1 Parallelization by Grid Partitioning

The parallelization of the (2D and 3D) multigrid algorithms follows the principle of *grid partitioning*: the discrete grid Ω_h is divided into (2-dimensional and 3-dimensional, respectively) subgrids Ω_h^i, each of which, together with some overlapping area along the artificial boundaries, is assigned to a particular process. (This leads to a natural partitioning of the coarser grids Ω_H. For a more detailed description see [6].)

The different multigrid levels are treated in a sequential order, parallelism is gained on each level within the multigrid components. This is easy for local grid operations such as point relaxation, intergrid transfer operations or residual calculations. Data exchanges are needed in order to update the overlap areas. Problems occur for non local operations such as line, plane or IBLU relaxations, see Sections 3.2 - 3.4.

Essential for the high parallelizability of Gauss-Seidel type relaxations is the fact that it is performed in a *multi-color* ordering of blocks. Within each halfstep, the ordering of the blocks (points, lines or planes) to be updated is unimportant - all blocks can be treated independently, which can be exploited to gain (asymptotically) full parallelism (also with respect to vectorizability, which is important to mention although the results presented here do not refer to that case).

We use the typical structure of a regular multigrid application on a local memory MIMD system, which is described in the following for the case that point relaxation is performed (the problem of parallelizing block relaxations is treated in Sections 3.2 - 3.4):

- The host process reads the global parameters for the algorithm and for the parallel configuration, creates the node processes and sends them the necessary initial information, including identification of the 'process neighbors', index range of their computational subdomain and parameters for the algorithm.

- **The node processes receive the initial information and start the calculation. After certain computational steps, for example, after relaxation steps and after residual restrictions to coarser grids, data are exchanged between process neighbors in order to update the overlap areas.**

- **The discrete grid is coarsened as far as possible such that all processors are kept busy, i.e. until the last grid is reached on which each process contains at least one inner point.**

- During the computation globally dependent results (like residual norms) are assembled treewise and sent to the host. This allows control over the convergence of the algorithm.

- After the computation every node sends its part of the solution to the host, where the results are assembled.

Routines of the SUPRENUM communications library ([5, 6]) are called for all process creation, mapping and communication tasks. This leads to a strict separation of the computation and the communication parts and, thus, to an easier and safer programming and to a great portability of the code (cf. the remarks at the end of Section 1).

The grid partitioning as described above restricts the number of multigrid levels if one sticks to a fixed partitioning during the application: the coarsest grid is the finer the more processes are used. This may affect the multigrid efficiency.

It is possible to avoid this situation by changing the number of active processes during the distributed application. In our codes, this facility has been integrated. It is supported by routines of the SUPRENUM communications library. These *agglomeration* (resp. *deagglomeration*) routines perform the collection of a distributed application from a given number of processes to a smaller number of processes setting some processes idle (resp. the reverse task). Of course, the collection and the redistribution of data cost communication.

3.2 Parallel Line Relaxation

Performing *line relaxation* requires the solution of (many) independent tridiagonal systems. We consider the case when *alternating line relaxation* is used as a smoother for the multigrid solution of a 2D equation (either directly for a given 2D problem or within plane relaxation for a 3D equation). Then, for at least one direction, we have the situation that the lines, i.e. the corresponding tridiagonal systems, are distributed over several processes.

Consequently, we concentrate on the following problem: given a number $J \geq 1$ of tridiagonal systems with N unknowns each, find a parallel algorithm for solving these systems on P processes assuming that process k $(1 \leq k \leq P)$ knows only a certain number N_k of equations of each of the J systems.

We briefly sketch the main ideas and refer to [7], [8], [9] for the details.

First one should note that a global re-distribution of the data (such that each process receives $m = [J/P]$ or $m+1$ *complete* systems and applies a standard solver) can only be efficient if J is large enough and if the data transfer rate of the communication system is sufficiently high in comparison with the computational speed (see e.g. [9]).

Thus we assume that the systems need to remain distributed during the solution procedure. Concretely, we assume a splitting as indicated in Fig. 1 a) (here, $N = 15$ and $P = 4$). For

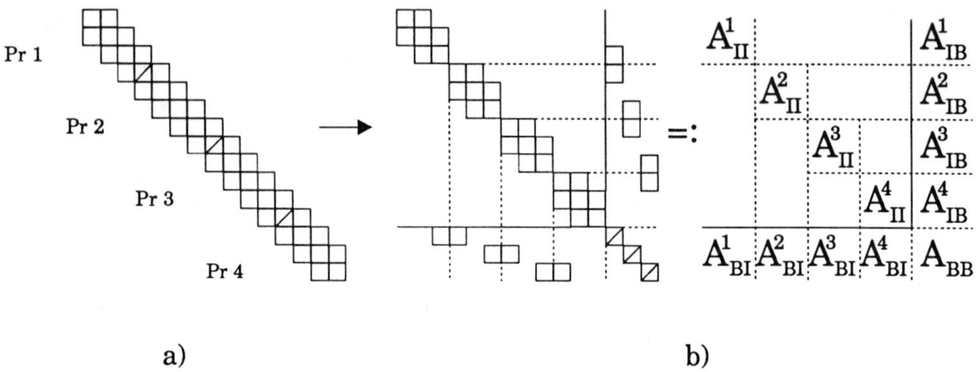

Figure 1: Splitting of a tridiagonal system

$k = 2, .., P$, the first equation of process k coincides with the last equation of process $k-1$. We call the equations shared by two neighboring processes (marked by cross-lines in Figure 1) '*interface*' or '*boundary*' equations, the remaining ones are the *inner* equations, being known to only one particular process. The parallel solver is based on the following observation. The *inner* unknowns of two *different* processes are not directly coupled to each other. Thus, elimination operations only involving *inner* equations may be performed *fully in parallel* by the different processes. Only at the end of the elimination process, the interface equations are handled, which requires communication.

This gives an idea rather than an exact formal description of the algorithm. However, it becomes clear that implicitely the unknowns are re-numbered: inner equations first, interface equations last. Explicitely, this yields a matrix as shown in Figure 1 b). Here, the decoupled 'inner' blocks can be seen clearly (denoted by A_{II}^k for process k, 'II' indicating that the couplings of inner unknowns to inner unknowns are represented). Moreover, process k holds the $(P-1 \times M_k)$ matrix A_{BI}^k (M_k = number of inner unknowns), containing the couplings of the boundary with the inner unknowns, and the transpose $A_{IB}^k = (A_{BI}^k)^T$. Only the A_{BB} part, a $(P-1 \times P-1)$ diagonal matrix, is non local. Process 1 (P) knows the first (last) entry, respectively, whereas process k contains the entries $k-1$ and k ($k = 2, .., P-1$).

Now each process can apply a standard elimination (and the corresponding backward substitution) algorithm in order to transform its 'II-block' and to eliminate the 'BI-part'. *Cyclic*

reduction **or *Gauss algorithm* are possible. Obvious modifications, see [7], of the standard (single process) methods have to be introduced, due to the elimination of the BI-blocks which do not occur on one process.**

The single process cyclic reduction (Gauss) algorithm needs ~ 17 (~ 8, respectively) operations per unknown. The advantage of cyclic reduction consists in its vectorizability, Gauss algorithm being purely sequential.

The situation is different in the parallel case, when these algorithms are applied to the *renumbered* matrix shown in Fig. 1 b). The operation count of reduction is unchanged, whereas Gauss algorithm looses its advantage. The reason is that, in contrast to the single process case, *fill in* occurs during the elimination. The fill in occuring in the BI-part of an *interior* process (i.e. a process with number k, $2 \leq k \leq P-1$) is shown in Fig. 2. $M_k - 1$ additional

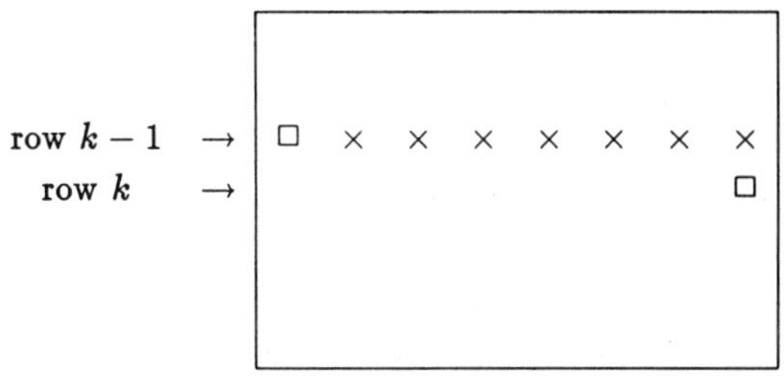

Figure 2: $(P-1 \times M_k)$ matrix A^k_{BI} with fill in locations marked by $\times$'s

eliminations are performed in this part. Note that fill in does *not* occur in process 1, and that process P can avoid fill in by eliminating in reversed (bottom to top) order. This gives a particularly efficient Gauss algorithm on 2 processes.

The most complicated component of the parallel solver is the handling of the interface (BB) part. Both cyclic reduction and Gauss algorithm produce fill in transforming A_{BB} into a *tridiagonal* matrix $\tilde{A}_{BB}$, distributed across the processes as the original A_{BB} (each process contains at most 2 rows).

Now we have reduced the original system to a small tridiagonal system for the interface unknowns. Assuming that this system has been solved and each process knows the values of its interface unknowns, the inner unknowns can be solved for by a fully parallel backsubstitution corresponding to the chosen elimination algorithm. For sufficiently large N, the size of the interface system is negligible, so that we have an *asymptotically fully parallel solver with* 17 *operations per unknown.*

Nevertheless, a careful handling of the interface system is neccessary. For an extensive investigation see e.g. [9]. To explain the most important principle, we assume that the time needed to transfer a message of l bytes between two processes is

$$t(l) = \alpha + \beta \cdot l. \tag{7}$$

α is a fixed *startup* overhead, β the transfer time for 1 byte. This (idealizing) assumption satisfactorily approximates the reality for many MIMD machines (e.g. SUPRENUM, iPSC2).

Recall that actually J ($\gg 1$ eventually) *independent* interface systems have to be solved. For each of them, a sequence of data transfers between certain processes has to be performed. Of course, the same communication structure can (and should) be applied to all of them. Then, simultaneous messages with *the same sending and receiving process* occur for each system (or, more generally, for each 'block' to be relaxed). In this situation, it is important to respect the following

> Principle:
> Simultaneous messages occuring for all of the blocks are collected and sent as a single message. In this way, the number of *startups* is kept to a minimum. (8)

Many variants of the interface phase, respecting this principle, are possible. Let us sum up the main properties valid for all of the implemented algorithms.

Given J tridiagonal systems with N unknowns each, P processes,

- the elimination and backsubstitution arithmetic solving for the $J \cdot (N - P + 1)$ *inner* unknowns (with $\sim$17 or $\sim$8 operations per unknown for $P > 2$ or $P \leq 2$, resp.), is fully parallel,
- the communication in the interface phase, based on the *spanning tree* structure, requires $O(\log_2 P)$ startups and, depending on the variant, the transfer of $O(J)$ or $O(J \cdot \log_2 P)$ real numbers,
- depending on the variant, $O(J)$ or $O(J \cdot \log_2 P)$ operations per process are needed in solving for the interface unknowns.

Concrete efficiencies of the line relaxation routines, measured on the Intel iPSC2, are documented in [9]. Their efficiency is reflected also, to some extent, by the results in Section 3 (Table 9).

Remark: Here, the problem of adequately defining and interpreting the quantities introduced in (6) becomes evident. In order to define speedup and efficiency of the line relaxation routines, one could take for $T(1)$ the CPU time of single process cyclic reduction. Then, $E(P)$ would approach 1 for increasing problem size and could be seen as *amount of parallelizability* of cyclic reduction. On the other hand, for $P = 1$ (usually) Gauss algorithm is taken (this is provided in our programs). Then, S measures the acceleration achieved by the parallel program compared with the fastest single process algorithm ('numerical speedup'), which is restricted to about $8P/17$ on P processes.

3.3 Parallel IBLU Smoothing

See [10] for more details.
We consider the 2D case of (4) which means that L_h is *block tridiagonal*, every block being a banded matrix. IBLU is based on a regular splitting

$$L_h = K_h - R_h \tag{9}$$
$$K_h = M_h D_h M_h^T, \tag{10}$$

where M_h (D_h) is block lower triangular (block diagonal, resp.) and $R_h \geq 0$. The corresponding relaxation is defined by

$$u_h^{(i+1)} = u_h^{(i)} + K_h^{-1}(f_h - L_h u_h^{(i)}). \tag{11}$$

For each such iteration one linear system with the matrix K_h has to be solved, which is easily done with help of the block Cholesky decomposition (10). It is essential to define M_h and D_h in (10) such that

- M_h and D_h are sparse and their calculation is cheap (which is not true for the *exact* (block) Cholesky decomposition),
- an application of M_h^{-1} and D_h^{-1} is cheap,
- the spectral radius of the iteration matrix $I - K_h^{-1}L_h$ is small.

A well known way (in the single process case) of achieving this is to use *incomplete block decompositions*, see e.g. [1]. The main ideas are easily explained again with help of Fig. 1, where the matrix in Fig. 1 a) now represents the whole operator L_h. Each of the □'s symbolizes a banded matrix, corresponding to a grid line. (In case of a five point discretization, the block main diagonal consists of tridiagonal matrices, whereas the off diagonal blocks are diagonal matrices).

In an *exact* block Cholesky decomposition, the *pivots* are *matrices* which have to be inverted. Since usually such inverses are full, exact inversions cannot be worthwile. The essential of an *incomplete* decomposition is to approximate these inverses by *banded* matrices (coinciding with the corresponding *exact* inverse within the band and being zero outside; for an $(m \times m)$ banded matrix, this can be achieved by an $O(m)$ algorithm, see [1]). In this way it is easy to modify the block Cholesky decomposition so that each occurring block is sparse. We call such a decomposition **standard IBLU** if no fill in is allowed: for each block of the decomposition matrices the same sparsity pattern is chosen as occurring in the corresponding block of L_h. (E.g. the D_h blocks are chosen tridiagonal.)

We now consider the parallel case. As for plain (non block) tridiagonal systems (see Section 3.2) the problem occurs that the standard block Cholesky algorithm is not parallel. Thus, we make use of the same technique as explained in Section 3.2: we apply standard IBLU to the *renumbered* matrix Fig. 1 b). Geometrically this means that we assume a grid partitioning into (vertical or horizontal) stripes. The lines separating two stripes are the *interfaces*, marked by cross-lines in Fig. 1. Now, fully in parallel and in direct analogy to the plain tridiagonal case, we can apply standard IBLU to the *inner* parts. (One could also use the incomplete block analogon of cyclic reduction which, however, is known to be unstable eventually.) The main problem is an appropriate definition of the decomposition for the 'BI' parts and for the interface system. According to the 'no fill in rule' it is natural to define *standard* IBLU so that the original matrix pattern (Fig. 1 b)) is maintained. Fill in within the BI matrices, see Fig. 2, and in the BB part is neglected (which means that it is put into the remainder matrix R_h). This makes the decomposition and the application of K_h^{-1} highly parallel. The crucial point, however, is that this parallel preconditioner K_h differs from the one coming out on a single process. Due to the fact that *more* fill in is neglected in the parallel algorithm, one intuitively expects a worse quality of the preconditioner. Neglecting fill in within the BI part, see Fig. 2, heuristically means an insufficient handling of the couplings (or, more physically, the flow of information) from the interface lines to the interior of the stripes. This should become particularly critical if these couplings are very strong.

Let us demonstrate this problem for the anisotropic operator L obtained by setting $a \equiv \epsilon$ for some constant $\epsilon > 0$, $b \equiv 1$ in (1) ('ϵ problem'). If ϵ is large and the interface lines are vertical, the above problem should occur. In Table 3 we list results obtained on the iPSC2 (without vectorization), for an example with $\epsilon = 100$ on a (192×192) grid. Parallel IBLU as explained above was used as smoother in multigrid with V(1,1) cycling, full weighting and bilinear interpolation. Agglomeration was applied, so that the coarsest grid contained one single inner point. The dependence of the convergence rates and the CPU times in seconds (for the solution up to some accuracy) on P processes is shown, for several numbers P.

P	convergence rate	CPU time
1	.03	325
2	.02	153
4	.59	640
8	.62	398
16	.71	325
32	.79	326

Table 3: Dependence on P of multigrid with standard IBLU smoothing

The expected effect is confirmed drastically. No speedup can be obtained on any number $P \geq 4$ of processes. Only the case $P = 2$ is very beneficial: here the fill in problem can be avoided when the second process performs its eliminations in reversed order, see Section 3.2.

To get more insight into what can be done, we have to analyze the remainder matrix R_h. The first fill in block in the BI part (marked by the leftmost '$\times$' in Fig. 2) is neglected by standard IBLU and thus becomes an entry of R_h. In case of the ϵ problem, this remainder block is easily calculated as

$$E_1 = \epsilon^2 D_1^{-1}, \tag{12}$$

where $D_1 = [-1\ (2+2\epsilon)\ -1]$ is the first (tridiagonal) block in the diagonal of the inner matrix. Some easy analysis shows that for fixed l the entries in the l-th diagonal (i.e. those with location (i,j), $|i-j| = l$) are $O(\epsilon^{(1-l)})$ ($\epsilon \to \infty$). In particular, the diagonal is $O(\epsilon)$, which explains the bad behavior of K_h. On the other hand, the entries outside the innermost three diagonals are small. Thus, we could include the central tridiagonal band of E_1 into the matrix pattern. (This tridiagonal choice is not arbitrary: it corresponds to the tridiagonal approximation of inverses used in the *inner* decomposition. Thus, heuristically, the remainder entries in this BI block should have the same order of size as the 'usual' remainders in the inner parts.) Inductively following the above arguments, tridiagonal blocks may be included into the matrix pattern in all the locations where fill in occurs (i.e. in the BI part, see Fig. 2 and in the first sub- and super- blockdiagonal of the BB matrix). We want to call this variant (parallel) **full pattern IBLU**. Before commenting on this algorithm somewhat more extensively, in Table 4 we show its success for the example treated in Table 3.

P	convergence rate	CPU time	speedup S	superior to standard IBLU by a factor of
1	.03	325		
2	.02	153	2.1	
4	.02	133	2.4	4.8
8	.02	75	4.3	5.3
16	.02	48	6.8	6.8
32	.02	41	8.0	8.0

Table 4:
Analogue of Table 3 with full pattern IBLU

Comments:

- The convergence rates are now independent of the number of processes.
- The speedup cannot be optimal since the total amount of arithmetic is higher in the parallel than in the single process case. Each relaxation costs about 1.6 times more than in the standard case, a complete V(1,1) cycle is approximately 1.5 times more expensive. Thus, assuming the same convergence speed as on one process, parallel full pattern IBLU should achieve a speedup of about $2 \cdot P/3$ on $P > 2$ processes. In our example, this is nearly reached for $P = 4$, where the inner domains are sufficiently large.
- The speedup compared to standard IBLU is remarkable.

- The BB system was handled by incomplete block cyclic reduction, being parallel to some extent. The stability of the algorithm was not influenced by this.
- Agglomeration was incorporated.
- We tested parallel IBLU for many different sizes of ϵ in the ϵ problem (always underlying the same (vertical) stripe partitioning). For small ϵ, yielding weak coupling of the different stripes, standard IBLU is superior to full pattern IBLU; for $\epsilon = 1$ (isotropic case), both are about equal. Thus, one could always choose the 'right' stripe decomposition in order to get weak couplings across the interfaces. However, in general we do not know much about where the anisotropies occur (e.g. for non constant, strongly varying a and b in (1)). We look upon the ϵ problem as a (hard) test of how well 'bad' anisotropies are handled.

  In all our examples, with remarkable regularity we observed the same as in Table 4: by full pattern IBLU, the single process convergence rates were re-established for all numbers of processes.

Mixed derivatives

When using parallel standard IBLU for the 'τ problem' (2), the same difficulties as for the ϵ problem arise. Remainder blocks in the BI matrices, analogous to (12), with non negligible entries occur. An additional difficulty is due to the fact that the *dominant* entries in the i-th fill in block in A_{BI} are in the $i+1$-st sub- or super-diagonal (depending on the discretization). We have to include these block locations into the IBLU pattern, with *variable* sparsity pattern for the single blocks. We have programmed the choice of these patterns such that the fill in blocks are *banded* (with some small bandwidth) and, respecting this restriction, the sum of the absolute values of all entries is maximal. (This way, the right choice comes out also in case of the ϵ problem.)

In Table 5 below we show the results obtained for $\tau = 1.9$, on a 192×192 grid. V(1,1) multigrid with IBLU smoothing was applied. The comments on Table 4 can be adopted nearly word by word. We observed the beneficial effects of full pattern IBLU for all tested sizes of τ (these effects were even the stronger, the larger $\tau < 2$).

Table 6 shows the analogue of Table 1 for $P = 4$ processes. The benefits of IBLU for $\tau \simeq 2$ are evident also in the parallel case, although the speedup is not optimal (see the comments above).

3.4 Parallel Plane Relaxation; Numerical 3D Results

Until now, the development of parallel multigrid codes has been mainly restricted to the 2-dimensional case. For the implementation of 3D codes, it may be reasonable to use (parts of) codes for corresponding 2D problems.

	Standard IBLU		Full pattern IBLU			
P	rate	CPU time	rate	CPU time	speedup S	superior to standard IBLU by a factor of
1	.01	370	.01	370		
2	.01	190	.01	190	1.9	
4	.30	377	.01	148	2.5	2.5
8	.30	199	.01	82	4.5	2.4
16	.36	127	.01	51	7.3	2.5
32	.44	110	.01	43	8.6	2.6

Table 5:
IBLU smoothing in a τ problem, $\tau = 1.9$.

Smoother	τ=1		τ=1.5		τ=1.9		τ=1.99	
	ρ	CPU time	ρ	CPU time	ρ	CPU time	ρ	CPU time
point, V(2,1)	.03	47	.12	62	.57	238	.94	2604
altern. line, V(1,1)	.08	85	.15	113	.66	461	.95	4691
IBLU, V(1,1)	.03	183	.02	170	.01	148	.01	148

Table 6:
Analogue of Table 1 for 4 processes.

In order to illustrate the efficient incorporation of a 2D multigrid code into the smoothing components of a 3D multigrid program, we explain the implementation of (x, y)-plane relaxation, making use of a parallel 2D multigrid code.

One zebra (x, y)-plane relaxation step consists of two halfsteps, each of which is followed by a data exchange in the z-direction. Every relaxation halfstep is performed by a 2D method.

The overall amount of data that has to be exchanged in the parallel 2D algorithm (within the planes) can not be decreased. However, following 'Principle' (8), we can save startups by collecting the messages for the exchanges from all planes incorporated.

In order to show the importance of this principle, we explored two different versions of parallel plane relaxation halfsteps (for more details see [3]).

1. Version

Do *not* collect messages (according to (8)) and choose the easiest possible way: for each plane to be relaxed, call a 2D multigrid solver. The communication tasks are performed for each plane separately.

2. Version

Use 'collective' messages according to (8). After each 2D relaxation (restriction, interpolation), applied simultaneously to all incorporated planes, the data to be exchanged are, as far as possible, collected (from all those planes) and sent in a single message.

Using the second version instead of the first one leads to a considerably increased multiprocessor efficiency of the 3D multigrid solver on machines with a relative high startup time. With help of assumption (7), it is straightforward to roughly estimate the multiprocessor efficiencies of both versions. In Table 7 this has been done by inserting for α, β the values relevant for the SUPRENUM machine. The increase of multiprocessor efficiency obtained by switching from the first to the second version is clearly reflected. In our multigrid package the second version has been realized.

Size of Finest Grid	Processes	Version 1. E (in %)	Version 2. E (in %)
$32 \times 32 \times 32$	$4 \times 4 \times 4$	18	50
$64 \times 64 \times 64$	$4 \times 4 \times 4$	41	77

Table 7:
Multiprocessor efficiency E (obtained by simulation), relaxation type: 3D alternating plane/ 2D alternating line.

The Tables 8 – 10 contain numerical results obtained with the 3D multigrid program for different grid sizes and different process numbers on the iPSC2 (without vectorization). We measured the times needed to perform one V(1,1)-Cycle, using three grid levels and ten relaxation steps on the coarsest grid. (Agglomeration was not used in these examples.)

Due to the memory restrictions on the iPSC2, the program could not be run on a single process for the grid sizes considered. Hence, we estimated $T(1)$ (defined at the beginning of Section 3) and the multiprocessor efficency E by extrapolating the single process computational time from the solution time spent on eight processes.

Comments:

- It is important to remark that the numerical results in the Tables 8-10 merely give a measurement of the *efficiency of parallelization* of the algorithms, not including an evaluation of the *numerical efficiency* of the different methods. Given a concrete differential equation, an appropriate parallel relaxation method can basically be chosen following the same criteria as in the sequential case (cf. Section 2).

- The tables show a satisfactory efficiency of parallelization for reasonable ratios 'problem size/processor number'. The results obtained for the $64 \times 64 \times 8$ grid problem give an impression of the efficiencies that can be expected on the $64 \times 64 \times 64$ grid.

Size of Finest Grid	Processes	$T(1)$ (in sec)	E (in %)
$32 \times 32 \times 32$	$2 \times 2 \times 2$	19.458	60.3
	$4 \times 2 \times 2$		43.1
	$4 \times 4 \times 2$		30.7
$64 \times 64 \times 8$	$4 \times 2 \times 1$	17.887	70.8
	$8 \times 2 \times 1$		55.4
	$8 \times 4 \times 1$		43.6

Table 8:
Multiprocessor efficiency, relaxation type: 3D point.

Size of Finest Grid	Processes	$T(1)$ (in sec)	E (in %)
$32 \times 32 \times 32$	$2 \times 2 \times 2$	27.603	63.3
	$2 \times 4 \times 2$		49.4
	$4 \times 2 \times 2$		40.9
	$2 \times 4 \times 4$		37.3
	$4 \times 4 \times 2$		30.0
$64 \times 64 \times 8$	$2 \times 4 \times 1$	25.431	75.5
	$4 \times 2 \times 1$		63.3
	$2 \times 8 \times 1$		62.2
	$8 \times 2 \times 1$		48.6
	$4 \times 8 \times 1$		40.9

Table 9:
Multiprocessor efficiency, relaxation type: 3D x-line.

- A comparison of point and line relaxation shows: as long as the lines to be relaxed are distributed over $P \leq 2$ processes, line relaxation is more expensive but has a better multiprocessor efficiency (when the same or an equivalent grid partitioning is underlying). The reason is that line relaxation requires more arithmetical work but not significantly more communication. The situation is different when the lines belong to $P \geq 4$ processes. In this case, line relaxation has lower multiprocessor efficiency than point relaxation. This is explained by the remark end of Section 3.2.

- The degree of parallelism is lower for plane relaxation than for point relaxation. This reflects the fact that plane relaxation requires more 'very coarse grid visits'. Visits on very coarse grids decrease the calculation/communication ratio, because the volume/surface ratios of the subgrids are smaller on coarser grids. In *every* plane relaxation step *all* coarser grids are visited – independent of the grid level one has started on.

Size of Finest Grid	Processes	$T(1)$ (in sec)	E (in %)
$32 \times 32 \times 32$	$2 \times 2 \times 2$	49.831	55.6
	$2 \times 2 \times 4$		44.7
	$4 \times 2 \times 2$		32.2
	$4 \times 2 \times 4$		23.8
$64 \times 64 \times 8$	$4 \times 2 \times 1$	43.912	61.1
	$8 \times 2 \times 1$		42.5
	$8 \times 4 \times 1$		27.8

Table 10:
Multiprocessor efficiency, relaxation type: 3D (x,y)-plane/2D point

4 Future Generalizations

The described parallel multigrid codes will be developed further:

1. Their vectorizability will be exploited on the SUPRENUM machine where more memory is available that on the VX boards of the iPSC2.
2. They will be extended for the solution of more general elliptic equations.
3. IBLU smoothing will be tested in 3D (cf. footnotes in Section 2).

References

[1] A. Axelsson: *Incomplete block preconditioning - the ultimate answer?* J. Comp. Appl. Math. 12/13 (1985), 3-18.

[2] L. Bomans, D. Roose and R. Hempel: *The Argonne/GMD macros in FORTRAN for portable parallel programming and their implementation on the Intel iPSC/2*, Parallel Computing 15 (1990), 119-132

[3] U. Gärtel: *Parallel multigrid solver for 3D anisotropic elliptic problems*, Proceedings of the First European Workshop on Hypercube and Distributed Computers, Rennes 1989, F.Andre, J.P.Verjus (eds.), North-Holland, 1989.

[4] W. Hackbusch: *Multi–Grid Methods and Applications*, Springer Series in Computational Mathematics, 4, Springer–Verlag, Berlin, 1980

[5] R. Hempel: *The SUPRENUM Communications Subroutine Library for grid-oriented problems*, Argonne National Laboratory Report ANL-87-23, Argonne, 1987

[6] R. Hempel, A. Schüller: *Experiments with parallel multigrid algorithms using the SUPRENUM communications subroutine library*, GMD-Studien 141, GMD St. Augustin, 1988

[7] A. Krechel, H.J. Plum, K. Stüben: *Solving tridiagonal linear systems in parallel on local memory MIMD machines*, Arbeitspapiere der GMD 372, GMD St. Augustin, 1989.

[8] A. Krechel, H.J. Plum, K. Stüben: *Parallel solution of tridiagonal linear systems*, Proceedings of the First European Workshop on Hypercube and Distributed Computers, Rennes 1989, F.Andre, J.P.Verjus (eds.), North-Holland, 1989.

[9] A. Krechel, H.J. Plum, K. Stüben: *Parallelization and vectorization aspects of the solution of tridiagonal linear systems*, Parallel Computing 14 (1990), 31-49.

[10] Krechel, A., Plum, H.J.: *A parallel block preconditioned conjugate gradient method*, Arbeitspapiere der GMD 459, GMD St.Augustin, 1990,

To appear in: *Notes on Numerical Fluid Mechanics*, Vieweg (Proceedings of the GAMM Seminar Kiel, 1990).

[11] J. Linden, B. Steckel and K. Stüben: *Parallel multigrid solution of the Navier-Stokes equations on general 2D-domains*, Arbeitspapiere der GMD 294, GMD St. Augustin, 1988

[12] Y. Saad, M. H. Schultz: *Parallel implementations of preconditioned conjugate gradient methods* Research Report YALEU/DCS/RR-425, Yale University, 1985

[13] K. Solchenbach, C.-A. Thole and U. Trottenberg: Parallel multigrid methods: Implementation on local memory multiprocessors and applications to fluid dynamics, Arbeitspapiere der GMD 264, GMD St. Augustin, 1987

[14] K. Stüben: *Parallel multigrid for general, block-structured grids*, to appear

[15] C.-A. Thole, U. Trottenberg: *Basic smoothing procedures for the multigrid treatment of elliptic 3D-operators*, Arbeitspapiere der GMD 141, GMD St. Augustin, 1985

[16] C.-A. Thole: *A short note on standard parallel multigrid algorithms for 3D-problems*, SUPRENUM Report 3, SUPRENUM GmbH, Bonn, 1987

[17] U. Trottenberg: *SUPRENUM – a MIMD system for multilevel scientific supercomputing*, SUPRENUM Report 2, Bonn, 1987

Parallel Multigrid Methods on Sparse Grids

Michael Griebel
Technische Universität München
Institut für Informatik
Arcisstr 21, Postfach 20 24 20, D-8000 München

Abstract

This paper deals with multigrid methods on two-dimensional sparse grids and their vectorization. First we will introduce sparse grids and discuss briefly their properties. Sparse grids contain only O(n ld n) grid points in contrast to the usually used $O(n^2)$-grids whereas, for a sufficiently smooth function, the accuracy of the representation is only slightly deteriorated from $O(n^{-2})$ to $O(n^{-2}$ ld n). We sketch the main features of a multigrid method that works on these sparse grids and discuss its vectorization and parallelization aspects. Additionally, we present the results of numerical experiments for an implementation of this algorithm on the CRAY-Y-MP.

Sparse grids: Properties and discretization

Let us first address the question: What is a sparse grid? This can be answered best by the simple example on the unit square shown in figure 1.

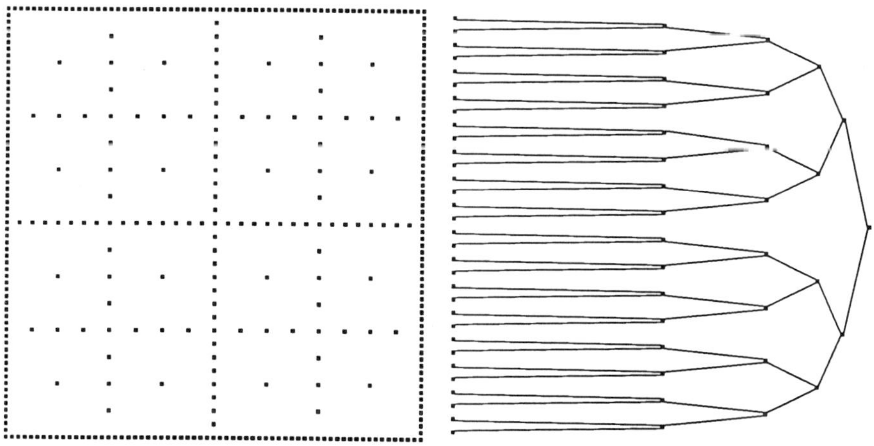

Figure 1: The sparse grid $\Omega^s_{6,6}$ and its tree structure.

We see here the sparse grid $\Omega^s_{6,6}$ with boundary grid size $h = 2^{-6}$. Despite of its sparsity, this grid contains some sort of structure that can be described by a binary tree. Every node of the tree is related to a grid line in x-direction. The level of a node corresponds to the number of points on the associated grid line. Boundary grid lines have to be treated separately. Of coarse, every grid line itself can be represented by a binary tree where every node corresponds to a sparse grid point. If we sum the grid points of the tree of grid lines, we see directly the first property of two dimensional sparse grids:

Property 1: The number of sparse grid points is only of the order O(n ld n) in contrast to $O(n^2)$ for the usually used full grids.

It can be shown (see [ZEN90]) that sufficiently smooth functions are represented on sparse grids with nearly the same accuracy as on full grids. For that purpose, we introduce a special hierarchical basis for the FE-space of the full grid $\Omega_{k,k}$.

Let $\Omega_{i,j}$ be the equidistant rectangular grid on the unit square $\Omega = [0,1]\times[0,1]$ with mesh width $h_i = 2^{-i}$ in x- and $h_j = 2^{-j}$ in y-direction. Let moreover $S_{i,j}$ be the space of piecewise bilinear functions on grid $\Omega_{i,j}$. To simplify the discussion we assume that all functions vanish on the boundary.

The finite element space $S_{i,j}$ of piecewise bilinear functions on grid $\Omega_{i,j}$ that satisfy the boundary condition can be decomposed by $S_{i,j} = \sum_{s=1}^{i} \sum_{t=1}^{j} T_{s,t}$, where $T_{s,t}$ denotes the subspace of $S_{s,t}$ vanishing on all grid points corresponding to the grids of $S_{s-1,t}$ and $S_{s,t-1}$. We can introduce bilinear basis functions in $T_{s,t}$ that are uniquely described by their non-overlapping rectangular support of size $1/2^{s-1}$ in x- and $1/2^{t-1}$ in y-direction. This results in a special hierarchical basis. The supports of the first few $T_{s,t}$ can be seen in figure 2.

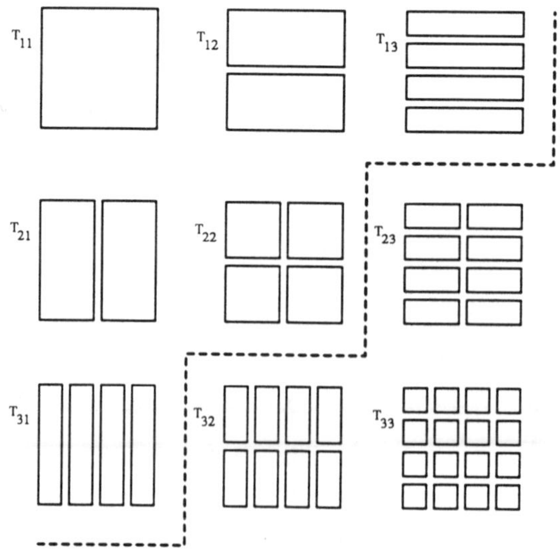

Figure 2: The supports of the first few $T_{s,t}$.

Then, every function $u \in S_{i,j}$ is represented by

$$u = \sum_{s=1}^{i} \sum_{t=1}^{j} u_{s,t}\,, \qquad \text{where } u_{s,t} \in T_{s,t},\ s=1,..,i,\ t=1,..,j.$$

Define now $|u| = \| \frac{\partial^4 u}{\partial_x^2 \partial_y^2} \|_\infty$. We learned in [ZEN90] (see (2.5)) that for the coefficients $u_{s,t}$ the estimation

$$\| u_{s,t} \|_\infty \le 4^{-s-t-1}\ |u|$$

is valid if the function u is sufficiently smooth so that $|u|$ is bounded.

Now we consider the full grid space $S_{k,k}$. Here, the subspace $T_{k,k}$ with a quite big dimension 2^{2k-2} and a rather small contribution to the error of magnitude $4^{-2k-1}|u|$ is included, but, for example, the space $T_{1,k+1}$ with relatively small dimension 2^k and a big contribution to the error of magnitude $4^{-k-3}|u|$ is not contained.

It would be more advantageous to take into account only those subspaces where the contribution to the error is equal or larger than some prescribed tolerance and to omit the rest. This leads to a triangular scheme of subspaces $T_{i,j}$ with $i+j \le k+1$ (see dashed line in figure 2), and results in an approximation space $S^s_{k,k}$ that corresponds to the sparse grid $\Omega^s_{k,k}$ instead of the full grid $\Omega_{k,k}$.

For the sparse grid space $S^s_{k,k}$, it can be proved that the L_∞- norm of the interpolation error is bounded by $\frac{1}{48} h^2 (\text{ld } h^{-1} + \frac{4}{3}))\ |u|$. For a complete proof see [ZEN90]. This gives directly the following property 2.

Property 2: If we represent a sufficiently smooth function on a sparse grid, then the order of accuracy deteriorates from the usual order $O(n^{-2})$ for the full grid representation only slightly to the order $O(n^{-2}\ \text{ld } n)$.

The features of the HB-Block-MG-algorithm

In the following we present a multigrid algorithm for the solution of Poisson problems, that works on sparse grids. This algorithm has a convergence rate which is nearly independent of the grid size like standard multigrid methods on full grids (see the experiments in [GRI90/a]). Due to the sparse grid approach, it has in practice an overall operation count of only $O(n\ \text{ld } n)$. Additionally, this algorithm is perfectly suited for parallelization and vectorization.

To some extend, the algorithm generalizes the ideas of the hierarchical basis multigrid method that is due to Yserenant and Bank [BAN88], and therefore we call it

Hierachical-Basis-Block-Multigrid algorithm (HB-Block-MG).

Before we sketch the main features of the algorithm, let us first mention another important property of sparse grids that concerns its implementation. A sparse grid $\Omega^s_{k,k}$ with boundary grid size $h = 2^{-k}$ contains the standard grids $\Omega_{i,j}$, i+j=k, with the mesh sizes $h_x = 2^{-i}$ and $h_y = 2^{-j}$ in the x- and y-direction. This is shown in figure 3.

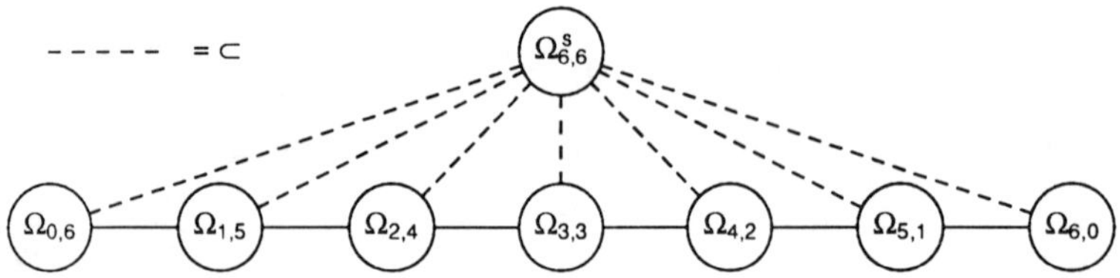

Figure 3: The sparse grid $\Omega^s_{6,6}$ as the union of a sequence of full grids.

This allows simple arrays as data structures for sparse grids, and will be exploited in a vector implementation of our algorithm.

Assume now for simplicity that k is even and k = 2D. Then we see that especially the full and uniform grid $\Omega_{D,D}$ with grid size $H = 2^{-D} = \sqrt{h}$ is contained. This is shown in figure 4. The separation of the sparse grid points into the points of $\Omega_{D,D}$ and the set of the remaining grid points is the basic idea of our algorithm. This approach works analogously in the case of an odd value for k.

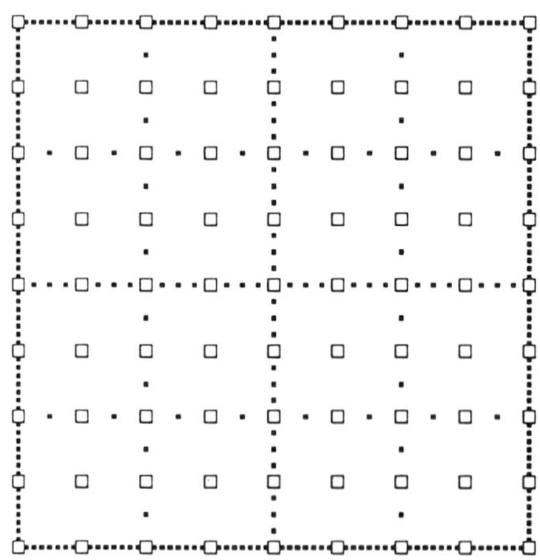

Figure 4: The sparse grid $\Omega^s_{6,6}$ (• and □) and its full uniform subgrid $\Omega_{3,3}$ (□), $h = 2^{-6}$.

For the coarse grid problem on $\Omega_{D,D}$, we use a standard MG-algorithm with red-black Gauss-Seidel relaxation as smoothing procedure.

The remaining grid points belong to 2^D stripe subdomains for the x-direction and 2^D stripe subdomains for the y-direction that are mutually decoupled and independent, i.e., the interior of the intersection of two different subdomains contains no grid points at all. See figure 5. Thus, the problems on the different stripe subproblems can be treated fully in parallel.

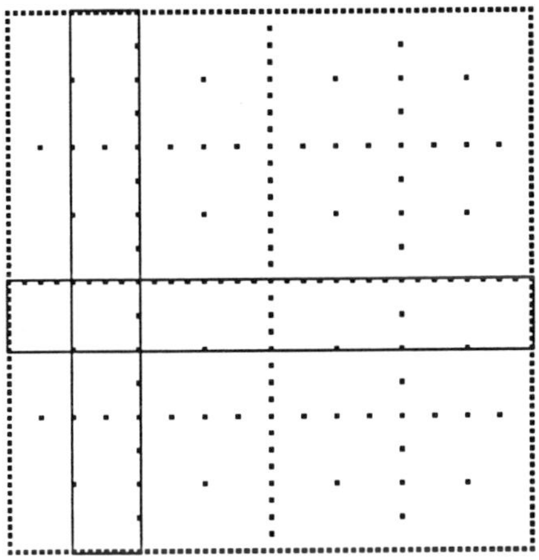

Figure 5: Stripe subdomains in x- and y-direction.

The local grids on the subdomains can be regarded as sparse grids with 2^D+1 points on the boundary, where the x- or y-axis is scaled by a factor 2^{-D}. See figure 6.

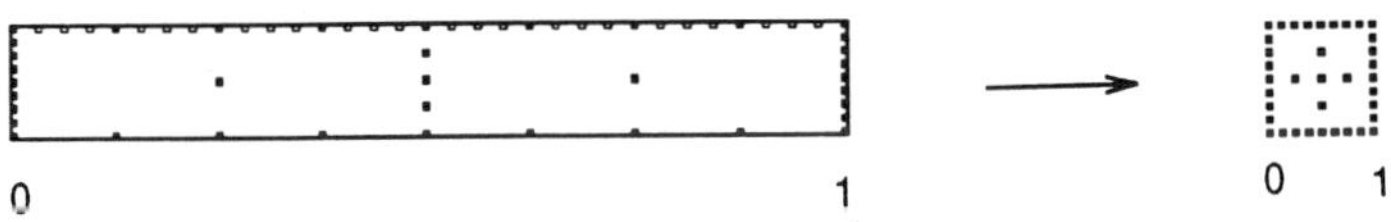

Figure 6: The sparse grid of a subdomain and scaling.

However on the subdomains, the finite element stencils for the Laplace operator show a strong coupling in one direction and a weak coupling in the other direction. Therefore, we get nearly one-dimensional stencil operators on the stripes. Here we use a semicoarsening approach, i.e. we subdivide the stripes and their sparse grids recursively into still thinner stripes.

This approach assigns the grid points that might belong to several full grids of figure 3 uniquely to their specific coarsest level with respect to the partial ordering of the grids in figure 7. Each interior point of the stripe subdomains will be relaxed only on its assigned level according to the HB-MG-method of Yserentant and Bank [BAN88]. Thus we get the (nearly) exact solution of the stripe problems already after one cycle.

All together, we have now a two-level method: We have a root level with grid $\Omega_{D,D}$ where we use a standard MG-algorithm. And we have an additional level with all remaining grid points that belong to the stripe subdomains. Here we use for the nearly one-dimensional problems a semicoarsening approach together with the HB-MG-algorithm that has been extended to sparse grid problems.

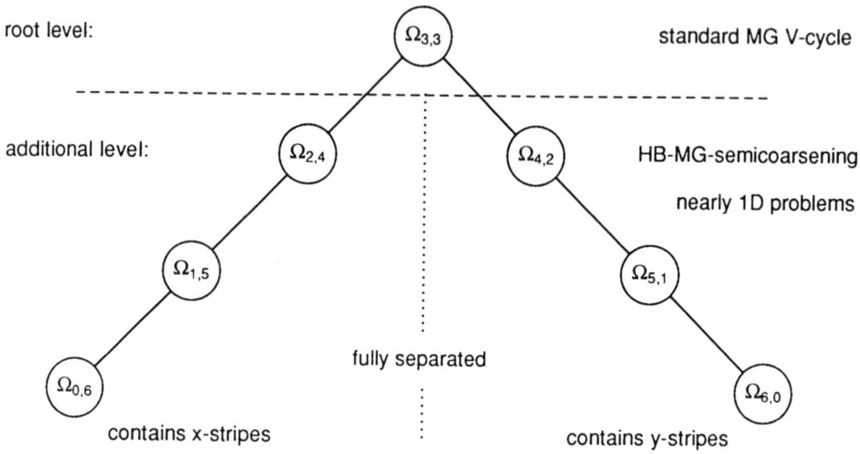

Figure 7: Partial ordered sequence of full grids.

The whole algorithm has been implemented in a way that it runs on the partial ordered sequence of full grids $\Omega_{i,j}$, i+j = k of figure 7. The values at coarser grid points serve as Dirichlet conditions for the finer level points. The accumulation of the right hand side of the root level problem and the coupling of the two levels is done according to the hierarchical basis method. All together, we need just simple arrays as data structures, and this allows efficient vectorization. For details of the implementation and the complete code see [GRI90/b].

Numerical experiments

In the following we present the results of numerical examples using a vectorized version of the HB-Block-MG algorithm. We run the code on the CRAY-Y-MP4/432 of the Leibniz Rechenzentrum in Munich. This machine has 8 vector registers of 64 words per CPU and a cycle time of 6 nano seconds.

As we are mainly interested in the vectorization performance and not in auto-tasking questions, we used only one CPU. The maximal mega flop rate that can be theoretically reached if an addition is always followed by a multiplication and no subroutine calls occur is 333 MFlops/sec. Practically however, a rate of 333/2 ≅ 166 mega flops is fairly good. The maximal vector length that can be achieved on our machine is 64.

We applied the options -o vector and -o novector of the cft77 compiler to produce vectorized and sequential code. In the following experiments we used 10 overall cycles of the new two level algorithm. On its root level we employed a standard (1,1)-V-cycle. 10 cycles are sufficient to reduce the error to the machine accuracy. The performance of the code was measured by the utility PERFTRACE with an accuracy of 95 %. As Dirichlet model problem we solved $\Delta u = f$ with the exact solution $u = \sin(\pi x)\sinh(\pi y) / \sinh(\pi)$.

The following table 1 and the figure 8 show the behavior of the execution time.

	D=	2	3	4	5	6	7	8	9
	bound.-h=	1/16	1/64	1/256	1/1024	1/4096	1/16384	1/65536	1/262144
add	vec	7.99E-5	5.297E-4	2.26E-3	1.024E-2	4.71E-2	2.37E-1	1.07E+0	4.99E+0
	seq	2.49E-4	2.83E-3	1.85E-2	1.02E-1	5.18E-1	2.52E+0	1.19E+1	5.46E+1
root	vec	9.38E-4	2.61E-3	5.89E-3	1.31E-2	3.06E-2	8.44E-2	2.684E-1	9.87E-1
	seq	9.13E-4	3.37E-3	1.18E-2	4.39E-2	1.70E-1	6.68E-1	2.65E+0	1.06E+1
alg	vec	1.02E-3	3.14E-3	8.15E-3	2.33E-2	7.77E-2	3.21E-1	1.34E+0	5.98E+0
	seq	1.16E-3	6.20E-3	3.04E-2	1.46E-1	6.88E-1	3.19E+0	1.45E+1	6.52E+1

Table 1: Time (sec.) for 10 cycles of the algorithm on the CRAY-Y-MP using one CPU.

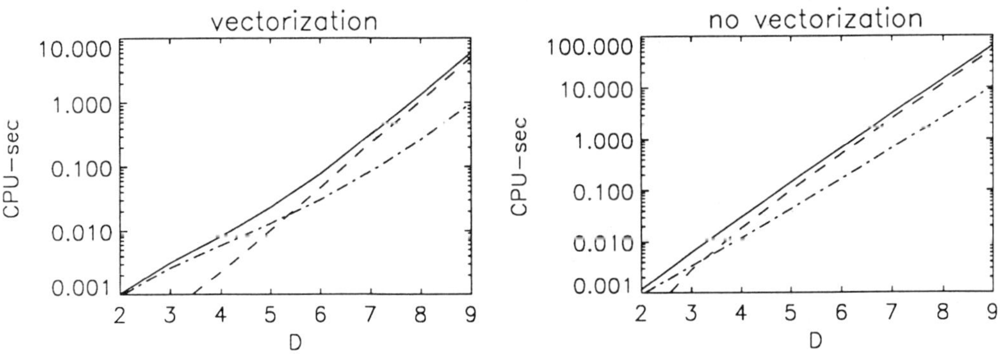

Figure 8: Time for 10 cycles of the algorithm on the CRAY-Y-MP using one CPU.

Here and in the following, we denote the part of the algorithm that runs on the additional level of the remaining grid points by *add*, and indicate it by a dashed line. The part that runs on the root level is denoted by *root* and indicated by a dash pointed line. The whole algorithm is denoted by *alg* and indicated by a bold line.

As the parameter D is coupled by $h = 2^{-2D}$ with the grid size h, our table contains only every second possible grid. But the cases of a grid size with $h = 2^{-2D-1}$ behave similar. Note that it was not difficult to compute even problems with grid sizes of $h = 2^{-18}$ in moderate time. In general, this is impossible for full grid problems because of storage limitations.

Using the vectorized code, we measured for the grid size h = 2^{-8} 0.0771 CPU-seconds, and for h = 2^{-10} we get 0.321 CPU-seconds. We see that our algorithm is an extremely fast solver. This is basically due to the reduced number of unknowns respectively grid points of only O(n log n) for the sparse grid and to the very good vectorization properties of our algorithm.

The following tables 2 and 3 and the figure 9 show the achieved speedup and the achieved average vector length of our algorithm.

D=	2	3	4	5	6	7	8	9
bound.-h=	1/16	1/64	1/256	1/1024	1/4096	1/16384	1/65536	1/262144
add	3.11	5.35	8.20	9.97	10.99	10.63	11.10	10.94
root	0.97	1.29	2.01	3.37	5.55	7.92	9.90	10.72
alg	1.14	1.98	3.73	6.27	8.85	9.92	10.86	10.91

Table 2: Achieved speedup.

D=	2	3	4	5	6	7	8	9
bound.-h=	1/16	1/64	1/256	1/1024	1/4096	1/16384	1/65536	1/262144
add	2.17	8.52	22.27	47.61	62.01	63.92	63.97	63.99
root	1.43	2.80	5.54	10.97	21.76	42.51	56.11	61.37
alg	1.74	4.81	11.27	23.66	42.81	57.76	62.33	63.53

Table 3: Achieved average vector length.

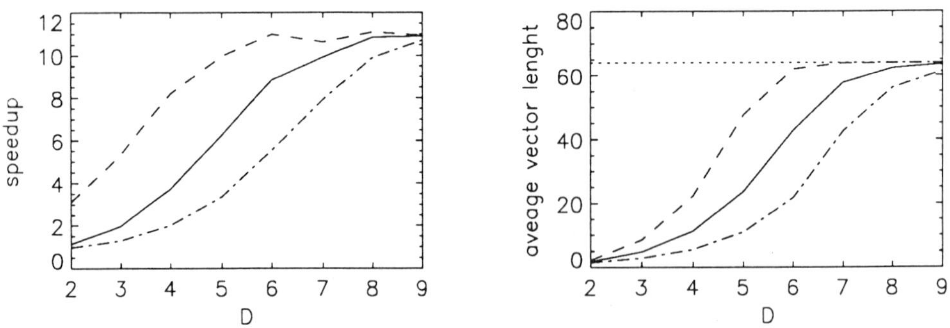

Figure 9: Achieved speedup and average vector length.

The speedup is approaching the factor 11 which is a fairly good result on the CRAY. Note that the speedup for the additional level reaches its maximum very fast. Already for the case of h = 2^{-8}, we get here a value of 8.2, and for h = 2^{-10} we get a value of 9.97. The speedup for the root level part of the algorithm reaches slower its maximum. But still the overall algorithm shows a very good performance.

The same behavior can be seen for the achieved average vector length. The vector length is approaching the optimal value 64. For the additional level this value is reached very fast. Already for $h = 2^{-8}$, we get here a vector length of 22.27, and for $h = 2^{-10}$ we have 47.61. The vector length for the root level is naturally smaller because we face here only small problems. But the overall algorithm still shows a very good average vector length.

The following tables 4 and 5 and the figure 10 show the number of flops of the algorithm and the rate of mega flops per second.

D=	2	3	4	5	6	7	8	9
bound.-h=	1/16	1/64	1/256	1/1024	1/4096	1/16384	1/65536	1/262144
add	3.37E+3	2.90E+4	2.45E+5	1.44E+6	7.51E+6	3.68E+7	1.74E+8	8.08E+8
root	6.84E+3	2.64E+4	1.29E+5	5.81E+5	2.44E+6	9.97E+6	4.03E+7	1.62E+8
alg	1.02E+4	5.54E+4	3.74E+5	2.02E+6	9.95E+6	4.68E+7	2.15E+8	9.66E+8

Table 4: Number of FLOPS, sequential code has nearly the same numbers.

	D=	2	3	4	5	6	7	8	9
	bound.-h=	1/16	1/64	1/256	1/1024	1/4096	1/16384	1/65536	1/262144
add	vec	42.27	54.73	108.37	140.92	159.35	155.26	163.30	161.92
	no vec	13.27	10.07	13.05	14.00	14.38	14.51	14.63	14.67
root	vec	7.29	10.13	21.97	44.51	79.84	118.16	150.44	163.83
	no vec	7.49	7.84	10.93	13.22	14.40	14.93	15.19	15.28
alg	vec	10.03	17.66	45.91	86.90	128.07	145.50	160.63	161.57
	no vec	8.73	8.88	12.22	13.75	14.39	14.60	14.74	14.77

Table 5: MFLOPS/sec.

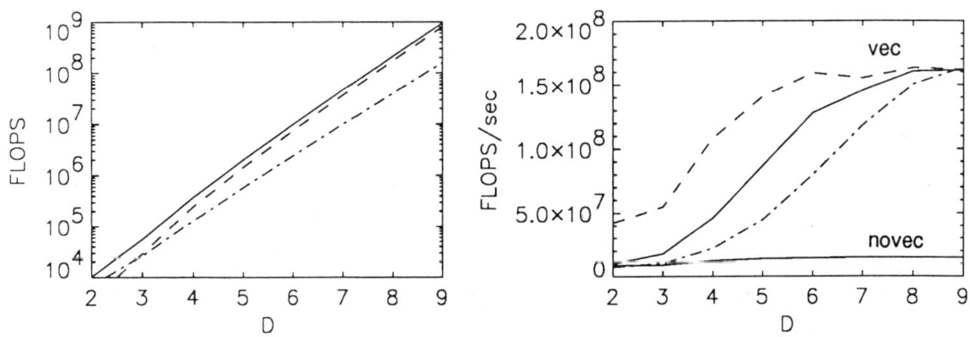

Figure 10: Number of FLOPS and MFLOPS/sec.

The mega flop rate of the vectorized code shows the same behavior as the vector length. It is approaching a value of 161 which is nearly optimal. For the additional level this value is reached very fast. Already for $h = 2^{-8}$, we get here a mega flop rate of 108.37 and for $h = 2^{-10}$ we have 140.92. The rate for the root level is naturally smaller, but the overall algorithm still has a very good mega flop rate.

From these results we see the good vectorization properties of our algorithm.

In the last figure 11 we compare our sparse grid algorithm with a standard MG-code that runs on the full $O(h^2)$-grids. Here we plot the accuracy of the computed discrete solution versus the CPU-time needed to execute 10 cycles. The figure shows the results for sequential and vectorized codes.

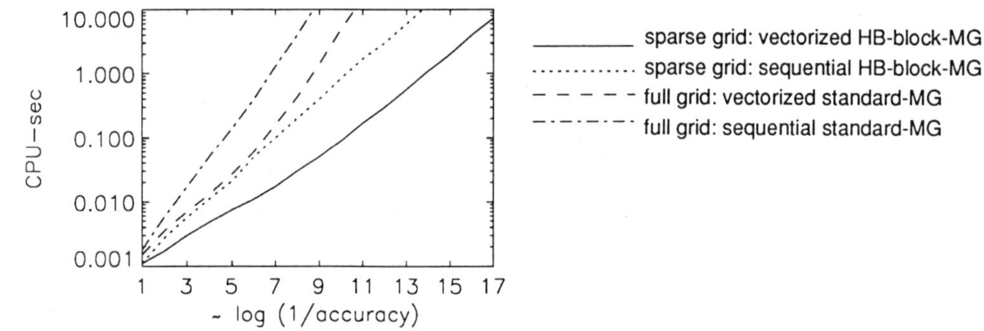

Figure 11: Comparison of the sparse grid algorithm with a standard MG-algorithm.

We see clearly that we need substantially less time for the sparse grid algorithm than for the standard full grid algorithm to obtain a solution with a prescribed accuracy. For a fixed time we get a substantially more accurate solution with a (much finer) sparse grid.

This shows once more the superiority of the sparse grid approach. The accuracy of the computed solution on the full $O(n^2)$-grid is of the order $O(n^{-2})$. But the accuracy of the computed solution on the sparse $O(n \text{ ld } n)$-grid is of the order $O(n^{-2} \text{ ld } n)$.

Concluding remarks

We presented a really fast algorithm for the solution of two-dimensional Poisson problems on sparse grids. It is superior to any MG-method for the standard grid because it needs only $O(n \text{ ld } n)$ operations in contrast to $O(n^2)$ of the full grid approach. The accuracy of the solution deteriorates only slightly from $O(n^{-2})$ to $O(n^{-2} \text{ ld } n)$. Despite of the used sparse grid, the algorithm is very well suited for parallelization and especially vectorization.

For three-dimensional problems sparse grids can be derived in a similar way. It can be shown that then the number of grid points is only of the order $O(n\,(\mathrm{ld}\,n)^2)$ instead of $O(n^3)$ for the usual full grids. The accuracy deteriorates only slightly from $O(n^{-2})$ to $O(n^{-2}\,(\mathrm{ld}\,n)^2)$. For Poisson's problem on the unit cube, there exists already a sequential, fully adaptive HB-MG-code written by H. Bungartz. Details will be published in a future paper.

Beyond adaptivity, even the resolution of curved domain boundaries has been treated successfully. But the combination of adaptivity and vectorization or parallelization is still an open question for future research.

References

[BAN88] Bank R., Dupont T., Yserentant H., "The Hierarchical Basis Multigrid Method", Numer. Math. No. 52, p. 427-458, 1988.

[GRI90/a] Griebel M., "A Parallelizable and Vectorizable Multi-Level Algorithm on Sparse Grids", Proc. Conf. GAMM-Workshop, Kiel 1990, Notes on Numerical Fluid Mechanics, Vieweg-Verlag, 1990.

[GRI90/b] Griebel M., "Parallel Multigrid Methods on Sparse Grids", to appear as technical report, TU München, in the series TUM INFO.

[ZEN90] Zenger C., "Sparse grids", Proc. Conf. GAMM-Workshop, Kiel 1990, Notes on Numerical Fluid Mechanics, Vieweg-Verlag, 1990.

International Series of Numerical Mathematics, Vol. 98, © 1991 Birkhäuser Verlag Basel

Analysis of multigrid methods for general systems of PDE

by

Bertil Gustafsson*
Per Lötstedt**

Abstract

Most iteration methods for solving boundary value problems can be viewed as approximations of a time-dependent differential equation. In this paper we show that the multigrid method has the effect of increasing the time-step for the smooth part of the solution leading back to an increase of the convergence rate. For the non-smooth part the convergence is an effect of damping. Fourier analysis is used to find the relation between the convergence rate for multigrid methods and single grid methods. The analysis is performed for general partial differential equations and an arbitrary number of grids.

1. Introduction

The convergence analysis of multigrid methods is usually based on the assumption that the problem is solved exactly on the coarsest grid. In this way high convergence rates are often predicted, at least for elliptic model problems. The situation is different for large scale real life problems, where the geometry and the structure of the grid is such that a grid coarse enough to permit exact solutions is never reached. Instead the smoothing operator (for example Jacobi, Gauss-Seidel, Conjugate gradient, Runge-Kutta iteration) applied on the finer grids is also used on the coarsest grid, and the number of grids is usually low, typically two, three or four. The convergence rates observed for this kind of computations are often lower than the ones predicted by too simplified model problems.

The traditional way of performing convergence analysis is to estimate the magnitude of the eigenvalues of the iteration matrix M. The analysis is in most cases based on Fourier modes (see for example [1], [7], [8]), which means that the solutions are assumed to be periodic in space or that the domain is unbounded. If ξ is the wave number, h is the fine grid step-size, the differential equation has no source terms and $\hat{M}(h\xi)$ is the symbol of M, there is always one eigenvalue $\lambda(h\xi)$ of $\hat{M}(h\xi)$ with $\lambda(0) = 1$. If a fixed number of iterations with the fine grid smoother is used also on the coarse grids, the eigenvalues for the modified symbol $\hat{M}$ are continuous functions of $h\xi$. This means that the convergence rate is arbitrarily low for small values of $h\xi$. It was shown in [2] that for low wave numbers the two-grid procedure plays the rôle of scaling up the time-variable compared to what it would have been for the single-grid method.

* Dept of Scientific Computing, Uppsala University, Sweden
**Dept of Scientific Computing, Uppsala University, and
SAAB-SCANIA, Linköping, Sweden

In this paper we give some results concerning the consistency of multigrid methods with a time-dependent differential equation where the time variable is scaled up compared to the corresponding single grid method (cf. [5], [6]). However, the properties of the time-dependent differential equation obtained in this way can be used only for the smooth part of the solution. The remaining part of the discrete solution is completely independent of the differential equation. In our analysis we take into account the interaction between the two parts. We use Fourier analysis to derive precise results concerning the behaviour of the low and the high wave number parts of the solution. The analysis is carried out for discretizations of general constant coefficient differential equations of arbitrary order for a V-cycle on an arbitrary number of grids in two space dimensions.

The results discussed here will be presented in a more detailed form with full proofs in a forthcoming paper.

2. Consistency with a time-dependent system

In this section we shall present a theorem which shows that a full multigrid iteration is consistent with a time-dependent differential equation where the time variable is scaled compared to the equation which corresponds to a single grid iteration. We begin by giving the notation.

We shall use $L+1$ grids $\{G_l\}_{l=0}^{L}$ where G_L is the finest one. For convenience it is assumed that the step–size h_l on grid G_l is equal in all directions. Q_l is the difference or finite volume approximation of the linear differential operator

$$P = \sum_{\nu} A_\nu(x)\left(\frac{\partial}{\partial x^1}\right)^{\nu_1}\cdots\left(\frac{\partial}{\partial x^d}\right)^{\nu_d},\ A_\nu \in \mathbf{R}^s x \mathbf{R}^s,\ \nu = (\nu_1,\ldots,\nu_d),\ x = (x^1,\ldots,x^d)^T, \tag{2.1}$$

on the grid G_l, $l = 0, 1,\ldots,L$, and we seek the solution to

$$Q_L u = f, \tag{2.2}$$

where u and f are vector functions with s components. The restriction operator from G_l to G_{l-1} is r_l, and the prolongation operator from G_{l-1} to G_l is p_l, $l = 1, 2,\ldots,L$. On each grid G_l there is an iterative method

$$R_l(u,f) := S_l u + T_l f \rightarrow u\ , \tag{2.3}$$

which is applied p times before and q times after the coarse grid corrections. On the coarsest grid we use $p+q$ iterations.

Consistency of the iteration method requires

$$T_l Q_l = I - S_l\ . \tag{2.4}$$

When analyzing the error and its convergence to zero, it is sufficient to consider the case $f = 0$. Let n be the number of the iteration. Then the multigrid V-cycle can be written as, cf. [3, Lemma 7.1.4],

$$\begin{aligned} &M_{-1} = I, \\ &M_l = S_l^q(I-p_l(I-M_{l-1})Q_{l-1}^{-1}r_lQ_l)S_l^p, \quad l = 0,1,...,L, \\ &u^{n+1} = M_L u^n. \end{aligned} \tag{2.5}$$

(If the exact solution is computed on G_0 then $M_0 = 0$ in the recursion (2.5) for $l \geq 1$.)

From now on we use the notation $u^{(l)}$ for a grid-function defined on G_l. Consider the time-dependent problems

$$\frac{\partial u^{(l)}}{\partial t} + Q_l u^{(l)} = f^{(l)} \quad , \qquad l = 0,1,...,L, \tag{2.6}$$

and introduce the time–steps Δt_l, $l = 0, 1,...,L$. We also use the notation

$$\Delta t = \Delta t_L \, , \; h = h_L \; , \quad \alpha_l = \frac{\Delta t_l}{\Delta t} \, , \quad l = 0,1,...,L.$$

The idea is to consider the iteration $R_l(u^{(l)},f^{(l)})$ as one time step Δt_l in a solution procedure of (2.6). If the whole multigrid cycle (2.5) is considered as one time-step Δt, we want to relate it to the time-dependent problem (2.6).

Instead of (2.6) we could of course consider the more general systems

$$\frac{\partial u^{(l)}}{\partial t} + D_l Q_l u^{(l)} = D_l f^{(l)}$$

where D_l are non–singular operators. But this is just a preconditioning of the original system e.g. with "local time-stepping" as in [5]. For convenience we assume that the preconditioner is already included in Q_l.

We first make

<u>Definition 2.1</u> The prolongation operator p_l and the restriction operator r_l are consistent if for all smooth functions $u(x)$

$$\begin{aligned} &(p_l u)(x) = u(x) + O(h_l) \quad , \quad x \in G_l \; , \; l = 1,2,...,L, \\ &(r_l u)(x) = u(x) + O(h_{l-1}) \; , \quad x \in G_{l-1} \; , \; l = 1,2,...,L. \end{aligned} \tag{2.7}$$

By deriving the explicit form of M_L one can prove

Theorem 2.1 Assume that

1) Q_l is consistent with the differential operator P, $l = 0,1,\dots,L$.
2) The iteration (or smoothing) operator S_l is consistent with (2.5), i.e.

$$S_l = I - \Delta t_l Q_l + O(\Delta t_l^2). \tag{2.8}$$

3) The operators $\{p_l, r_l\}_{l=1}^{L}$ are consistent.

If the multigrid iteration (2.5) is considered as one time–step Δt in a time–dependent procedure, it is consistent with

$$\frac{\partial u}{\partial t} + (p+q) \sum_{l=0}^{L} \alpha_l Pu = 0. \tag{2.9}$$

For $p = 1, q = 0, \alpha_l = 2^{L-l}, l = 0,1,\dots,L$, it is consistent with

$$\frac{\partial u}{\partial t} + (2^{L+1}-1)Pu = 0 .$$

Remark When writing $O(\Delta t_l^2)$ in (2.8) it is tacitly understood that S_l is applied to smooth functions.

The theorem shows that on a fixed number of grids the iteration formula converges to the modified time-dependent equation (2.9) as $\Delta t \to 0$. In practice this means that for first order systems we can expect the waves corresponding to low wave numbers to move $(\sum_{l=0}^{L} \alpha_l)$ times faster by using the multigrid procedure instead of a single grid solver. Alternatively, we can of course consider the procedure as an increase in the time-step for the original system (2.6), and this interpretation applies to all types of operators P. Note that we have assumed consistency also on the coarsest grid G_0. The practical implication of this is that G_0 must be fine enough such that the low frequency part of the solution can be represented. If there are only two points, say, in each direction of G_0, the theorem has no meaning. This does not mean that one should avoid very coarse grids if the geometry of the computational domain permits it. On the contrary, it may accelerate the convergence as a result of stronger damping.

3. Fourier analysis of the multigrid cycle

In this section we consider the constant coefficient case and we shall use Fourier analysis as our main tool of investigation. For the sake of notational simplicity the analysis is carried out for two space dimensions but the results generalize to any finite number d of dimensions. All variables are associated with level l except when the level is explicitly written as a subindex or superindex on the variable. The norm in the sequel is the Euclidean vector norm and the subordinate spectral matrix norm.

The main result is that two effects are responsible for the convergence: the amplification of the time scale for low wave number modes and damping of intermediate and high wave number modes.

The Fourier representation of the solution is

$$u(x) = \int_{-\infty}^{\infty}\int_{-\infty}^{\infty} e^{i\xi\cdot x}\,\hat{u}(\xi)d\xi_1 d\xi_2\,,$$

$$x = (x^1,x^2)^T\,,\ \xi = (\xi_1,\xi_2)^T\,.$$

We are interested in the solution of the Cauchy problem at discrete points $x_{\mu\upsilon}$ on level l

$$x_{\mu\upsilon} = x_0 + h\begin{pmatrix}\mu\\ \upsilon\end{pmatrix},\ (\mu,\upsilon)\in \mathbf{Z}\times\mathbf{Z}\,,$$

$$\mathbf{Z} = \{\text{the integer numbers}\}.$$

After some transformations the expression for the grid function $u_{\mu\upsilon}$ is

$$u_{\mu\upsilon} = \int_0^{2\pi/h_l}\int_0^{2\pi/h_l} \exp(i(\xi'_1\mu+\xi'_2\upsilon)h)\hat{u}'(\xi')d\xi'_1\, d\xi'_2. \qquad (3.1)$$

Henceforth, we drop the primes on ξ and $\hat{u}$ and use the form (3.1).

We introduce the notation $\eta_l = h_l\xi$ and make

Assumption 3.1

The Fourier–transform $\hat{S}_l$ of the smoothing operator S_l has the form

$$\hat{S}_l = I - \Delta t_l\,\hat{H}_l + O(\Delta t_l(\Delta t_l + \|\eta_l\|)).$$

The matrix $\hat{H}_l$ is bounded, and if S_l satisfies (2.8) then $\hat{H}_l = \hat{Q}_l$.

The V–cycle applied at level l introduces a coupling between 2^{2l} different wave–numbers and we collect the corresponding Fourier coefficients in the vector $\tilde{u}$. The transformed iteration is

$$\tilde{u}^{n+1} = \tilde{M}\tilde{u}^n,$$

where $\tilde{M}$ is a matrix of order $s2^{2l}\times s2^{2l}$. Denote the wave number domain in (3.1) by C_l, see Fig. 3.1. The wave numbers in C_l are coupled by $\tilde{M}$ such that for each ξ in C_0 the Fourier coefficients of

$$\xi' = \xi + 2\pi/h_0\begin{pmatrix} j \\ k \end{pmatrix}, \quad \xi' \in C_l,$$

$$(j,k) \in I_m \times I_m, \quad I_m = [0,1,\cdots m-1], \quad m = 2^l,$$

are found in $\tilde{u}^n$. If $\|\xi\|$ is small, the Fourier coefficients are related to smooth solutions. By periodicity this relationship holds also if either one or both of ξ_1, ξ_2 is in the neighbourhood of $2\pi/h_l$. We denote by C_*^l this composite domain in the (ξ_1,ξ_2)–plane, see Fig. 3.1. With $\xi \in C_*^l$ we place the corresponding Fourier coefficients first in the vector $\tilde{u}$ and order $\tilde{M}$ in the same way. We need a second assumption.

<u>Assumption 3.2</u>
The lower right submatrix $\tilde{M}_{II}$ of $\tilde{M}$ of order $s(2^{2l}-1)\text{x}\, s(2^{2l}-1)$ has simple eigenvalues μ_j satisfying

$$|\mu_j| \le \theta < 1, \quad \xi \in C_*^0.$$

Under consistency assumptions and a few other very natural technical assumptions one can prove

<u>Theorem 3.1</u>
Let $\xi \in C_*^0$ and let the assumptions be satisfied. The transformed multigrid iteration matrix for a V-cycle at level l after n iterations is

$$\tilde{M}^n = \begin{pmatrix} I - n\delta t\,\hat{H} + O(\Delta t_0 + \|\eta_0\|) & O(\Delta t_0) \\ O(\Delta t_0\|\eta_0\|) & \tilde{M}_{II}^n + O(\Delta t_0\|\eta_0\|) \end{pmatrix},$$

where

$$\delta t = (p+q)\sum_{j=0}^{l} \Delta t_j = (p+q)\Delta t \sum_{j=0}^{l} \alpha_j,$$

$$\|\tilde{M}_{II}^n\| \le c_{II}\,\theta^n, \quad \theta < 1.$$

Alternatively, the upper left corner of $\tilde{M}^n$ can be written

$$T_0\,\mathrm{diag}(\{\exp(-\sigma_k n\delta t - i\tau_k n\delta t)\}_{k=0}^{s})T_0^{-1} + O(\Delta t_0 + \|\eta_0\|),$$

where

$$\sigma_k = \sigma_k(\xi,h) = \mathrm{Re}\,\lambda_k(\hat{H}), \quad \tau_k = \tau_k(\xi,h) = \mathrm{Im}\,\lambda_k(\hat{H}),$$

and T_0 is the eigenvector matrix of $\hat{H}$.

The theorem shows that the multigrid V-cycle scales up the time (or pseudo time) variable for the smooth part of the solution. For the other wave-numbers, which become coupled to ξ through the multigrid iteration, the convergence is achieved through damping.

If ξ does not belong to C_*^0 such that $\tilde{u}$ does not include any components related to smooth solutions, then we simply assume that

$$\|\tilde{M}(\xi)\| \leq \theta < 1 .$$

This guarantees that the whole non-smooth part of the solution is damped out independently of the grid size.

Let $\hat{U}_L(n\Delta t,\xi)$ be the solution in Fourier space obtained after n steps on grid L with the smoothing operator S, and let $\hat{u}^n_{MG}(\xi)$ be the multigrid solution after n cycles. Then one can show

Theorem 3.2
Let the assumptions in this section be fulfilled. Then

$$\hat{u}^m_{MG}(\xi) = \hat{U}_L(n\delta t_L, \xi) + O(\Delta t_0 + \|\eta_0\|), \quad \xi \in C_*^L,$$

$$\|\hat{u}^n_{MG}(\xi)\| \leq c_L(\xi)\theta^n + O(\Delta t_0^{1+1/\nu}), \quad \xi \in C_L \setminus C_*^L,$$

where υ is the order of the differential equation and $c_L(\xi)$ is a constant which depends on $\hat{u}^0$.

As an illustration of the theoretical results in this paper the eigenvalues of $\tilde{M}$ in Theorem 3.1 are plotted for a one-dimensional example in fig. 3.2 and three two-dimensional examples in Fig. 3.3.

In Fig. 3.2 the model equation to be solved is

$$u_x = f .$$

The equation is discretized by a cell-centered finite volume scheme with additional 4:th order artificial viscosity, see [4]. The smoothing iteration is a three stage Runge-Kutta method and $\Delta t_l = h_l$. The number of presmoothing steps is 1 and there are no postsmoothing iterations.

The symbol of the restriction operator is

$$\hat{r} = \cos(\xi\, h_l/2),$$

and $\hat{p} = \hat{r}$.

In the left column $|\lambda_j(\tilde{M})|$, $j = 1,2,\cdots m$, $m = 2^L$, is plotted as a function of $\eta_L = \xi h_L$, where $\eta_L \in [0, 2\pi/m]$.

In the right column of Fig. 3.2 we plot the factor

$$\text{Im}\log \lambda_j(\tilde{M})/\zeta\Delta t_L,$$

$$\zeta = \begin{cases} \xi & \text{if } \xi \in [0,\pi/h_0], \\ \\ 2\pi/h_0 - \xi & \text{if } \xi \in (\pi/h_0, 2\pi/h_0]. \end{cases}$$

It follows from [2] and Theorem 3.1 that this is the speed–up factor $\gamma = \sum_{j=0}^{L} \alpha_j$ of the slowly varying Fourier mode with $\lambda_1 \approx 1$ for small ξ and ξ close to $2\pi/h_0$.

The results are in accordance with the theory in Theorem 3.1. There is one eigenvalue λ_1 in in each of the three cases, $L = 0,1,2$, for which $|\lambda_1| \approx 1$ when ξ is in the neighbourhood of 0 or $2\pi/h_0$, and furthermore

$$\gamma = 2^{L+1} - 1 .$$

For the other eigenvalues

$$|\lambda_j(\tilde{M})| \le \theta < 1, \quad j = 2,\cdots m,$$

for all ξ.

The two-dimensional model example is

$$a\frac{\partial u}{\partial x^1} + b\frac{\partial u}{\partial x^2} = f . \tag{3.2}$$

The discretization method, the smoothing procedure and the multigrid strategy are the same as above and there are two grids, $L = 1$. Isolines of

$$\max_j |\lambda_j(\tilde{M}(\xi))|$$

are plotted in Fig. 3.3. The coefficients in (3.2) in Fig. 3.3a are $a = b = 1$. When $\xi \in C_*^L$, i.e. the squares in the corners of the wave number domain C_L, the maximum of $|\lambda_j|$ is close to 1 and exactly 1 at the corners. In the interior of C_L we have

$$\max_j |\lambda_j| \leq \theta < 1 .$$

The coefficients in (3.2) in Fig 3.3.b are $a = 1$ and $b = 0$. We still have good damping properties in the interior and $\max |\lambda_j|$ is close to 1 in the corners. There are no "grid alignment" effects with no damping at all in large parts of the wave number domain. This is sometimes a problem with upwind discretizations of (3.2) [7]. The reason is that the artificial viscosity here is isotropic and independent of a and b.

On the other hand, if we scale the artificial viscosity term in the x^1 direction by a and in the x^2 direction by b, then we have an obvious "grid alignment" problem as in Fig 3.3.c. The multigrid method does not reduce the amplitude of the modes with e.g. η_1 small and $\eta_2 \approx \pi/2$, simply because there is no artificial viscosity in the x^2 direction.

References

1. Brandt A: Multi-level adaptive solutions to boundary-value problems. Math. Comp. 31 (1977), pp 333-390.
2. Gustafsson B and Lötstedt P: Analysis of the multigrid method applied to first order systems. Proc. of the Fourth Copper Mountain Conf. on Multigrid Methods. Eds. J Mandel et. al., SIAM, Philadelphia (1989).
3. Hackbusch W: Multigrid methods and applications, Springer Verlag, Berlin, Heidelberg, (1985).
4. Jameson A: Solution of the Euler equations for two-dimensional transonic flow by a multigrid method, Appl. Math. Comp., 13 (1983), pp. 327-355.
5. Jameson A: Computational transonics, Comm. Pure Appl. Math., XLI (1988), pp. 507-549.
6. Jespersen D C: A time-accurate multiple-grid algorithm, AIAA paper 85-1493-CP, (1985).
7. Mulder W A: A new multigrid approach to convection problems, J. Comp. Phys., Vol. 83 (1989), pp. 303-323.
8. Stüben K and Trottenberg U: Multigrid methods: Fundamental algorithms, model problem analysis and applications. Multigrid methods, Proc., Köln-Porz (1981), Eds. W Hackbusch and U Trottenberg. Lecture notes in Mathematics 960, Springer, Berlin.

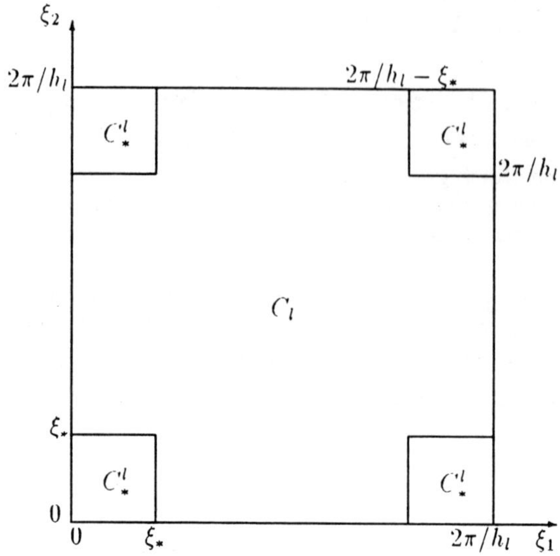

Fig. 3.1. The definition of C_*^l as a subset of $C_l = [0, 2\pi/h_l] \times [0, 2\pi/h_l]$ with wave numbers corresponding to slowly varying modes.

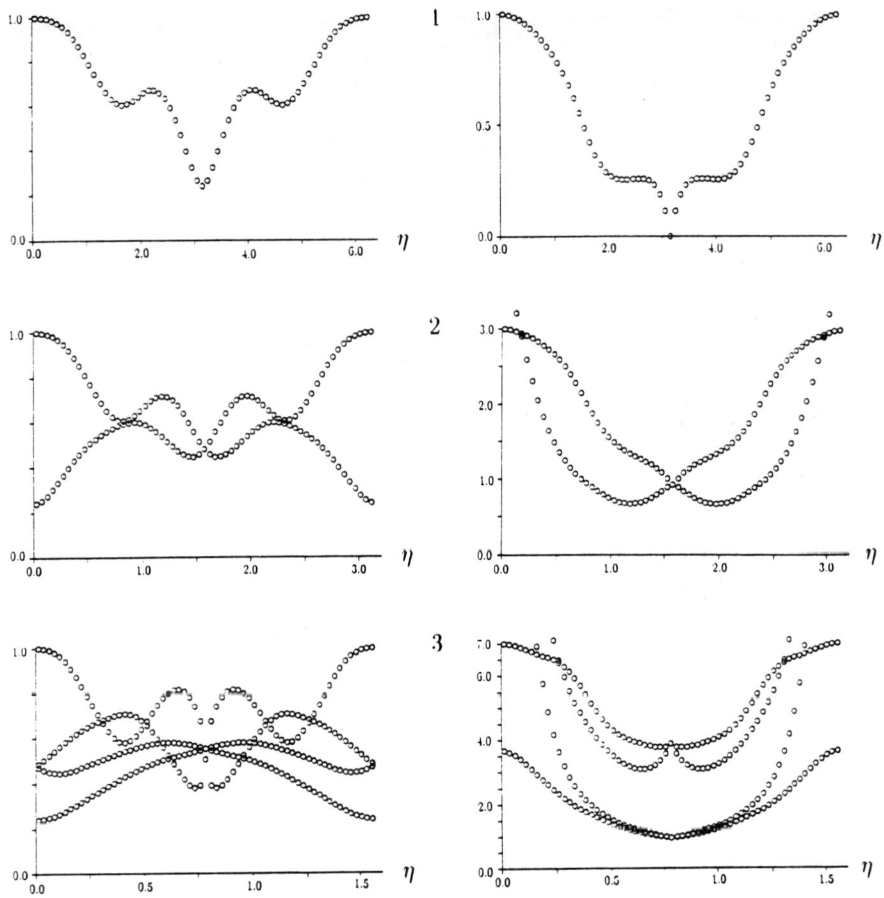

Fig. 3.2. The eigenvalues $\lambda_j(\eta)$ of the Fourier transform of the multigrid matrix $\hat{M}$ are displayed for the model equation $u_x = f$ and Runge-Kutta iteration on 1, 2 and 3 grids. To the left $|\lambda_j|$ is plotted and to the right $Im \log \lambda_j / \zeta \Delta t_l$ shows the speed up factor for $|\lambda_j|$ close to 1.

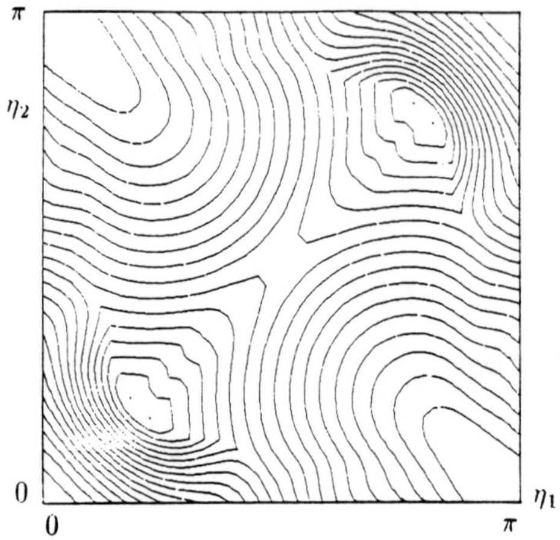

Fig. 3.3. Isolines of the maximum of $|\lambda_j(\eta)|$ for each η of the Fourier transform of the multigrid matrix $\hat{M}$ are displayed for the equation (3.2) and Runge-Kutta iteration on 2 grids. The maximum is 1 in the corners and less than 1 in the interior of the η domain.

Fig. 3.3a. $a = 1, b = 1$.

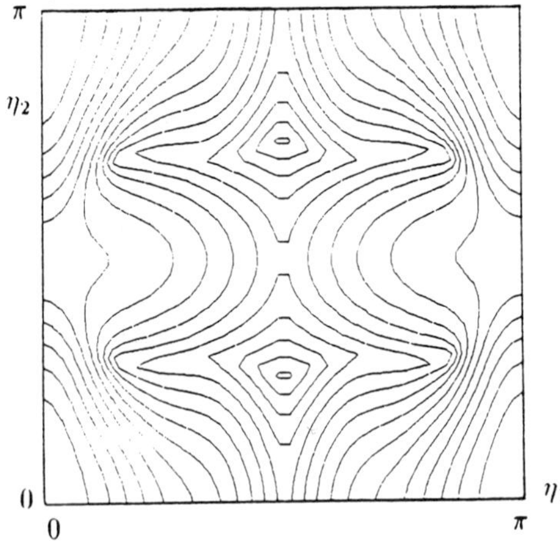

Fig. 3.3b. $a = 1, b = 0$. Isotropic artificial viscosity.

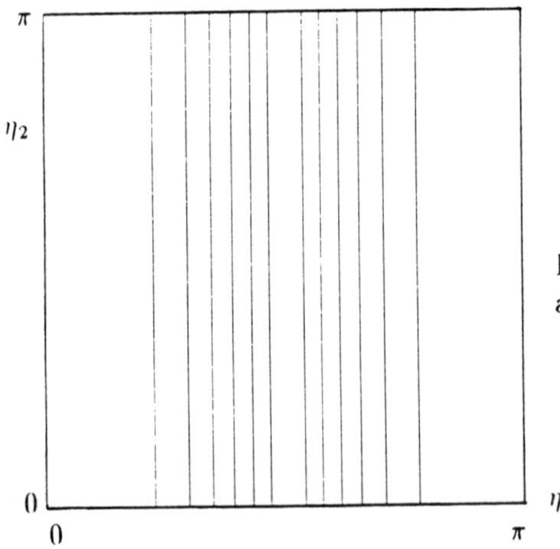

Fig. 3.3c. $a = 1, b = 0$. Anisotropic artificial viscosity.

On the convergence of the nonlinear multigrid algorithm for potential operators

S. Hengst G. Telschow

Karl-Weierstraß-Institut für Mathematik,Mohrenstr. 39, Berlin 1086

Abstract

Among numerical methods to solve large-scale algebraic equations systems (which may result from the approximation of two-or three-dimensional integral or differential equations), the multi-grid-methods have been rapidly developed during the last 10 years. These methods have the property that the necessary effort is proportional to the number of unknown parameters. In the present paper we give results on a non-linear multigrid method that supports an efficient solution of non-linear problems without linearization. Thereby the linear convergence of the method can be shown under rather general assumptions. For the proofs we use mappings with potentials (cf. Daniel [1]). Under stronger assumptions we show for a two-level-algorithm the rate of convergence to be independent of the level of refinements. Our results continue the investigation done by Hackbusch and Reusken [2-5].

Introduction

Let $H_k, k \geq 0$ be Hilbert spaces. The inner product on the H_k is denoted by $(\cdot, \cdot)$ and the corresponding norms by $\| \ \|$.
Moreover, let $\phi_k : H_k \Rightarrow R^1$ be some given differentiable functional.
Then

$$\phi_k^v(u) = \phi_k(u) - (\phi_k'(u), u - v) - \phi_k(v)$$

defines a functional on H_k for any k and $v \in H_k$.
Obviously $\phi_k^v(v) = 0$ and $(\phi_k^v)'(v) = 0$.
In what follows we suppose the gradients to be strongly monotone and Lipschitz continuous, i.e. there exist numbers $l_k \geq m_k > 0$ that for all $u, v \in H_k$

$$(\phi_k'(v) - \phi_k'(u), u - v) \geq m_k \| v - u \|^2$$

and

$$\| \phi_k'(v) - \phi_k'(u) \| \leq l_k \| v - u \|$$

holds.
Furthermore we have to introduce certain parameters.

Hence, let $0 < \beta_k < 1, \psi_k > 0, \delta_k \neq 0$ real numbers, $\sigma_k > 0$. Assume that there are given $\nu_k^1, \nu_k^2 \in N$ (number of pre- and post-relaxation) with $\nu_k^1 + \nu_k^2 > 0$ and $\tilde{u} \in H_k$. The given mappings $p_k : H_{k-1} \Rightarrow H_k$ we call prolongations.
The restrictions $r_k : H_k \Rightarrow H_{k-1}$ are given by the adjoint mappings of p_k, i.e. for all $u \in H_k$ and $v \in H_{k-1}$:

$$(u, p_k v) = (r_k u, v).$$

The multigrid method to solve $\phi_k'(u) = \phi_k'(v)$ will be described recursively by the following mappings:
Smoothing mappings $R_k : H_k \times H_k \Rightarrow H_k$ by

$$R_k(u,v) = u - \beta_k \frac{2}{l_k}(\phi_k^v)'(u).$$

Coarse-grid corrections $C_k : H_k \times H_k \Rightarrow H_k$ by

$$\begin{aligned} C_k(v,u) &= u - \psi_k \frac{2}{l_k} \frac{p_k(y - \tilde{u}_{k-1})}{\sigma_k} \quad \text{with} \\ y &= \begin{cases} w & \text{if} \quad k = 1 \\ F_{k-1}^{\delta_{k-1}}(w, \tilde{u}_{k-1}) & \text{if} \quad k > 1 \end{cases} \end{aligned}$$

where w means the exact solution of the coarse-grid equation

$$\phi_{k-1}'(w) = \phi_{k-1}'(\tilde{u}_{k-1}) + \sigma_k r_k(\phi_k'(u) - \phi_k'(v)).$$

The multigrid algorithm is now defined through the mapping

$$F_k(v,u) = R_k^{\nu_k^2}(v, C_k(v, R_k^{\nu_k^1}(v,u))).$$

Global convergence

Lemma 1 gives a general estimation used in the sequel.
Lemma 1: Take $\beta \geq 0$ and $u, v, y \in H_k$.
Then the following holds

$$\phi_k^v(u - \beta\frac{2}{l_k}y) \leq \phi_k^v(u) - \beta\frac{2}{l_k}(((\phi_k^v)'(u), y) - \beta \parallel y \parallel^2).$$

Proof: Using the Lipschitz condition for $(\phi_k^v)'$ and $\xi = \beta\frac{2}{l_k}$ the Taylor expansion with respect to ξ yields:

$$\begin{aligned} \phi_k^v(u - \xi y) &= \phi_k^v(u) - \xi \int_0^1 ((\phi_k^v)'(u - t\xi y), y)dt \\ &= \phi_k^v(u) - \xi((\phi_k^v)'(u), y) - \xi \int_0^1 ((\phi_k^v)'(u - t\xi y) - (\phi_k^v)'(u), y)dt \\ &\leq \phi_k^v(u) - \xi((\phi_k^v)'(u), y) + \xi^2 \frac{l_k}{2} \parallel y \parallel^2 \\ &= \phi_k^v(u) - \beta\frac{2}{l_k}((\phi_k^v)'(u), y) + \beta^2 \frac{2}{l_k} \parallel y \parallel^2 \\ &= \phi_k^v(u) - \beta\frac{2}{l_k}((\phi_k^v)'(u), y)) - \beta \parallel y \parallel^2). \end{aligned}$$

The elementary Lemma 1 has an interesting application concerning the relaxation R_k.
Lemma 2: The following holds:

$$\phi_k^v(R_k(v,u)) \leq q_k \phi_k^v(u) \qquad \text{with} \qquad q_k = 1 - \beta_k(1-\beta_k)(\frac{2m_k}{l_k})^2 < 1$$

Proof: We verify that Lemma 1 may be applied with $y = (\phi_k^v)'(u)$ and $\beta = \beta_k$.
So we have

$$\phi_k^v(u - \beta_k \frac{2}{l_k}(\phi_k^v)'(u)) \leq \phi_k^v(u) - \delta' \parallel (\phi_k^v)'(u) \parallel^2$$

with $\delta' = \beta_k(1-\beta_k)\frac{2}{l_k}$.
Also we get

$$m_k \parallel u - v \parallel^2 \leq (\phi_k'(u) - \phi_k'(v), u - v) \leq \parallel (\phi_k^v)'(u) \parallel \parallel u - v \parallel .$$

So

$$m_k \parallel u - v \parallel \leq \parallel (\phi_k^v)'(u) \parallel .$$

Furthermore

$$\begin{aligned} \phi_k^v(u) &= \phi_k^v(u) - \phi_k^v(v) \\ &= \int_0^1 ((\phi_k^v)'(v + t(u-v)), u - v) dt \\ &= \int_0^1 (\phi_k'(v + t(u-v) - \phi_k'(v), u - v) dt \\ &\leq \frac{l_k}{2} \parallel u - v \parallel^2 . \end{aligned}$$

Consequently, we obtain

$$\begin{aligned} \phi_k^v(u - \beta_k \frac{2}{l_k}(\phi_k^v)'(u)) &\leq \phi_k^v(u) - \delta' \parallel (\phi_k^v)'(u) \parallel^2 \\ &\leq \phi_k^v(u) - \delta' m_k^2 \parallel u - v \parallel^2 \\ &\leq \phi_k^v(u) - \delta' m_k^2 \frac{2}{l_k} \phi_k^v(u) \\ &= (1 - \delta' m_k^2 \frac{2}{l_k}) \phi_k^v(u). \end{aligned}$$

The following Lemma shows that the direction given by the coarse grid correction C_k makes the functional ϕ_k decrease.
Lemma 3: Take $u, v \in H_k$ and $\tilde{u}, w \in H_{k-1}$ with

$$\phi_{k-1}'(w) = \phi_{k-1}'(\tilde{u}) + \sigma_k r_k(\phi_k'(u) - \phi_k'(v)).$$

Let $y \in H_{k-1}$ be such that $\phi_{k-1}^w(y) \leq \phi_{k-1}^w(\tilde{u})$.
Then

$$\left((\phi_k^v)'(u), \frac{p_k(y - \tilde{u})}{\sigma_k}\right) \geq 0.$$

Proof: Note that

$$\begin{aligned}\phi_{k-1}^{w}(y) &= \phi_{k-1}^{w}(\tilde{u}) + \int_0^1 ((\phi_{k-1}^{w})'(\tilde{u}+t(y-\tilde{u})), y-\tilde{u})dt \\ &\geq \phi_{k-1}^{w}(\tilde{u}) + ((\phi_{k-1}^{w})'(\tilde{u}), y-\tilde{u})\end{aligned}$$

and so we have

$$((\phi_{k-1}^{w})'(\tilde{u}), y-\tilde{u}) \leq \phi_{k-1}^{w}(y) - \phi_{k-1}^{w}(\tilde{u}) \leq 0.$$

This results in

$$\begin{aligned}\left(\phi_k'(u) - \phi_k'(v), \frac{p_k(y-\tilde{u})}{\sigma_k}\right) &= \frac{1}{\sigma_k}(r_k(\phi_k'(u) - \phi_k'(v)), y-\tilde{u}) \\ &= -\frac{1}{\sigma_k^2}(\phi_{k-1}'(\tilde{u}) - \phi_{k-1}'(w), y-\tilde{u}) \\ &= -\frac{1}{\sigma_k^2}((\phi_k^{w})'(\tilde{u}), y-\tilde{u}) \\ &\geq 0\end{aligned}$$

Now we can give the first theorem on the convergence of the multigrid method. The linear convergence holds under general assumptions. However, the rate of convergence cannot be proven to be independent of the level of refinement k.

Theorem 4: Assume that $\psi_k \parallel \frac{p_k(y-\tilde{u}_{k-1})}{\sigma_k} \parallel^2 \leq ((\phi_k^v)'(u), \frac{p_k(y-\tilde{u}_{k-1})}{\sigma_k})$.
Then the following holds:

$$\phi_k^v(F_k(v,u)) \leq Q_k \phi_k^v(u) \tag{1}$$

$$\parallel F_k^i(v,u) - v \parallel \leq \sqrt{\frac{l_k}{m_k}}\left(\sqrt{Q_k}\right)^i \parallel u - v \parallel \qquad \text{where} \quad Q_k < 1. \tag{2}$$

Proof: Applying Lemma 1 with $y = \frac{p_k(y-\tilde{u}_{k-1})}{\sigma_k}$ and $\beta = \psi_k$ results in:

$$\begin{aligned}\phi_k^v(C_k(v,u)) &= \phi_k^v(u) - \psi_k \frac{2}{l_k}((\phi_k^v)'(u), \frac{p_k(y-\tilde{u}_{k-1})}{\sigma_k}) - \psi_k \parallel \frac{p_k(y-\tilde{u}_{k-1})}{\sigma_k} \parallel^2) \\ &\leq \phi_k^v(u).\end{aligned}$$

Using Lemma 2 we get

$$\begin{aligned}\phi_k^v(F_k(v,u)) &= \phi_k^v(R_k^{\nu_k^2}(v, C_k(v, R_k^{\nu_k^1}(v,u)))) \\ &= q_k^{\nu_k^2}\phi_k^v(C_k(v, R_k^{\nu_k^1}(v,u))) \\ &\leq q_k^{\nu_k^2}\phi_k^v(R_k^{\nu_k^1}(v,u)) \\ &\leq q_k^{\nu_k^1+\nu_k^2}\phi_k^v(u) \\ &= Q_k\phi_k^v(u)\end{aligned}$$

with $Q_k = q_k^{\nu_k^1+\nu_k^2} < 1$.
This proves (1).
Also we have

$$\phi_k^v(F_k^i(v,u)) \leq Q_k^i \phi_k^v(u).$$

From the Lipschitz condition and the strong monotonicity follows:

$$\begin{aligned}\phi_k^v(u) &= \int_0^1 ((\phi_k^v)'(v+t(u-v)), u-v)dt \\ &= \int_0^1 (\phi_k'(v+t(u-v)) - \phi_k'(v), u-.v)dt \\ &\leq \frac{l_k}{2} \parallel u-v \parallel^2\end{aligned}$$

and

$$\begin{aligned}\phi_k^v(u) &= \int_0^1 (\phi_k'(v+t(u-v)) - \phi_k'(v), u-v)dt \\ &\geq \frac{m_k}{2} \parallel u-v \parallel^2 .\end{aligned}$$

So we have

$$\frac{m_k}{2} \parallel F_k^i(v,u) - v \parallel^2 \leq \phi_k^v(F_k^i(v,u)) \leq \phi_k^v(F_k^i(v,u)) \leq Q_k^i \phi_k^v(u) \leq Q_k^i \frac{l_k}{2} \parallel u-v \parallel^2$$

and finally

$$\parallel F_k^i(v,u) - v \parallel \leq \sqrt{\frac{l_k}{m_k}} \left(\sqrt{Q_k}\right)^i \parallel u-v \parallel$$

To get the rate of convergence independent of k additional assumptions are necessary. Hence, let the Lipschitz constants be bounded, i.e. there is an L with $l_k \leq L$. Moreover we suppose a complementary condition to hold. To it we define the exact coarse-.grid correction $\tilde{C}_k$ by

$$\tilde{C}_k(v,u) = u - \psi_k \frac{2}{l_k} \frac{p_k(w - \tilde{u}_{k-1})}{\sigma_k}$$

with w as the exact solution of the coarse-grid equation

$$\phi_{k-1}'(w) = \phi_{k-1}'(\tilde{u}_{k-1}) + \sigma_k r_k(\phi_k'(u) - \phi_k'(v)).$$

Then we assume that there is an M independent of k with

$$\parallel (\phi_k^v)'(\tilde{C}_k(v,u)) \parallel^2 \geq M \phi_k^v(\tilde{C}_k(v,u))$$

for any $u \in H_k$. The following theorem gives the convergence of the two-level method defined by the mapping

$$\tilde{F}_k(v,u) = R_k(v, \tilde{C}_k(v,u)).$$

Theorem 5: The following holds:

$$\phi_k^v(\tilde{F}_k(v,u)) \leq \tilde{q}_k \phi_k^v(u)$$

with $\tilde{q}_k = 1 - \beta_k(1-\beta_k)\frac{2}{l_k}M \leq 1 - \beta_k(1-\beta_k)\frac{2}{L}M \leq 1$.

Proof: Let $\tilde{u} = \tilde{C}_k(v,u)$. Using Lemma 1 and δ' from Lemma 2 we get

$$\begin{aligned}\phi_k^v(\tilde{F}_k(v,u)) &= \phi_k^v(R_k(v,u)) \\ &= \phi_k^v(\tilde{u}) - \delta' \parallel (\phi_k^v)'(\tilde{u}) \parallel^2 \\ &\leq (1-\delta' M)\phi_k^v(\tilde{u}) \\ &\leq (1-\delta' M)\phi_k^v(u).\end{aligned}$$

The last inequality can be proven as in theorem 4.

References

[1] James W. Daniel. *The Approximate Minimization of Functionals.* Prentice-Hall, Englewood Cliffs,N.J., 1971

[2] W.Hackbusch and A.Reusken. *Analysis of a damped nonlinear multilevel method.* 1988

[3] W.Hackbusch and A.Reusken. *On global multigrid convergence for nonlinear problems.* In W.Hackbusch,editor, Robust Multigrid Methods, Vieweg-Verlag Braunschweig, 1988

[4] Arnold Reusken. *Convergence of the multigrid full approximation scheme for a class of elliptic mildly nonlinear boundary value problems.* Numerische Mathematik, 52:251-277,1988

[5] Arnold Reusken.*Convergence of the multilevel full approximation scheme including V-cycle.* Numerische Mathematik, 53:687-699,1988

A MULTIGRID ALGORITHM WITH TIME-DEPENDENT, LOCALLY REFINED GRIDS FOR SOLVING THE NONLINEAR DIFFUSION EQUATION ON A NONRECTANGULAR GEOMETRY.

W. Joppich - Institut für Methodische Grundlagen (F1/T),
Gesellschaft für Mathematik und Datenverarbeitung mbH,
Schloß Birlinghoven, D-5205 Sankt Augustin,
Federal Republic of Germany

ABSTRACT: The numerical solution of the diffusion equation in VLSI process simulation leads to large systems of nonlinear equations which have to be solved at every time step. For this purpose, a multigrid (MG) algorithm with locally refined grids is constructed. It is demonstrated that the method used here yields typical MG algorithm convergence rates, and reduces the amount of work considerably. The local refinements are controlled by an estimation of the discretization error which is calculated within the nonlinear MG method (FAS) and requires no extra computational work.

1 Description of the problem and the discretization

The design of semiconductor devices provides many, highly interesting problems for numerical analysis. The MG application which is presented here arises from the process simulation which in general has to simulate the complete set of all processing steps during the fabrication of modern silicon devices. One main result is the concentration of doping material which is responsible for the physical properties of the silicon device. To calculate the final distribution of impurities is one of the most time consuming tasks of process simulation.

The nonlinear model problem (1)-(3) together with an initial condition $N(x,y,0)$, which governs the redistribution of an impurity during the diffusion in an oxidizing ambient, has to be solved for the concentration $N(x,y,t)$ with a concentration-dependent diffusion coefficient $D(N)$ and – due to the oxidation of silicon – on a varying nonrectangular domain $\Omega(t)$:

$$\nabla\Big(D(N)\nabla N\Big) = \frac{\partial N}{\partial t} \text{ in } \Omega(t), \tag{1}$$

$$\frac{\partial N}{\partial \vec{n}} = 0 \text{ at } B_1(t), B_2(t) \text{ and } B_3(t), \tag{2}$$

$$D(N)\frac{\partial N}{\partial \vec{n}} - (k-m)\dot{U}_n N = 0 \text{ at } B_4(t) \text{ for } t > 0. \tag{3}$$

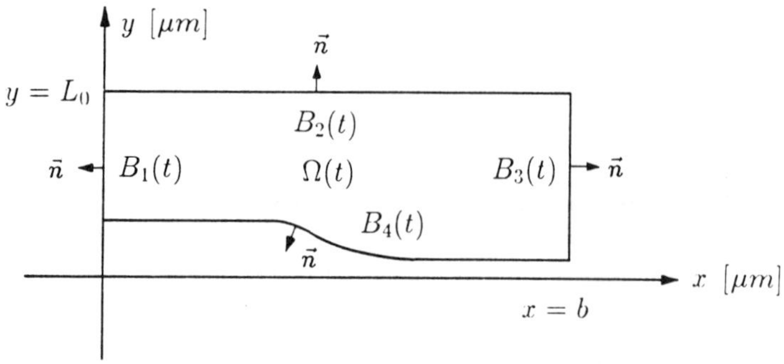

Figure 1: Description of the initial boundary value problem

$\vec{n}$ in (1)-(3) denotes the outward normal vector to the boundary of $\Omega(t)$. k is the segregation coefficient. The amount of consumed silicon to produce a volume unit of SiO_2 is given by m. For every $t \geq 0$ the position $(x, y) = (x, mU(x,t))$ of the Si-SiO_2 interface is determined by a two-dimensional oxidation model. Then $\dot{U}_n$ stands for the oxide growth rate normal to $B_4(t)$. A more detailed description of the problem is contained in Maldonado [5] and Joppich [3].

For the time discretization, implicit schemes of first and second order are used: backward-Euler 'BE' and Crank-Nicolson 'CN', respectively. The space discretization uses Cartesian grids. The discretization of (1) and (2) on an equispaced grid with the aid of 'ghost points', which are eliminated later, furnishes a five-point-stencil in $\bar{\Omega} - B_4$. To approximate the boundary condition (3) on the same grid with the same order of consistency, $\frac{\partial N}{\partial \vec{n}}$ is expanded in a Taylor series for points on B_4. This results in a seven-point-formula for the boundary condition. For a more detailed description of the discretization see Joppich [3].

Without mapping the physical domain onto a rectangle the physical domain is also the computational domain. Having in mind the use of MG methods in combination with local refinements, this approach is based upon the idea of treating local problems locally and maintains simple grid structures and uncomplicated five-point-formulas in the main part of the computational domain. This makes the chosen MG approach effective for quite general situations, not just for the considered problem of impurity diffusion on the so called 'bird's beak' geometry of Figure 1.

2 A multigrid algorithm with local refinements

Local refinement may be considered as an important feature that should be included in every algorithm for process and device simulation, because this algorithmic detail is strongly connected with the aim of reducing the amount of computational work. Within the MG approach 'local refinements' are realized usually on a composite mesh which is the union of a sequence of grids with decreasing mesh sizes. We call G_h a 'refinement' or 'patch' if $G_h \subset \mathcal{G}_h \cap (\Omega \cup \partial\Omega)$, where $\mathcal{G}_h$ is the infinite mesh of stepsize h. Grids that cover the complete computational domain $G_h = \mathcal{G}_h \cap (\Omega \cup \partial\Omega)$ are called 'global' grids.

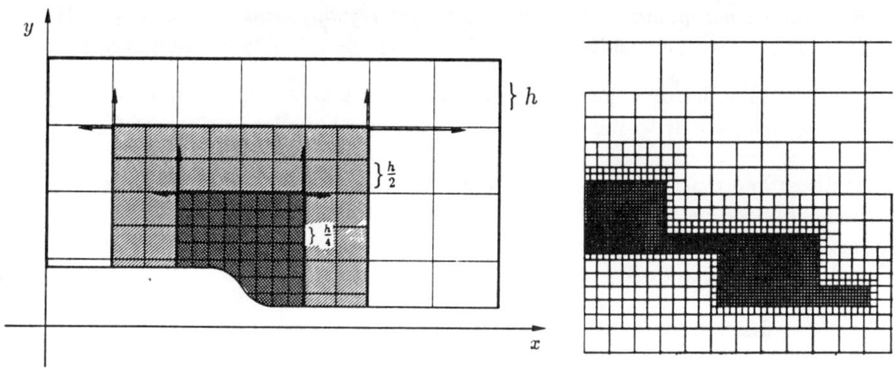

Figure 2: Global grid with two refinement levels and a refined grid which is used in practice

Generally, several disjoint refinements of arbitrary shape within the domain are possible. As usual there is a trade-off between minimizing the number of unknowns and technical complexity. A reasonable approach for this model problem is a restriction to one simply connected refinement together with some additional technical conventions. Thus, the patches are nested rectangular grids all having the same orientation. Additionally, the boundary lines of a fine grid coincide with lines of the next coarser one, and the mesh sizes are decreased by a factor of $\frac{1}{2}$, from one grid to the next finer one. A composite mesh for the given application will, under these conditions, look like the left one in Figure 2. The arrows indicate the varying position of the interior boundaries of refinements due to the time dependency of the problem. In this way, the patches may blow up to global grids. Dropping the technical convention of 'nested rectangles' (see the right mesh in Figure 2) enables refined meshes which can be handled in the same way and the treatment of interior boundaries remains relatively inexpensive. These meshes are used in practice.

The principle of MG on a composite mesh is more easily presented by formulating the corresponding two grid scheme for a linear or nonlinear, scalar elliptic boundary value problem $Lu = f$ on $\Omega \cup \partial\Omega$. Let $L_h u_h = f_h$ be the corresponding discrete problem on $G_h = \mathcal{G}_h \cap (\Omega \cup \partial\Omega)$. The set G_h° of 'regular points' consists of all those points of G_h where the operator L_h (including the discrete boundary operator) is defined. Included in this definition is the case of points on the physical boundary $\partial\Omega$ belonging to G_h°. Within the grid sequence $\{G_h\}_{h\in\mathcal{H}}$, indexed by $\mathcal{H} = \{h_k | h_k = h_0/2^k, k = 0, ..., M-1\}$, there should exist at least one global grid and at least one refinement $G_h \neq \phi$. In the sequel, for a given grid $G_h := G_{h_k} (k \in \{1, ..., M-1\})$ the next coarser grid will be marked by G_{2h} or G_H with $H := 2h$, instead of writing $G_{h_{k-1}}$.

To solve the discrete problem on the composite grid $\Gamma_{\mathcal{H}} := \bigcup_{h\in\mathcal{H}} G_h$ the discretization with the finest possible mesh size has to be used. In this context, G_H-points of two consecutive grids G_h and G_H will have different functions:

1. Those *regular points* of G_H which are also *regular points* of G_h are used to calculate *corrections* to the approximate solution on G_h°.

2. For those *regular points* of G_H which are *not regular points* of G_h, G_H is the finest discretization level. A *solution* to the discrete problem has to be computed there.

3. To evaluate the discrete operator L_h at all regular points of G_h it is necessary to determine approximations to the solution at all 'interior boundary points' of G_h.

Satisfying these two aims - computing an approximate solution as well as corrections to an approximate solution on the same level of discretization - becomes possible by the use of a slightly modified FAS (Full Approximation Scheme) within the MLAT (MultiLevel Adaptive Technique) approach due to Brandt [1].

With $w_h^{(0)}$ as first approximation to the solution u_h of the problem (4) let $w_h^\nu = RELXLV^\nu(w_h^{(0)}, L_h, f_h, u_H)$ be the result of ν relaxations for

$$\begin{aligned} L_h u_h &= f_h \text{ on } G_h^\circ \text{ and} \\ u_h &= \hat{I}_H^h u_H \text{ on } G_h - G_h^\circ. \end{aligned} \tag{4}$$

Defining

$$F_H := \begin{cases} L_H \hat{I}_h^H u_h + r_H & on \quad G_H^\circ \cap G_h^\circ \\ f_H & on \quad G_H^\circ - (G_H^\circ \cap G_h^\circ) \end{cases} \tag{5}$$

simplifies the formulation of the FAS two grid scheme for composite grids:

$$\begin{cases} (i) & Pre-smoothing & \bar{w}_h := w_h^{\nu_1} = RELXLV^{\nu_1}(w_h^{(0)}, L_h, f_h, u_H) \\ (ii) & Computation\ of\ residuals & r_h := f_h - L_h \bar{w}_h \\ (iii) & Restriction\ of\ residuals & r_H := I_h^H r_h \\ (iv) & Exact\ solution\ of\ the & L_H u_H = F_H \quad from\ (5)\ using\ \bar{w}_h\ instead\ of\ u_h \\ & coarse\ grid\ problem & \bar{u}_H := u_H \\ (v) & Computation\ and\ inter- & v_H := u_H - \hat{I}_h^H \bar{w}_h \\ & polation\ of\ the\ correction & v_h := I_H^h v_H \\ (vi) & Correction & \tilde{w}_h := \bar{w}_h + v_h \\ (vii) & Post-smoothing & w_h^{(1)} := RELXLV^{\nu_2}(\tilde{w}_h, L_h, f_h, \bar{u}_H) \end{cases} \tag{6}$$

The formal index H marks coarse grid quantities on grid G_H. L_H is a reasonable discretization of L on G_H, for instance analogous to L_h. $\hat{I}_h^H$ and I_h^H denote restrictions which map fine grid functions onto coarse grid functions. $\hat{I}_H^h$ and I_H^h stand for prolongation (interpolation) operators which calculate fine grid values from coarse grid values. Linearization is done locally, using nonlinear relaxation methods for smoothing.

The coarse grid correction equation, step (iv) of (6), for points of $G_h^\circ \cap G_H^\circ$ is given by

$$L_H u_H = L_H \hat{I}_h^H w_h^{\nu_1} + I_h^H (f_h - L_h w_h^{\nu_1}) \tag{7}$$

with some restriction $\hat{I}_h^H$ which may be different from I_h^H. This equation will be used again in section 3.

For G_h a refinement, the interior boundary $G_h - G_h^\circ$ has to be treated carefully. The determination of function values on interior boundaries requires cubic interpolation $\hat{I}_H^h$ in order to maintain the order of consistency. The aforementioned conventions simplify these interpolations. Other techniques like 'interpolation by relaxation of the interior boundaries' are presented in Ritzdorf [6]. Both approaches have similar properties with regard to accuracy and convergence.

3 The determination of the refinement

The *a priori* selection of refinement areas is extremely difficult. This is the reason for adaptive strategies which determine the critical regions as the solution process progresses. There are essentially two approaches: the first one uses physically based information of the problem to refine grids, while the second one, minimizing the local discretization error, is numerically motivated and is, from the numerical point of view, more reliable. The principle is extendable to systems of equations, to higher order equations and to problems in more dimensions. Even for nonlinear cases, the FAS-scheme (6) offers a numerically cheap possibility to estimate the local discretization error with the aid of two consecutive grids. Rewriting the coarse grid equation (7), with u_h instead of $w_h^{\nu_1}$, as

$$L_H u_H = L_H \hat{I}_h^H u_h + I_h^H f_h - I_h^H L_h u_h = \tau_h^H + f_H \tag{8}$$

introduces the 'relative local discretization error of G_H and G_h' $\tau_h^H := L_H \hat{I}_h^H u_h - I_h^H L_h u_h$. For mesh sizes $H = 2h$ and second order of consistency, the local discretization error τ_H and the quantity τ_h^H are related by $\tau_H \doteq \frac{4}{3}\tau_h^H$. Because we have $\tau_H \doteq \tau_h + \tau_h^H$ the relative local discretization error τ_h^H equals the discretization errors of interest, τ_h and τ_H on G_h and G_H, respectively, up to high order terms ($\doteq$) and a factor depending on the mesh size ratio and the order of consistency. τ_h^H is an inherent information of the FAS scheme and is cheaply analyzed on the coarse grid G_H in order to determine where to refine the h-grid by a $\frac{h}{2}$-patch.

One condition to control the refinements is

$$\mid \tau_h^H(x,y) \mid \leq \varepsilon_{ref} * \tau^{sup} \quad \text{with} \quad \tau^{sup} := \sup_{(x,y)\in G_H} \mid \tau_h^H(x,y) \mid, \tag{9}$$

with a reasonable $\varepsilon_{ref} < 1$. Modifications of this condition are possible. If (9) is violated for some $\tau_h^H(x,y)$ the point (x,y) should be within a refinement. A proper choice for second order consistency and mesh size ratios of $\frac{1}{2}$ is $\varepsilon_{ref} = \frac{1}{4}$ because the local discretization error decreases by a factor of $\frac{1}{4}$ as the meshsize is reduced by $\frac{1}{2}$.

Having found a condition for the decision 'where' to refine a grid the next question is 'when' to adapt the grid. As we don't have exact solutions but only approximations the aforementioned quantities change within the solution process in time as well as within the MG-iteration. This raises the question: when is the local discretization error approximated well enough by τ_h^H for the grid adaption to be performed? In other words: when is the approximate solution accurate enough to ensure a good approximation of the local discretization error by τ_h^H? This in fact occurs only after some iterations and is demonstrated by the values of Table 4 in section 7. Therefore, for the model problem considered, the τ_h^H analysis is performed at the end of each time step in order to fix the grid for the following time step.

4 Smoothing factors depending on the time step size

Classical single grid methods such as ADI or SOR are not used here, although they can be efficient even for nonlinear problems; since the convergence speed of these methods slows down with increasing time steps. Indeed, this application needs to reach a given time $t > 0$ with a numerical effort as small as possible, which means for a given algorithm 'with time steps as large as possible'. To verify that the well known MG properties for every time t and especially for large Δt are maintained, a smoothing analysis with respect to time step size has been carried out under the usual assumptions like 'infinite grid, neglected boundary conditions, ideal coarse grid correction operator' and with some additional conditions for this problem. For several values of $t > 0$ and different Δt the smoothing analysis was done for nonlinear relaxation schemes with local linearization by one Newton step. To cover the whole variety of values at different locations, the smoothing factors are calculated for all inner points of the computational domain.

The results for the analyzed lexicographic point Gauss-Seidel-Newton method (PGSN) and lexicographic column Gauss-Seidel-Newton relaxation (CGSN) show an asymptotic behaviour for different $t > 0$ and increasing Δt. Upper bounds are the corresponding smoothing rates for the discrete Laplacian Δ_h. This is demonstrated by Figure 3 where $\max_{x \in G_{h_6}} \tilde{\mu}(h_6, \Delta t)$ is shown [1].

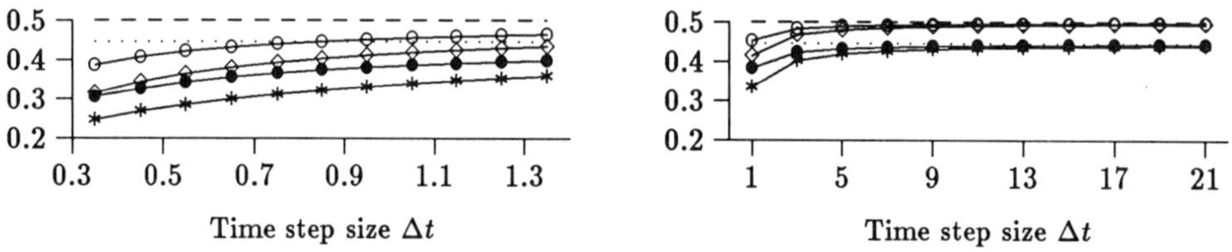

Figure 3: Smoothing rates for different Δt on G_{h_6}, t=0.05 (left) and t=80.0 (right), $\circ \hat{=}$ PGSN,BE, $\bullet \hat{=}$ CGSN,BE, $\diamond \hat{=}$ PGSN,CN, $* \hat{=}$ CGSN,CN

The left part of Figure 3 shows the maximum of all computed smoothing factors at the beginning of the simulation while the right part of the picture represents the values for an increased time. The asymptotic behaviour with respect to Δt is obvious. As expected CN is superiour to BE and the corresponding values of CGSN are better than those of PGSN.

G_{h_6}	PGSN-lex			CGSN-lex		
$t = 80 \quad \Delta t = 15$	CN	BE	*difference*	CN	BE	*difference*
$(\tilde{\mu}(h_6))^3$	0.1196	0.1220	0.0024	0.0835	0.0863	0.0028
ρ_{av} -(V(2,1))	0.1032	0.1075	0.0043	0.0811	0.0815	0.0004
difference	0.0164	0.0145	0.0019	0.0024	0.0048	0.0024

Table 1: Empirical convergence rates ρ_{av} compared with estimated values

Smoothing analysis is a tool which is not only applicable to idealized cases or problems of only theoretical interest. This is demonstrated by comparing the empirical convergence rates ρ_{av} and estimated convergence rates. For different mesh sizes, at a time of 80 minutes with $\Delta t = 15$ minutes for a $V(2,1)$-cycle, the empirical convergence rates are in good agreement

[1] A value for 'h_6' is given in section 7.

with the predicted convergence rate obtained from the third power of the smoothing rates. Table 1 contains values for PGSN and CGSN with CN and BE scheme on grid G_{h_6} which show the quality of the smoothing analysis in this case.

Another experience of numerical experiments is the fact that the often recommended red-black relaxation for this time-dependent problem does not show the behaviour which is known for Helmholtz-type equations. This type of relaxation turns out to be well suited only for for large t and on fine grids. An explanation may be the 'red-black induced' coupling of high and low frequencies of this stiff problem.

5 Time step size control

The process simulation has to provide an accurate doping profile at every time $t > 0$. To justify the implicit approach using an 'elliptic' solver for every time step, Δt should be as large as possible to reduce the computational effort. Therefore, the time step size is chosen by a time stepping procedure which is based upon well-known extrapolation techniques. Large time steps may cause convergence problems for the 'elliptic' solver used, although they are necessary for the algorithm to be competitive with methods like ADI and SOR. The smoothing analysis shows that good smoothing properties will not disappear with increasing time step size. On the other hand it is known that the Crank-Nicolson scheme for nonlinear operators has to satisfy a stability condition. In this context it is of interest to know whether this condition might be too restrictive.

An estimation of the stability condition may be derived by considering the Crank-Nicolson scheme which leads to a stability condition of the form

$$\frac{|1 + \frac{\Delta t}{2}\lambda_{lm}(N_h^n)|}{|1 - \frac{\Delta t}{2}\lambda_{lm}(N_h^{n+1})|} < 1,$$

with solution-dependent and time-dependent eigenvalues λ_{lm} of L_h. Replacing the space operator $L_h(N_h^n)$ for instance by an approximation $\Lambda_h(N_h^n) := \widehat{D(N_h^n)}\Delta_h$ the stability requirement can be estimated. Using $\widehat{D(N_h^n)} = max_{x \in G_h} D(N_h^n)$ a numerical experiment shows that the estimated stability condition will not restrict the time step size too much.

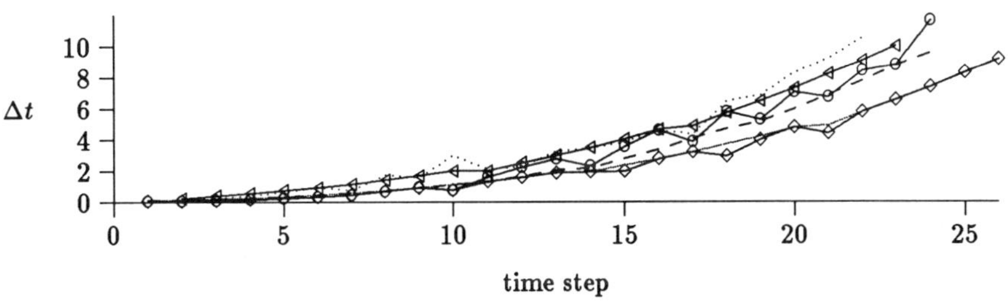

Figure 4: $\circ \hat{=}$ MGI-43[2]cntrl, $\diamond \hat{=}$ MGI-43 cntrl/stab, $\triangleleft \hat{=}$ MGI-43 stab

[2]MGI-43 $\hat{=}$ three cycles per time step on grid G_{h_4}

Figure 4 evidently shows that time step size control (cntrl) and stability condition (stab) lead to Δt's which are of a similar size. A simulation controlled by the time stepping procedure ($\circ$) proposes time step sizes which are close to the estimated CN stability condition (dashed line). Another experiment using only Δt due to the stability condition ($\triangleleft$) shows the time step size proposed by the step size algorithm (dotted line) again in close agreement. From the numerical point of view a combination of both methods is recommended ($\diamond \hat{=}$ min(stab,cntrl)).

For t small, the theoretical advantage of CN as against BE was not observed in practice. This prompted the analysis of the influence of nonlinearity and boundary movement on the overall behaviour of the algorithm.

The following diagram separates the influence of both phenomena, shown for CN. The upper left corner gives the number of time steps, used for a 1000 minutes simulation with fixed domain, implying 'no oxidation' and constant diffusion coefficient, i.e. a 'linear problem'. Horizontally, the movement of the boundary is added to the simulation experiment. Vertically the nonlinearity complicates the problem. In combination with BE as well as in combination with CN the nonlinearity causes more problems than the time dependent domain. With CN the nonlinearity is much more restrictive than for BE.

$$\begin{array}{ccc} (D_0, \quad \Omega_0) : 15 & \xrightarrow{\quad +4 \quad} & (D_0, \quad \Omega(t)) : 19 \\ +13 \downarrow & \text{CN} & \downarrow +18 \\ (D(N), \Omega_0) : 28 & \xrightarrow[\quad +9 \quad]{} & (D(N), \Omega(t)) : 37 \end{array}$$

Figure 5: CN – Number of time steps used with increasingly complex problems

6 Multigrid Components

Although the basic idea of the MG-technique is quite simple, the construction of efficient methods requires some careful tuning of the algorithmic components of which each multigrid solver is composed. For the model problem under consideration, Joppich [3] contains a detailed discussion of both the multigrid components and results for different components.

The results of the following section are based on a nonlinear Gauß-Seidel iteration using a full weighting (I_h^H) which is modified near the moving boundary $B_4(t)$. The discretized boundary condition at $B_4(t)$ is treated separately. The FAS restriction ($\hat{I}_h^H$) and the transfer of the corrections (I_H^h) are standard (injection, bilinear interpolation, respectively). Using the FMG algorithm to solve the problem for every time step, bi-cubic interpolation is in this case not the best choice to interpolate approximate solutions, even with only global grids: FMG has to start on fine grid levels and additional cycles become necessary in order to obtain convergent solutions. Monotone interpolation methods with sufficiently high order of accuracy, like the one of Carlson and Fritsch [2], permit standard FMG procedures: starting on the coarsest level without additional cycles the algorithm converges and the amount of work is reduced considerably.

By diffusion, the doping concentration redistributes and the region of interest will change position and shape. Grid functions have to be calculated carefully on new refinement areas which have been created by the last grid adaption. This calculation can be done by interpolation of coarse grid values because the grids are nested. Using standard bicubic polynomial interpolation which is sufficient in many applications leads to oscillations which will not disappear automatically. 'Monotone' interpolation methods overcome this difficulty. So they are not only used as FMG interpolation but also to calculate the right hand side on 'new refinements' and they are suitable for the correct treatment of the interior boundaries, which means 'with sufficiently high order to maintain consistency'. The influence of different interpolation techniques with sufficiently high approximation order will be discussed in Joppich and Lorentz [4].

7 Numerical experiments with local refinements

Using the presented principles in an MG algorithm for the model problem (1)-(3) of section 1, on a given geometry (see Figure 1), of initial interest is the overall behaviour of the algorithm. From the MG point of view it is interesting to know whether the expected h-independent convergence is achieved. Another point of interest is the amount of computational work which is necessary to provide a good discretization error estimate for the grid adaption. Before briefly discussing these aspects with the aid of some numerical results, the conditions for the underlying experiments should be mentioned.

Starting with a coarsest grid G_{h_1} ($h_1 = \frac{1}{6}$) the grid sequence generally continues up to G_{h_6} ($h_6 = \frac{1}{192}$) and in some cases up to G_{h_7} ($h_7 = \frac{1}{384}$). The numerical effort is measured in 'work units (WU)' where $1WU$ stands for one relaxation sweep over the finest grid, considered as a global grid. The numerical experiments simulate diffusion steps until $t = 80$ minutes. For this situation, Table 2 contains information about the number of time steps needed (#), the largest time step used (Δt_{max}) and the total number of WU (ΣWU), including the effort for time step control (one Δt step and two $\frac{\Delta t}{2}$ steps).

To find a lower bound for the efficiency gain by MG with local refinements, the ε_{ref} in (9) is chosen such that the refinements nearly degenerate to global grids at the end of the experiment. Each grid sequence used consists of at least two global grids (grid$_{glob}$) and at least two refinement levels κ. In the following tables, for a given number of global grids 'grid$_{glob}$' abbreviations like 'MGI-jk-κ' stand for k MG iterations using a finest mesh size h_j and $\kappa = j -$grid$_{glob}$ refinement levels.

In Table 2 MG iterations with only one cycle per time step show the lowest number of work units ΣWU. On the other hand, the considerably larger number of time steps indicates a poor approximation of the discretization error in time and as a consequence a non-optimal time stepping. Comparing results with two cycles per time step shows the improved behaviour. It should be noted that the total amount of computational work ΣWU is not doubled with two

MG iterations per time step. The number of time steps for different grid sequences is approximately constant and will not change significantly with an increased number of MG cycles. The same behaviour is observed with standard FMG-cycles using monotone interpolation methods wherever interpolation is required.

$j = 4,5,6$ $\kappa = j-\text{grid}_{glob}$	G_{h_4}			G_{h_5}			G_{h_6}		
method	#	Δt_{max}	ΣWU	#	Δt_{max}	ΣWU	#	Δt_{max}	ΣWU
$\text{grid}_{glob} = 2$									
MGI-j2-κ	19	14.3	263	21	13.4	234	21	16.9	195
MGI-j1-κ	22	10.2	159	26	11.7	152	29	9.1	159
$\text{grid}_{glob} = 3$									
MGI-j2-κ				23	13.8	271	23	22.1	228
MGI-j1-κ				25	11.8	150	28	12.0	145

Table 2: Simulations on locally refined grids, overall behaviour

As already mentioned, the h-independent convergence rate is of practical and theoretical interest. Usually, this value is approximated by the defect reduction after a certain number of cycles. Table 3 presents such values for the model problem with two global grids and four refinement levels for the sixth and seventh MG iteration (ρ_6 and ρ_7) using red-black relaxation. The empirical behaviour is in good accordance with theoretical expectations. The reduction is approximately the same over all grids and the similarity of ρ_6 and ρ_7 is due to the asymptotic behaviour.

$t = 80 \quad \Delta t = 15$	G_{h_2}	G_{h_3}	G_{h_4}	G_{h_5}	G_{h_6}
$\rho_6 - L_2$	*[3]	0.0419	0.0418	0.0417	0.0421
$\rho_7 - L_2$	*	*	0.0421	0.0420	0.0422

Table 3: Defect reduction on a sequence of grids with four refinement levels

The numerical effort is tightly coupled with the size of the refinement area. A natural question is whether a better approximation of the discretization error by τ_h^H might reduce the area of the patches and as a consequence might reduce the total work. But as τ_h^H becomes a really good approximation of the discretization error only after some iterations the gain of smaller refinements might be lost by these additional computations. The history of τ^{sup} and

[3] * marks the fact that the total defect reduction is about 10^{-15}. In this case of attained machine accuracy, the actual defect reduction rate differs from the asymptotic behaviour. This situation is reached first on coarse grids.

$\|\tau_h^H\|_{L_2}$ for an increasing number of cycles offers an orientation for choosing an acceptable number of MG iterations before adapting the grid.

Table 4 shows the development of relative deviations of both quantities from those values which are adopted at convergence. The grid sequence used, consists of G_{h_1} - G_{h_7}. τ_h^H has been analyzed on the two finest grid levels G_{h_5} and G_{h_6} where it is defined. The values which are adopted at convergence are approximated up to 10^{-5} within four or five MG iterations. Nevertheless, for more than three iterations a significant shrinking of the patch area that justifies the additional computational work was not observed.

$grid_{glob} = 2$	G_{h_5}		G_{h_6}	
method	τ^{sup}	$\|\tau_h^H\|_{L_2}$	τ^{sup}	$\|\tau_h^H\|_{L_2}$
MGI-71-5	43.877	2.668	4.116	6.832
MGI-72-5	1.112	0.029	0.058	0.093
MGI-73-5	0.025	0.005	0.003	0.004
MGI-74-5	0.004	0.0	0.0	0.0

Table 4: Relative deviations of τ_h^H-values, finest grid used: G_{h_7}

To summarize this section, there are essentially three results. The use of local refinements does not disturb the expected MG convergence properties. Two or three cycles are sufficient for a good approximation of the discretization error, whereas more than three cycles are not justified, from the grid adaption point of view. With two MG cycles a satisfactory time stepping is possible. Similar experiments using FMG give the same result for the standard FMG-procedure.

8 Conclusions

The presented MG approach for solving an initial boundary value problem arising from process simulation shows its potential to solve more general problems. Therefore, the future work will be to increase the class of application fields.

The use of local refinements for systems of equations with more complex interior boundaries in combination with parallelization will pose questions which are not yet answered. One of these questions has to deal with the use of 'point minimizing' refinements or refinements 'composed of simply structured patches' in order to get an optimal efficiency. When treating systems of equations, the control quantity τ_h^H for every equation should be supplemented by an information which takes the coupling of the system into account. For a collective point relaxation with Newton-like linearization the local Jacobi matrix may give such information about the local coupling of the system which can be used for grid refinement.

Important for every code to be used in practice are reliable termination criteria which determine -for a given accuracy with respect to time and space- the iterative work, the time stepping, the mesh sizes and the number of refinement levels. Because there is not yet much experience with 'parabolic multigrid' this approach will be investigated especially for systems of nonlinear equations. Perhaps the questions of how to find and how to use coupling information, as well as reliable termination criteria, will be answered by this approach.

References

[1] Brandt, A. *Multigrid techniques: 1984 guide with applications to fluid dynamics*, GMD-Studie No. 85, GMD, Sankt Augustin, 1984

[2] Fritsch, F. N., Carlson, R. E. *Monotone Piecewise Cubic Interpolation*, SIAM J. NUMER. ANAL. Vol. 17, No. 2, [119–138], April 1980

[3] Joppich, W., *Mehrgitterverfahren für Diffusionsprobleme der Prozeßsimulation*, Dissertation, Universität Bonn, 1990

[4] Joppich, W., Lorentz, R. A., *High-order, Positive, Monotone and Convex Multigrid Interpolation*, in preparation, 1990

[5] Maldonado, C. D., *ROMANS II : A Two-Dimensional Process Simulator for Modeling and Simulation in the Design of VLSI Devices*, Applied Physics (Solids and Surfaces), A31 [119–138], 1983

[6] Ritzdorf, H., *Lokal verfeinerte Mehrgitter-Methoden für Gebiete mit einspringenden Ecken*, Diplomarbeit, Universität Bonn, 1984

[7] Stüben, K., Trottenberg, U., *Multigrid Methods: Fundamental Algorithms, Model Problem Analysis and Applications*, Lecture Notes in Mathematics, 960 [1–176], 1982

International Series of Numerical Mathematics, Vol. 98, © 1991 Birkhäuser Verlag Basel

Multigrid Methods for Hyperbolic Equations

Edgar Katzer

Institute of Informatics and Applied Mathematics,
Christian-Albrechts University, D-2300 K I E L, F. R. G.

Abstract

A multigrid method for solving hyperbolic partial differential equations is presented. A second order conservative discretisation of the differential equation is obtained by a cell vertex scheme which has been introduced by Ni [1]. For the smoothing iteration a truncated Lax-Wendroff scheme is applied. A frequency decomposition approach uses multiple coarse grid corrections and avoids the need for numerical dissipation. The coarse grid system is a modified Galerkin product. Numerical results for a two-grid iteration for a linear periodic initial value problem are presented. Convergence rates of 0.5 are observed. Convergence deteriorates when the characteristics are aligned with the grid.

1. Introduction

Multigrid methods are efficient tools for solving partial differential equations. In contrast to the many results on elliptic equations (see Brandt and Dinar [2], Hackbusch and Trottenberg [3] and Hackbusch [4] for further references), the applications on non-elliptic and hyperbolic equations are not satisfactory. Especially the optimal convergence rates observed with elliptic problems are not obtained with non-elliptic problems (see the review of Kroll [5]). Several partial differential systems of practical importance, for example the Euler equations of gas dynamics, are not elliptic or are singular perturbed problems, e. g. the Navier-Stokes equations. It should be noted, that I am not recommending the use of multigrid methods for *purely* hyperbolic problems, because then alternative procedures could possibly be applied. For example the method of characteristics reduces the complexity of the problem by one dimension and leads to a very efficient algorithm. But for a nonlinear mixed elliptic/hyperbolic problem a multigrid method with high efficiency in the elliptic and in the hyperbolic regime is required. This motivates our research on hyperbolic equations.

As a linear hyperbolic system may be decoupled into a system of scalar equations, we restrict ourself to a scalar model problem with constant coefficients:

$$\mathcal{L}\, u = a_1\, \partial_x u + a_2\, \partial_y u = 0 \tag{1}$$

$$a_1 \neq 0 \;, \quad a_2 > 0 \;, \quad a_1^2 + a_2^2 = 1$$

on

$$\Omega = [\,0,1\,]^2 \quad , \qquad u \in C^1_{x-per}(\Omega) \quad .$$

We will analyse a periodic initial value problem:

$$u(x,0) = u_0(x) \quad , \qquad u(1,y) = u(0,y) \qquad \text{(periodic!)} \quad .$$

Furthermore we define a time dependent augmented problem:

$$\partial_t u = -\mathcal{L}u \tag{2}$$

and assume, that the solution of (1) is an asymptotic solution ($t \to \infty$) of the augmented unsteady problem.

2. Discretisation

A second order cell vertex scheme, similar to those of Ni [1], Hall [6], Jameson [7], and Kroll et al. [8], is applied for discretising the differential operator (1). The discrete equations yields:

$$L\,u_l = f_l \quad . \tag{3}$$

Here, u_l and f_l are real (or complex) grid functions on equidistant grids:

$$\mathcal{U}_l = \{u_l : \Omega_l \to \mathbf{C}\} \quad , \qquad \mathcal{F}_l = \{f_l : \Omega_l^* \to \mathbf{C}\}$$

$$\Omega_l = \{(x_j, y_k) = (jh_l, kh_l)\,,\ 1 \le j,k \le n_l\}$$

$$\Omega_l^* = \Omega_l - (\tfrac{1}{2}h_l, \tfrac{1}{2}h_l) = \{(x - h_l/2\,,\ y - h_l/2)\,|\,(x,y) \in \Omega_l\}$$

$$h_l = 1/n_l = 2^{-l} \quad , \qquad l \in \mathbf{N} \quad .$$

Grid points of Ω_l are vertices of quadrilateral cells with cell centers indicated by points in Ω_l^* (see figure 3). We define difference operators on grid functions:

$$\begin{aligned}
(\Delta_x u)(x,y) &= u(x+\tfrac{1}{2}h_l, y) - u(x-\tfrac{1}{2}h_l, y)\\
(\Delta_y u)(x,y) &= u(x, y+\tfrac{1}{2}h_l) - u(x, y-\tfrac{1}{2}h_l)\\
(I_x u)(x,y) &= \tfrac{1}{2}(u(x+\tfrac{1}{2}h_l, y) + u(x-\tfrac{1}{2}h_l, y))\\
(I_y u)(x,y) &= \tfrac{1}{2}(u(x, y+\tfrac{1}{2}h_l) + u(x, y-\tfrac{1}{2}h_l)) \quad .
\end{aligned}$$

The differential operators are approximated by:

$$\partial_x \simeq \delta_x = \tfrac{1}{h_l} I_y \Delta_x \quad , \qquad \partial_y \simeq \delta_y = \tfrac{1}{h_l} I_x \Delta_y \tag{4}$$

$$\Delta_{xy} = \Delta_y \Delta_x$$

and we obtain:

$$L = a_1\,\delta_x + a_2\,\delta_y \quad . \tag{5}$$

Notice, that for simplicity the subscript l is omitted and we will not distinguish between operators defined on different domains $\mathcal{U}_l$ and $\mathcal{F}_l$. This will not cause any confusion, because it will always be clear on which domains the operators are defined.

3. Smoothing

The discrete system (3) may easily be solved by a marching procedure in y-direction. Furthermore, for a nonperiodic initial value problem, a Gauß-Seidel iteration solves the system with one iteration, provided that the sweep direction is aligned with the characteristic direction. These results would not be representative for systems of partial differential equations which may have different characteristic directions or may be elliptic. For these reasons, we will analyse Jacobi type smoothing procedures only. As the augmented problem (2) is hyperbolic with respect to time, smoothing iterations are constructed by a time marching procedure. Discretisation of the time derivative:

$$\partial_t \simeq (u^{n+1} - u^n)/\tau$$

yields a semi-discrete version of the Lax-Wendroff scheme:

$$u^{n+1} = u^n - \tau \mathcal{L} u^n + \tfrac{1}{2}\tau^2 \mathcal{L}^2 u^n \quad . \tag{6}$$

Also, semi-discrete Runge-Kutta time marching procedures (Jameson [7], Jameson et al. [9]) may be defined as polynomials in spacial operators:

$$\begin{aligned} u^{n+1} &= S^{RK} u^n \\ S^{RK} &= \textstyle\sum_{i=0}^{m} \alpha_i(\tau) \mathcal{L}^i \quad . \end{aligned} \tag{7}$$

Here, numerical dissipation is neglected and the coefficients $\alpha_i(\tau)$ have to fulfil several stability and consistency conditions. The time marching approach motivates the construction of Jacobi type smoothing iterations:

$$S = I - \omega_1 Q_1 + \omega_2 Q_2 - \omega_3 Q_3. \tag{8}$$

with damping coefficients $\omega_1, \omega_2, \omega_3 \in \mathbf{R}$. The choice of:

$$\begin{aligned} \omega_1 &= \tau \quad , \quad & Q_1 &= I_{xy} L \\ \omega_2 &= \tfrac{1}{2}\tau^2 \quad , \quad & Q_2 &= L^{\mathsf{T}} L \end{aligned} \tag{9}$$

gives the Lax-Wendroff scheme proposed by Ni [1]. Stability requires:

$$\tau \le h_l / \sqrt{a_1^2 + a_2^2} = h_l \quad .$$

Operator Q_3 defines numerical dissipation. A particular choice is a fourth order dissipation term:

$$Q_3 = \tfrac{1}{16} {\Delta_{xy}}^{\mathsf{T}} \Delta_{xy} \quad .$$

Other dissipation operators have been proposed by Ni [1], Jameson et al. [7, 9] and Decker and Turkel [10].

Notice, that the transposed Matrix L^{T} approximates $\mathcal{L}$ and the resulting scheme (8) is consistent with the semi-discrete Lax-Wendroff scheme (6).

A local mode analysis gives the amplification rates

$$\begin{aligned} S\, e^{\theta_1,\theta_2} &= \lambda^{\theta_1,\theta_2}(S)\, e^{\theta_1,\theta_2} \\ \lambda^{\theta_1,\theta_2}(S) &= 1 - i\omega_1\lambda_1 - \omega_2\lambda_2^2 - \omega_3\lambda_3 \\ \lambda_1 &= \lambda_2 \cos(\theta_1/2)\cos(\theta_2/2) \\ \lambda_2 &= 2(a_1 \sin(\theta_1/2)\cos(\theta_2/2) + a_2 \sin(\theta_2/2)\cos(\theta_1/2)) \\ \lambda_3 &= \sin^2(\theta_1/2)\sin^2(\theta_2/2) \end{aligned} \tag{10}$$

of the frequency modes:

$$e^{\theta_1,\theta_2}(x_j, y_k) = \exp i(\theta_1 j + \theta_2 k)$$

$$\theta_1 = 2\pi\kappa_1 h_l \ , \qquad \theta_2 = 2\pi\kappa_2 h_l \in]-\pi, \pi\,]$$

$$\kappa_1, \kappa_2 \in \mathbf{Z} \bmod n_l \quad .$$

The amplification rates are reduced when $\omega_1 = 0$ and the imaginary part vanishes. The resulting scheme will be called **truncated Lax-Wendroff scheme:**

$$S = I + \omega_2\, Q_2 \tag{11}$$

and is a distributed Jacobi iteration with distribution operator L^{T}. (For details on distributive iterations see [2] and Wittum [11].) Clearly the resulting truncated scheme is no longer consistent with the time marching approach, but this is not essential, unless we are interested in the unsteady problem.

For the case $a_1 = a_2$, figure 1 shows the modulus of the amplification rates for the Ni-Lax-Wendroff scheme (9) and the truncated scheme without numerical dissipation ($\omega_3 = 0$). The region of good smoothing rates ($|\lambda^{\theta_1,\theta_2}(S)| \leq 1/2$) is hatched. The truncated scheme has a larger smoothing region. Similar results are observed for $a_1 \neq a_2$.

The region of low smoothing rates consists of three parts:

1. The low frequency range: $\theta_1 \approx \theta_2 \approx 0$.

2. Numerical characteristic modes: $\lambda_2 \approx 0$.
These are modes which propagate approximately orthogonal to the characteristic direction (a_1, a_2) (see also Brandt [12]):

$$\lambda_2 = \left\langle \begin{pmatrix} a_1 \\ a_2 \end{pmatrix}, \begin{pmatrix} \sin(\theta_1/2)\cos(\theta_2/2) \\ \sin(\theta_2/2)\cos(\theta_1/2) \end{pmatrix} \right\rangle \approx 0 \quad . \tag{12}$$

3. The high frequency range: $|\theta_1| \approx |\theta_2| \approx \pi$.

Results of Gustavson and Lötstedt [13] and the author's experience suggest, that low frequency errors are mainly reduced by convection when a time marching approach is used. Thus, with the Lax-Wendroff scheme, the low frequency errors are propagated out of the domain, provided that nonreflecting boundary conditions are applied. When a multigrid approach is used, low frequency errors are reduced by the coarse grid correction.

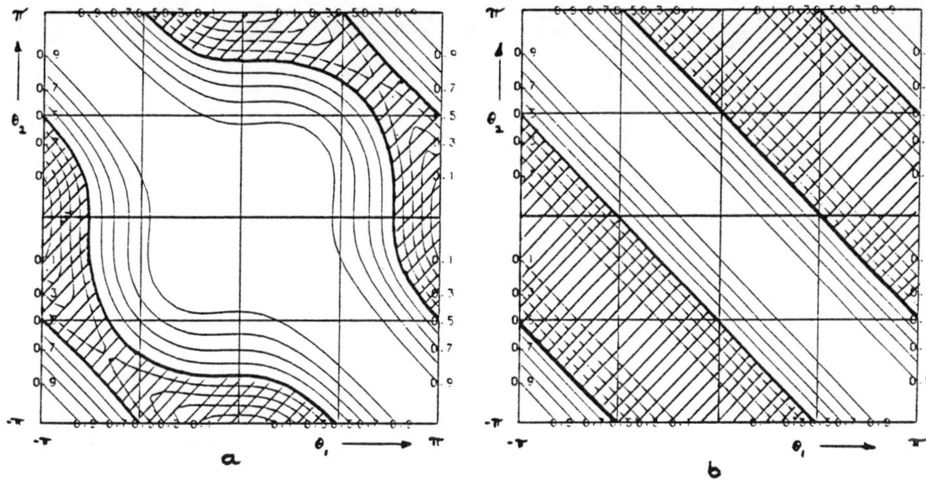

Figure 1: Amplification rates for a) Ni-Lax-Wendroff and b) truncated scheme.

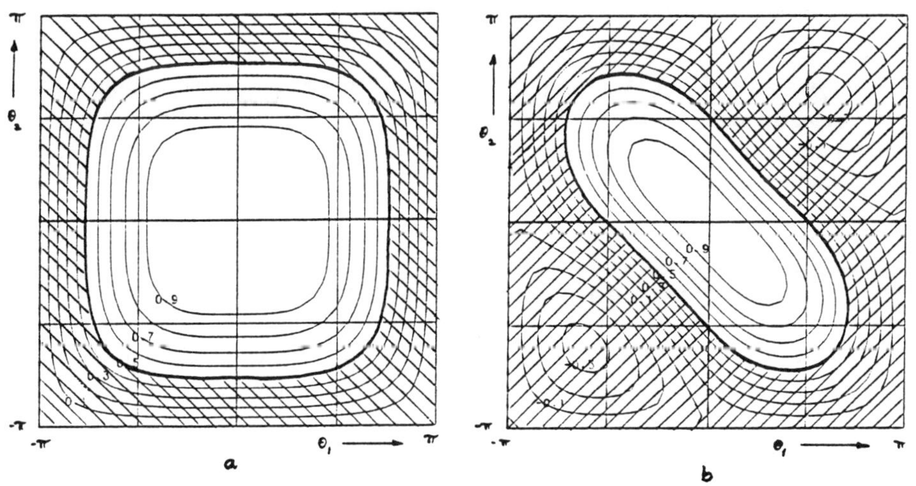

Figure 2: Same as in figure 1 but with numerical damping ($\omega_3 = 1$).

Then there is no need for error propagation with a time accurate iteration and we may omit the term Q_1 in (8) and obtain the truncated scheme (11) which has better smoothing rates.

For consistency reasons, low fequency errors ($\sin\theta \approx \theta, \cos\theta \approx 1$) propagating orthogonal to the characteristics are decoupled. They cannot be reduced by a local smoothing iteration. For higher frequencies, the sine and cosine functions modify the scalar product (12) and the numerical characteristics are different from the continuous case.

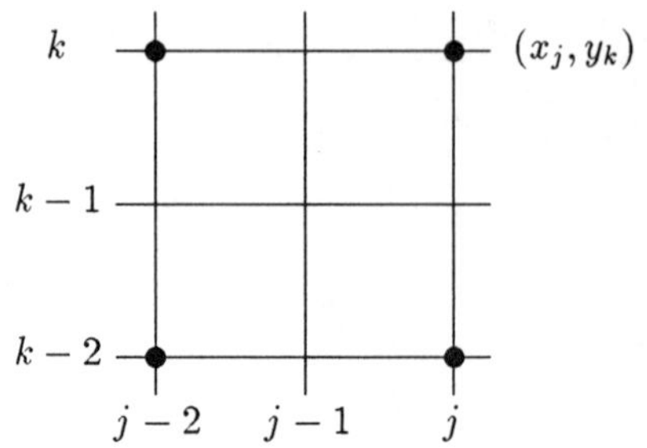

Figure 3: Coarse-grid points • and fine-grid cells.

Bad damping of high frequency modes is commonly observed when first order derivatives are approximated with second order accuracy. This results in a decoupling of grid points with even from those with odd indices $(j + k)$. This even-odd decoupling is usually remedied by adding numerical dissipation. Figure 2 shows amplification rates for the truncated scheme with numerical dissipation $(\omega_3 = 1)$. The region of good smoothing is largely increased and the resulting iteration may be used for the smoothing step in multigrid schemes. The disadvantage of dissipation is, that the limit of the smoothing iteration is no solution of the discrete system.

A frequency decomposition approach introduced by Hackbusch [14,15] avoids the need for numerical dissipation. Here, a second coarse grid correction is applied for reducing the high frequency errors. Furthermore, multiple coarse grid corrections will be applied to damp characteristic mode errors.

4. Frequency decomposition approach

Multiple coarse grid corrections are applied for reducing low and high frequency modes which are not damped by the smoothing iteration. Low frequency errors are removed by a standard coarse grid correction. As the coarse grid cells are a union of four adjacent fine grid cells (figure 3), the coarse grid defect is defined as an average of four fine grid defects:

$$\begin{aligned} r_1 &: \mathcal{F}_l \rightarrow \mathcal{F}_{l-1} \\ r_1 &= R^l_{l-1}\, I_{xy} \quad . \end{aligned}$$

The restriction is represented by the following stencil:

$$r_1 = \frac{1}{4}\begin{bmatrix} 1 & 1 \\ 1 & 1 \end{bmatrix} \quad .$$

which gives the weights of the fine grid cells. Here R^l_{l-1} is the trivial restriction (trivial injection). For the prolongation operator we simply use bilinear interpolation:

$$p_1 : \mathcal{U}_{l-1} \rightarrow \mathcal{U}_l$$

with stencil:

$$p_1 = \frac{1}{4}\begin{bmatrix} 1 & 2 & 1 \\ 2 & 4 & 2 \\ 1 & 2 & 1 \end{bmatrix} \quad .$$

High frequency modes with $|\theta_1| \approx |\theta_2| \approx \pi$ are reduced by a specially designed coarse grid operator similar to those introduced by Hackbusch [14,15]. Following his notation, the operators are indexed by number 4. The restriction is:

$$r_4 = R_{l-1}^{l} \, \Delta_{xy} \quad .$$

$$r_4 = \frac{1}{4}\begin{bmatrix} -1 & 1 \\ 1 & -1 \end{bmatrix}$$

and the prolongation is an alternating sign prolongation with stencil:

$$p_1 = \frac{1}{4}\begin{bmatrix} 1 & -2 & 1 \\ -2 & 4 & -2 \\ 1 & -2 & 1 \end{bmatrix} \quad .$$

A modified Galerkin product is used as coarse grid operator and will be presented below.

The two coarse grid operators given above, are not sufficient for reducing errors with periodicity $2h_l$. This may be illustrated by special defects shown in figure 4. Errors, which result in such defects, vanish under the restrictions r_1 and r_4. Consequently they are not reduced by the coarse grid correction. For reducing these errors, a second coarse grid is used which is shifted by h_l in x-direction:

$$\Omega_{l-1}^{+} = \Omega_{l-1} + (h_l, 0) \quad . \tag{13}$$

Corresponding grid functions and restrictions and prolongations are introduced:

$$r_i^{+} : \mathcal{F}_l \rightarrow \mathcal{F}_{l-1} \quad , \qquad r_i^{+} \simeq r_i$$

$$p_i^{+} : \mathcal{U}_{l-1} \rightarrow \mathcal{U}_l \quad , \qquad p_i^{+} \simeq p_i$$

With coarse grid systems L_1, L_4, L_1^{+}, L_4^{+} defined below, we obtain coarse grid corrections:

$$\begin{aligned} G_i &= p_i L_i^{-1} r_i L \\ G_i^{+} &= p_i^{+} (L_i^{+})^{-1} r_i^{+} L \ , \quad i \in \{1,4\} \end{aligned} \tag{14}$$

and apply a two-grid iteration:

$$M_l^{TG} = (I - G_1^{+} - G_4^{+})\, S\, (I - G_1 - G_4)\, S \quad . \tag{15}$$

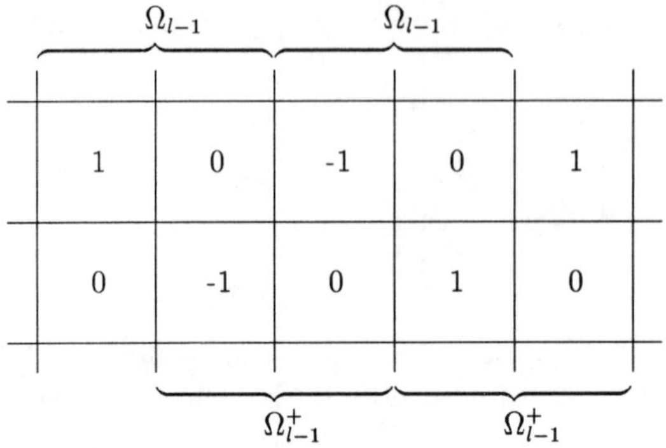

Figure 4: Defects on fine grid and coarse grids.

Clearly, more smoothing steps and post smoothing could be applied. The Galerkin coarse grid system is obtained by:

$$\tilde{L}_1 = r_1 L p_1 = a_1\,\delta_x + a_2\,\delta_y$$

and

$$\tilde{L}_4 = r_4 L p_4 = a_2\,\delta_x + a_1\,\delta_y \quad .$$

The coarse grid system $\tilde{L}_1$ is a consistent approximation to $\mathcal{L}$ on the coarse grid, but $\tilde{L}_4$ is consistent with a different equation.

With the Galerkin approach, the coarse grid corrections G_1 and G_4 are projections and their mixed products vanish:

$$G_1 G_4 = G_4 G_1 = G_1^+ G_4^+ = G_4^+ G_1^+ = 0 \quad .$$

Therefore additive and multiplicative coarse grid corrections are equivalent:

$$(I - G_1)(I - G_4) = I - G_1 - G_4 \quad . \tag{16}$$

The additive approach allows the calculation of the coarse grid operators in parallel and is more adapted for parallel computation.

The Galerkin coarse grid system gives good convergence rates for the two-grid iteration when $|a_1| = |a_2|$ but bad convergence or even divergence when $|a_1| \neq |a_2|$. The reason is a spatial instability of the coarse grid system. Therefore modified coarse grid operators are introduced:

$$\begin{aligned} L_1 &= \tilde{L}_1 + \frac{\epsilon}{h_{l-1}}\Delta_{xy} \\ L_4 &= \tilde{L}_4 + \frac{\epsilon}{h_{l-1}}\Delta_{xy} \quad . \end{aligned}$$

a_1	a_2	$h_l = 1/8$	$h_l = 1/16$	$h_l = 1/32$
1.0	1.0	0.481	0.495	0.499
-1.0	1.0	0.481	0.495	
0.4	0.6	0.752	0.831	0.845
0.6	0.4	0.752	0.831	
1	2	0.877	0.909	0.923
1	3	0.952	0.973	0.978
1	4	0.975	0.990	0.995
4	1	0.975	0.990	
1	9	0.992	1.00	

Table 1: Convergence rates for two-grid iteration, truncated Lax-Wendroff scheme.

Here, L_1 is still consistent with $\mathcal{L}$, but is only a first order approximation on the coarse grid:

$$L_1 = \mathcal{L} + \mathcal{O}(h_{l-1}) \quad , \qquad \text{on} \quad \Omega_{l-1} \quad .$$

Stability of L_1 and L_4 is obtained when:

$$\epsilon = \begin{cases} \frac{1}{2}|a_1 - a_2| & \Longleftarrow \quad a_1 a_2 > 0 \\ -\frac{1}{2}|a_1 + a_2| & \Longleftarrow \quad a_1 a_2 < 0 \end{cases} \quad . \tag{17}$$

The results in the following section are obtained using this modified Galerkin coarse grid system.

5. Results

Convergence rates for the two-grid iteration have been determined numerically and are presented in table 1. Here, the coefficients a_1 and a_2 are not normalised. The damping parameters are:

$$\omega_1 = \omega_3 = 0$$

$$\omega_2 = \frac{\sqrt{a_1^2 + a_2^2}}{4 \max\{a_1^2, a_2^2\}}$$

Good convergence rates of order 0.5 are obtained for the case $|a_1| = |a_2|$. For $|a_1| \neq |a_2|$, the convergence rates increase and they deteriorate when a_1 or a_2 approaches infinity. This should have been expected, because then the discrete problem (3) becomes singular when the (normalised) coefficient a_1 or a_2 is 0. Many authors have observed convergence rates of order 0.95 for the Euler equations. Our present results, which are well within this range and are better when $|a_1| \approx |a_2|$, are therefore quite satisfactory. Further improvement of the two-grid iteration requires the incorporation of additional coarse grid corrections which are specially designed to reduce errors aligned with the coordinate directions. Here, we plan to follow the ideas of Hackbusch [14,15].

6. Conclusion

A two-grid method for a hyperbolic model problem is presented. Smoothing properties of Ni's Lax-Wendroff scheme [1] for a time dependent augmented problem are analysed. With the time marching iteration, low frequency errors are primarily propagated through the computational domain. But there is no need for a time accurate iteration, because these erors are reduced by the coarse grid correction. A new truncated Lax-Wendroff smoothing iteration is obtained by discarding a first order term of the Ni scheme. We obtain a Jacobi type distributive smoothing iteration which is no more consistent with a time marching approach but has improved smoothing properties.

A tendency for decoupling of errors on even and odd grid points is usually remedied by adding numerical dissipation. Here, we propose an alternative approach. Multiple coarse grid corrections are defined, which are specially designed to reduce even-odd error components. These corrections are also used for the reduction of characteristic error modes. A modified Galerkin approach with improved stability is applied for the coarse grid system.

Numerical results for a two-grid iteration show convergence rates of 0.5 when the characteristics are parallel with the grid diagonals. Convergence deteriorates, when the characteristics are aligned with the grid lines. Future plans concern improvement of the convergence and applications to systems of first order partial differential equations.

7. References

[1] Ni, Ron-Ho: *A multiple-grid scheme for solving the Euler equations.* AIAA Paper 81-1025, 1981, or AIAA J. 20 (1982), pp. 1565-1571.

[2] Brandt, A.; Dinar, N.: *Multigrid solutions to elliptic flow problems.* in: Parter, S. V. (ed.): Numerical methods for partial differential equations. New York, N. Y.: Academic Press, 1979, pp. 53 - 147.

[3] Hackbusch, W. Trottenberg, U. (eds.): *Multigrid methods.* Köln, 1981. Proceedings: (Lecture Notes in Mathematics, 960.) Berlin, Heidelberg, New York: Springer, 1982.

[4] Hackbusch, W.: *Multi-grid methods and applications.* Berlin, Heidelberg, New York: Springer, 1985.

[5] Kroll, N.: *Einsatz von Mehrgittertechniken zur Berechnung stationärer reibungsloser Strömungen auf der Basis der Eulerschen Bewegungsgleichungen - Literaturstudie. (The use of multigrid methods based on the Euler equations for calculating steady inviscid flows. - A literature study.)* Braunschweig, (Germany): Institut für Entwurfsaerodynamik; German Aerospace Research Center, internal Report: DFVLR-IB 129 - 85/9, 1985.

[6] Hall, M. G.: *Cell-vertex multigrid schemes for solution of the Euler equations.* in: Morton, K., W.; Baines, M., J.; (eds.): Proc. Numerical Methods for Fluid Dynamics, II. Reading, 1985. Oxford: Clarendon Press, 1986, pp. 303 - 345.

[7] Jameson, A.: *Computational transonics.* Comm. Pure Appl. Math. 41 (1988), pp. 507 - 549.

[8] Kroll, N.; Radespiel, R.; Rossow, C.-C.: *Experiences with explicit time-stepping schemes for supersonic flow fields.* in: Wesseling, P.(ed.): Eighth GAMM-Conf. Numer. Methods in Fluid Mechanics. Delft, 1989. Proceedings: (Notes on Numer. Fluid Mechanics 29). Braunschweig: Vieweg, 1990, pp. 252 - 261.

[9] Jameson, A.; Schmidt, W.; Turkel, E.: *Numerical solution of the Euler equations by finite volume methods using Runge-Kutta time stepping schemes.* AIAA paper 81-1259, 1981.

[10] Decker, N. H.; Turkel, E.: *Multigrid for hypersonic inviscid flows.* Third Int. Conf. on Hyperbolic Problems, 1990. ICASE Rep. No. 90-54.

[11] Wittum, G.: *Multi-grid methods for Stokes and Navier-Stokes equations. Transforming smoothers - algorithms and numerical results.* Numerische Mathematik 54 (1989), pp. 543-563.

[12] Brandt, A.: *General PDE systems at vanishing ellipticity and other projects.* Third European Conf. on Multigrid Methods, Bonn, 1990.

[13] Gustafsson, B.; Lötstedt, P.: *Convergence of the multigrid method for systems of first order differential equations.* Third European Conf. on Multigrid Methods, Bonn, 1990.

[14] Hackbusch, W.: *A new approach to robust multi-grid solvers.* in: McKenna, J.; Temam, R. (ed.): ICIAM '87: First Int. Conf. on Industrial and Appl. Math.. Paris, 1987. Proc.: Philadelphia, PA: SIAM, 1988, pp. 114 - 126.

[15] Hackbusch, W.: *The frequency decomposition multi-grid algorithm.* in: Hackbusch, W. (ed.): Fourth GAMM-Seminar Kiel, 1988: Robust multi-grid methods. Proc.: (Notes on Numer. Fluid Mechanics 23). Braunschweig: Vieweg, 1988, pp. 96 - 104.

Low-Diffusion Rotated Upwind Schemes, Multigrid and Defect Correction for Steady, Multi-Dimensional Euler Flows

by

Barry Koren

Abstract: *Two simple, multi-dimensional upwind discretizations for the steady Euler equations are derived, with the emphasis lying on both a good accuracy and a good solvability. The multi-dimensional upwinding consists of applying a one-dimensional Riemann solver with a locally rotated left and right state, the rotation angle depending on the local flow solution. First, a scheme is derived for which smoothing analysis of point Gauss-Seidel relaxation shows that despite its rather low numerical diffusion, it still enables a good acceleration by multigrid. Next, a scheme is derived which has not any numerical diffusion in crosswind direction, and of which convergence analysis shows that its corresponding discretized equations can be solved efficiently by means of defect correction iteration with in the inner multigrid iteration the first scheme. For the steady, two-dimensional Euler equations, numerical experiments are performed for some supersonic test cases with an oblique contact discontinuity. The numerical results are in good agreement with the theoretical predictions. Comparisons are made with results obtained by a standard, grid-aligned upwind scheme. The grid-decoupled results obtained are promising.*

Note: *This work was supported by the European Space Agency, through Avions Marcel Dassault - Bréguet Aviation.*

1. Introduction

In the present paper, for the steady, 2-D Euler equations in a cell-centered finite volume context, we develop a multi-D upwind method with some appropriate balance between crosswind diffusion and efficiency. The steady equations will be solved directly. Herewith, for obtaining a good efficiency, we rely on nonlinear multigrid (multigrid-Newton) iteration. As the smoothing technique for the multigrid iteration, we prefer to apply point Gauss-Seidel relaxation, using the exact differential matrices (exact Newton). The latter requires the cell face fluxes to be continuously differentiable. If multigrid does not easily meet our standards, we will not try to repair it, but - instead - we will rely on defect correction iteration as an additional iteration. The multi-D upwind schemes to be considered here will be very simple schemes only. They will use neither decoupling of the Euler equations (as in [5,10]), nor rotated fluxes (as in [1,8]). The schemes will be based on *rotated left and right states solely*. Per cell face, just as with grid-aligned upwind schemes, only a single numerical flux is computed: that normal to the cell face. The numerical flux function to be applied should allow a good resolution of both oblique shock waves and oblique contact discontinuities, fixing choices to flux difference splitting schemes. Given the good experience with Osher's scheme [9] in combination with multigrid - (exact) Newton [4], here we will apply the latter flux difference splitting scheme.

The contents of the paper is as follows. First, in section 2, to set a frame of reference, on the basis of a linear, scalar model equation, an accuracy analysis and a simple solvability analysis are performed for the standard, grid-aligned first-order upwind scheme. Then, two multi-D upwind schemes are derived. Next, in section 3, on the basis of the same model equation, solvability properties of the two schemes derived in section 2 are analyzed. These analyses will be Fourier analyses. Finally, in section 4, the theoretical results found in the previous sections are verified for some steady, fully supersonic Euler flows on the unit square.

2. Derivation of grid-decoupled upwind schemes

We will try to derive positive, grid-decoupled upwind schemes which have a low crosswind diffusion (preferably zero), and which (hopefully) still have enough artificial diffusion in characteristic direction to preserve good solvability. An important property required from these schemes is that their stencils are as compact as possible. The motivation for compactness is to avoid: non-consistent boundary condition treatments and - if possible - non-positivity. Compactness will result in a close relationship with the grid-aligned first-order upwind scheme, which therefore will be used as the main reference for comparison. Another property strived for is continuous differentiability, this because of the intended application of exact Newton iteration. The schemes will be investigated on the basis of the linear, scalar, 2-D model equation

$$\cos\phi\frac{\partial u}{\partial x}+\sin\phi\frac{\partial u}{\partial y}=0, \quad 0\leqslant\phi\leqslant\pi/2, \tag{2.1}$$

with ϕ the angle made by the characteristic direction and the x-axis. Discretization of (2.1) on a square, cell-centered finite volume grid yields

$$\cos\phi(u_{i+\frac{1}{2},j}-u_{i-\frac{1}{2},j})+\sin\phi(u_{i,j+\frac{1}{2}}-u_{i,j-\frac{1}{2}})=0, \tag{2.2}$$

where the non-integer indices refer to the cell faces in between the (integer-indexed) cell centers.

In the grid-aligned first-order upwind scheme, given the being positive of $\cos\phi$ and $\sin\phi$, for the cell face states we take

$$\begin{pmatrix}u_{i+\frac{1}{2},j}\\ u_{i,j+\frac{1}{2}}\end{pmatrix}=\begin{pmatrix}u_{i,j}\\ u_{i,j}\end{pmatrix}, \quad 0\leqslant\phi\leqslant\pi/2. \tag{2.3}$$

Similar choices are made for $u_{i-\frac{1}{2},j}$ and $u_{i,j-\frac{1}{2}}$. Substituting these cell face states into (2.2) and applying next truncated Taylor series expansions, with h as the mesh size, the following modified equation may be derived:

$$\cos\phi\frac{\partial u}{\partial x}+\sin\phi\frac{\partial u}{\partial y}-\frac{h}{2}\left[\cos\phi\frac{\partial^2 u}{\partial x^2}+\sin\phi\frac{\partial^2 u}{\partial y^2}\right]=O(h^2), \quad 0\leqslant\phi\leqslant\pi/2. \tag{2.4}$$

By transformation to characteristic coordinates, (2.4) becomes

$$\frac{\partial u}{\partial s}-\frac{h}{2}\left[(\cos^3\phi+\sin^3\phi)\frac{\partial^2 u}{\partial s^2}-2\cos\phi\sin\phi(\cos\phi-\sin\phi)\frac{\partial^2 u}{\partial s\partial n}+\right.$$
$$\left.\cos\phi\sin\phi(\cos\phi+\sin\phi)\frac{\partial^2 u}{\partial n^2}\right]=O(h^2), \quad 0\leqslant\phi\leqslant\pi/2. \tag{2.5}$$

From (2.5) now, it appears that for the grid-aligned first-order upwind scheme, zero-crosswind diffusion (i.e. both $\cos\phi\sin\phi(\cos\phi-\sin\phi)=0$ and $\cos\phi\sin\phi(\cos\phi+\sin\phi)=0$) occurs only in case of $\phi=0$ or $\phi=\pi/2$, i.e. in case of grid-alignment of the characteristic direction.

For the sake of comparisons to be made hereafter, in Fig. 2.1a we give the distributions of the diffusion coefficients from (2.5), over the complete range of ϕ considered. In here and also in the following, μ_{ss}, μ_{sn} and μ_{nn} denote the coefficients of $\partial^2 u/\partial s^2$, $\partial^2 u/\partial s \partial n$ and $\partial^2 u/\partial n^2$, respectively.

As opposed to the poor accuracy properties, the smoothing properties of point Gauss-Seidel relaxation applied to the grid-aligned first-order upwind system are known to be good. The fact that these smoothing properties are at least not bad is reflected by the fact that the discretization is positive, which clearly appears from the stencil

$$\begin{array}{c} j \\ j-1 \\ \end{array} \begin{array}{|c|c|} \hline -\cos\phi & \cos\phi+\sin\phi \\ \hline & -\sin\phi \\ \hline i-1 & i \end{array}, \quad 0 \leqslant \phi \leqslant \pi/2. \tag{2.6}$$

2.1. A positive, continuously differentiable scheme

The most compact, grid-decoupled upwind schemes use

$$\begin{pmatrix} u_{i+\frac{1}{2},j} \\ u_{i,j+\frac{1}{2}} \end{pmatrix} = \begin{pmatrix} \delta_1(\phi)u_{i,j} + [1-\delta_1(\phi)]u_{i,j-1} \\ \delta_2(\phi)u_{i,j} + [1-\delta_2(\phi)]u_{i-1,j} \end{pmatrix}, \quad 0 \leqslant \delta_1(\phi), \delta_2(\phi) \leqslant 1, \quad 0 \leqslant \phi \leqslant \pi/2. \tag{2.7}$$

The coefficients $\delta_1(\phi)$ and $\delta_2(\phi)$, whose choice determines the scheme, are taken in the range [0,1] in order to prevent unphysical features such as e.g. a negative density. The stencil corresponding with (2.7) reads

$$\begin{array}{c} j \\ j-1 \\ \end{array} \begin{array}{|c|c|} \hline -\delta_1\cos\phi+(1-\delta_2)\sin\phi & \delta_1\cos\phi+\delta_2\sin\phi \\ \hline -(1-\delta_1)\cos\phi-(1-\delta_2)\sin\phi & (1-\delta_1)\cos\phi-\delta_2\sin\phi \\ \hline i-1 & i \end{array}, \quad 0 \leqslant \phi \leqslant \pi/2. \tag{2.8}$$

A natural requirement imposed to the discretization is that of symmetry with respect to $\phi = \pi/4$; in formula:

$$\delta_2(\phi) = \delta_1(\pi/2 - \phi), \quad 0 \leqslant \phi \leqslant \pi/2. \tag{2.9}$$

The modified equation corresponding with (2.7) reads in characteristic coordinates

$$\frac{\partial u}{\partial s} - \frac{h}{2}\Bigg[\Big[(1+\cos\phi\sin\phi)(\cos\phi+\sin\phi) - 2(\delta_1\cos\phi+\delta_2\sin\phi)\cos\phi\sin\phi\Big]\frac{\partial^2 u}{\partial s^2} +$$

$$2(\cos\phi-\sin\phi)\Big[1+\cos\phi\sin\phi-(\delta_1\cos\phi+\delta_2\sin\phi)(\cos\phi+\sin\phi)\Big]\frac{\partial^2 u}{\partial s\partial n} +$$

$$2\cos\phi\sin\phi\Big[(\delta_1-\tfrac{1}{2})\cos\phi+(\delta_2-\tfrac{1}{2})\sin\phi\Big]\frac{\partial^2 u}{\partial n^2}\Bigg] = O(h^2), \quad 0 \leqslant \phi \leqslant \pi/2. \tag{2.10}$$

Following (2.10), $\mu_{sn}=0$ and $\mu_{nn}=0$ lead to respectively

$$\delta_1\cos\phi+\delta_2\sin\phi = \frac{1+\cos\phi\sin\phi}{\cos\phi+\sin\phi}, \quad 0 \leqslant \phi \leqslant \pi/2, \tag{2.11a}$$

$$\delta_1\cos\phi+\delta_2\sin\phi = \tfrac{1}{2}(\cos\phi+\sin\phi), \quad 0 \leqslant \phi \leqslant \pi/2, \tag{2.11b}$$

which clearly is an inconsistent system of equations, leading to the following theorem:

THEOREM (2.1)

No most compact grid-decoupled upwind scheme exists for which both $\mu_{sn}=0$ and $\mu_{nn}=0$.

Next, concerning crosswind diffusion, we only require μ_{nn} to be zero, i.e. we assume (2.11b) to hold solely. Further, following (2.8), positivity can be expressed as:

$$(\delta_1\cos\phi+\delta_2\sin\phi)\begin{pmatrix}1\\1\\1\\-1\end{pmatrix}>\begin{pmatrix}0\\ \sin\phi\\ \cos\phi\\ -\cos\phi-\sin\phi\end{pmatrix},\quad 0\leqslant\phi\leqslant\pi/2. \tag{2.12}$$

It appears that the system (2.11b)-(2.12) is inconsistent as well, leading to the theorem:

THEOREM (2.2)
No most compact grid-decoupled upwind scheme exists which is positive and for which $\mu_{nn}=0$.

Finally, we require $\mu_{sn}=0$ to hold (i.e. (2.11a)) in combination with (2.12). It can be verified that this system is consistent. Assuming the form $\delta_1(\phi)=(\cos\phi+a\sin\phi)/(\cos\phi+\sin\phi)$, a being a constant, with symmetry requirement (2.9) we get $\delta_2(\phi)=(\sin\phi+a\cos\phi)/(\sin\phi+\cos\phi)$. Substitution of these forms of $\delta_1(\phi)$ and $\delta_2(\phi)$ into (2.11a) yields $a=½$, so

$$\begin{pmatrix}u_{i+½,j}\\u_{i,j+½}\end{pmatrix}=\frac{1}{\cos\phi+\sin\phi}\begin{pmatrix}(\cos\phi+½\sin\phi)u_{i,j}+½\sin\phi\ u_{i,j-1}\\(\sin\phi+½\cos\phi)u_{i,j}+½\cos\phi\ u_{i-1,j}\end{pmatrix},\quad 0\leqslant\phi\leqslant\pi/2, \tag{2.13}$$

which gives the (positive) stencil

j	$\dfrac{-\cos^2\phi}{\cos\phi+\sin\phi}$	$\dfrac{1+\cos\phi\sin\phi}{\cos\phi+\sin\phi}$	, $0\leqslant\phi\leqslant\pi/2$, (2.14)
$j-1$	$\dfrac{-\cos\phi\sin\phi}{\cos\phi+\sin\phi}$	$\dfrac{-\sin^2\phi}{\cos\phi+\sin\phi}$	
	$i-1$	i	

and the modified equation

$$\frac{\partial u}{\partial s}-\frac{h}{2}\left[\frac{1+\cos\phi\sin\phi}{\cos\phi+\sin\phi}\frac{\partial^2 u}{\partial s^2}+\frac{\cos\phi\sin\phi}{\cos\phi+\sin\phi}\frac{\partial^2 u}{\partial n^2}\right]=O(h^2),\quad 0\leqslant\phi\leqslant\pi/2. \tag{2.15}$$

In Fig. 2.1b we give the graph of the present scheme's diffusion properties. The scheme's crosswind diffusion is significantly lower than that of the grid-aligned first-order upwind scheme (Fig. 2.1a). The lower crosswind diffusion in combination with the properties of positivity and continuous differentiability makes that scheme (2.13) is possibly more appropriate for our present multigrid purposes than the grid-aligned first-order upwind scheme (2.3).

2.2. A zero-crosswind diffusion scheme

Because of the limitation of the most compact, grid-decoupled upwind schemes with respect to the elimination of all crosswind diffusion (the limitation expressed by THEOREM (2.1)), in this section, we will consider wider stencils. To start with, we consider a situation with small ϕ. Still striving for compactness, the extrapolation is done from the nearest lines connecting two neighboring cell center states, also avoiding negative coefficients herewith:

$$\begin{pmatrix}u_{i+½,j}\\u_{i,j+½}\end{pmatrix}=\begin{pmatrix}\delta_1(\phi)u_{i,j}+[1-\delta_1(\phi)]u_{i,j-1}\\ \delta_2(\phi)u_{i-1,j}+[1-\delta_2(\phi)]u_{i-1,j+1}\end{pmatrix},\quad 0\leqslant\delta_1(\phi),\delta_2(\phi)\leqslant1,\ 0\leqslant\phi\leqslant\phi_{up}, \tag{2.16}$$

with the (small) upper bound ϕ_{up} not yet fixed. With (2.16), we find the following modified equation

$$\frac{\partial u}{\partial s}-\frac{h}{2}\Bigg[\left[\cos^3\phi+2(1-\delta_1)\cos^2\phi\sin\phi+2\cos\phi\sin^2\phi+(2\delta_2-1)\sin^3\phi\right]\frac{\partial^2 u}{\partial s^2}+$$
$$2\left[(1-\delta_1)\cos^3\phi+(\delta_1+2\delta_2-2)\cos\phi\sin^2\phi-\sin^3\phi\right]\frac{\partial^2 u}{\partial s\partial n}+$$
$$\cos\phi\sin\phi\left[(2\delta_1+2\delta_2-3)\cos\phi-\sin\phi\right]\frac{\partial^2 u}{\partial n^2}\Bigg]=O(h^2),\quad 0\leqslant\phi\leqslant\phi_{up}. \tag{2.17}$$

From this, it follows that no crosswind diffusion occurs for $\delta_1(\phi)=1$, $\delta_2(\phi)=\frac{1}{2}(1+\tan\phi)$, $0\leqslant\phi\leqslant\pi/4$, where the indicated ϕ-range (with $\phi_{up}=\pi/4$) is that for which negative coefficients are just avoided. Taking for the remaining subrange the symmetrical counterpart of (2.16):

$$\begin{pmatrix}u_{i+\frac{1}{2},j}\\ u_{i,j+\frac{1}{2}}\end{pmatrix}=\begin{pmatrix}\delta_1(\phi)u_{i,j-1}+[1-\delta_1(\phi)]u_{i+1,j-1}\\ \delta_2(\phi)u_{i,j}+[1-\delta_2(\phi)]u_{i-1,j}\end{pmatrix},\quad \pi/4\leqslant\phi\leqslant\pi/2, \tag{2.18}$$

with symmetry condition (2.9) we get $\delta_1(\phi)=\frac{1}{2}(1+\text{cotan}\phi)$, $\delta_2(\phi)=1$, $\pi/4\leqslant\phi\leqslant\pi/2$. Summarizing, we have derived as expressions for the cell face states:

$$\begin{pmatrix}u_{i+\frac{1}{2},j}\\ u_{i,j+\frac{1}{2}}\end{pmatrix}=\begin{pmatrix}u_{i,j}\\ \frac{1}{2}(1+\tan\phi)u_{i-1,j}+\frac{1}{2}(1-\tan\phi)u_{i-1,j+1}\end{pmatrix},\quad 0\leqslant\phi\leqslant\pi/4, \tag{2.19a}$$

$$\begin{pmatrix}u_{i+\frac{1}{2},j}\\ u_{i,j+\frac{1}{2}}\end{pmatrix}=\begin{pmatrix}\frac{1}{2}(1+\text{cotan}\phi)u_{i,j-1}+\frac{1}{2}(1-\text{cotan}\phi)u_{i+1,j-1}\\ u_{i,j}\end{pmatrix},\quad \pi/4\leqslant\phi\leqslant\pi/2, \tag{2.19b}$$

as corresponding stencils:

	$i-1$	i
$j+1$	$\frac{1}{2}\sin\phi(1-\tan\phi)$	
j	$\sin\phi(\tan\phi-\text{cotan}\phi)$	$\cos\phi$
$j-1$	$-\frac{1}{2}\sin\phi(1+\tan\phi)$	

, $0\leqslant\phi\leqslant\pi/4$, (2.20a)

	$i-1$	i	$i+1$
j		$\sin\phi$	
$j-1$	$-\frac{1}{2}\cos\phi(1+\text{cotan}\phi)$	$\cos\phi(\text{cotan}\phi-\tan\phi)$	$\frac{1}{2}\cos\phi(1-\text{cotan}\phi)$

, $\pi/4\leqslant\phi\leqslant\pi/2$, (2.20b)

and as modified equations:

$$\frac{\partial u}{\partial s}-\frac{h}{2}\frac{1}{\cos\phi}\frac{\partial^2 u}{\partial s^2}=O(h^2),\quad 0\leqslant\phi\leqslant\pi/4, \tag{2.21a}$$

$$\frac{\partial u}{\partial s}-\frac{h}{2}\frac{1}{\sin\phi}\frac{\partial^2 u}{\partial s^2}=O(h^2),\quad \pi/4\leqslant\phi\leqslant\pi/2. \tag{2.21b}$$

A geometric interpretation of scheme (2.19) is given in [3]. In Fig. 2.1c, we give the graph with its diffusion coefficients. Unfortunately, from (2.20) it appears that the scheme is non-positive. A favorable property of the scheme is its simplicity.

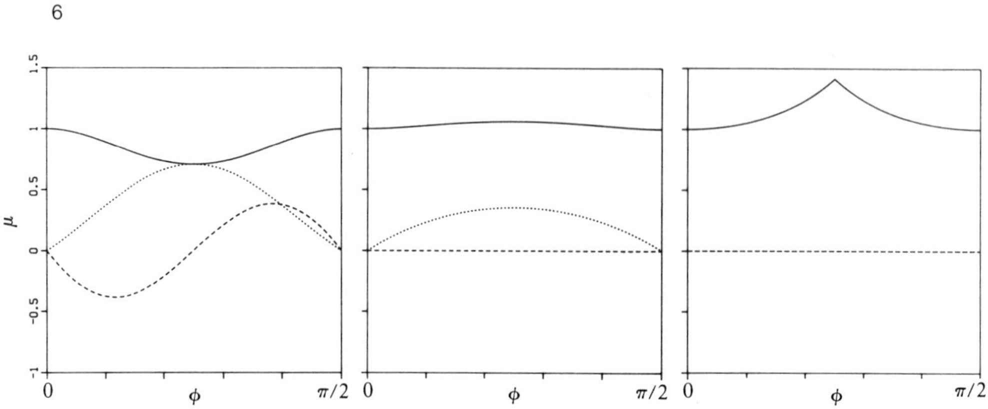

a. Grid-aligned first-order upwind scheme (2.3).

b. Continuously differentiable scheme (2.13).

c. Zero-crosswind diffusion scheme (2.19).

Fig. 2.1. Diffusion coefficients modified equation
(μ_{ss}: —— , μ_{sn}: ------ , μ_{nn}: ....).

3. Analysis of multigrid iteration and defect correction iteration

In the present section, we will analyze some possible solution methods for the two grid-decoupled schemes derived in the previous section. The analyses will be done again for model equation (2.1) on a square, cell-centered finite volume grid. First, the smoothing behavior of point Gauss-Seidel relaxation will be investigated, both for scheme (2.13) and scheme (2.19). (Though we do not expect the latter scheme to be sufficiently dissipative to allow a successful application of multigrid iteration, for sake of certainty we do consider its smoothing behavior.) The smoothing properties found will be compared with those of the grid-aligned first-order upwind scheme (2.3). If analysis shows that multigrid iteration applied directly to the zero-crosswind diffusion operator does not work, to solve the corresponding system of discretized equations, we will rely on defect correction iteration. We already have good experience with defect correction iteration, in efficiently solving higher-order grid-aligned Euler and Navier-Stokes equations (see [2,6] and [7], respectively).

3.1. Smoothing analysis of point Gauss-Seidel relaxation

Four different relaxation sweep directions are considered: downwind, upwind and twice crosswind, each of those four with the i-loop as the inner sweep-loop, and each for the complete range of ϕ considered: $[0,\pi/2]$. To apply smoothing analysis, denoting the number of sweeps performed by n, we introduce: (i) the iteration error

$$\delta_{i,j}^n \equiv u_{i,j}^* - u_{i,j}^n, \tag{3.1}$$

with $u_{i,j}^*$ the converged numerical solution in finite volume i,j, and (ii) the Fourier form

$$\delta_{i,j}^n = D\rho^n e^{\mathrm{i}(\theta_1+\theta_2)}, \tag{3.2}$$

with D constant, ρ the amplification factor, and $\theta_1 \equiv \omega_1 h$ and $\theta_2 \equiv \omega_2 h$, ω_1 and ω_2 being the error mode in i- and j-direction, respectively. In Fig. 3.1, results of the smoothing analysis are given for successively: grid-aligned first-order upwind scheme (2.3), continuously differentiable scheme (2.13) and zero-crosswind diffusion scheme (2.19). In here, the smoothing factor ρ_s is defined as

$$\rho_s \equiv \sup|\rho(\theta_1,\theta_2)|, \quad (|\theta_1|,|\theta_2|) \in \{[0,\pi]\times[0,\pi] \,|\, |\theta_1| \in [\pi/2,\pi] \vee |\theta_2| \in [\pi/2,\pi]\}. \tag{3.3}$$

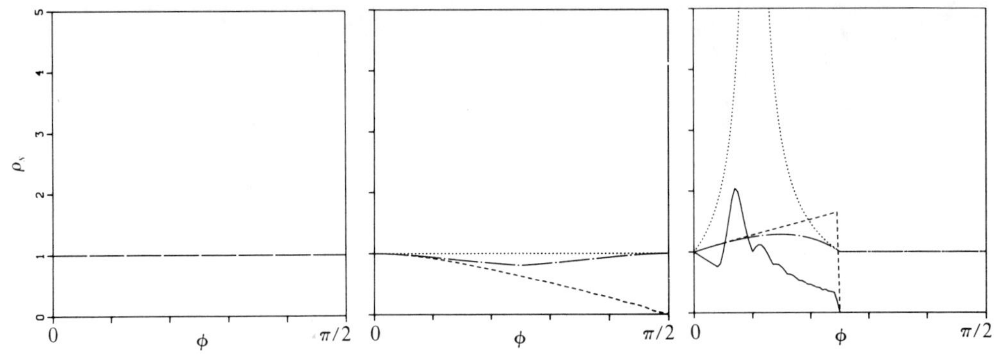

a. Grid-aligned first-order upwind scheme (2.3).

b. Continuously differentiable scheme (2.13).

c. Zero-crosswind diffusion scheme (2.19).

Fig. 3.1. Smoothing factors point Gauss-Seidel relaxation, with i-loop as inner loop (downwind; : ——, upwind; : ------, crosswind up; : ------, crosswind down; : ....).

It appears that the zero-crosswind diffusion scheme does not enable a successful application of multigrid with point Gauss-Seidel relaxation as the smoother (Fig. 3.1c). Notice that even for the corresponding downwind relaxation sweep, smoothing is not guaranteed over the complete range of ϕ considered. As opposed to this, for the positive, continuously differentiable scheme we do have smoothing (Fig. 3.1b).

To find a suitable solution method for zero-crosswind diffusion scheme (2.19), in the following, we will study the convergence properties of defect correction iteration with positive, continuously differentiable scheme (2.13) as the approximate scheme in the inner multigrid iteration. For comparison, a similar study will also be made with grid-aligned first-order upwind scheme (2.3) as the approximate scheme.

3.2. *Convergence analysis of defect correction iteration*

Denoting the zero-crosswind diffusion operator by L_h^+, defect correction iteration reads

$$L_h(u_h^{n+1})=L_h(u_h^n)-L_h^+(u_h^n), \quad n=0,1,...,N, \tag{3.4}$$

with L_h denoting the positive operator to be inverted, and with the index n denoting the iteration counter. From (3.4), it is clear that the closer the resemblance between the target operator L_h^+ and the approximate operator L_h, the better the convergence of the defect correction iteration. Hence, in this respect, the best convergence of (3.4) is expected from the approach with (2.13) as the approximate scheme. Introducing, as before, the iteration error (3.1) in its Fourier form (3.2), we can write for the amplification factor

$$\rho(\theta_1,\theta_2)=1-L_h^{-1}(\theta_1,\theta_2)L_h^+(\theta_1,\theta_2). \tag{3.5}$$

In Fig. 3.2, for each of the two approximate schemes (2.3) and (2.13), and for successively $\phi=0.1\pi, 0.2\pi, 0.3\pi$ and 0.4π, we give the distributions of the convergence factor ρ_c;

$$\rho_c \equiv |\rho(\theta_1,\theta_2)|, \quad (|\theta_1|,|\theta_2|)\in\{[0,\pi]\times[0,\pi]\}. \tag{3.6}$$

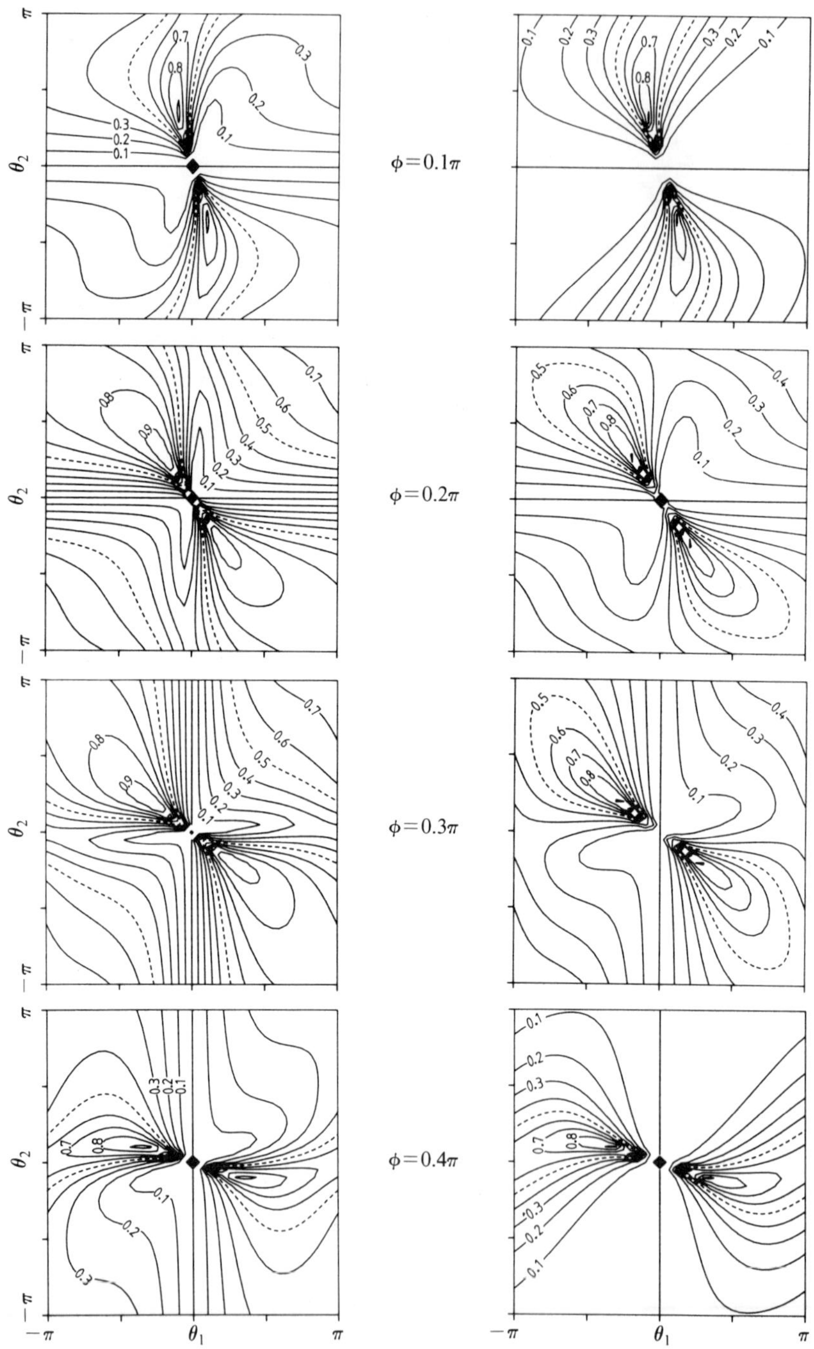

a. Grid-aligned first-order upwind scheme (2.3) as approximate scheme.

b. Continuously differentiable scheme (2.13) as approximate scheme.

Fig. 3.2. Convergence factor distributions defect correction iteration.

In all graphs of Fig. 3.2, the dashed iso-lines correspond with $\rho_c=0.5$. It appears that, for all four values of ϕ considered, scheme (2.13) as the approximate scheme gives better convergence factor distributions indeed.

4. Numerical results

To investigate some theoretical results found in the previous sections, in this section, for a perfect gas with $\gamma=1.4$, we perform numerical experiments for a set of fully supersonic Euler flows with oblique contact discontinuity (Fig. 4.1a). The discontinuity is considered for the flow angles $\phi=0.1\pi, 0.2\pi, 0.3\pi$ and 0.4π. The flows are computed on the 32×32-grid given in Fig. 4.1b. In all cases - for simplicity - at each of the four boundaries, the exact solution is imposed (overspecification). The multigrid method applied for all cases is nonlinear multigrid with V-cycles, and with per level a single pre- and post-relaxation sweep only. In all cases, the coarsest grid considered is a 2×2-grid. Further, in all cases we take as the initial solution: the solution with $q=q^L$ (the q^L from Fig. 4.1a) uniformly constant over the complete domain. For both grid-decoupled upwind schemes derived before, as the angle to be considered at each cell face, we take the streamline angle. In [3], we give a physically proper way of computing this angle.

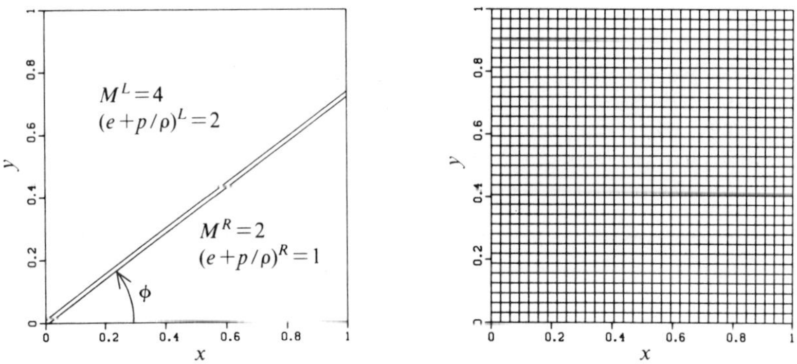

a. Flow with contact discontinuity ($\phi=0.1\pi, 0.2\pi, 0.3\pi, 0.4\pi$, $p=1$).

b. Finest grid (32×32).

Fig. 4.1. Test case on unit square.

4.1. *Results positive, continuously differentiable scheme*

In Fig. 4.2 we first give, on top of each other, the reference results obtained for $\phi=0.1\pi, 0.2\pi, 0.3\pi$ and 0.4π with the grid-aligned first-order upwind scheme; in Fig. 4.2a the multigrid convergence histories and in Fig. 4.2b the enthalpy $(e+p/\rho)$ distributions. The iso-enthalpy values considered in these and all following enthalpy distributions are: $1.1, 1.2, 1.3, \cdots, 1.9$. Because of the severe smearing of the grid-aligned first-order upwind scheme, hardly any distinction can be made between the four solutions. Notice that the layers along $x=1$ and $y=1$, in Fig. 4.2b and all following enthalpy graphs, are only due to the overspecification.

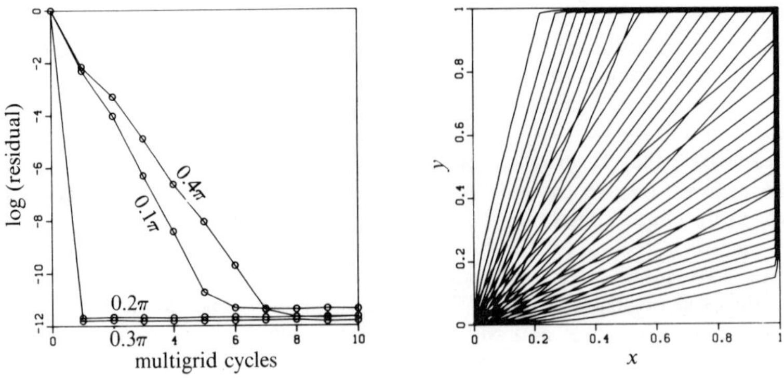

a. Multigrid convergence histories. b. Enthalpy distributions.

Fig. 4.2. Results grid-aligned first-order upwind scheme.

In Fig. 4.3 we give the similar results obtained with the grid-decoupled, positive, continuously differentiable scheme. Though not as very fast as the reference convergence in Fig. 4.2a, fortunately, the present scheme's multigrid convergence is still fast. Though clearly more accurate than the reference distributions in Fig. 4.2b, the present enthalpy distributions are still insufficiently accurate.

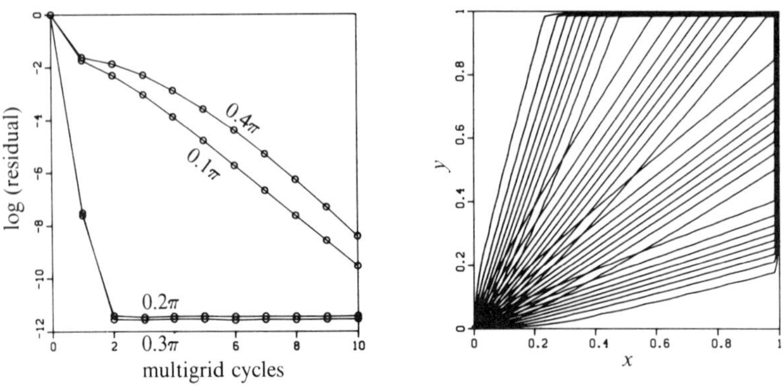

a. Multigrid convergence histories. b. Enthalpy distributions.

Fig. 4.3. Results positive, continuously differentiable scheme.

4.2. Results zero-crosswind diffusion scheme

In Fig. 4.4 we give the enthalpy distributions for the zero-crosswind diffusion scheme, as obtained after successively 1,2 and 10 defect correction cycles (with per defect correction cycle a single nonlinear multigrid cycle only). The distributions in Fig. 4.4c appear to be almost free of crosswind diffusion. (In [3] we also give the enthalpy distributions obtained with a third-order accurate, grid-aligned upwind scheme. All four enthalpy distributions in Fig. 4.4c appear to be even less diffusive than those of that third-order accurate scheme.) Though the being non-positive of the zero-crosswind diffusion scheme allows solutions with spurious oscillations, its distributions are still monotone. Concerning the convergence rate of the defect correction iteration, we are of the opinion that this is good.

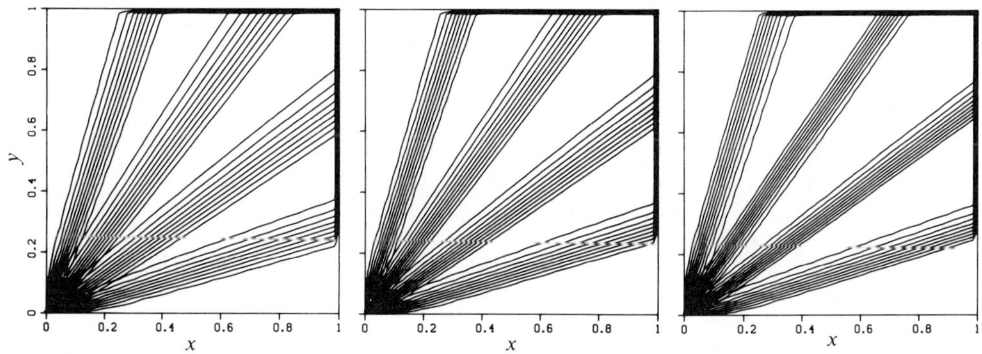

a. After *1st* cycle. b. After *2nd* cycle. c. After *10th* cycle.

Fig. 4.4. Convergence history defect correction iteration, enthalpy distributions zero-crosswind diffusion scheme.

5. Conclusions

In the present paper, for multi-D Euler equations, two promising multi-D upwind schemes have been derived; a positive, continuously differentiable scheme and a zero-crosswind diffusion scheme. Both schemes are based on a one-dimensional Riemann solver. Their multi-D nature is simply realized through a local, solution-dependent rotation of the left and right Riemann state, allowing to keep the number of numerical flux computations per cell face at one only. Good efficiency is strived for by means of nonlinear multigrid iteration and (if necessary) defect correction iteration. The accuracy and efficiency of the numerical results obtained are promising. One important result obtained is that when applying the positive, continuously differentiable scheme, for flows with contact discontinuities, the performance of nonlinear multigrid with point Gauss-Seidel relaxation as the smoother, is very good. Another important result is that, again for flows with contact discontinuities, the solutions of the zero-crosswind diffusion scheme appear to be practically free of crosswind diffusion indeed. Moreover, their computation by means of defect correction iteration, with the grid-decoupled, positive, continuously differentiable scheme as the approximate scheme, is efficient.

REFERENCES

1. DAVIS, S.F., A rotationally biased upwind difference scheme for the Euler equations, *J. Comput. Phys.* **56** (1984), 65-92.
2. HEMKER, P.W., Defect correction and higher order schemes for the multi grid solution of the steady Euler equations, *Proceedings 2nd European Conference on Multigrid Methods, Cologne, 1985,* Lecture Notes in Mathematics, No. 1228, Springer, Berlin, 1986, 149-165.
3. HEMKER, P.W., KOREN, B. AND VAN LEER, B., Low-diffusion upwind schemes, multigrid and defect correction for steady, multi-dimensional Euler flows, Center for Mathematics and Computer Science, Amsterdam, NM-Report (in preparation).
4. HEMKER, P.W. AND SPEKREIJSE, S.P., Multiple grid and Osher's scheme for the efficient solution of the steady Euler equations, *Appl. Numer. Math.* **2** (1986), 475-493.
5. HIRSCH, CH., LACOR, C. AND DECONINCK, H., Convection algorithms based on a diagonalization procedure for the multidimensional Euler equations, *AIAA Paper 87-1163* (1987).
6. KOREN, B., Defect correction and multigrid for an efficient and accurate computation of airfoil flows, *J. Comput. Phys.* **77** (1988), 183-206.
7. KOREN, B., Multigrid and defect correction for the steady Navier-Stokes equations, *J. Comput. Phys.* **87** (1990), 25-46.
8. LEVY, D.W., POWELL, K.G. AND VAN LEER, B., An implementation of a grid-independent upwind scheme for the Euler equations, *AIAA Paper 89-1931* (1989).
9. OSHER, S. AND SOLOMON, F., Upwind difference schemes for hyperbolic systems of conservation laws, *Math. Comput.* **38** (1982), 339-374.
10. ROE, P.L., Discrete models for the numerical analysis of time-dependent multidimensional gas dynamics, *J. Comput. Phys.* **63** (1986), 458-476.

Barry Koren
Center for Mathematics and Computer Science
P.O. Box 4079
1009 AB Amsterdam, The Netherlands

International Series of Numerical Mathematics, Vol. 98, © 1991 Birkhäuser Verlag Basel

MULTIGRID CONVERGENCE ACCELERATION FOR COMPLEX FLOW INCLUDING TURBULENCE

by

Fue-Sang Lien and Michael A. Leschziner

Abstract: The paper reports a study on the performance of variants of the *Full Approximation Multigrid Scheme* in the computation of complex recirculating flows, both laminar and turbulent. The MG variants are implemented into a three-dimensional, non-orthogonal, collocated finite-volume procedure in which the Cartesian velocity components and the pressure are determined via a pressure-correction algorithm. Convection is approximated by three methods: a first-order hybrid scheme combining the central and upwind approximations, the quadratic upstream-weighted QUICK scheme and the TVD-type MUSCL scheme. Three turbulence models are considered: a low-Re and a high-Re variant of the two-equation k-ε eddy-viscosity model, and a Reynolds-stress-transport closure, all implemented within a non-orthogonal grid environment. Multigrid performance is investigated for four cases: a laminar flow in a 2D plane constriction, a laminar flow in a 3D skewed lid-driven cavity, a turbulent flow in a constricted pipe, computed with two eddy-viscosity models, and a turbulent flow behind a 2D backward-facing step, computed with a Reynolds-stress-transport model within a strongly distorted non-orthogonal mesh. Convergence acceleration is shown to be significant in turbulent conditions but not of the same order as that for laminar flows.

1. Introduction

Rapid advances in computer hardware in recent years have given decisive impetus to the development and application of general numerical algorithms for complex turbulent flows of industrial significance. Three-dimensional general-geometry codes, based on both finite-volume and finite-element strategies, have emerged and combined with complex turbulence closures which account, on grounds of physical realism, for transport of turbulence. Such codes are now being applied to problems as complex as jet-engine combustors, complete aircraft configurations and nuclear-reactor cores.

The ability of any practical code to yield quantitatively meaningful statements rests principally on two factors: the accuracy of the numerical approximation in conjunction with the density of the grid supporting the solution, and the realism of the mathematical models describing pertinent physical processes such as turbulence, heat transfer, combustion and multiphase interaction. While the latter issue is paced by limitations to human insight, the former is mainly constrained by computer-resource limitations in the form of insufficient memory and CPU time (at acceptable cost). Although both constraints are being progressively eroded by computer-hardware developments, the rate at which the latter diminishes is disproportionally slow because the rate of convergence of conventional iterative schemes for highly coupled, non-linear systems decreases exponentially with increasing node number, the exponent being of order 2-3. There is thus a strong need to explore promising convergence-acceleration techniques in the context of geometrically and physically complex, practically relevant conditions. The multigrid method is one such promising technique, and it is its efficient implantation into a non-orthogonal finite-volume algorithm incorporating a range of advanced turbulence closures which is the focus of this study.

The principles of the MG method were established for linear systems in the late 1970s and early 1980s, principally by Brandt (1977), and Hackbusch, Stüben & Trottenberg (1982). The central property of the method is its ability to return optimal convergence rates, in terms of computing effort, which depend linearly on the number of grid points. Much work has been done in the 1980s on extending the technique to the Euler and Navier Stokes equations. Studies directed at laminar flow in simple rectilinear geometries by Lonsdale (1984), Vanka (1986), Thompson & Ferziger (1988), Gaskell & Lau (1987), Barcus et al (1988) and Becker et al (1989), among others, have demonstrated that the favourable convergence properties observed in linear cases carry over to the NS set, and speed-up factors of the order 10-100 have been reported in the literature, depending on grid density, type of flow and the precise manner of MG implementation. Very few studies have ventured into the area of turbulent flow, however, and none appear to have explored MG performance in elliptic-flow algorithms combining turbulent-transport models with non-orthogonal grid systems. Following a rather discouraging foray into turbulence by Phillips et al (1985), Peric et al (1989) reported attaining speed-up factors of the order 30 when applying a *Full Approximation Scheme* with a 320x64 grid to a turbulent backward-facing-step flow within a collocated finite-volume scheme which incorporated a high-Re k-ε eddy-viscosity model. This effort reflects, in effect, the status of multigrid acceleration for turbulent recirculating flow at the time the present study was initiated.

The present paper reports on the incorporation of variants and the *Full Approximation Scheme* into a three-dimensional, non-orthogonal, collocated finite-volume framework, and investigates their performance for a range of options, among them grid disposition (skewness and non-uniformity), type of convection scheme and type of turbulence closure. In the last area, MG performance is examined for high-Re and low-Re versions of the k-ε two-equation model and for a Reynolds-stress transport model which solves differential transport equations for all pertinent Reynolds stresses.

2. The Continuum Description

The numerical approximation takes as its starting point the Reynolds-averaged versions of the Navier-Stokes equations and the mass-conservation equation. The numerical approach adopted is to integrate the equations, prior to the approximation process, over non-orthogonal finite volumes, whilst retaining the Cartesian velocity decomposition within the non-orthogonal framework. In view of this choice, the most expedient form of the continuum equation for the momentum components U_i is vectorial and conservative, as given below:

$$\nabla\cdot\rho\vec{U}\,U_i+\nabla\cdot\left(\mu\,\mathbf{grad}\,U_i+\rho\overline{u_iu_1}\,\vec{i}+\rho\overline{u_iu_2}\,\vec{j}+\rho\overline{u_iu_3}\,\vec{k}\right)=-\mathbf{grad}\,P \tag{1}$$

where $\vec{i}, \vec{j}, \vec{k}$ are the Cartesian unit vectors and the subscript 'i' denotes Cartesian co-ordinate directions.

Three turbulence models have been used in following tests to determine the Reynolds stresses $\overline{u_iu_j}$. Two evaluate the stresses via the eddy-viscosity concept:

$$-\rho\overline{u_i u_j} = \mu_t\left(\frac{\partial U_i}{\partial x_j}+\frac{\partial U_j}{\partial x_i}\right) - \frac{2}{3}\rho k\delta_{ij} \tag{2}$$

the eddy viscosity being related to the turbulence energy k and its rate of dissipation ε via:

$$\mu_t = \rho C_\mu \frac{k^2}{\varepsilon} \tag{3}$$

The distributions of k and ε are determined by solving modelled differential transport equations of the form:

$$\nabla\cdot(\rho\vec{U}\phi - \nabla\cdot\mu_t/\sigma_\phi\, grad\phi) = S^\phi \tag{4}$$

where ϕ=k or ε, and S^ϕ depends on the model variant. Here, use is made of the *high-Reynolds-number* and *low-Reynolds-number* k-ε models of Jones and Launder (1972), and the associated forms of S^ϕ may be found in the references given. The essential difference between the model variants lies in the way in which the near-wall region is handled. The high-Re version does not account for turbulence/viscosity interaction and requires the use of log-law-based, semi-emperical relations to bridge the near-wall region $y_+<100$. The low-Re version, in contrast, permits integration upto the wall but requires a very fine near-wall resolution to account for the steep variations of the properties in this region, the adverse result being very high cell-aspect ratios.

The third model used is Gibson & Launder's Reynolds-stress-transport closure (1978). This consists of a set of modelled transport equations, of the form (4), with ϕ being $\overline{u_i u_j}$, the components of the Reynolds-stress tensor. This set is augmented by an equation for ε which extracts turbulence energy, isotropically, from the three normal stresses $\overline{u_1u_1}$, $\overline{u_2u_2}$ and $\overline{u_3u_3}$. Space constraints do not permit the model to be presented in detail. Suffice it to mention that each stress-transport equation represents a balance between stress convection, diffusion, production, dissipation and redistribution. The principal merit of this rather complex approach is that it is able to capture the subtle interaction between streamline curvature and turbulence without *ad hoc* modifications. This ability is essentially rooted in the fact that stress-generation levels, which differ greatly from stress to stress, are represented exactly in the model. The particular version used in this study is applicable to high-Re conditions only, and log-law-based relations, similar to those used in conjunction with the high-Re k-ε model, must be applied in the semi-viscous near-wall region. A recent review focusing on the performance of Reynolds-stress closures in complex flow conditions is given by Leschziner (1990).

3. Discretization and Basic Smoothing Algorithm

Equation (4), for any pertinent intensive property, is integrated over the volume shown in Fig. 1 to yield, after application of the Gauβ Divergence Theorem, a balance of face fluxes and volume-integrated net source which may be expressed, vectorially, as:

$$\int_s(\rho\vec{U}\phi + \mu_t/\sigma_\phi\, grad\phi)\cdot\vec{n}\, dS = \int_{vol} S^\phi\, dVol \tag{5}$$

The convection fluxes through cell faces are approximated either by the Hybrid (upwind/central-) differencing scheme (Patanker, 1980) or by the quadratic upstream-weighted scheme QUICK of Leonard (1979) or by the TVD-type MUSCL scheme of van Leer (1979). The diffusive fluxes are approximated by central differences. Introduction of the approximations into equation (5) ultimately leads to a weighted-average formula of the form

$$A_P\phi_P + \sum A_j\phi_j = S^\phi \tag{6}$$

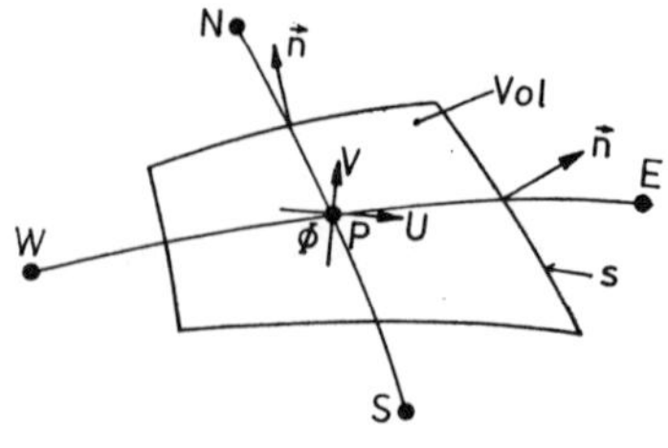

Fig. 1 Finite volume and storage arrangement

where j denotes all neighbours of node 'P' used in approximating the fluxes. When ϕ stands for the momentum components, S^ϕ consists of pressure differences across the cell 'P', the principal contributions arising from the immediate neighbours W,E,S and N. In the particular case of an equidistant mesh and with central differencing used to approximate pressure gradients, equation (6) may be written:

$$\begin{aligned} A_P U_P + \sum A_j U_j &= D^U(P_W - P_E) + S^U \\ A_P V_P + \sum A_j V_j &= D^V(P_S - P_N) + S^V \end{aligned} \tag{7}$$

Given the correct pressure field, equations (7) yield the velocity field. To determine P, and hence U and V, an iterative *guess-and-correct* procedure is used as the smoother, constituting the SIMPLE algorithm of Patankar (1980). This involves the following nine steps:

(i) 'Guess' a pressure field;

(ii) With initial conditions for U,V and turbulence parameters assumed, solve equations (7) for the nodal velocities;

(iii) Interpolate for cell-face velocities {special interpolation practices are here essential to maintain solution smoothness, as demonstrated by Rhie & Chow (1983)};

(iv) Substitute these into the volume-integrated mass-conservation equations and determine the mass residuals for all cells;

(v) Use the linearised and truncated versions of equations (7), together with the interpolation of step (iii) to derive relationships between cell-face

$$\begin{aligned} A_P U'_P &= D^U(P'_W - P'_E) \\ A_P V'_P &= D^V(P'_S - P'_N) \end{aligned} \tag{8}$$

velocity corrections and nodal pressure corrections;

(vi) Substitute cell-face velocity corrections in terms of pressure corrections into the mass-conservation equations, imposing the requirement of vanishing mass residuals obtained in step (iv). This yields a pressure-correction equation of the form:

$$A_P P'_P + \sum A_j P'_j = R^m_P \tag{9}$$

in which R^m is the mass residual for cell P;

(vii) solve equation (9) and add pressure corrections to the prevailing pressure values;

(viii) solve equation (6) for any pertinent turbulence parameter, determining the eddy-viscosity field, if appropriate (special interpolation practices are also needed here in the case of the Reynolds-stress model in order to avoid seriously destabilising oscillations);

(ix) return to step (ii).

Any equation of the form (6) is solved by a line-implicit ADI-type scheme. To enhance stability and smoothness, under-relaxation is employed in all equations, the factors being 0.6 for velocity, 0.3 for pressure and 0.7 for turbulence parameters.

4. Basic FAS V-Cycle

In all applications to be presented, variants of the *Full Approximation Scheme* have been used. For the purpose of conveying the essential features of the adopted multigrid algorithm, it suffices to consider a two-level arrangement, denoted by 'k' and 'k-1', as shown in Fig. 2. A central feature of the arrangement is that four fine-grid cells always combine to form a coarse-grid volume. As a consequence, the coarse-cell residuals arise from an appropriate summation of four fine-cell residuals. By similar arguments, mass fluxes through coarse-cell faces may be obtained by summation of the mass fluxes through pairs of fine-cell faces coincident with the associated coarse-cell faces.

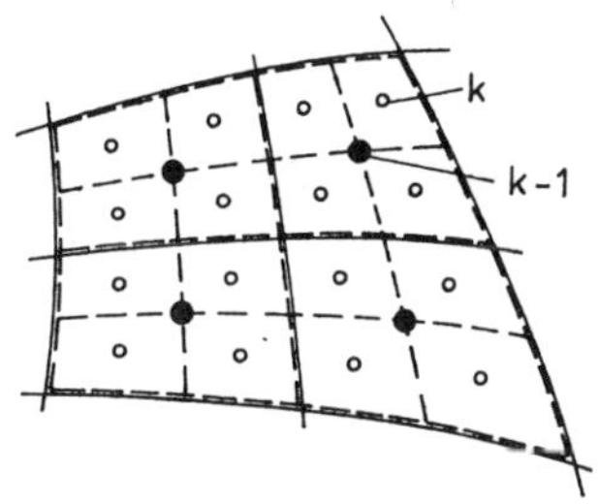

Fig. 2 Two-level finite-volume arrangement

The generic FAS V-cycle starts with typically 2 SIMPLE iterations on the fine k-grid, giving the approximate solutions $\tilde{U}_k$, $\tilde{V}_k$ and $\tilde{P}_k$ with residuals $\tilde{R}^U_k$, $\tilde{R}^V_k$ and $\tilde{R}^m_k$. Within any SIMPLE iteration, the U and V equations are solved once, while the P' equation is relaxed 15 times, the objective being to reduce the mass residuals below a pre-set norm.

Next, the velocities are restricted to the coarse (k-1) grid using bi-linear interpolation operators, while their residuals are restricted by summation:

$$\tilde{U}_{k-1} = I_k^{k-1}\,\tilde{U}_k \qquad \tilde{R}^U_{k-1} = \sum\nolimits_k^{k-1}\tilde{R}^U_k$$
$$\tilde{V}_{k-1} = I_k^{k-1}\,\tilde{V}_k \qquad \tilde{R}^V_{k-1} = \sum\nolimits_k^{k-1}\tilde{R}^V_k \qquad (10)$$

In order to assemble the coarse-grid equations, the sources $\tilde{S}^\phi_k$ would, ordinarily, also need to be restricted. However, because the sources are, in general, dependent on the associated variables, it is inappropriate to do so. Rather, to maintain consistency between the sources and the associated variable fields, it is preferable to re-evaluate the sources from the restricted variable fields, thus yielding $\tilde{S}^U_{k-1}$ $\tilde{S}^V_{k-1}$. Also for reasons of consistency, the residual $\tilde{R}^m_{k-1}$ is not restricted or lumped but re-evaluated from the restricted coarse-grid velocities $\tilde{U}_{k-1}$ and $\tilde{V}_{k-1}$. Finally, the coefficients $\tilde{A}_k$ must be assembled on the coarse grid. This is done here in two parts: first, the convective fragment, being a cell-face mass flux, is restricted by summation; second, the diffusive contribution, being a function of turbulence parameters which are computed as part of the iterative sequence, are not restricted but evaluated on the coarse grid from coarse-grid variables.

The coarse-grid equations now arise as:

$$\{\hat{A}_{k-1}\}^U \hat{U}_{k-1} = \{\tilde{A}_{k-1}\}^U \tilde{U}_{k-1} + \hat{S}^U_{k-1} - \tilde{S}^U_{k-1} + \tilde{R}^U_{k-1} \tag{11}$$
$$\{\hat{A}_{k-1}\}^V \hat{V}_{k-1} = \{\tilde{A}_{k-1}\}^V \tilde{V}_{k-1} + \hat{S}^V_{k-1} - \tilde{S}^V_{k-1} + \tilde{R}^V_{k-1}$$

in which '{A}U' and '{A}V' represent the left-hand sides of equations (7) for 'U' and 'V', respectively, and '^' identifies coarse-grid quantities. Next, starting with

$$\hat{S}^U_{k-1} - \tilde{S}^U_{k-1} = 0 \qquad \hat{S}^m_{k-1} - \tilde{S}^m_{k-1} = 0 \tag{12}$$
$$\hat{S}^V_{k-1} - \tilde{S}^V_{k-1} = 0 \qquad \hat{P}_{k-1} = 0$$

SIMPLE iterations (usually 1) are performed to obtain the coarse-grid solutions. Coarse-grid pressure corrections (actually, corrections of corrections) thereby obtained are fed into the sources of the coarse-grid momentum equations.

There now follows the prolongation process:

$$\tilde{U}_k \leftarrow \tilde{U}_k + I^k_{k-1}(\hat{U}_{k-1} - \tilde{U}_{k-1})$$
$$\tilde{V}_k \leftarrow \tilde{V}_k + I^k_{k-1}(\hat{V}_{k-1} - \tilde{V}_{k-1}) \tag{13}$$
$$\tilde{P}_k \leftarrow \tilde{P}_k + I^k_{k-1}\hat{P}_{k-1}$$

which is terminated by smoothing SIMPLE iterations (usually 2) on the fine grid.

Two special modifications are necessary in relation to turbulence-model equations to maintain the realisability constraints k, ε, $\overline{u_i u_i} > 0$. The first involves under-relaxing the coarse-grid solutions according to:

$$\Delta\hat{\Phi}_{k-1} \leftarrow (\hat{\Phi}_{k-1} - I^{k-1}_k \tilde{\Phi}_k)\,\alpha^{\Phi} \qquad \alpha^{\Phi} < 1 \tag{14}$$

while the second entails conditioning the prologation process via:

$$\tilde{\Phi}_k \leftarrow |\tilde{\Phi}_k + I^k_{k-1}\Delta\hat{\Phi}_{k-1}| \tag{15}$$

or

$$I^k_{k-1}\Delta\hat{\Phi}_{k-1} = \delta\phi^+ + \delta\phi^- \qquad \tilde{\Phi}_k \leftarrow \frac{\tilde{\Phi}_k + \delta\phi^+}{\tilde{\Phi}_k - \delta\phi^-}\tilde{\Phi}_k \tag{16}$$

where $\delta\phi^+$ and $\delta\phi^-$ are unconditionally positive and negative fragments, respectively.

5. MG Cycle Options

Three types of cycles feature in comparisons presented in Section 6. The first is a fixed V-cycle, as shown in Fig. 3, with prescribed iteration numbers on each level. The second is a *Full Multigrid* cycle in which each V-cycle is preceded by an interpolation of the converged solution on the finest mesh of the previous V-cycle to the finest mesh of the following V-cycle.

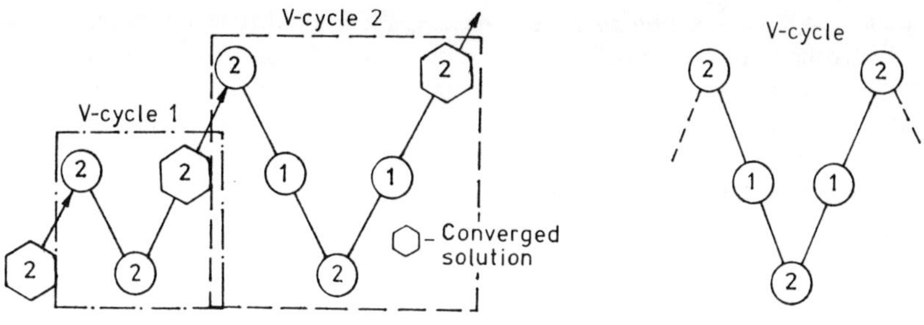

Fig. 3: typical FMG-cycle and V-cycle

In the third cycle, the iteration number at each level is controlled by the behaviour of the Eucledian norm,

$$\overline{R^2} = \sum_{ij} \{(R_{ij}^U)^2 + (R_{ij}^V)^2 + (R_{ij}^m)^2\} /3\ IxJ \tag{17}$$

here IxJ is the total number of cells in the grid level being considered.

The rate of decline of the above norm is tracked, and restriction to the level (k-1) is initiated when

$$R_k^{(n)} / R_k^{(n-1)} \succ \eta \tag{18}$$

where 'n' is the iteration counter and η is a pre-set value. On reaching the convergence criterion

$$R_k^{(n)} \preceq \epsilon_k \tag{19}$$

the coarse-grid corrections are prolongated to the finer level (k+1).

6. Performance Comparisons

The study, in its entirety, has encompassed a wide range of parametric features, investigating, *inter alia,* the dependence of the convergence rate on grid skewness, grid non-uniformity, flow-type, Reynolds number, dimensionality (2D,3D), convection scheme, turbulence modelling, cycle type, level of convergence criterion and vector vs. scalar execution. Space constraints do not permit more than a fraction of the investigation to be exposed. Four cases are chosen here to illustrate the scope of the study and its outcome. In all examples, a solution is held to have converged when the absolute sum of the momentum and mass residuals fell below 0.1% of corresponding momentum- and mass-flux scales. In the case of open, through-flow geometries, these scales arise as the inlet fluxes, while for closed cavities, the lid velocity and the cavity depth were used to derive the scales. In most cases, careful comparisons with other calculations or with experimental data (for turbulent flows) have been performed. These are not included here for lack of space.

The first example examines the dependence of MG performance on the cycle type and convection approximation in the case of a separated laminar flow in a sinusoidal diffuser at Re=100, as shown in Fig. 4. A sequence of four equidistant grids ranging from 32x8 to 256x64 lines has been adopted in conjunction with the Hybrid, QUICK and MUSCL schemes. In each case, two FMG variants have been employed, one with fixed iteration number on every level - denoted by FMG-V, and the other - denoted by FMG-R, subject to criterion (18). Computational requirements are summarised in Table 1 giving, for each scheme separately, expenditure figures in terms of ratios of single-grid-to-multigrid (sg/mg) CPU time and Work Units (WU). The main features deserved to be highlighted are, first, the insensitivity of CPU and WU to the convection scheme; second, the exceptionally high WU ratios - conveying a somewhat distorted message, however, due to the fact that, here, WU only measures the very few iterations on the finest grid of the FMG; and third, the fact that the residual-driven cycle is not significantly superior, in terms of CPU, to the fixed iteration cycle. This last outcome has been found to apply in other geometries examined in the present study.

The next case is a 3D lid-driven equal-sided cavity in which two parallel side walls are inclined at 63° to the horizontal. Calculations were performed for Re-numbers in the range 100-400, and the one chosen here, 100, is quite representative of other values in respect of MG efficiency. Table 2 gives information on WU requirements for 3 types of MG cycles with grids of upto 40^3 nodes. Dependence on the convection scheme has been observed to be relatively minor, with QUICK yielding efficiencies upto 25% above those arising from the Hybrid scheme. Although the grids used in this case are coarse, speedup ratios are quite respectable. A particularly encouraging observation is that even with a very coarse 24-node grid, speedup values of the order 3 have been attained. The virtual constancy of WU is indicative of the MG method performing in an optimal manner.

Attention is now turned to the more challenging turbulent cases. The first is a turbulent flow at Re=15000 through a circular pipe containing a sinusoidal constriction, as shown in Fig. 5. Results reported here have been obtained with the high-Re and low-Re k-ε models outlined in Section 2. The simple FAS V-cycle variant has been tested with the finest grids employed containing 200x40 and 200x56 lines for the high-Re and low-Re models, respectively. The maximum cell-aspect ratio arising with the second model is 120. The speedup ratios are recorded in Table 3 and seen to be far lower than those in laminar conditions, and this is due to a number of factors. First, convergence characteristics are observed to be strongly dependent on flow type and geometry. In this particular case, convergence is seriously hindered by the strong acceleration provoked by the constriction. Second, the grid is skewed and highly stretched, particularly in the case of the low-Re model. Third, the grid across the flow is relatively coarse. Finally, MG efficiency in turbulent flows appears to be quite sensitive to seemingly small details in the manner in which the turbulence-model equations are coupled, within the MG sequence, to the basic aerodynamic equation set. Although speedup ratios are here modest, variations in WU arising from the high-Re k-ε model (in which grid non-uniformities are less severe) are minor, indicating that the multigrid scheme is implemented correctly.

We finally present initial results obtained with the FAS V-cycle for a turbulent flow behind a 90° backward-facing step at Re=110000, computed with a deliberately distorted non-orthogonal grid and the Reynolds-stress-transport model. The grid and streamfunction plots resulting from the high-Re k-ε model and the stress closure are given in Fig. 6. While comparisons with experimental data are not included here, it is noted in passing that the stress model produces a larger recirculation zone which is in accord with experimental observations. MG performance is summarised in Table 4 which compares CPU and WU figures obtained with the k-ε model and the stress variant for relatively coarse grids. Although, speed-up ratios are, here again, significantly lower than in laminar conditions, it is clear that the savings obtained are encouraging, considering the relative coarseness and disposition of the grid and the complexity of the physical situation.

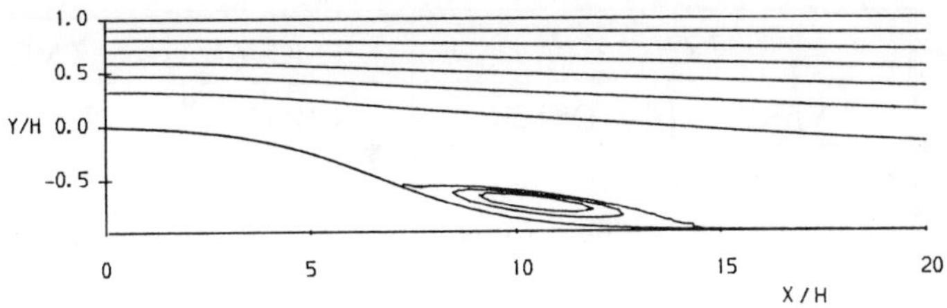

Fig. 4: Case 1: laminar flow in plane constriction

	HYBRID						QUICK		
	FMG-V			FMG-R			FMG-V		
GRID	sg/mg CPU	WU	sg/mg WU	sg/mg CPU	WU	sg/mg WU	sg/mg CPU	WU	sg/mg WU
32x8	1.0	4.8	4.6	1.1	6.0	3.7	1.0	4.8	5.2
64x16	5.9	5.0	16.4	6.5	4.5	18.0	5.7	5.0	17.0
128x32	15.3	2.0	144.0	16.6	4.5	64.0	15.5	2.0	158.5
256x64	64.9	2.0	552.5	70.8	3.5	315.7	63.6	2.0	558.5.

	QUICK			MUSCL					
	FMG-R			FMG-V			FMG-R		
GRID	sg/mg CPU	WU	sg/mg WU	sg/mg CPU	WU	sg/mg WU	sg/mg CPU	WU	sg/mg WU
32x8	1.1	6.3	4.0	1.1	4.8	5.4	1.1	6.0	4.3
64x16	6.5	4.5	18.9	6.1	5.0	17.0	7.0	4.5	18.9
128x32	17.9	3.5	90.6	16.7	2.0	158.0	18.2	3.5	90.3
256x64	70.2	3.5	319.1	68.3	2.0	558.5	74.4	3.5	319.1

Table 1: MG performance for Case 1: 2D plane constriction

	V-CYCLE				FMG-V		FMG-R	
	HYBRID		QUICK		HYBRID		HYBRID	
CV	CPU sg/mg	WU sg/mg	CPU sg/mg	WU sg/mg	CPU sg/mg	WU sg/mg	CPU sg/mg	WU sg/mg
24x24x24	3.3	3.7	3.0	3.7	3.3	4.4	4.5	6.0
32x32x32	5.7	6.3	5.0	5.9	5.7	7.5	7.6	9.9
40x40x40	9.1	9.1	6.7	7.5	10.3	11.7	12.6	15.0

Table 2: MG performance for Case 2: 3D skewed cavity

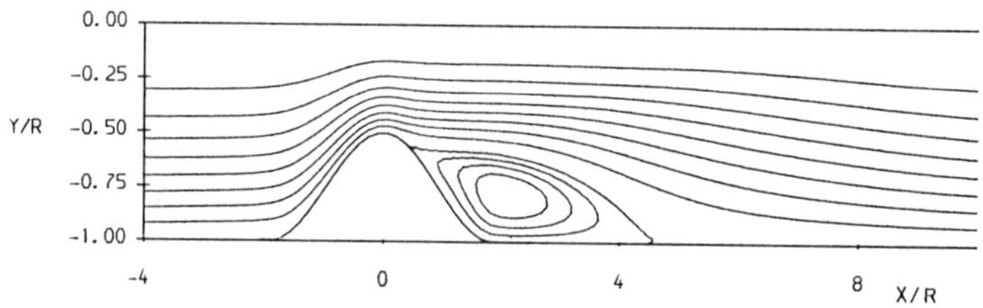

Fig. 5: Case 3: turbulent flow in circular constriction

	High-Re k~ε				Low-Re k~ε			
	SG	V-CYCLE			SG	V-CYCLE		
CV	WU	CPU sg/mg	WU	WU sg/mg	WU	CPU sg/mg	WU	WU sg/mg
80x16	189	1.26	94.4	2.00	296	1.71	117.7	2.52
120x24	417	2.77	94.4	4.42	529	2.62	142.9	3.70
160x32	639	4.19	97.1	6.58	1013	3.92	183.9	5.51
200x40	933	5.90	110.3	8.46	1255	4.26	224.9	5.58

Table 3: MG performance for Case 3: 2D circular pipe

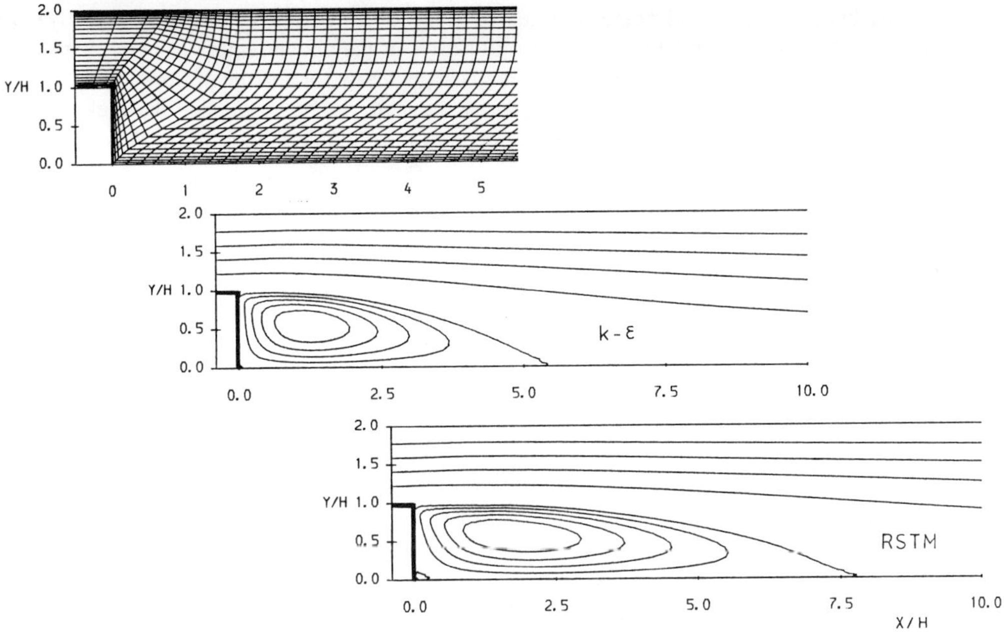

Fig. 6: Case 4: turbulent flow behind backward-facing step

	High-Re k~ε				RSM			
	SG	V-CYCLE			SG	V-CYCLE		
CV	WU	CPU sg/mg	WU	WU sg/mg	WU	CPU sg/mg	WU	WU sg/mg
40x8	135	1.22	71.2	1.90	153	1.74	49.3	3.11
80x16	390	2.60	83.9	4.65	446	2.65	96.5	4.62
120x24	790	3.59	126.1	6.26	808	3.27	149.0	5.42
160x32	1329	5.44	139.3	9.54	1094	3.44	191.0	5.72

Table 4: MG performance for case 4: 2D backward-facing step

7. Conclusions

Some aspects of a wider investigation into multigrid convergence acceleration for complex, practically relevant flows have been introduced. Considerable performance variations have been observed, the most influential factors being flow geometry and type. CPU savings in turbulent flow are significant, but far less dramatic than in some laminar flows. This is not merely due to the addition of a highly non-linear turbulence-transport model to the coupled equation set, but also a result of strongly increased levels of grid compression and convective processes. To ascertain the relative contribution of each of the above factors to performance variability, an extensive parametric study within a given geometry for a range of flow conditions would be required. The main message conveyed by the study is, however, that the multigrid method offers real benefits even if skewed grids are used in conjunction with the most complex turbulence model currently available.

8. References

1. Barcus, M., Peric, M. and Scheuerer, G., Notes on Numerical Fluid Mechanics, 20, Vieweg Verlag, Braunschweig, (1988), 9-16.
2. Becker, C., Ferziger, J.H., Peric, M. and Scheuerer, G., Vieweg Verlag, Braunschweig, (1989), 30-40.
3. Brandt, A., Math. Comput., 31, No. 138, (1977), 333-390.
4. Gaskell, P.H., Lau, A.K.C. and Wright, N.G., Two efficient solution strategies for use with high order discretization schemes in simulation of fluid flow problems, Proc. 5th Conf. on Num. Meth. in Laminar and Turbulent Flow, Montreal (1987), 210.
5. Gibson, M.M. and Launder, B.E., J. Fluid Mech., 85, (1978), 491-511.
6. Hackbusch, W. and Trottenberg, U.(eds.) Multigrid Methods, Lecture Notes in Mathematics, 960, (Spring, Berlin, 1982).
7. Jones, W.P. and Launder, B.E., Int. J. Heat Mass Transfer, 16 (1972), 301-313.
8. van Leer, B., J. Comp. Phys., 32, (1979), 101-136.
9. Leonard, B.P., Comp. Meths. Appl. Mech. Eng., 19, (1979), 59-98.
10. Lonsdale, G., Solution of a rotating Navier-Stokes problem by a non-linear multigrid algorithm, Report No 105, Dept. of Mathematics, University of Manchester, (1985)
11. Leschziner, M.A., J. Wind Engineering and Industrial Aerodynamics, 35 (1990), 21-47.
12. Patankar, S.V., Numerical Heat Transfer and Fluid Flow, Hemisphere Publishing Co., McGraw Hill, (1980).
13. Peric, M., Ruger, M. and Scheuerer, G., A finite volume multigrid method for calculating turbulent flows, 7th Sym. on Turb. Shear Flows, Stanford Univ., Aug. (1989), 7.3.1.-7.3.6.
14. Phillips, R.E., Miller, R.F. and Schmidt, F.W., A multilevel-multigrid algorithm for turbulent recirculating flows, Proc. 5th Sym. on Turb. Shear Flows, (1985), 20.21-20.25.
15. Rhie, C.M. and Chow, W.L., AIAA J., 21, (1983), 1525.
16. Thompson, M.C. and Ferziger, H.J., An adaptive multigrid solution technique for the steady state incompressible Navier-Stokes equations, *Computational Fluid Dynamics*, G. de Vahl Davies and C. Fletcher (Ed), Elsevier Science Publishers B.V. (North-Holland), 1988, 715-724
17. Vanka, S.P., Block-implicit multigrid calculation of flows, Comp. Meth. Appl. Mech. Eng., 59,(1986) 29-48.

Acknowledgement

The work reported herein was supported by Grant GR/E26808 awarded to the investigators by the UK Science and Engineering Research Council. Thanks are due to Mrs. Irene Bowker for her professional preparation of the manuscript.

Time Accurate Multigrid Solutions of the Navier-Stokes Equations

M. Meinke, D. Hänel
Aerodynamisches Institut, RWTH Aachen
Wüllnerstr. zw. 5 und 7, 5100 Aachen, Germany

Abstract

The time dependent two-dimensional Navier-Stokes equations for compressible laminar flows are solved with an explicit Runge-Kutta time stepping scheme. The influence of the direct FAS multigrid method on the time accuracy of the numerical solution is investigated for several unsteady viscous flow problems. Even in highly unsteady flows, like the self induced unsteady flow around a circular cylinder, up to 50% of computer time can be saved without significant loss of accuracy.

1 Introduction

The numerical simulation of unsteady viscous flows requires sufficient resolution of all existing scale lengths. Typical small scales in space in viscous flows are usually the boundary layer thickness or the diameter of small scale vortices, which develop e. g. behind a self-induced separation point. Scale lengths in time can exist in several orders of magnitude, dependent on the nature of the unsteady flow. Unsteady flows due to a movement of a rigid wall, for instance, can have much larger scale lenghts than those appearing in flows with an instability.

The numerical algorithm for an accurate and efficient prediction of unsteady flows should be properly chosen. In [2,12] it is shown that, among other influences, the amount of numerical dissipation in the discretization scheme can have a large influence on the unsteady solution of flows with small scale vortices.

For steady state problems Jameson [8] and other authors applied the multigrid method to an explicit solution scheme for the Euler equations. The efficiency of the scheme could be improved considerably. A similar approach but for unsteady problems is done in this paper. The so called direct multigrid method was formulated for the use with an explicit solution scheme for the unsteady Navier-Stokes equations. The direct multigrid method uses the Full Approximation-Storage (FAS) concept in space and time. Consequently the physical time also advances on the coarser grids. In principle it can be applied to explicit and implicit methods. The use of the direct multigrid method in implicit schemes is of advantage, when the time step cannot be chosen as large as permissible due to stability limits, e. g. in factorization schemes. For explicit schemes the benefit of the direct method is twofold: first computational work is saved on the coarser grids, secondly the time steps can be chosen larger on the coarse grids due to larger spatial steps. The influence of the multigrid method vanishes in a convergent steady state solution, because the discretization error preserves the fine grid accuracy also on the coarse grids. In transient time-accurate solutions, however, an influence remains. The coarse grid solution is corrected by the discretization error, but the latter is not updated during the time steps on the coarse grids. Therefore the direct method has an influence on the unsteady solution, which increases with larger temporal changes in the flow and the number of time steps carried out on the coarse grids.

A careful formulation of the multigrid algorithm is necessary to achieve time accurate solutions. To investigate the influence of different parameters of the multigrid method on the accuracy and to study the possible increase of the efficiency, a couple of different unsteady flows problems are simulated.

2 Governing Equations and Spatial Discretization

The governing equations are the time-dependent Navier-Stokes equations for a compressible fluid. In the following the equations are presented for a twodimensional curvilinear coordinate system, here (ξ, η, t). In conservative form the equations read:

$$\widehat{Q}_t + (\widehat{F} - \frac{1}{Re}\widehat{S})_\xi + (\widehat{G} - \frac{1}{Re}\widehat{T})_\eta = 0 \tag{1}$$

The vector $\widehat{Q}$ is the vector of the conservative variables $Q = (\rho, \rho u, \rho v, \rho E)^T$ multiplied by the Jacobian, i. e. $\widehat{Q} = J\,Q$, where $J = x_\xi y_\eta - x_\eta y_\xi$, which has the meaning of a volume. The Euler fluxes $\widehat{F}, \widehat{G}$ and the viscous terms $\widehat{S}, \widehat{T}$ are the contravariant components, e. g. $\widehat{F} = F y_\xi - G x_\xi$, which can be interpreted as the physical fluxes multiplied by the normal surface element. The gas under consideration is assumed to be perfect, and the flow to be laminar.

To preserve the conservative properties in the discretized space, equations (1) are applied to a finite control volume. In general a node-centered arrangement of the control volume (Fig. 1a) was used in the present study. But for comparison a cell vertex formulation (Fig. 1b) is considered, as well. For both cases the result of the discretization is a set of difference equations approximating the Navier-Stokes equations:

$$\frac{\Delta\widehat{Q}}{\Delta t} + Res(Q) = 0 \tag{2}$$

with $\Delta\widehat{Q}/\Delta t$ representing the discrete time derivative, defined later, and $Res(Q)$ corresponds to the discretized steady-state operator

$$Res(Q)_{i,j} = \delta_\xi(\widehat{F} - \frac{1}{Re}\widehat{S}) + \delta_\eta(\widehat{G} - \frac{1}{Re}\widehat{T}) \tag{3}$$

For the node-centered scheme the evaluation of the fluxes at the cell interfaces $\Gamma_{i\pm\frac{1}{2},j}$ and $\Gamma_{i,j\pm\frac{1}{2}}$ of the control volume is required. The difference operators δ_ξ and δ_η are then defined as:

$$\delta_\xi f = \frac{f_{i+\frac{1}{2},j} - f_{i-\frac{1}{2},j}}{\Delta\xi}, \qquad \delta_\eta f = \frac{f_{i,j+\frac{1}{2}} - f_{i,j-\frac{1}{2}}}{\Delta\eta}$$

In the cell vertex scheme the control volume for the Euler fluxes consists of the four neighbouring cells with a common vertex (i, j). For this case the operators δ_ξ and δ_η read:

$$\begin{aligned}\delta_\xi f &= \sum_{k,l=0}^{1} \frac{f_{i+k,j+\frac{1}{2}-l} - f_{i+k-1,j+\frac{1}{2}-l}}{\Delta\xi} \\ \delta_\eta f &= \sum_{k,l=0}^{1} \frac{f_{i+\frac{1}{2}-k,j+l} - f_{i+\frac{1}{2}-k,j+l-1}}{\Delta\eta}\end{aligned}$$

The viscous fluxes in the cell vertex scheme are integrated over a smaller control volume, which is defined by the center points of the four cells $(i \pm \frac{1}{2}, j \pm \frac{1}{2})$.

The essential properties of this method are determined by the definition of the fluxes. The Euler fluxes are approximated either by central- or upwind differences, the viscous terms in all cases by central differences. The different discretizations will be explained briefly in the following for one component.

In the cell vertex scheme the Euler fluxes are computed at the vertices and then averaged in order to obtain the values on the the cell faces, e. g.:

$$\widehat{F}_{i\pm\frac{1}{2},j} = \frac{1}{2}(\widehat{F}_{i\pm 1,j} + \widehat{F}_{i,j})$$

In the case of central differencing in the node centered scheme, the Euler fluxes on the cell interfaces $\Gamma_{i\pm\frac{1}{2},j}$ are determined as functions of the averaged conservative variables:

$$\widehat{F}_{i\pm\frac{1}{2},j} = \widehat{F}(\frac{Q_{i\pm 1,j} + Q_{i,j}}{2})$$

For the upwind discretization the Euler fluxes are split according to the sign of the eigenvalues by van Leer's splitting concept [11]. This results in

$$\widehat{F}_{i\pm\frac{1}{2},j} = \widehat{F}^{+}(Q^{+}_{i\pm\frac{1}{2},j}) + \widehat{F}^{-}(Q^{-}_{i\pm\frac{1}{2},j})$$

Since the splitting has been derived for a Cartesian grid, its use for curvilinear coordinate systems is provided by rewriting the components of the transformed Euler fluxes in such a way that every term of the reformulated components belongs to a locally Cartesian grid. This procedure was applied to the contravariant flux components as defined in Eq. (3). With the Cartesian velocities u, v, the speed of sound a and the definitions:

$$\begin{aligned} \bar{W} &= u\omega_x + v\omega_y \\ \widehat{a} &= a\,|\nabla\omega| \\ |\nabla\omega| &= \sqrt{(\omega_x^2 + \omega_y^2)} \end{aligned}$$

the general fluxes $\widehat{P}^{\pm}$ for subsonic flow i. e. for $|\bar{W}/\widehat{a}| < 1$ result in:

$$\widehat{P}^{\pm} = J^{-1} \begin{pmatrix} \pm\rho\widehat{a}/4\,(\bar{W}/\widehat{a} \pm 1)^2 \equiv \widehat{P}_1^{\pm} \\ \widehat{P}_1^{\pm}\,[u + \frac{p\omega_x}{\rho a}(-\bar{W}/\widehat{a} \pm 2)] \\ \widehat{P}_1^{\pm}\,[v + \frac{p\omega_y}{\rho a}(-\bar{W}/\widehat{a} \pm 2)] \\ \widehat{P}_1^{\pm}\,H_t \end{pmatrix} \tag{4}$$

where H_t is the total enthalpy. Substituting $\omega = \xi$ one determines $\widehat{F}^{\pm} = \widehat{P}^{\pm}$ with $\widehat{U} = \bar{W}$, and in a similar way the other components of Eq. (3).

The Euler fluxes are approximated according to Gudonov's approach, i. e. at each cell interface a Rieman problem has to be solved. For the terms of the split Euler fluxes MUSCL type differencing (Monotonic Upstream Centered Schemes for Conservation Laws) is used [10]. First the values of the conservative variables Q of the grid points are extrapolated to the cell boundaries $\Gamma_{i\pm\frac{1}{2},j}$. The extrapolation in ξ-direction yields for example:

$$Q^{+}_{i+\frac{1}{2},j} = Q_{i,j} + \delta Q_{i,j}\,, \qquad Q^{-}_{i+\frac{1}{2},j} = Q_{i+1,j} - \delta Q_{i+1,j} \tag{5}$$

for the forward flux $\widehat{F}^{+}_{i+\frac{1}{2},j}$, and for the backward flux $\widehat{F}^{-}_{i+\frac{1}{2},j}$, resp.. Then the fluxes are evaluated by the extrapolated variables:

$$\widehat{F}^{\pm}_{i\pm\frac{1}{2},j} = \widehat{F}^{\pm}(Q^{\pm}_{i\pm\frac{1}{2},j}) \tag{6}$$

A more accurate splitting for viscous flows than the original concept by van Leer is used in the present study, [9]. In this case the variables ρ, p, and $\bar{W}$ are updated as in Eq. (6), but the velocities tangential to the cell interfaces, for example $\widehat{V}$, is updated according to the sign of the normal velocity $\bar{W} = \widehat{U}$:

$$\widehat{V}^{\pm} = \widehat{V}(\tilde{Q}) \qquad \begin{cases} \tilde{Q} = Q^{+} : \widehat{U} \geq 0 \\ \tilde{Q} = Q^{-} : \widehat{U} < 0 \end{cases} \tag{7}$$

Choosing:

$$\delta Q_{i,j} = \left[\left(\frac{\overrightarrow{S}}{\overrightarrow{S} + \overleftarrow{S}} \right) \varphi_{i,j} \left(\frac{\overrightarrow{\Delta Q}\,\overleftarrow{S} + \overleftarrow{\Delta Q}\,\overrightarrow{S}}{\overrightarrow{S} + \overleftarrow{S}} \right) \right]_i \tag{8}$$

and $\varphi = 1$, a piecewise linear distribution over the cell is assumed resulting in a second-order accurate approximation. The quantities $\overrightarrow{\Delta Q}$, $\overleftarrow{\Delta Q}$ in Eq. (8) represent forward and backward differences in the conservative variables Q, and $\overrightarrow{S}$, $\overleftarrow{S}$ describe the distance between two neighbouring nodal points, e. g.:

$$\overrightarrow{\Delta Q_{i,j}} = Q_{i+1,j} - Q_{i,j} \qquad \overleftarrow{\Delta Q_{i,j}} = Q_{i,j} - Q_{i-1,j}$$

$$\overrightarrow{S} = [\overrightarrow{\Delta x}^2 + \overrightarrow{\Delta y}^2]^{\frac{1}{2}} \qquad \overleftarrow{S} = [\overleftarrow{\Delta x}^2 + \overleftarrow{\Delta y}^2]^{\frac{1}{2}}$$

3 Method of Solution

3.1 Runge-Kutta Time-Stepping Scheme

The integration in time of the differential equations can be carried out with an explicit Runge-Kutta method. At present a method is used, which has been succesfully applied to the Euler- and Navier-Stokes equations by Jameson [8] and other authors. A general solution scheme for a N-step Runge-Kutta method reads:

$$\begin{aligned} Q^{(0)} &= Q^n \\ &\vdots \\ \Delta Q^{(l)} &= -\alpha_l \Delta t\, Res(Q^{(l-1)}) / J \\ Q^{(l)} &= Q^{(0)} + \Delta Q^{(l)} \\ &\vdots \\ Q^{n+1} &= Q^N \end{aligned} \quad \left. \right\} \; l = 1, \ldots, N \tag{9}$$

Herein a 5-step Runge-Kutta scheme was adapted for a maximum Courant number which is 4.0 for the central and 3.5 for the upwind scheme. The coefficients in the Runge-Kutta steps are chosen as $\alpha_l = (0.25, 0.1667, 0.375, 0.5, 1)$ for the central difference scheme and $\alpha_l = (0.059, 0.14, 0.273, 0.5, 1)$ for the upwind scheme.

For the central discretization of the Euler fluxes artificial damping terms are necessary to supress high frequency error components. A fourth order difference of the following form is added to the steady state operator:

$$\delta_\xi(\frac{J}{\Delta t}\epsilon\,\Delta_\xi^3 Q) + \delta_\eta(\frac{J}{\Delta t}\epsilon\,\Delta_\eta^3 Q)$$

with

$$\Delta_\xi Q = Q_{i+1,j} - Q_{i,j}\,, \qquad \Delta_\eta Q = Q_{i,j+1} - Q_{i,j}$$

The parameter ϵ controls the amout of numerical dissipation. In the flow simulations computed here, it was chosen between $0.0039 < \epsilon < 0.00196$. A damping term of second order, which is often used for shock capturing is not needed here, because all flows considered here are subsonic.

3.2 Direct Multigrid Method

In the following the direct multigrid method, used in this paper, is decribed in more detail.

3.2.1 Full Approximation Storage Multigrid Concept

For nonlinear equations Brandt [3] has proposed the Full Approximation Storage (FAS) multigrid concept. This is adopted in the present solution scheme for the Navier-Stokes equations. Considering a grid sequence $G_k, k = 1, \ldots, m$ with the step sizes $h_k = 2h_{k+1}$ a finite difference approximation on the finest grid G_m may be

$$L_m Q_m = 0 \tag{10}$$

where $L_m Q_m$ corresponds to the discretized conservation equation, Eq. (2). If the solution on the fine grid is sufficiently smooth with respect to the high frequency solution components, Eq. (10) may be approximated on a coarser grid G_{k-1} by a modified difference approximation:

$$L_{k-1} Q_{k-1} = \tau_{k-1}^m \tag{11}$$

where τ is the "fine to coarse defect correction" and refered to as the "discretization error". It maintains the truncation error of the fine grid G_m on the coarser grids G_{k-1} and is defined by

$$\tau_{k-1}^m = \tau_k^m + L_{k-1}(I_k^{k-1} Q_k) - I\!I_k^{k-1}(L_k Q_k) \tag{12}$$

where I_k^{k-1} and $I\!I_k^{k-1}$ are restriction operators from grid G_k to G_{k-1} which are applied to the variable Q_k and the difference approximation $L_k Q_k$, as well. These operators can be used as injection or full weighting operators and will be explained later.

The solution sequence used in this paper corresponds to the so-called V-cycle. Starting with the solution of Eq. (10) on the finest grid G_m, repeatedly the transfer Eq. (12) to the coarser grids G_k is carried out and Eq. (11) is solved, until the coarsest grid is reached. Then the variables are interpolated back to the finest grid with some solution steps on the coarser grid levels in between.

According to the FAS scheme only the correction between the "old" fine grid solution Q_{k+1}^{old} and the "new" coarse grid solution Q_k is interpolated to the fine grid to update Q_{k+1}^{old}:

$$Q_{k+1}^{new} = Q_{k+1}^{old} + I_k^{k+1}(Q_k - I_{k+1}^k Q_{k+1}^{old}) \tag{13}$$

Herein I_k^{k+1} is the interpolation operator, where so far bilinear interpolation is used.

3.2.2 Restriction Operators and Coarse Grid Boundary Conditions

In the node-centered and cell vertex scheme, as sketched in Fig. 1, the nodes on the coarser grid coincide with the fine grid nodes. Therefore point-to-point injection is usually used for the conservative variables Q, (not $\widehat{Q}$), and in some cases also used for the restriction of the residuals (= flux times normal surface).

$$I_k^{k-1} Q_k = Q_k, \qquad I\!I_k^{k-1} Res_k(Q_k) = Res_k(Q_k) \tag{14}$$

Alternatively, a weighted restriction over the neighbouring nine points can be used for the residual:

$$I\!I_k^{k-1} Res_k(Q_k) = \sum_{9\text{cells}} \beta \; Res_k(Q_k) \tag{15}$$

where β is either a constant weighting factor, e. g. $\beta = (\frac{1}{4}, \frac{1}{8}, \frac{1}{16})$, or proportional to the cell volume $\beta = J_k / \sum J_k$.

For time accurate computations a correct FAS treatment of the boundary conditions is required. Consider a boundary approximation on the finest grid G_m:

$$C_m Q_m + g_m = 0 \tag{16}$$

which may be for example a Neumann boundary condition. If Eq. (16) is applied directly to the coarser grids, a large truncation error would occur. A coarse grid correction is therefore defined similar to and consistent with the FAS procedure, Eq. (11). The boundary approximation on the coarse grid G_{k-1} then reads:

$$C_{k-1}\,Q_{k-1} + g_{k-1} = \tau_{B\,k-1}^{\;m} \tag{17}$$

where $$\tau_{B\,k-1}^{\;m} = \tau_{B\,k}^{\;m} + C_{k-1}\,(I_k^{k-1} Q_k) + g_{k-1} - I_k^{k-1}\,(C_k Q_k + g_k) \tag{18}$$

3.2.3 Formulation of the Discretization Error

Employing the FAS-multigrid procedure, Eq. (11), to the Runge-Kutta scheme, Eq. (9), an intermediate Runge-Kutta step (l) on a coarser grid reads:

$$\begin{aligned} \Delta Q_{k-1}^{(l)} &= -\alpha_l\,\Delta t_{k-1}\,(\,Res_{k-1}\,Q_{k-1}^{(l-1)} - \tau_{k-1}^{m}\,)\,/\,J_{k-1} \\ Q_{k-1}^{(l)} &= Q_{k-1}^{(0)} + \Delta Q_{k-1}^{(l)} \end{aligned} \tag{19}$$

The discretization error τ between the fine grid "m" and the coarse grid "$k-1$" is defined by Eq. (12) and can be split in two terms:

$$\tau_{k-1}^{m} = (\tau_{k-1}^{m})_{Res} + (\tau_{k-1}^{m})_{time} \tag{20}$$

The first part of the discretization error corrects the spatial accuracy and is defined by:

$$(\tau_{k-1}^{m})_{Res} = (\tau_k^m)_{Res} + Res_{k-1}(\,I_k^{k-1}\,Q_k) - II_k^{k-1}\,Res_k(\,Q_k) \tag{21}$$

Approximating the time derivative of Q in the Runge-Kutta algorithm by $\Delta Q/\Delta t$ the second term can be written as:

$$\begin{aligned} (\tau_{k-1}^{m})_{time} = (\tau_k^m)_{time} &+ \left[I_k^{k-1}\,(\,Q_k(\,t^{\tilde n}\,) - Q_k(\,t^{\tilde n} - \Delta t_{k-1}\,))\right] J_{k-1}/\Delta t_{k-1} \\ &- II_k^{k-1}\left[(\,Q_k(t^{\tilde n}) - Q_k(\,t^{\tilde n} - \Delta t_k))\,J_k\,/\Delta t_k \right] \end{aligned} \tag{22}$$

where $t^{\tilde n}$ is the time when the grid level is changed. If the time steps Δt_k and Δt_{k-1} on the fine and on the coarse grid are different, then the variables on several fine grid levels have to be stored to establish time-accurate correction on the coarser grids.

Assuming the same time steps on all grids, computer memory can be saved and Eq. (22) can be simplified to:

$$(\tau_{k-1}^{m})_{time} = (\tau_k^m)_{time} + I_k^{k-1}\left[Q\,(\,t^{\tilde n}\,) - Q\,(\,t^{\tilde n} - \Delta t\,)\right]\,(\,J_{k-1} - II_k^{k-1} J_k\,)\,/\Delta t \tag{23}$$

The consideration of steady-state solutions does not require time accuracy and therefore allows the usage of Eq. (23) even though local time stepping may be used.

In the present definition of the discretization error the residual corresponds to the boundary integral of the fluxes over cell interfaces, Eq. (3). If the residual in the form of the difference operator is used, i. e. the present definition divided by the volume J, the formulation of the algorithm and restriction must be properly adapted.

In the following considerations a node-centered scheme is used, if not otherwise specified. In all cases the transfer from coarse to fine grids was carried out by bilinear FAS interpolation according to Eq. (13).

4 Results

In the present study the multigrid method is applied to an explicit Runge-Kutta time stepping scheme for unsteady solutions of the Navier-Stokes equations. To preserve the accuracy of the finest grid with the multigrid method the discretization error has to correct the larger truncation errors in time and space on the coarser grids. To reduce the amount of required computer storage the same time step Δt was chosen on all grid levels. Then the simpler formulation of the discretization error in time Eq. (23) can be used.

As a first test case the flow about an infinite oscillating flat plate, also called Stoke's second problem, is calculated, for which an exact solution is available. Several computations were carried out with different V-cycles and a different number of grid levels. In Fig. 2a the velocity profiles of the single grid and the analytical solution show good agreement at all time levels. In Fig. 2b the deviation between numerical and exact solution is presented. The deviation is defined by $\frac{1}{N}\sum_{i=1}^{N}(u_{i\,exact} - u_{i\,num.})^2$, where u is the velocity parallel to the plate and N is the number of grid points normal to the wall. It is plotted against the work units of the single grid solution devided by the work units used for the actual solution. This factor roughly corresponds to the speed-up achieved with the multigrid procedure. As expected the deviation increases with the amount of work done on the coarse grids, due to the frozen discretization error. Remarkable is the fact that V-cycles with two grid levels show smaller deviations than V-cycles of comparable computational effort with three grid levels.

In the previous simulation the flow field was resolved with 17 grid points normal to the plate and about 650 time steps per period of oscillation. To simulate a more realistic case, the flowfield about a plate of finite length is computed. The exact solution for the plate of infinite length still can be used as a reference solution, provided the Reynolds number is large enough. Fig. 3a shows the velocity profiles at three different positions on the plate. They still agree very well with the analytical solution. Reducing the resolution of the flow field to 9 grid points normal to the wall and to 25 time steps per period, the results in Fig. 3b are obtained. The deviation from the analytical solution presented here, is larger than in Fig. 2b, due to the coarser resolution of the flow field. The influence of the number of grid levels, however, is comparable to that in Fig. 2b. For a given speed-up the deviation is smaller, when less grid levels are used. In highly unsteady flows the number of grid levels should therefore be restricted to two. The sequence of time steps on the coarse grids also influences the accuracy. The best results are obtained with a V-cycle, in which the discretization error is updated as often as possible and one time step is done on all intermediate grid levels between two successive interpolations and restrictions.

The results of a simulation of a self induced vortex separation behind a circular cylinder are shown in Fig. 4. An impression of the flow structure is given in Fig. 4a by the computed streaklines. The domain of integration and the boundary conditions are given in Fig. 4b. Two different spatial resolutions were used in the simulation. Fig. 4c and Fig. 4d show the deviation of the Strouhal number in percent of the multigrid from the single grid solution. The number of grid points used in Fig. 4c is 177x113 and 225x225 in Fig. 4d., where the first number denotes the number of grid points in circumferential and the second in radial direction. Comparing the Strouhal number of the two single grid solutions, they differ by about 4%, which shows that the flow field is not sufficiently resolved with the coarser grid. The deviation of the multigrid solutions at a speed up factor of about 2 amounts to 2% for the coarse and ca. 0.5% for the finer grid. The errors induced by the multigrid method obviously become smaller in better resolved flow fields, since the approximation on the coarser grids becomes better and the discretization error smaller. The use of the direct multigrid method is therefore better suited for flow fields, where either the discretization error is small due to a good flow resolution, or where the changes in the discretization error per time step are small due to small temporal changes in the flow.

To examine the influence of different restriction operators the flow instability of a jet was simulated. Fig. 5a shows the domain of integration with the initial and boundary conditions. This flow develops highly unsteady, small scale structures for large Reynolds numbers. In [2] it is shown that small variations in the numerical scheme can change the solution remarkably. The simulation with the multigrid scheme is therefore not done with the intention to produce an exact solution, but to make changes in the restriction operator clearly visible. Fig. 5b-e show streaklines of the computed flow. The streaklines consist of 10000 particles, which were initially distributed on the two discontinuities of the flowfield. The flow structure is shown after about 500 time steps with the Runge-Kutta scheme. Two main vortices dominate the flow, in addition smaller structures are visible in the big vortices. All multigrid solutions reproduce the main structure but they differ to a great extent in the details. The multigrid solutions with a weighted average or volume-weighted restriction come closest to the single grid solution, where the simple injection operator shows larger deviations. The volume-weighted restriction operator therefore seems to be of advantage in unsteady flow computations.

For a quantitative analysis of the influence of the multigrid procedure the flow around a circular cylinder at Reynolds numbers of 100 and 200 is computed with different V-cycles and compared to the single grid solution. The flow fields computed with a single and a multigrid algorithm are shown in Fig. 6a,b and compare very well. The small differences are mainly due to the fact that the physical times do not coincide exactly. The Strouhal numbers of all solutions lie in the range of experimental data. The drag- and the lift coefficients of the single grid solutions are in agreement with numerical results of [13]. The evaluation of the multigrid solutions in Table 1 and Table 2 shows small deviations up to 2% compared with the single grid solution at a maximum effictive speed up factor of about two. This shows that accurate solutions can be obtained with the direct multigrid method in unsteady flows.

5 Conclusions

A direct multigrid method for an explicit multistage Runge-Kutta scheme was formulated for the unsteady Navier-Stokes equations. The applicability of this method has been verified for several unsteady flow problems. The numerical experiments have shown that the number of grid levels should be minimized to minimize the additional errors introduced by the frozen discretization error.

The influence of the frozen discretization error becomes smaller when either the discretization error itself is or the temporal changes of the error are small. Therefore the multigrid method is well suited for flow problems which can be sufficiently resolved in time and/or space. Up to 50% of computer time could be saved without significant loss of accuracy even in highly unsteady flows. The simulation of unsteady flows at higher Reynolds numbers and with highly stretched grids is the matter of work in progress.

References

[1] Brandt, A.: *Multi-level adaptive solutions to boundary-value problems,* Mathematics of Computation, vol. 31, No. 138, pp. 333-390, (1977).

[2] Hänel, D.: *Effects of Numerical Dissipation in Solutions of the Navier-Stokes equations,* 3rd International Conference on Hyperbolic Problems, Uppsala, 1990.

[3] Brandt, A.: *Guide to multigrid development,* In: Lecture Notes in Mathematics vol. 960., pp. 220-312, Springer Verlag Berlin, (1981).

[4] Radespiel, R., Swanson, R.C.: *An investigation of cell centered and cell vertex multigrid schemes for the Navier-Stokes equations,* AIAA paper 89-0548, (1989).

[5] Hall, M.G.: *Cell-vertex multigrid schemes for solutions of the Euler equations,* In: Num. Meth. for Fluid Dynamics, part II, Morton, K.W., Baines, M.J. (Edit.), Larendon Press, Oxford, (1986).

[6] Shaw, G., Wesseling, P.: *Multigrid method for the compressible Navier-Stokes equations,* Rep. of the Dep. of Math. and Inf., Nr. 86-13, Univ. Delft, (1986).

[7] Anderson, W. K., Thomas, J.L., Rumsey, C. L.: *Extension and application of flux-vector splitting to calculation on dynamic meshes,* AIAA paper 87-1152 CP (1987).

[8] Jameson, A.: *Solution of the Euler equations for two dimensional transonic flow by a multigrid method,* Presented at the International Multigrid Conference, Copper Mountain, (1983).

[9] Schwane, R., Hänel, D.: *An implicit flux-vector splitting scheme for the computation of viscous hypersonic flow,* AIAA-paper No. 89-0274, (1989).

[10] van Leer, B.: *Towards the ultimate conservative difference scheme V. A second-order sequel to Godunov's method,* J. Comp. Phys., vol. 32, pp. 101-136, (1979).

[11] van Leer, B.: *Flux-vector splitting for the Euler equations,* Lecture Notes in Physics vol. 170, pp. 507-512, (1982).

[12] Meinke, M., Hänel, D.: *Simulation of Unsteady Flows,* 12th International Conference on Numerical Methods in Fluid Dynamics, Oxford 1990

[13] Braza, M., Chassaing P., Ha Minh, H.: *Numerical study and physical analysis of the pressure and velocity fields in the near wake of a circular cylinder*, Journal of Fluid Mechanics vol. 165, pp. 79-130, (1986).

[14] Berger, E., Wille, R.: *Periodic Flow Phenomena*, Annual Review of Fluid Mechanics vol. 4, pp. 313-340, (1972).

[15] Friehe, C.: *Vortex shedding from cylinders at low Reynolds numbers*, Journal of Fluid Mechanics vol. 100, pp. 237-241, (1980).

6 Figures

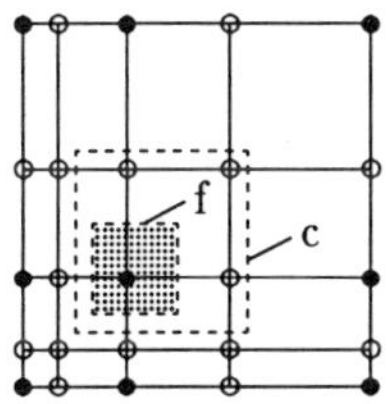

Fig. 1a Node-centered scheme

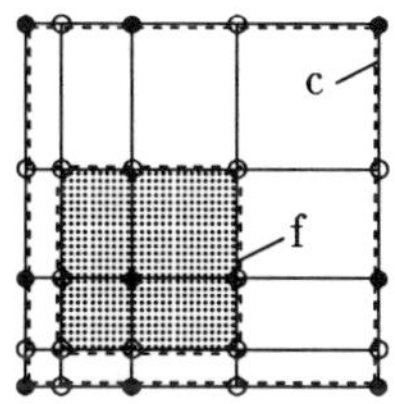

Fig. 1b Cell vertex scheme

Fig. 1 Representation of fine (f) and coarse grid (c) control volumes for the node-centered and cell vertex scheme

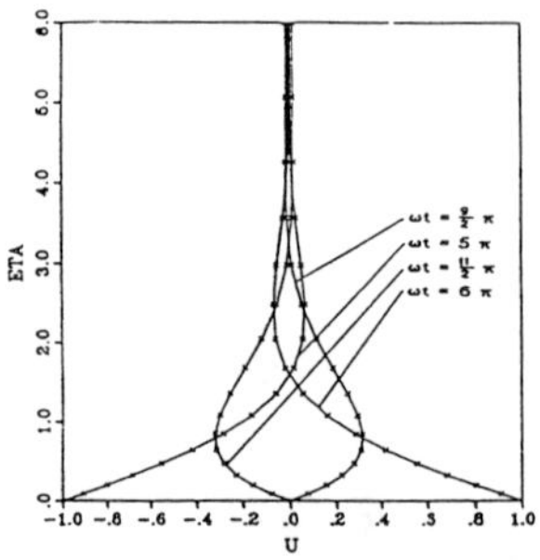

Fig. 2a Velocity profiles for different time levels, 17x17 grid points

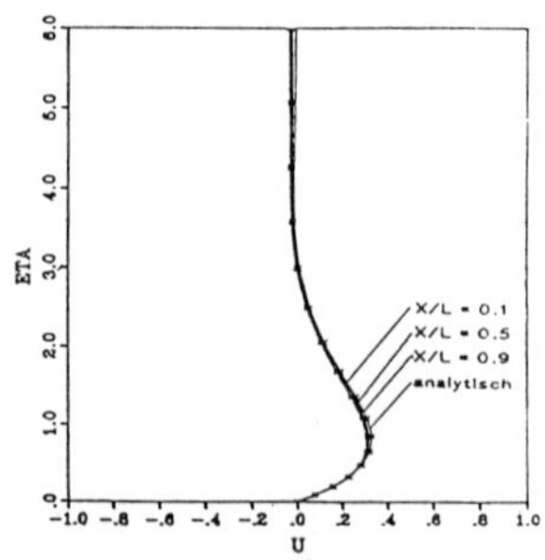

Fig. 3a Velocity profiles for different positions x, 65x17 grid points, Re=100

× single grid solution, — analytical solution

Unsteady flow about an oscillating flat plate of infinite (Fig. 2) and finite (Fig. 3) length, Navier-Stokes solution with the modified Flux-Vector Splitting, horizontal velocity component u as a function of the dimensionless wall distance ETA

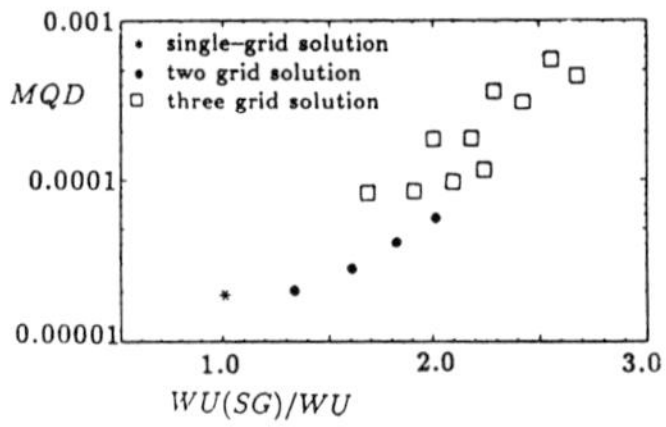

Fig. 2b Deviation from the analytical solution, 17x17 grid points

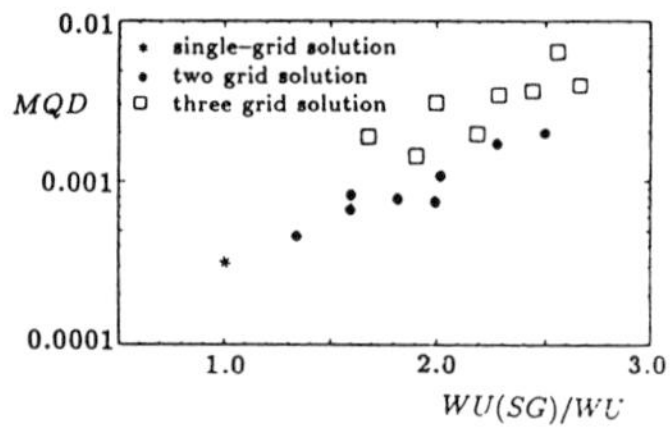

Fig. 3b Deviation from the analytical solution, 65x9 grid points, Re=200

$WU(SG)$ = work units of the single grid solution, WU = work units of the multigrid solution

MQD=Mean quadratic sum of the deviation

Unsteady flow about an oscillating flat plate of infinite (Fig. 2) and finite (Fig. 3) length, deviation from the analytical solution for different V-cycles

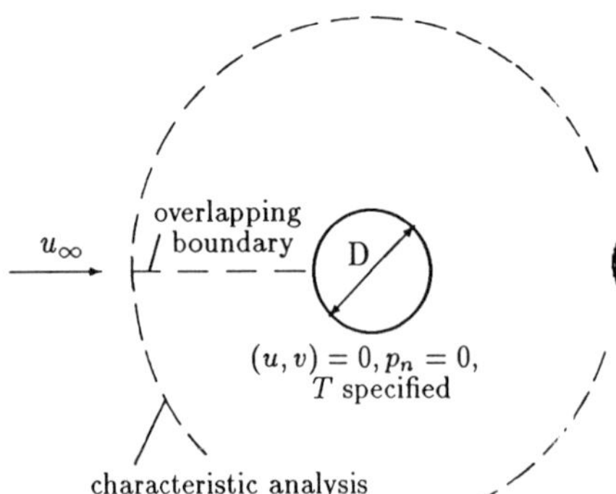

Fig. 4a Domain of integration

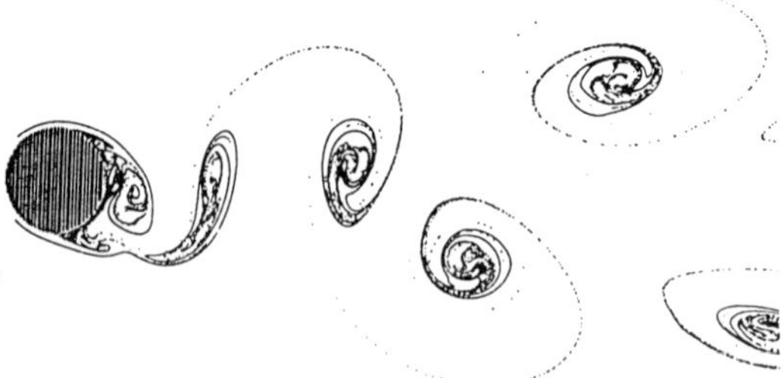

Fig. 4b Computed streaklines (cell-vertex scheme)

Fig. 4 Unsteady flow around a circular cylinder

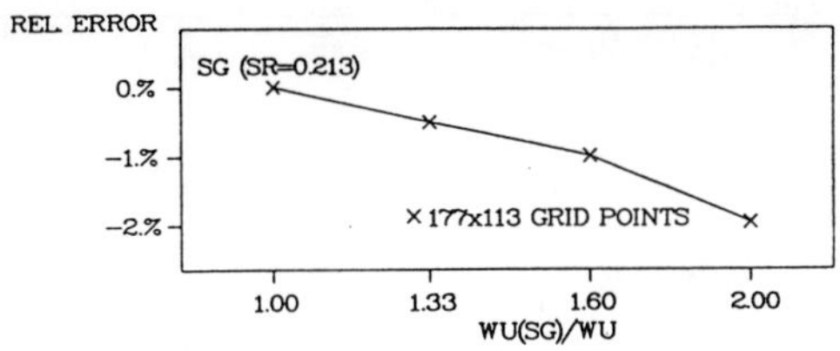

Fig. 4c Solution with 177x113 grid points

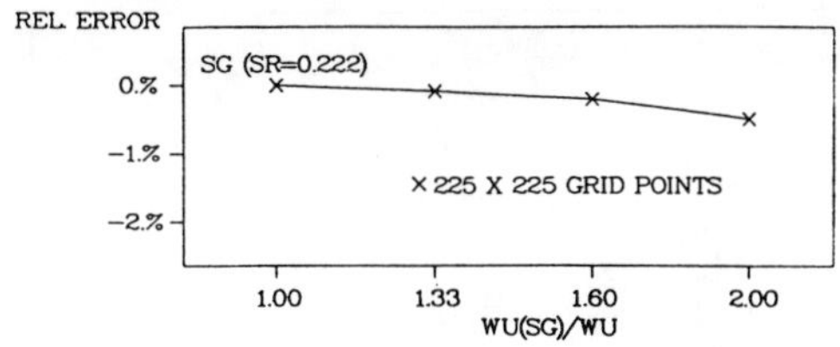

Fig. 4d Solution with 225x225 grid points

Fig. 4 Unsteady flow around a circular cylinder, Deviation of the Strouhal number, $Sr = \frac{fD}{u_\infty}$, of the multigrid from the single grid solution in percent, Navier-Stokes solution with the node-centered scheme and central differences, Ma=0.3, Re=500

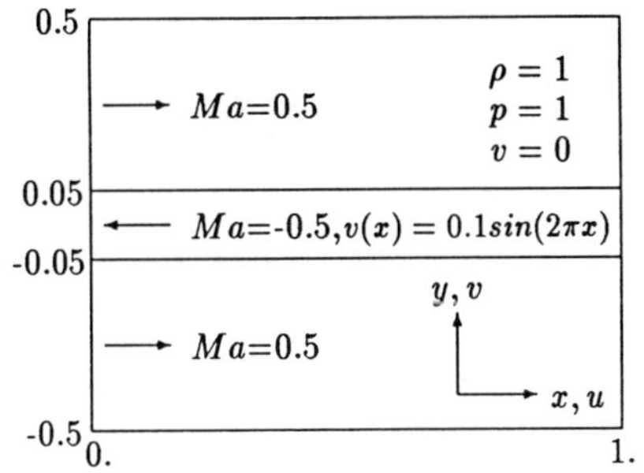

Fig. 5a Domain of integration.
Boundary conditions:
periodic boundary conditions for $x = 0.$ and $x = 1.$,
characteristic analysis normal to the boundary for $y = -0.5$ and $y = 0.5$

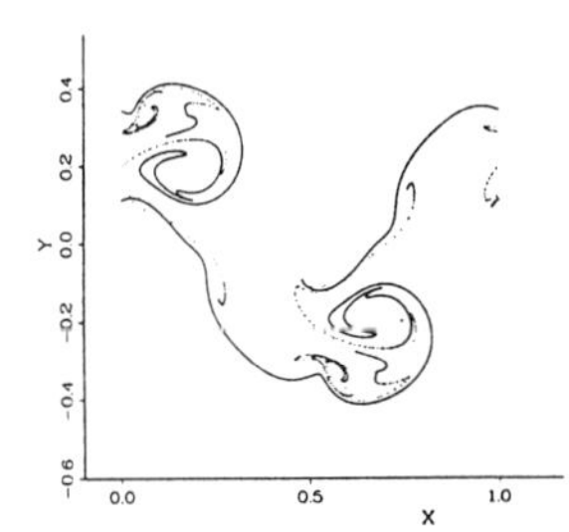

Fig. 5b single grid solution

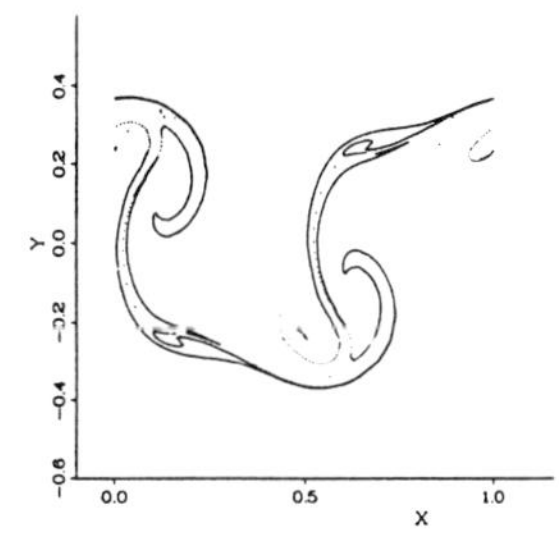

Fig. 5c multigrid solution, simple injection

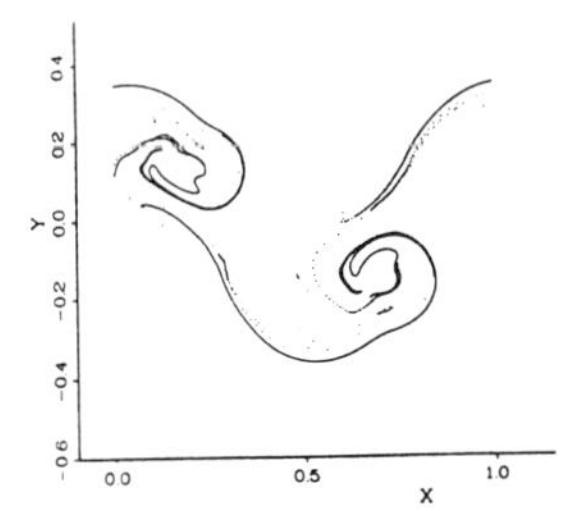

Fig. 5d multigrid solution, volume weighted restriction

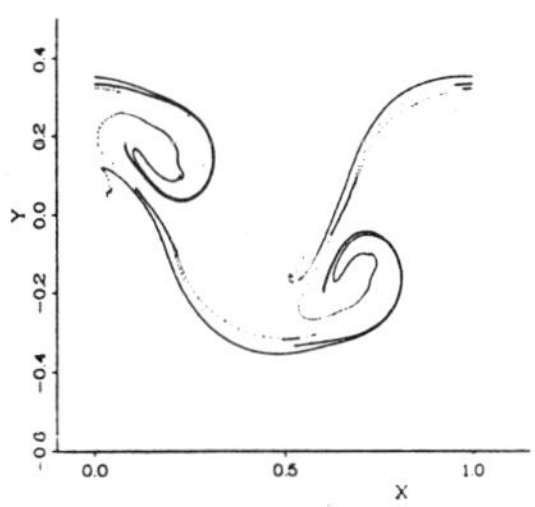

Fig. 5e multigrid solution, weighted average

Fig. 5 Flow instability of a jet, Navier-Stokes solution with the node-centered scheme and central differences, 113x113 grid points, Re=10000, streaklines for t=2 and different restriction operators

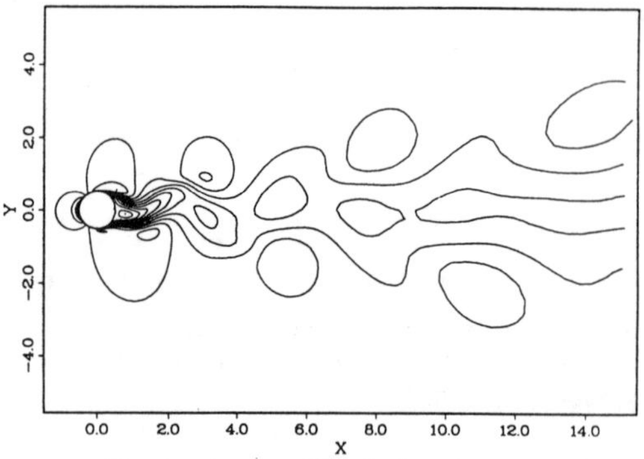

Fig. 6a single grid solution

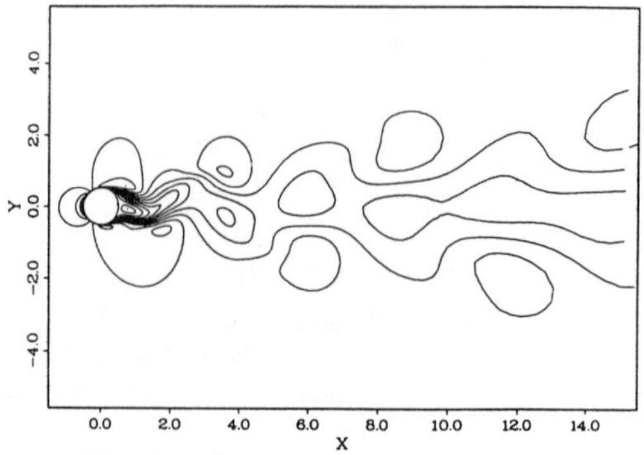

Fig. 6b multigrid solution

Fig. 6 Unsteady flow around a circular cylinder, Navier-Stokes solution with the node-centered scheme and central differences, Re = 100, Ma=0.3, 225x225 grid points, lines of constant x-momentum

	Sr	$c_{a\,max} - c_{a\,min}$	$c_{w\,max}$	$c_{w\,min}$
exp. data [15]	0.159-0.166	–	–	–
num. data [13]	–	0.60	1.38	1.35
single grid	0.1629	0.676	1.390	1.367
$WU(SG)/WU$ = 1.33	0.1622	0.677	1.390	1.370
$WU(SG)/WU$ = 1.50	0.1610	0.677	1.389	1.370
$WU(SG)/WU$ = 2.50	0.1608	0.674	1.385	1.370

Table 1

	Sr	$c_{a\,max} - c_{a\,min}$	$c_{w\,max}$	$c_{w\,min}$
exp. data [14]	0.182-0.192	–	–	–
num. data [13]	–	1.55	1.42	1.34
single grid	0.1933	1.324	1.419	1.334
$WU(SG)/WU$ = 1.33	0.1924	1.323	1.413	1.330
$WU(SG)/WU$ = 1.60	0.1919	1.332	1.410	1.327
$WU(SG)/WU$ = 2.00	0.1914	1.330	1.408	1.328
$WU(SG)/WU$ = 2.29	0.1908	1.340	1.412	1.331
$WU(SG)/WU$ = 2.50	0.1917	1.332	1.410	1.325

Table 2

Table 1 and Table 2: Unsteady flow around a circular cylinder, Navier-Stokes solution with the node-centered scheme and central differences, Re = 100 for Table 1, Re = 200 for Table 2, Ma=0.3, 225x225 grid points, comparison of Strouhal number, lift (c_a) and drag (c_w) coefficients

MESH–ADAPTIVE SOLUTION OF THE NAVIER–STOKES EQUATIONS

J. A. Michelsen
Dept. Fluid Mech., Techn. Univ. of Denmark, build. 404, DK–2800, Lyngby, Denmark.

ABSTRACT

The laminar, incompressible Navier–Stokes equations in 2D and axisymmetry are solved by finite volume technique on fixed or dynamically adaptive BFC meshes. The solution algorithm is of velocity–pressure coupled type. The scheme employs either the SIMPLE method, in which one Poisson equation for pressure correction is solved per time step, or the PISO method in which two Poisson equations are solved in each time step.

Steady–state solutions are accellerated by use of full multigrid (FMG) and a full approximation storage (FAS) multigrid scheme. A linear multigrid scheme is employed for the Poisson system(s).

It is suggested to employ the PISO method instead of the normally preferred SIMPLE method as a basic smoother. Convergence data show that the extra workload of solving a second Poisson system is more than balanced by a higher convergence rate for steady state problems.

INTRODUCTION

Multigrid schemes for the incompressible Navier–Stokes equations have been reported by a number of authors. Ghia et al. [1] presented an FAS scheme for the vorticity streamfunction formulation, while Barcus [2] and Barcus et al. [3] preferred a correction grid (CG) scheme for the primitive variables, using the SIMPLE algorithm of Patankar and Spalding [4]. Vanka [5] and Dick [6] present solution methods employing point collective relaxation of all solution components.

All of the above mentioned schemes can be viewed as relaxation procedures, whether of timestepping or iterative nature. As is the case for relaxation methods for linear systems, each of the schemes have certain convergence and smoothing properties. The present work is focused on velocity–pressure coupled methods, such as the above mentioned SIMPLE and the *P*ressure–*I*mplicit–by–*S*plit–*O*perator (PISO) method devised by Issa [7] to treat, in particular, unsteady problems. The smoothing action of the two schemes is discussed and convergence data are compared.

In order to facilitate the treatment of non–linear source terms, viz. the Coriolis and centrifugal forces, an FAS scheme is preferred instead of the less complicated CG scheme.

CONSERVATION LAWS

Two dimensional plane flows and axisymmetric swirling flows are considered in this study. Assuming laminar and incompressible flow, the conservation laws for mass and momentum are written in a fixed, general non–orthogonal frame (ξ,η,θ)

$$Q_t + (F+F_p)_\xi + (G+G_p)_\eta + H_\theta = (V_1+V_2)_\xi + (W_1+W_2)_\eta + X_\theta, \qquad (1)$$

where $(\cdot)_\xi$ denotes partial differentiation with respect to ξ. Here, quantities Q, F, G, V, W,

H, and X are defined below.

For plane flow, θ refers to the third (redundant) dimension in the 3D Cartesian frame, while for axisymmetric flow, θ is the circumferental direction. As seen below, H_θ and X_θ vanish for plane flow. The coordinate mapping

$$x = x(\xi,\eta),\ y = y(\xi,\eta) \tag{2}$$

in the plane case, or in the meridian plane in the axisymmetric case, defines the transformation Jacobian

$$J^* = x_\xi y_\eta - x_\eta y_\xi, \tag{3}$$

while the corresponding coordinate mapping in (ξ,η,θ) defines the Jacobian

$$J = x^\alpha \cdot J^*\ , \tag{4}$$

with α zero for plane flow and unity for axisymmetric flow. For axisymmetric flow, x is the radial direction, while y is the axial direction.

The conserved quantities are

$$Q = J\cdot\begin{Bmatrix} \rho \\ \rho u \\ \rho v \\ \rho w \end{Bmatrix}, \tag{5}$$

where (u,v) are the velocity components in the (x,y) directions, respectively. For axisymmetry, this corresponds to the radial and axial velocity components, respectively. The tangential momentum component ρw is only conserved for axisymmetric flow. ρ represents the fluid density.

F and G are the convective fluxes, while F_p and G_p represent the pressure terms

$$F = x^\alpha[y_\eta(u-\dot{x})-x_\eta(v-\dot{y})]\cdot\begin{Bmatrix} \rho \\ \rho u \\ \rho v \\ \rho w \end{Bmatrix},\quad G = x^\alpha[-y_\xi(u-\dot{x})+x_\xi(v-\dot{y})]\cdot\begin{Bmatrix} \rho \\ \rho u \\ \rho v \\ \rho w \end{Bmatrix}, \tag{6}$$

$$F_p = x^\alpha P\cdot\begin{Bmatrix} 0 \\ y_\eta \\ -x_\eta \\ 0 \end{Bmatrix},\quad G_p = x^\alpha P\cdot\begin{Bmatrix} 0 \\ -y_\xi \\ x_\xi \\ 0 \end{Bmatrix}. \tag{7}$$

Here, $(\dot{x},\dot{y})$ represents the mesh–speed in the case of dynamically–adaptive mesh generation. Using the present surface integral form of the pressure terms, an extra pressure contribution arises with the θ–curvature. The H_θ term contains this extra pressure term as well as the Coriolis and centrifugal force contributions

$$H_\theta = \alpha J^*\cdot\begin{Bmatrix} 0 \\ \rho w^2 - P \\ 0 \\ \rho u w \end{Bmatrix}. \tag{8}$$

The diffusive fluxes are split into the ∇^2 part and the part proportional to the frame, or mesh, skewness

$$V_1 = \frac{\mu x^\alpha}{J^*}\cdot(x_\eta^2+y_\eta^2)\cdot\begin{Bmatrix}0\\u_\xi\\v_\xi\\w_\xi\end{Bmatrix},\ W_1 = \frac{\mu x^\alpha}{J^*}\cdot(x_\xi^2+y_\xi^2)\cdot\begin{Bmatrix}0\\u_\eta\\v_\eta\\w_\eta\end{Bmatrix}, \tag{9}$$

$$V_2 = -\frac{\mu x^\alpha}{J^*}\cdot(x_\xi x_\eta+y_\xi y_\eta)\cdot\begin{Bmatrix}0\\u_\eta\\v_\eta\\w_\eta\end{Bmatrix},\ W_2 = -\frac{\mu x^\alpha}{J^*}\cdot(x_\xi x_\eta+y_\xi y_\eta)\cdot\begin{Bmatrix}0\\u_\xi\\v_\xi\\w_\xi\end{Bmatrix}, \tag{10}$$

where μ is the dynamic viscosity. For axisymmetric flow, the θ–curvature results in an extra diffusive term

$$X_\theta = \alpha\mu J^*\cdot\begin{Bmatrix}0\\-2\frac{w}{x}\\0\\y_\eta w_\xi - y_\xi w_\eta - \frac{w}{x}\end{Bmatrix}. \tag{11}$$

ADAPTIVE MESH GENERATION

The Brackbill–Saltzman [8] system, which is preferred for the present use, comprises a set of quasi–linear Euler equations for the minimization of a weighted sum of measures for global mesh smoothness, orthogonality, and density distribution

$$b_1 x_{\xi\xi}+b_2 x_{\xi\eta}+b_3 x_{\eta\eta}+a_1 y_{\xi\xi}+a_2 y_{\xi\eta}+a_3 y_{\eta\eta} = \lambda_c wJ^2\cdot\frac{\partial w}{\partial x}, \tag{14}$$

$$a_1 x_{\xi\xi}+a_2 x_{\xi\eta}+a_3 x_{\eta\eta}+c_1 y_{\xi\xi}+c_2 y_{\xi\eta}+c_3 y_{\eta\eta} = \lambda_c wJ^2\cdot\frac{\partial w}{\partial y}, \tag{15}$$

with coefficients given by

$$a_i = a_{si} + \lambda_c w^2 a_{ci} + \lambda_o a_{oi},\ \text{etc.}, \tag{16}$$

where s,c, and o refer to *smoothness*, *concentration*, and *orthogonality*. λ_c and λ_o are user–specified parameters, which shift the emphasis from smoothness to concentration, respectively orthogonality.

The majority of the boundary points in the mesh are assigned a Cauchy–like boundary condition, which specifies the points to follow the internal points, while sliding along the boundary contour, always preserving mesh orthogonality on the boundary Γ

$$\forall(\xi,\eta)\in\Gamma : f[\underline{x}(\xi,\eta)]=0 \wedge x_\xi x_\eta+y_\xi y_\eta=0\ . \tag{17}$$

In order to render the mesh problem well–posed, a limited number of boundary mesh–points are assigned Dirichlet conditions. In order to ensure the presence of mesh points on possible sharp corners, such mesh–points are always assigned Dirichlet conditions.

Since the mesh speed $(\dot{x},\dot{y})$ in (6) is found as the difference between coordinates at time levels $n+1$ and n, the mesh and flow problems are uncoupled during a time step.

Some examples of dynamically–adaptive meshes, controlled by a function of derivatives of a scalar flow variable, have been reported, notably by Saltzman & Brackbill [9], who based the mesh control on the pressure gradient. For incompressible viscous flow, solution adaptation would be based on the velocity field rather than on the pressure field. Hence, an invariant of the velocity vector field is sought. Being the simplest invariant of the strain tensor, the dissipation–like term

$$w = U_{i,j} \cdot (U_{i,j} + U_{j,i}) \tag{18}$$

is selected as a control function.

In zones of high velocity–gradients emphasis is almost totally on mesh concentration. Consequently, mesh smoothness is ultimately controlled by the distribution of the control function w in those zones. Thus, for practical purposes, the control function must be smoothed. This is done here by propagating/damping the control function away from the high resolution zones, keeping in mind the mesh expansion factor.

For dynamically adaptive mesh generation, (14–15) is to be solved once for each time step. In practice, a single relaxation sweep on the mesh generator system for each time step is adequate.

VELOCITY–PRESSURE COUPLING

The momentum equations (1) are discretized, using central difference representation of diffusive terms and upwind differences for the convective terms. Following the concept of Rhie [10], all solution variables are stored at the cell center. The linearized momentum equations are written

$$\frac{\rho J^n}{\Delta t} \cdot (\underline{u}_P^{n+1} - \underline{u}_P^n) + a_P \underline{u}_P^{n+1} = \sum_{nb=E,N,W,S} a_{nb} \underline{u}_{nb}^{n+1} + \underline{S} + \underline{S}_P, \tag{19}$$

$$a_P = \sum_{nb=E,N,W,S} a_{nb} \quad . \tag{20}$$

Here, $\underline{u}$ represents the velocity vector $\{u,v,w\}^T$. The coefficient subscipts refer to the compass notation. Convection and the ∇–square part of the diffusion contribute to the coefficients a, while the diffusive terms proportional to cell skewness goes into the source term $\underline{S}$. The pressure term is represented by $\underline{S}_P$. The present evaluation of the diagonal coefficient a_P ensures consistency, also with moving mesh–points. For the present purpose, the time step is local. Based on the local cell size and velocity, a timestep is taken such that the CFL number equals a fixed value on the order 2 – 4.

In order to construct an equation for pressure, consider the mass conservation, for incompressible flow

$$C_{\xi e} - C_{\xi w} + C_{\eta n} - C_{\eta s} = 0, \; C_\xi = \rho x^\alpha [y_\eta u - x_\eta v], \; C_\eta = \rho x^\alpha [x_\xi v - y_\xi u], \tag{21}$$

The appropriate momentum equations are now substituted for the cell face velocities in (21), for example

$$u_e^{n+1} = \frac{\frac{\rho J}{\Delta t} u_e^n + \sum a_{nb} u_{nb}^{n+1} + S_u + [y_\eta (P_P - P_E) - y_\xi (P_{se} - P_{ne})] x_e^\alpha}{a_P + \frac{\rho J}{\Delta t}}, \tag{22}$$

where a_P is here understood to be the diagonal coefficient for a fictitious control volume centered at the east cell face. The substituted momentum equations are interpolated between cell centers in order to obtain expressions involving cell center values. Following the practice by Rhie, the pressure term is treated by interpolating only the coefficient part

$$u_e^{n+1} = \overline{\left(\frac{\frac{\rho J}{\Delta t} \cdot u_P^n + \sum a_{nb} u_{nb}^{n+1} + S}{a_P + \frac{\rho J}{\Delta t}} \right)}_e + \overline{\left(\frac{1}{a_P + \frac{\rho J}{\Delta t}} \right)}_e \cdot [y_\eta \cdot (P_P - P_E) - y_\xi \cdot (P_{se} - P_{ne})] \cdot x_e^\alpha, \tag{23}$$

where overbar denotes linear interpolation between two cell centers. This practice precludes the well–known pressure uncoupling normally encountered with non–staggered variable allocation.

Substituting the right–hand–side of (23), and equivalent expressions for the other face velocities, into the mass conservation (21) yields a Poisson equation for pressure

$$A_P P_P = \sum A_{nb} P_{nb} + S_m, \; A_P = \sum A_{nb}. \tag{24}$$

PREDICTOR–CORRECTOR SCHEME

The above momentum and pressure equations (19–24) are employed in a predictor–corrector scheme. In a time step, a velocity field $\underline{u}^*$ is obtained as solution to the momentum equations (19), tentatively using the pressure field P^* from the former time level. In the subsequent corrector step, the pressure equation is applied. Insertion of $\underline{u}^*$ into the substituted momentum equations that go into the mass conservation creates an equation to correct the pressure and velocity

$$u_e^{**} = \overline{\left(\frac{\frac{\rho J}{\Delta t} \cdot u_P^n + \sum a_{nb} u_{nb}^* + S}{a_P + \frac{\rho J}{\Delta t}} \right)}_e + \overline{\left(\frac{1}{a_P + \frac{\rho J}{\Delta t}} \right)}_e \cdot [y_\eta \cdot (P_P^{**} - P_E^{**}) - y_\xi \cdot (P_{se}^{**} - P_{ne}^{**})] \cdot x_e^\alpha, \tag{25}$$

where P^{**} is the solution to the Poisson equation. The cell face velocities now satisfy the mass conservation (21). The cell center velocity field is corrected by replacing the tentative pressure P^* in (19) with P^{**}.

CHOICE OF BASIC SMOOTHER

In solving the pressure equation, it was not taken into account that in the term

$$\sum a_{nb} u_{nb}^* \tag{26}$$

in (25), the velocity field $\underline{u}^*$ should have been replaced by the corrected velocity field $\underline{u}^{**}$. If the mass defect S_m in (24) were a smooth field, the velocity correction $u' = u^{**}-u^*$ at the neighbour points would be of the same order as those at the point P. Hence, the pressure correction is overpredicted. In the SIMPLE algorithm, the pressure correction is thus underrelaxed

$$P^{n+1} = P^* + \alpha_P \cdot (P^{**}-P^*) \,, \tag{27}$$

where the underrelaxation factor α_P is very small, often on the order 0.2.

If, on the other hand, the mass defect contained largely rough components, the (rough part of the) pressure correction would not be so severely overpredicted as the low value of the underrelaxation parameter expresses. Hence, even if the convergence properties of the SIMPLE algorithm are fairly good, it is believed that its smoothing properties are degraded by the inherent pressure underrelaxation.

In the PISO method, the pressure correction is not underrelaxed. Instead, a second corrector step is performed. The recently corrected velocity field is inserted in the right hand side of (25), whereupon the new Poisson system is solved. Besides paying more attention to the rough components by not underrelaxing pressure corrections, the mere insertion of $\underline{u}^{**}$ into the mass conservation (25) means that local information about the momentum balance is utilized twice instead of once.

The coefficient matrix is the same for the two Poisson systems and the decomposition from the first correction may be reused. Only modest residual reduction – say two or three orders of magnitude – is needed, and as a multigrid technique is employed, the saving of the second construction and decomposition of the coefficient matrix is not insignificant. All in all, the second corrector step increases the workload per timestep by only 30–40 percent compared to the SIMPLE workload. Thus, between the two, the PISO algorithm is believed to be the better choice for a basic multigrid smoother.

In the case of dynamically adaptive mesh generation, a timestep is started by computing the mesh control function w, based on the most recent approximation. One LGSF1F2 relaxation is performed on the mesh generator equations (14–15). The old and new mesh coordinates and the (local) time step define the mesh–speed in (6), and the time step is now completed by one PISO predictor–corrector step.

MULTIGRID TIMESTEPPING STRATEGY

The above PISO algorithm is employed in a FMG/FAS scheme. A finer mesh is constructed by standard refinement, i.e. the coarse control volumes are subdivided into two by two finer cells. As a consequence of the cell centered variable allocation, the coarse and finees mesh do not have any common points of variable allocation.

The two level algorithm is now outlined. After ν_h presmoothings on the fine mesh, approximate solutions $\underline{u}_h$ and P_h are obtained, where h refers to the fine mesh, while superscript H is employed for the coarse mesh. The discrete momentum equations display residuals $\underline{R}$, hence

$$a_P^h \underline{u}_P^h = \sum a_{nb}^h \underline{u}_{nb}^h + \underline{S}^h + \underline{S}_P^h - \underline{R}^h . \tag{28}$$

The approximate solution $\underline{u}^h, P^h$ is restricted by volume weighted sum, whereas the corresponding residual $\underline{R}^h$ is restricted by a simple sum over the four fine cells

$$\underline{u}_0^H = I_h^H \underline{u}^h = \Sigma J_i^h \underline{u}_i^h / J^H, \; J^H = \Sigma J_i^h, \; i = 1,2,3,4 \tag{29}$$

$$\Delta \underline{R} = I_h^H \underline{R}^h - \underline{R}_0^H = \Sigma \underline{R}_i^h - \underline{R}_0^H, \; i = 1,2,3,4 \tag{30}$$

where R_0^H is the residual based on the restricted solution. J^h and J^H represent volumes of fine and coarse cells, respectively.

The coarse–mesh momentum equations have the same structure as the fine–mesh equations, except that an extra term is included in residual calculations. The momentum equations employed on the coarse mesh are now written

$$a_P^H \underline{u}_P^H = \sum a_{nb}^H \underline{u}_{nb}^H + \underline{S}^H + \underline{S}_P^H + \Delta \underline{R}\,, \tag{31}$$

After the restriction, ν_H time–steps are taken on the coarse mesh. The approximate solution is subsequently prolonged back to the fine mesh

$$\underline{u}_0^h = \underline{u}^h + P_H^h \cdot (\underline{u}_i^H - \underline{u}_0^H), \tag{32}$$

where $\underline{u}^h$ is the fine–mesh solution before the restriction, $\underline{u}_i^H - \underline{u}_0^H$ is the correction obtained on the coarse mesh. Bilinear interpolation in physical space is employed for the prolongation.

In the FAS part, the levels may be visited in a variety of patterns, see fig. 6.3, of which the *V*–cycle and the *W*–cycle are the most common ones. For Navier–Stokes problems, owing to poor smoothing factors, fairly high smoothing counts ν are employed. As a consequence of the high non–linearity, short cycles such as the *V*–cycle is normally employed for Navier–Stokes problems. Summing up, a *V*–cycle will be used here, employing 2–4 presmoothings and 1–2 postsmoothings.

The starting guess on the finest level is obtained by a full multigrid scheme.

MULTIGRID POISSON SOLVER

The above FMG/FAS procedure was designed for convergence to steady–state of the momentum–transport problem. For each timestep – or iteration – the solution of a Poisson equation for the pressure correction is required twice. The application of a linear multigrid technique for this problem seems appropriate.

After presmoothings, the pressure equation displays residual mass sources R^h

$$A_P^h P_P^h - \sum A_{nb}^h P_{nb}^h = S_m^h - R^h. \tag{33}$$

The restriction is performed as

$$R^H = \Sigma \underline{R}_i^h\,, \; i = 1,2,3,4, \; P_0^H = 0, \; S_m^H = 0, \tag{34}$$

resulting in the coarse–mesh correction equation

$$A_P^H P_P^H - \sum A_{nb}^H P_{nb}^H = R^H. \tag{35}$$

After converging the coarse–mesh correction equation, the pressure correction is prolongated, using bilinear interpolation

$$P_0^h = P^h + P_H^h P_i^H. \tag{36}$$

Calculation of the coarse–mesh pressure coefficients is explained by a resistance analogy, where the change δC in mass flux in fig. 1 acts as current, the change of pressure drop acts as voltage, while the pressure coefficient assumes the role of the resistance

$$\delta C_{e1}^h = \frac{\delta(P_{P1}^h - P_{E1}^h)}{\Delta\xi^h \cdot A_{e1}^h}, \tag{37}$$

where $\Delta\xi^h$ is the fine–mesh width in transformed space. The coarse–mesh massflux change over the east face is now calculated as the 'current' through two parallel 'resistors'

$$\delta C_e^H = \left[(A_{e1}^h)^{-1} + (A_{e2}^h)^{-1}\right] \cdot \frac{\delta(P_P^H - P_E^H)}{\Delta\xi^H}, \tag{38}$$

where $\Delta\xi^H$ is the coarse–mesh width in transformed space. For standard coarsening, the ratio $\Delta\xi^H/\Delta\xi^h$ equals two. Thence, the coarse–mesh coefficients are calculated as

$$(A_e^H)^{-1} = \frac{1}{2} \cdot \left[(A_{e1}^h)^{-1} + (A_{e2}^h)^{-1}\right], \; etc. \tag{39}$$

For the Poisson system, a FMG/FAS cycle is employed, using a LGSF1F2 smoother. A *W* sawtooth cycle is employed, where one relaxation is preformed on the finest level, two relaxations on the next coarser level, etc. In the near future, the LGSF1F2 will be replaced by a multiblock ILU smoother.

EXAMPLES

Square driven cavity

Barcus et al. [3] present convergence results for the square driven cavity, employing the SIMPLE algorithm as a basic smoother in a correction grid (CG) scheme. The authors claim, that the CG scheme does not reduce performance relative to a FAS scheme.

The present results are obtained employing the PISO algorithm in an FAS scheme. Convergence data for the two schemes are compared in table 1 for *Re* numbers 100, 1.000, and 5.000. Both calculations uses 4 levels in a *V*–cycle. Remembering that the calculation of a PISO time step takes 1.4 times longer than the calculation of a SIMPLE iteration, the PISO scheme still converges between 1.8 and 2.8 times faster than the SIMPLE scheme.

Table 7.1

Driven cavity 64×64 CV 4 MG levels	$Re = 100$	$Re = 1.000$	$Re = 5.000$
CG/SIMPLE (Barcus et al.)	15 cycles 54 WU	17 cycles 61 WU	27 cycles 97 WU
FAS/PISO (present)	4 cycles 14 WU	7 cycles 25 WU	12 cycles 39 WU

Fig. 2 shows comparisons of convergence histories for $Re = 100$ and 1.000, using 4–level and single–level PISO time stepping. In both cases, 4 levels were used for the solution of the Poisson equations. Hence, the comparison reveals the isolated effectivity of the FAS time stepping scheme, which is an order of magnitude faster than the single–level time stepping scheme. Furthermore, it is noted, that the multigrid efficiency decreases with increasing Re number.

Axisymmetric lid–driven cavity

The axisymmetric lid–driven cavity, see fig. 3, was computed by Michelsen [11]. The Re number based on lid radius and edge velocity was 2.494, while the height to radius ratio was 2.5.

The left part of fig. 3 is a visualization of Escudier [12], while the right side shows computed streamlines for the secondary flow. The flow exhibits a double vortex–breakdown (*VB*). The computed location, shape, and size of the *VB* compares excellently to the visualization. The computed *VB* looks a bit narrow compared to the visualization. However, as Escudier notes, the distortion due to difference in refraction index is about 8 percent. Having this in mind, the agreement between visualization and computation is almost exact.

The computation was performed on 96×192 control volumes, employing 5 multigrid levels for both FAS scheme and Poisson solver. The convergence history is shown in fig. 4. Between three and four orders of magnitude residual reduction was accomplished within 7 FAS cycles, or 30 work units.

Laminar separated airfoil

For the NACA 64_2A015 airfoil, visualizations for $Re = 7.000$ and 5^0 angle of attack were presented by Werlé [13]. Fig. 5 shows that the leeside flow is massively separated. A large counter–rotating zone between the primary separation and the pressure side flow is seen in the visualization, as well as in the computation.

The present calculation employed a dynamically–adaptive mesh of 40×208 control volumes, using 3 multigrid levels. The mesh extented two cords away from the airfoil. Note especially the adaptation to the shear layer above the separated zone. Mesh adaption took part during the FMG part and in the three first FAS cycles, after which the mesh was fixed. The argument for mesh fixation was that mesh movements partially impaired convergence of the flow solver. The mesh speed is accounted for in the convective terms. However, the tentative pressure employed in the momentum predictor step is calculated in one mesh position and subsequently employed in the shifted mesh position.

After the mesh–fixation, the convergence rate is still lower than the rates obtained for the driven cavity flows, see fig. 6. This is in line with the findings from the driven cavity flows, that multigrid efficiency drops with increasing Re number.

Closure

The multigrid Poisson solver was employed for all of the above flow cases. For early computations, comparisons were made between single grid and multigrid Poisson solutions. Application of the multigrid Poisson solver reduced CPU demands for the corrector steps by 10–50 times. For only partial convergence of the pressure equation, as often used, the multigrid Poisson solver has a positive influence on the FAS performance, since residual mass sources of low wave–numbers are removed more efficiently.

The LGSF1F2 smoother has recently been replaced by an alternating direction zebra solver, with more or less the same performance.

For non–orthogonal meshes, the Poisson equation results in a 9–point molecule. At least for steady–state computations, the corner coefficients may be moved to the right–hand–side. This removes the source of divergence noted by Hemker [14], and makes the application of an ILU smoother highly relevant.

CONCLUSIONS

The present study is concerned with the application of multigrid techniques in an existing Navier–Stokes code. The code employs the PISO time stepping scheme on general curved meshes with cell–centered variable storage. Multigrid techniques are emplyed in two different levels. An FMG–FAS method is emplyed on the time step level, while a linear multigrid scheme is employed for the Poisson equations for pressure correction.

Comparison to computations using the SIMPLE scheme shows that the increased workload of a second corretor step in the PISO is more than compensated by a higher convergence rate.

Application of the FAS scheme instead of the simpler correction grid scheme does not improve convergence for two dimensional problems, while the solution of axisymmetric problems is greatly facilitated.

For all cases reported herein, application of multigrid techniques increased computational efficiency. The FAS efficiency consistently drops with increasing *Re* number, while efficiency of the multigrid Poisson solver is retained for all *Re* numbers and flow cases.

ACKNOWLEDGEMENT

This study is supported by the Danish Techn. Research Council under grant FTU/STVF 5.17.7.6.62.

REFERENCES

[1] Ghia U., Ghia K.N., Shin C.T., (1982), *J. Comp. Phys.*, 48, p387.

[2] Barcus M., (1987) , M. Sc. Thesis, LSTM, Erlangen.

[3] Barcus M., Peric M., Scheuerer G., (1987), *Proc. 7'th GAMM Conf. Num. Meth. Fluid Mech.*

[4] Patankar S.V., Spalding D.B., (1972), *Int. J. Heat Mass Transfer*, 15, p1787.

[5] Vanka S.P., (1986), *Comp. Meth. Appl. Mech. Eng.*, 55, p321.

[6] Dick E., (1990), To appear in *Proc. 3'rd Int. Conf. on Multigrid methods.*

[7] Issa R.I., (1985), *J. Comp. Phys.*, 61.

[8] Brackbill J.U., Saltzman J.S., (1982), *J. Comp. Phys.*, 46, p342.

[9] Saltzman J.S., Brackbill J.U., (1982), *Appl. Math. Comp.*, 10.

[10] Rhie C.M., (1981) Ph.D. Thesis, Univ. of Illinois, Urbane–Champaign.

[11] Michelsen J.A. (1989), *Dept. Fluid Mech., rep. AFM 89–08*, Danish Tech. Univ.

[12] Escudier M.P., (1984), *Experiments in fluids*, 2, p289.

[13] Werlé H., (1974), Publ. no. 156, ONERA, France.

[14] Hemker P.W.,(1983), *Num. analysis Proc.*, Lecture Notes in Math. 1066, Griffiths DF (ed), Springer.

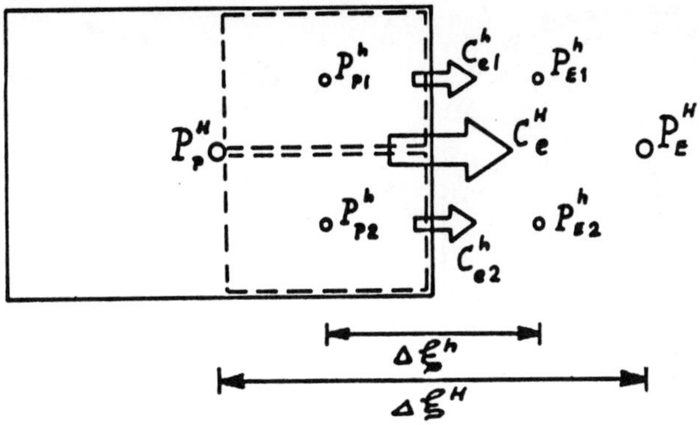

Fig. 1 Determination of pressure coefficients on coarse mesh.

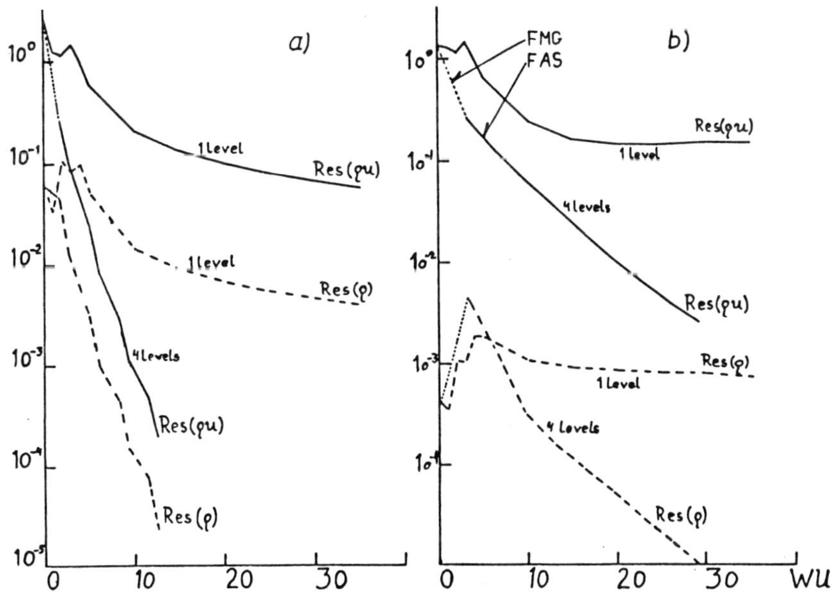

Fig. 2 Square driven cavity. Convergence of single—grid and 4—level FAS computations.

a) $Re = 100$, b) $Re = 1.000$.

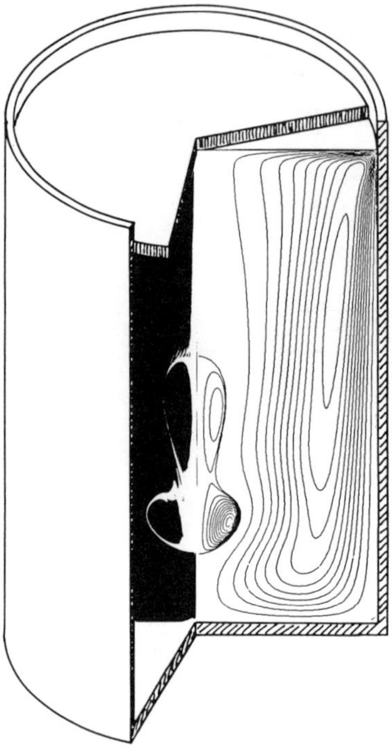

Fig. 3 Rotating lid system with double VB.
Left: visualization of Escudier (1984).
Right: computed streamlines. Re = 2.494, 96×192 CV.

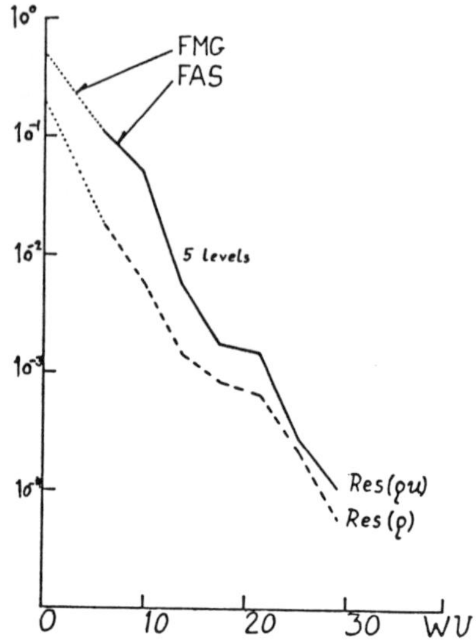

Fig. 4 Rotating lid system. $Re = 2.494$, $H/R = 2.5$.
Convergence of 5—level FAS.

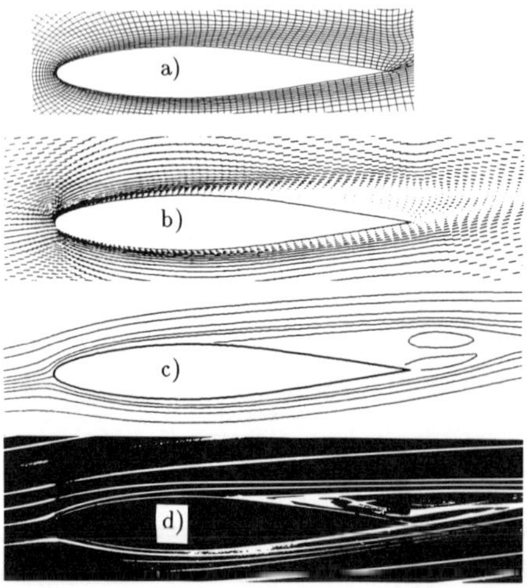

Fig. 5 NACA 64_2A015.
a) Mesh detail, b) velocity vectors,
c) streamlines, d) visualization by Werlé.

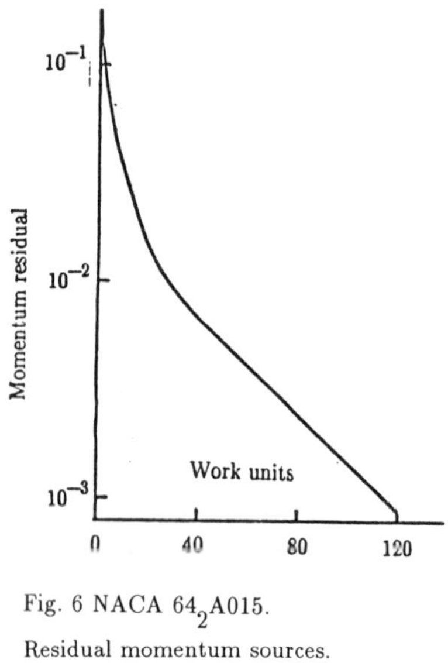

Fig. 6 NACA 64_2A015.
Residual momentum sources.
Convergence of 3—level FAS.

A two-grid analysis of the combination of mixed finite elements and Vanka-type relaxation

J. Molenaar
Centre for Mathematics and Computer Science
P.O. Box 4079, 1009 AB Amsterdam, The Netherlands

In this paper a two-grid algorithm is discussed for the mixed finite element discretization of Poisson's equation. The algorithm is based on a Vanka-type relaxation; the grid transfer operators are selected in accordance with the discretization. Local mode analysis is used to show that Vanka-type relaxation is an efficient smoother indeed. By studying the Fourier transform of the error amplification matrix we find that the canonical grid transfer operators are sufficiently accurate for grid independent convergence. However, this conclusion depends on the relaxation pattern used.

1. Introduction

The Mixed Finite Element (MFE) method is widely used for the discretization of second order elliptic systems. For the iterative solution of the discrete systems of equations multigrid methods are proposed. In [1] we presented a multigrid method for the MFE-discretization of the stationary semiconductor equations. This multigrid method employs a symmetric block Gauss-Seidel relaxation, as proposed by Vanka [2], which seems to be an efficient smoother. The prolongation and restriction in this multigrid algorithm are the canonical choice for the lowest order Raviart-Thomas elements that are used in the discretization. However, a priori it is not clear whether these canonical grid transfer operators are accurate enough to ensure grid independent convergence of the multigrid algorithm. Therefore we study two-grid algorithms for the model equation

$$\begin{aligned} div\ (\text{grad}\ u) &= 0, \quad \text{on } \Omega, \\ u &= 0, \quad \text{on } \partial\Omega. \end{aligned} \tag{1.1}$$

For ease of notation we only treat the case $\Omega \subset \mathbb{R}$. The Poisson equation (1.1) is discretized by the MFE-method (section 2). As the discretization is not stable if a sourceterm is present in (1.1) a quadrature rule is used which lumps the equations (cf. [3]). A two-grid algorithm for the discrete system is presented in section 3. We study the convergence behavior of this two-grid algorithm by Fourier analysis. The Fourier representations of the different operators in the two-grid algorithm are derived in section 4. By local mode analysis we show that Vanka-type relaxation is an efficient smoother for the discreet system indeed (section 5). In our analysis we include the use of a relaxation parameter as well as different relaxation patterns: lexicographical and red-black. In section 6 we study the error amplification matrix of the two-grid algorithm. Surprisingly we find that the required accuracy of the grid transfer operators is not only determined by the order of the differential equations (cf. [4,5]), but also depends on the relaxation pattern used. Our conclusions are summarized in the final section.

2. Discretization

To discretize the second order differential equation (1.1) by the mixed finite element method, we rewrite it as a system of two first order equations,

$$\sigma - \text{grad}\, u = 0, \tag{2.1a}$$

$$div\,\sigma = 0, \tag{2.1b}$$

$$u(0) = u(1) = 0. \tag{2.1c}$$

Let $L^2(\Omega)$ be the Hilbert space of square integrable functions on $\Omega \subset \mathbb{R}$ with inner product

$$(u,t) = \int_\Omega ut\, d\Omega,$$

and let $H(div, \Omega)$ be the Hilbert space defined by

$$H(div, \Omega) = \{\sigma | \sigma \in L^2(\Omega),\ div\sigma \in L^2(\Omega)\}$$

with norm

$$||\sigma||^2_{H(div,\Omega)} = ||\sigma||^2_{L^2(\Omega)} + ||div\sigma||^2_{L^2(\Omega)}.$$

By introduction of the product space $\Lambda(\Omega) = H(div,\Omega) \times L^2(\Omega)$ the weak formulation of (2.1) is: find $(\sigma,u) \in \Lambda(\Omega)$, such that

$$\int_\Omega \sigma \cdot \tau\, d\Omega + \int_\Omega u\, div\tau\, d\Omega = 0, \quad \forall \tau \in H(div, \Omega), \tag{2.2a}$$

$$\int_\Omega t\, div\sigma\, d\Omega = 0, \quad \forall t \in L^2(\Omega). \tag{2.2b}$$

To discretize (2.2) we decompose the domain Ω into a set Ω_h of N uniform cells Ω_h^i,

$$\Omega_h^i = \left[\frac{(i-1)}{h}, \frac{i}{h}\right], \qquad i = 1, \ldots, N, \tag{2.3}$$

of size $h = \dfrac{1}{N}$.

On this mesh lowest order Raviart-Thomas elements are defined, which span the subspace $\Lambda_h(\Omega) \subset \Lambda(\Omega)$. On each cell Ω_h^i the indicator function $e_h^i \in L^2(\Omega)$ is defined by

$$e_h^i(x) = \begin{cases} 1, & x \in \Omega_h, \\ 0, & x \notin \Omega_h. \end{cases} \tag{2.4}$$

For every edge at $x = jh$ of an interval Ω_h^i a piecewise linear function $\epsilon_h^j \in H(div, \Omega)$ is defined by

$$\epsilon_h^j(E_h^k) = \delta_{jk}, \quad k = 0, \ldots, N, \tag{2.5}$$

where δ_{jk} denotes the Kronecker delta. The discrete approximation (σ_h, u_h) of solution (σ, u) is

$$\begin{aligned} \sigma_h &= \sum_{j=0,N} \sigma_h^j \epsilon_h^j, \\ u_h &= \sum_{i=1,N} u_h^i e_h^i. \end{aligned} \tag{2.6}$$

To discretize the equation we proceed as usual and we replace $\Lambda(\Omega)$ in (2.2) by $\Lambda_h(\Omega)$ and use (ϵ_h^j, e_h^i) as the testfunctions. After division by h for proper scaling, the resulting algebraic system for $(\underline{\sigma_h}, \underline{u_h})^T$, i.e. the column vector of the coefficients $\{\sigma_h^j, u_h^i\}$ is

$$\begin{bmatrix} A_h & D_h \\ D_h^T & 0 \end{bmatrix} \begin{bmatrix} \underline{\sigma_h} \\ \underline{u_h} \end{bmatrix} = \begin{bmatrix} 0 \\ 0 \end{bmatrix}. \tag{2.7}$$

The matrix elements in this system are

$$(A_h)_{jk} = \frac{1}{h} \int_\Omega \epsilon_h^j \epsilon_h^k d\Omega = \begin{cases} \frac{1}{6}, & ,|j-k| = 1, \\ \frac{2}{3}, & ,j = k, \quad j \notin \{0, N\}, \\ \frac{1}{3}, & ,j = k, \quad j \in \{0, N\}, \\ 0, & ,\text{otherwise}, \end{cases} \tag{2.8a}$$

and

$$(D_h)_{ji} = \frac{1}{h}\int_{\Omega} e_h^i div \epsilon_h^j d\Omega = \begin{cases} -\frac{1}{h}, & ,j=i-1, \\ +\frac{1}{h}, & ,j=i, \\ 0, & ,\text{otherwise.} \end{cases} \tag{2.8b}$$

If a sourceterm is present in (1.1) this discretization is not stable in the sense that the matrix, obtained after elimination of σ_h, is necessarily a M-matrix. Therefore we change the discretization by approximating the integral in (2.8a) by a repeated trapezoidal rule,

$$(A_h)_{jk} \approx \frac{1}{2}\sum_{i=1,N} (\epsilon_h^j(ih)\epsilon_h^k(ih)+\epsilon_h^j((i-1)h)\epsilon_h^k((i-1)h)).$$

By this quadrature A_h is approximated by a diagonal matrix; effectively the matrix A_h is lumped. If σ_h is eliminated from the lumped system, we obtain a M-matrix indeed, even if a sourceterm is present.

In this paper we analyze two-grid algorithms for the solution of the linear system (2.7), both for the lumped and the non lumped case. In fact we treat the more general situation

$$(A_h)_{jk} = \begin{cases} \kappa, & |j-k|=1, \\ 1-2\kappa, & j=k, \quad j\notin\{0,N\}, \\ \frac{1}{2}-\kappa, & j=k, \quad j\in\{0,N\}, \\ 0, & \text{otherwise,} \end{cases} \tag{2.9}$$

which implies the lumped case ($\kappa=0$) as well as exact integration ($\kappa=\frac{1}{6}$).

3. Two-grid algorithm

In this section we discuss the different operators in a two-grid algorithm for the iterative solution of the system of equations (2.7). The discrete fine grid operator L_h: $\mathbb{R}^{2N+1}\rightarrow\mathbb{R}^{2N+1}$ is defined by the system (2.7),

$$L_h = \begin{bmatrix} A_h & D_h \\ D_h^T & 0 \end{bmatrix}. \tag{3.1}$$

The coarse grid is obtained by cellwise coarsening, i.e. by taking $H=2h$ in (2.3). This implies that the approximating subspaces are nested, $\Lambda_H(\Omega)\subset\Lambda_h(\Omega)$; hence the canonical grid transfer operators are available. The canonical prolongation P_h:$\mathbb{R}^{N+1}\rightarrow\mathbb{R}^{2N+1}$ is define on the space of coefficient vectors $(\sigma_H,u_H)^T$; the canonical restriction R_H:$\mathbb{R}^{2N+1}\rightarrow\mathbb{R}^{N}$ is the adjoint of P_h. The coarse grid operator is obtained by using the same discretization on the coarse grid Ω_H as on the fine grid Ω_h. If exact quadrature is used ($\kappa=\frac{1}{6}$), we find that L_H is the Galerkin approximation of L_h:

$$L_H = R_H L_h P_h. \tag{3.2}$$

As smoothing operator S_h:$\mathbb{R}^{2N+1}\rightarrow\mathbb{R}^{2N+1}$ we use symmetric block Gauss-Seidel relaxation. By this method all cells are scanned in some determined order, and in each cell Ω_h^i the three equations corresponding to the testfunctions $\epsilon_h^i,\epsilon_h^{i-1}$ and e_h^i, are solved for σ_h^{i-1}, σ_h^i and u_h^i. After each update a next cell is visited. In this paper we consider both lexicographical (SBGS) and red-black ordering (SBRB) ordering of the cells.

Finally we define the two-grid error amplification matrix

$$M_h^{\nu_2\nu_1} = S_h^{\nu_2}(I_h-P_h(L_H)^{-1}R_H L_h)S_h^{\nu_1}, \tag{3.3}$$

where $I_h:\mathbb{R}^{2N+1}\to\mathbb{R}^{2N+1}$ denotes the identity operator and ν_1,ν_2 the number of pre- and post relaxation sweeps, respectively.

4. Fourier analysis: the coarse grid correction

In order to derive Fourier representations of the different operators in the two-grid algorithm, we extend the domain to $\Omega=\mathbb{R}$ and omit the boundary conditions. The coefficient vectors $\underline{\sigma}_h$ and $\underline{u}_h$ are considered as gridfunctions defined on different discretization grids

$$\mathbb{Z}_{h,s} = \{(j-s)h | j\in\mathbb{Z}\}, \tag{4.1}$$

with

$$s = \begin{cases} 0, & \text{for } \underline{\sigma}_h, \\ \frac{1}{2}, & \text{for } \underline{u}_h. \end{cases}$$

The space of discrete L_2-functions on $\mathbb{Z}_{h,s}$, denoted by

$$L_{h,s}(\mathbb{Z}_{h,s}) = \{f_{h,s} | f_{h,s}\colon \mathbb{Z}_{h,s}\to\mathbb{C};\ h\sum_j |f_{h,s}((j-s)h)|^2<\infty\},$$

is a Hilbert space. The Fourier transform $FT(f_{h,s})=\hat{f}_{h,s}\colon T_h\to\mathbb{C}$ of a $L_{h,s}$-function is defined by

$$\hat{f}_{h,s}(\omega) = \frac{h}{\sqrt{2\pi}} \sum_{j\in\mathbb{Z}} e^{-i(j-s)h\omega} f_{h,s}(j-s)h), \tag{4.2}$$

with $T_h=(-\frac{\pi}{h},\frac{\pi}{h}]$. The inverse transformation is given by

$$f_{h,s}((j-s)h) = \frac{1}{\sqrt{2\pi}} \int_{\omega\in T_h} e^{i(j-s)h\omega}\hat{f}_{h,s}(\omega)d\omega. \tag{4.3}$$

By Parseval's equality the Fourier transformation operator $FT\colon L_{h,s}\to L_2(T_h)$ is an unitary operator.

Convolution or Toeplitz operators $B_h\colon L_{h,s}\to L_{h,s}$ are linear operators, generated by a gridfunction $b_{h,0}\in L_{h,0}$:

$$B_h f_{h,s}((j-s)h) = \sum_{k\in\mathbb{Z}} b_{h,0}(kh) f_{h,s}((j-k-s)h).$$

The Fourier transform $FT(B_h)=\hat{B}_h$ of a Toeplitz operator B_h is defined by

$$\hat{B}_h(\omega) = \hat{b}_{h,0}(\omega). \tag{4.4}$$

For example, the matrices A_h and D_h in (2.7) are Toeplitz operators with Fourier transforms

$$\hat{A}_h(\omega) = 1-4\kappa\sin^2(\frac{h\omega}{2}) \tag{4.5}$$

and

$$\hat{D}_h(\omega) = \frac{2i}{h}\sin(\frac{h\omega}{2}). \tag{4.6}$$

In order to obtain Fourier representations of the grid transfer operators we introduce the elementary prolongation $P^0_{h,s}\colon L_{H,s}\to L_{h,s}$,

$$(P^0_{h,s} f_{H,s})((j-s)h) = \begin{cases} f_{H,s}((\frac{j}{2}-s)h), & j \text{ even}, \\ 0, & j \text{ odd}, \end{cases} \tag{4.7}$$

and the elementary restriction $R^0_{H,s}\colon L_{h,s}\to L_{H,s}$,

$$(R^0_{H,s} f_{h,s})((j-s)H) = f_{h,s}((2j-s)h). \tag{4.8}$$

Using (4.2), (4.3) and (4.8) we find that the Fourier transforms of $f_{h,s}$ and $R^0_{H,s}f_{h,s}$ are related by

$$(R^0_{H,s}f_{h,s})(\omega) = e^{ish\omega} \sum_{p=0,1} e^{-ips\pi}\hat{f}_{h,s}(\omega+p\frac{\pi}{h}), \tag{4.9}$$

with $\omega \in T_H \subset T_h$. Notice that $R^0_{H,s}$ aliases two frequencies $\{\omega,\omega+\frac{\pi}{h}\} \in T_h$ with one frequency $\omega \in T_H$. The frequencies $\omega \in T_H$ are called low frequencies in T_h and the $\omega \in T_h / T_H$ are high frequencies. Now every $\omega \in T_h$ can be written as a 2-vector $(\omega+p\frac{\pi}{h})$, $p \in \{0,1\}$ on T_H. Hence for $\hat{f}_{h,s} \in L_2(T_h)$ we may also use the notation $\mathbf{f}_{h,s}$, where $\mathbf{f}_{h,s}$ is a 2-vector with entries $\hat{f}_{h,s}(\omega+p\frac{\pi}{h})$. Consistent with this notation, we write the Fourier transform $\hat{B}_{h,s}(\omega)$ of a Toeplitz operator as a 2×2-diagonal matrix $\hat{\mathbf{B}}_{h,s}(\omega)$ with entries $\hat{B}_{h,s}(\omega+p\frac{\pi}{h})$. Any restriction operator $R_{H,s}$, that is defined by an unique stencil, can be written as the combination of $R^0_{H,s}$ and a Toeplitz operator B_h. Hence the Fourier transform of $R_{H,s}$ is given by

$$\hat{\mathbf{R}}_{H,s}(\omega) = \hat{\mathbf{R}}^0_{H,s}(\omega)\hat{\mathbf{B}}_h(\omega), \tag{4.10}$$

where $\hat{\mathbf{R}}_{H,s}(\omega)$ and $\hat{\mathbf{R}}^0_{H,s}(\omega)$ are 1×2 matrices.

Analogous to the restriction we may write any prolongation $P_{h,s}$ as the combination of $P^0_{h,s}$ and a Toeplitz operator B_h. The Fourier transforms of $f_{h,s}$ and $P^0_{h,s}f_{H,s}$ are related by

$$(P^0_{h,s}f_{H,s})(\omega+p\frac{\pi}{h}) = \frac{1}{2}e^{-ish(\omega-p\frac{\pi}{h})}\hat{f}_{H,s}(\omega). \tag{4.11}$$

The Fourier transform of $P_{h,s}=B_hP^0_{h,s}$ is in matrix notation

$$\hat{\mathbf{P}}_{h,s}(\omega) = \hat{\mathbf{B}}_h(\omega)\hat{\mathbf{P}}^0_{h,s}(\omega), \tag{4.12}$$

where $\hat{\mathbf{P}}_{h,s}$ and $\hat{\mathbf{P}}^0_{h,s}$ are 2×1-matrices. Using (4.11) and (4.12) we compute the Fourier transform of the canonical prolongation P_h (see section 3) as

$$\hat{\mathbf{P}}_h(\omega) = \begin{bmatrix} \hat{\mathbf{P}}^\sigma_h(\omega) & 0 \\ 0 & \hat{\mathbf{P}}^u_h(\omega) \end{bmatrix} = \begin{bmatrix} \cos^2\frac{\theta}{2} & 0 \\ \sin^2\frac{\theta}{2} & 0 \\ 0 & \cos\frac{\theta}{2} \\ 0 & \sin\frac{\theta}{2} \end{bmatrix}, \tag{4.13}$$

with $\theta=h\omega$. The Fourier transform $\mathbf{R}_H(\omega)$ of the canonical restriction R_H is the transposed of $\hat{\mathbf{P}}_h(\omega)$.

The components of the Fourier transform of a grid transfer operator are trigoniometric functions of θ; this allows us to classify them according to their behavior in the case ω fixed and $h\to 0$. Suppose that the Fourier transforms of a prolongation and its adjoint restriction are given by $\hat{\Pi}_{h,s}(\omega)$ and $\hat{\Pi}^T_{h,s}(\omega)$, respectively. The low frequency order m_L of a grid transfer operator $\Pi_{h,s}$ is the largest number $m_L \geq 0$ such that

$$\hat{\Pi}_{h,s}(\omega) = 1+O(\theta^{m_L}), \quad \text{for} \quad h\to 0,\ \omega \in T_H.$$

The high frequency order m_H of $\Pi_{h,s}$ is the largest number $m_H \geq 0$ for which

$$\hat{\Pi}(\omega+\frac{\pi}{h}) = O(\theta^{m_H}), \quad \text{for} \quad h\to 0,\ \omega \in T_H.$$

For the canonical grid transfer operators we have: $m^\sigma_L=2$, $m^\sigma_H=2$, $m^u_L=2$ and $m^u_H=1$. To avoid large amplification of high frequencies, the high frequency order m_H should be at least equal to the order of the differential equation (cf. [4,5]). As we are considering a system of first order equations, we conclude that the canonical grid transfer operators are sufficiently accurate. In fact, the Fourier transform of the coarse grid correction operator M_h^{00} is given by

$$\hat{\mathbf{M}}_h^{00}(\omega) = \begin{bmatrix} +\sin^2\frac{\theta}{2} & -\cos^2\frac{\theta}{2} & 0 & 0 \\ -\sin^2\frac{\theta}{2} & +\cos^2\frac{\theta}{2} & 0 & 0 \\ \frac{h}{2i\sin\frac{\theta}{2}}\gamma_L(\frac{\theta}{2}) & \frac{h}{2i\sin\frac{\theta}{2}}\gamma_H(\frac{\theta}{2}) & \sin^2\frac{\theta}{2} & -\frac{1}{2}\sin\theta \\ \frac{h}{2i\cos\frac{\theta}{2}}\gamma_L(\frac{\theta}{2}) & \frac{h}{2i\cos\frac{\theta}{2}}\gamma_H(\frac{\theta}{2}) & -\frac{1}{2}\sin\theta & \cos^2\frac{\theta}{2} \end{bmatrix}, \tag{4.14}$$

with $\gamma_L(\theta) = \sin^2\theta(1-12\kappa\cos^2\theta)$ and $\gamma_H(\theta) = \cos^2\theta(1-12\kappa\sin^2\theta)$.

So, if it is assumed that $\inf_{\omega\in T_H}|\omega|$ is bounded away from zero by the boundary conditions, we see that all elements of $\mathbf{M}_h^{00}(\omega)$ remain bounded for $h\rightarrow 0$. This implies that errors in (σ_h,u_h) are not blown up by the coarse grid correction if $h\rightarrow 0$.

5. Fourier analysis: relaxation

In this section we derive the Fourier representation of the smoothing operators SBGS and SBRB. We start by treating the lexicographical ordering of the cells. In a single SBGS-sweep the σ_h^j are updated twice; so starting from initial values $\{\sigma_h^j,u_h^j\}$ SBGS yields new values $\{\bar{\sigma}_h^j,\bar{u}_h^j\}$, using intermediate values $\tilde{\sigma}_h^j$. If the cells are visited from left to right, we have in every cell Ω_h^i

$$\begin{aligned} &\frac{1}{h}(\tilde{\sigma}_h^i-\bar{\sigma}_h^{i-1}) = 0, \\ &\kappa\sigma_h^{i+1} + (1-2\kappa)\tilde{\sigma}_h^i + \kappa\bar{\sigma}_h^{i-1} + \frac{1}{h}(u_h^{i+1}-\bar{u}_h^i) = 0, \\ &\kappa\tilde{\sigma}_h^i + (1-2\kappa)\bar{\sigma}_h^{i-1} + \kappa\bar{\sigma}_h^{i-2} + \frac{1}{h}(\bar{u}_h^i-\bar{u}_h^{i-1}) = 0. \end{aligned} \tag{5.1}$$

Starting with a Fourier mode $\sigma_h^j=ae^{ijh\omega}$ and $u_h^j=be^{i(j-\frac{1}{2})h\omega}$ we see that $\tilde{\sigma}_h^j=\tilde{a}e^{ijh\omega}$, $\bar{\sigma}_h^j=\bar{a}e^{ijh\omega}$ and $\bar{u}_h^j=\bar{b}e^{i(j-\frac{1}{2})h\omega}$. After elimination of $\tilde{a}$ from (5.1) we obtain a relation between the components before and after relaxation,

$$\begin{bmatrix}\bar{a}\\ \bar{b}\end{bmatrix} = \hat{\mathbf{S}}_h^{GS}(\omega)\begin{bmatrix}a\\ b\end{bmatrix}, \tag{5.2}$$

with

$$\hat{\mathbf{S}}_h^{GS}(\omega) = \frac{e^{i\frac{\theta}{2}}}{2-2\kappa-(1-2\kappa)e^{-i\theta}}\begin{bmatrix}\kappa e^{i\frac{\theta}{2}}(1-e^{i\theta}) & \frac{1-e^{i\theta}}{h} \\ \kappa h(\kappa+(1-\kappa)e^{i\theta}) & e^{-i\frac{\theta}{2}}(\kappa+(1-\kappa)e^{i\theta})\end{bmatrix}. \tag{5.3}$$

The spectral radius $\rho(\cdot)$ and the spectral norm $\|\cdot\|_s$ of $\hat{\mathbf{S}}_h^{GS}(\omega)$ are respectively

$$\rho(\hat{\mathbf{S}}_h^{GS}(\omega)) = \left|\frac{1-2i\kappa\sin\theta}{2-2\kappa-(1-2\kappa)e^{-i\theta}}\right| \tag{5.4a}$$

and

$$\|\hat{\mathbf{S}}_h^{GS}(\omega)\|_S = (1+\kappa^2h^2)\frac{\left[\frac{4}{h^2}\sin^2\frac{\theta}{2}+1-4\kappa\sin\frac{\theta}{2}+4\kappa^2\sin^2\frac{\theta}{2}\right]^{\frac{1}{2}}}{|2-2\kappa-(1-2\kappa)e^{-i\theta}|}. \tag{5.4b}$$

By using (5.4a), the smoothing factor $\mu^{GS} = \sup\limits_{\frac{\pi}{2}\leq|\theta|\leq\pi} \rho(\hat{\mathbf{S}}_h^{GS}(\omega))$ is readily calculated:

$$\mu^{GS} = \begin{cases} (\frac{1}{5})^{\frac{1}{2}}, & \kappa=0, \\ (\frac{10}{29})^{\frac{1}{2}}, & \kappa=\frac{1}{6}, \end{cases} \tag{5.5}$$

independent of h.

From (5.4b) we see that $\|\hat{\mathbf{S}}_h^{GS}(\omega)\|_S$ becomes unbounded for $h\to0$. This is a consequence of the fact that, starting from an initial error $\sigma_h=0$, we find errors in $\overline{\sigma}_h$ of magnitude $O(h^{-1})$ for $h\to0$ (cf. 5.3). From the boundedness of $\rho(\hat{\mathbf{S}}_h^{GS}(\omega))$ we conclude that only in the first relaxation sweep errors in σ_h are blown up by SBGS. In order to obtain a measure of what happens in the first sweep, we introduce the scaled norm $\|\cdot\|_H$, which is defined by

$$\|\hat{\mathbf{A}}_h(\omega)\|_H = \|\hat{\mathbf{H}}_h\hat{\mathbf{A}}_h(\omega)\|_S, \tag{5.6}$$

with H_h: $\mathbb{R}^{2N+1}\to\mathbb{R}^{2N+1}$ a scaling operator,

$$H_h\begin{bmatrix}\sigma_h\\ -\\ u_h\\ -\end{bmatrix} = \begin{bmatrix}h\sigma_h\\ -\\ u_h\\ -\end{bmatrix}.$$

With respect to this norm we find for SBGS:

$$\|\hat{\mathbf{S}}_h^{GS}(\omega)\|_H = (1+\kappa^2h^2)\frac{(4\sin^2\frac{\theta}{2}+1-4\kappa\sin\frac{\theta}{2}+4\kappa^2\sin^2\frac{\theta}{2})^{\frac{1}{2}}}{|2-2\kappa-(1-2\kappa)e^{-i\theta}|}, \tag{5.7}$$

which is bounded for $h\to0$, indeed.

Vanka proposes underrelaxation for σ_h^j (cf. [2]) to improve the smoothing properties of SBGS. This can be analyzed by replacing $\tilde{\sigma}_h^j$ an $\overline{\sigma}_h^{j-1}$ in (5.1) by $\frac{1}{\alpha}(\tilde{\sigma}_h^j-(1-\alpha)\sigma_h^j)$ and $\frac{1}{\alpha}(\overline{\sigma}_h^{j-1}-(1-\alpha)\tilde{\sigma}_h^{j-1})$, respectively, where α denotes the relaxation parameter. For $\kappa=0$ the smoothing factor of this damped relaxation is easily derived and it is (independent of h) given by

$$\mu^{GS}(\alpha) = \max((\frac{1}{5})^{\frac{1}{2}},(1-\alpha)^2). \tag{5.8}$$

Fig. 5.1 shows a graph of $\mu^{GS}(\alpha)$ in the case $k=\frac{1}{6}$ (no lumping). Numerically we find an optimal smoothing rate $\mu^{GS}(\alpha_{opt})=0.369$ for $\alpha_{opt}=0.4583$.

The Fourier representation of SBRB relaxation is obtained by a similar method. As usual we write S_h^{RB} as the product of the partial step operators S_h^R and S_h^B,

$$\hat{\mathbf{S}}_h^{RB}(\omega) = \hat{\mathbf{S}}_h^R(\omega)\hat{\mathbf{S}}_R^B(\omega)$$

$$\hat{\mathbf{S}}_h^R(\omega) = \begin{bmatrix} s_1(\theta) & s_2(\theta+\pi) & s_3(\theta) & s_4(\theta+\pi)\\ s_2(\theta) & s_1(\theta+\pi) & s_4(\theta) & s_3(\theta+\pi)\\ s_5(\theta) & -is_5(\theta+\pi) & s_7(\theta) & -is_8(\theta+\pi)\\ is_5(\theta) & s_5(\theta+\pi) & is_8(\theta) & s_7(\theta+\pi)\end{bmatrix},$$

$$\left[\hat{\mathbf{S}}_h^B(\omega)\right]_{ij} = (-1)^{i+j}\left[\hat{\mathbf{S}}_h^R(\omega)\right]_{ij}, \tag{5.9}$$

with

$$s_1(\theta) = \kappa f(\theta) e^{i\frac{3\theta}{2}} \cos\frac{\theta}{2}, \qquad s_2(\theta) = -i\kappa f(\theta) e^{i\frac{3\theta}{2}} \sin\frac{\theta}{2},$$

$$s_3(\theta) = 2i \frac{f(\theta)}{h} \sin\theta \cos\frac{\theta}{2}, \qquad s_4(\theta) = 2\frac{f(\theta)}{h} \sin\theta \sin\frac{\theta}{2},$$

$$s_7(\theta) = e^{i\frac{\theta}{2}} \cos\frac{\theta}{2} + i(1-\kappa) f(\theta) \sin\theta, \quad s_8(\theta) = ie^{i\frac{\theta}{2}} \sin\frac{\theta}{2} + i(1-\kappa) f(\theta) \sin\theta,$$

$$s_5(\theta) = \frac{h\kappa}{2} e^{i\frac{3\theta}{2}} (1+(1-\kappa) f(\theta)),$$

and

$$f(\theta) = \frac{-1}{2-2\kappa+\kappa e^{-2i\theta}}.$$

We see that all elements of $\hat{\mathbf{S}}_h^{RB}(\omega)$ remain bounded for $h \to 0$ and ω fixed. However by doing so, we only consider the limit cases $|\theta| \to 0$, and $|\theta| \to \pi$; the spectral norm of $\hat{\mathbf{S}}_h^{RB}(\omega)$ becomes unbounded for $h \to 0$ and θ fixed. Numerical computation shows that the scaled norm of $\hat{\mathbf{S}}_h^{RB}(\omega)$ is bounded:

$$\sup_{0 \leq |h\omega| \leq \pi} \|\mathbf{S}_h^{RB}(\omega)\|_H < \infty, \quad \text{for } h \to 0.$$

As SBRB relaxation mixes low and high frequencies the smoothing factor μ^{RB} is defined by

$$\mu^{RB} = \sup_{\frac{\pi}{2} \leq |h\omega| \leq \pi} \rho(\hat{\mathbf{Q}} \hat{\mathbf{S}}_h^{RB}(\omega)), \tag{5.10}$$

where $\hat{\mathbf{Q}}$ denotes the operator that annihilates all low frequencies

$$\hat{\mathbf{Q}} = \begin{bmatrix} 0 & & & \\ & 1 & & \\ & & 0 & \\ & & & 1 \end{bmatrix}. \tag{5.11}$$

If underrelaxation of σ_h^i is taken into account, we obtain for $\kappa = 0$

$$\mu^{RB}(\alpha) = \max\left(\tfrac{1}{8}, (1-\alpha)^2\right), \tag{5.12}$$

independent of h. A plot of $\mu^{RB}(\alpha)$ for $\kappa = \frac{1}{6}$ is shown in Fig 5.2. In this case underrelaxation hardly improves the smoothing factor; numerically we find $\mu^{RB}(\alpha = 1) = 0.127$.

6. Fourier analysis: the two-grid algorithm

In the previous sections we have shown that appropriate norms of the coarse grid correction operator and the relaxation operator remain bounded in the limit case of vanishing meshsize. However, this does not imply that the scaled norm of the two-grid error amplification matrix is bounded, and hence that the convergence rate of the two-grid algorithm is mesh-independent. If $\kappa = 0$ both SBGS and SBRB relaxation without damping ($\alpha = 1$) eliminate σ_h, so we are solving a second order differential equation for u_h. Therefore we may expect that the canonical grid transfer operators P_h^u and R_H^u are not accurate enough.

We show that this is the case indeed by studying the two-grid algorithm with SBGS relaxation for $\kappa = 0$. If a single SBGS-sweep is used for pre- and post smoothing, the Fourier transform of M_h^{11} after n cycles is given by

$$(\hat{\mathbf{M}}_h^{11}(\omega))^n = \left[\frac{e^{2i\theta}}{4-e^{-2i\theta}}\right]^n \begin{bmatrix} 0 & M^{\sigma u} \\ 0 & M^{uu} \end{bmatrix}, \tag{6.1}$$

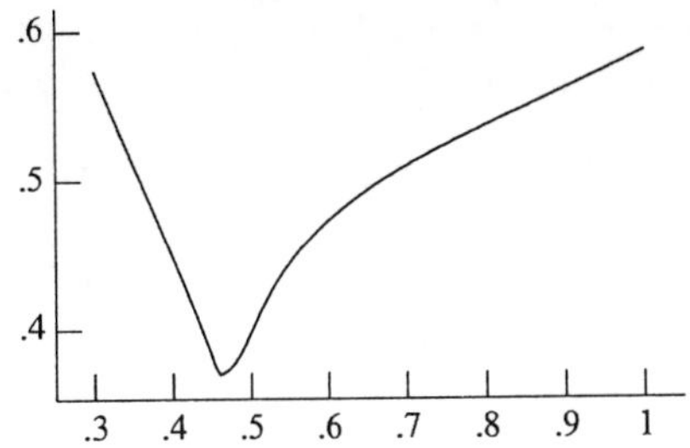

Fig. 5.1. Smoothing factor μ^{GS} depending on the relaxation parameter α

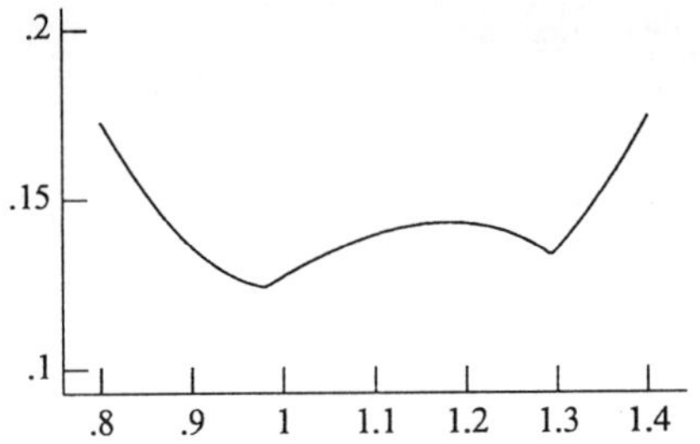

Fig. 5.2. Smoothing factor μ^{RB} depending on the relaxation parameter α

		(P_h, R_H)	$(\tilde{P}_h, \tilde{R}_H)$
$\kappa=0$	SBGS	∞	0.346
	SBRB	0.590	0.099
$\kappa=1/6$	SBGS ($\alpha=1$)	∞	0.531
	SBGS ($\alpha=\alpha_{opt}$)	∞	0.368
	SBRB	0.628	0.339

TABLE 6.1. Scaled norm of the two-grid error amplification matrix, $\sup_{0 \leq |\theta| \leq \frac{\pi}{2}} \|\hat{\mathbf{M}}_h^{11}(\omega)\|_H$.

	(P_h, R_H)		$(\tilde{P}_h, \tilde{R}_H)$	
ν	SBGS	SBRB	SBGS	SBRB
1	0.577	0.500	0.343	0.096
2	0.333	0.325	0,145	0.046
3	0.192	0.259	0.088	0.031
4	0.111	0.221	0.058	0.023

TABLE 6.2. Two-level convergence factor λ_ν, $\kappa=0$.

	(P_h, R_H)			$(\tilde{P}_h, \tilde{R}_H)$		
ν	SBGS ($\alpha=1$)	SBGS ($\alpha=\alpha_{opt}$)	SBRB	SBGS ($\alpha=1$)	SBGS ($\alpha=\alpha_{opt}$)	SBRB
1	0.657	0.477	0.457	0.476	0.391	0.264
2	0.429	0.211	0.367	0.311	0.277	0.178
3	0.281	0.147	0.314	0.236	0.212	0.136
4	0.184	0.093	0.278	0.177	0.166	0.111

TABLE 6.3. Two-level convergence factor λ_ν, $\kappa=\frac{1}{6}$.

with

$$n \text{ even}: M^{\sigma u} = \begin{bmatrix} -\frac{2}{ih}\sin\frac{\theta}{2} & 0 \\ 0 & \frac{2}{ih}\cos\frac{\theta}{2} \end{bmatrix}, \qquad M^{uu} = \begin{bmatrix} 1 & 0 \\ 0 & 1 \end{bmatrix},$$

$$n \text{ odd}: M^{\sigma u} = \begin{bmatrix} 0 & \frac{2}{ih}\cos\frac{\theta}{2} \\ \frac{2}{ih}\sin\frac{\theta}{2} & 0 \end{bmatrix}, \qquad M^{uu} = \begin{bmatrix} 0 & \cot\frac{\theta}{2} \\ \tan\frac{\theta}{2} & 0 \end{bmatrix}. \tag{6.2}$$

The two-grid algorithm exhibits a typical alternating convergence behavior. An initial high frequency mode $\underline{u}_h$ of amplitude b, causes a low frequency mode of amplitude $b\cot\frac{\theta}{2}$ after a single two-grid cycle; so if $h \to 0$ initial high frequency error modes in u_h are blown up. In the next cycle the large low frequency mode u_h is nicely removed by the coarse grid correction, although a small high frequency error mode is introduced. This alternating behavior is reflected by the scaled norm of $(\hat{\mathbf{M}}_h^{11}(\omega))^n$; for $h \to 0$ we find

$$\sup_{0 \le |h\omega| \le \frac{\pi}{2}} \|(\hat{\mathbf{M}}_h^{11}(\omega))^n\|_H = \begin{cases} (\frac{5}{3^n})^{\frac{1}{2}}, & n \text{ even}, \\ \infty, & n \text{ odd}. \end{cases}$$

By numerical computation we observe a similar alternating convergence behavior for $\kappa = \frac{1}{6}$, even though the coarse grid operator L_H satisfies Galerkin's relation.

The obvious cure is to use more accurate grid transfer operators. We introduce $\tilde{P}_h^u$ the linear interpolation operator for $\underline{u}_h$ and $\tilde{R}_H^u$ its adjoint. The Fourier transforms of these more accurate grid transfer operators are

$$\hat{\hat{\mathbf{P}}}_h(\omega) = \begin{bmatrix} \hat{\mathbf{P}}_h^{\sigma}(\omega) & 0 \\ 0 & \hat{\hat{\mathbf{P}}}_h^{u}(\omega) \end{bmatrix} = \begin{bmatrix} \cos^2\frac{\theta}{2} & 0 \\ \sin^2\frac{\theta}{2} & 0 \\ 0 & \cos^3\frac{\theta}{2} \\ 0 & \sin^3\frac{\theta}{2} \end{bmatrix}$$

and

$$\hat{\hat{\mathbf{R}}}_H(\omega) = \begin{bmatrix} \hat{\mathbf{R}}_H^{\sigma}(\omega) & 0 \\ 0 & \hat{\hat{\mathbf{R}}}_H^{u}(\omega) \end{bmatrix} = \left[\hat{\hat{\mathbf{P}}}_h(\omega)\right]^T,$$

so we have $m_L^u = 2$ and $m_H^u = 3$. Although it is not necessary to use $\tilde{P}_h^u$ for keeping the scaled norm of the error amplification matrix bounded, it is introduced to avoid similar problems with the amplification operator of the residuals. In table 6.1 we show values for $\sup_{0 \le |h\omega| \le \frac{\pi}{2}} \|\hat{\mathbf{M}}_h^{11}(\omega)\|_H$ for the different possible two-grid algorithms. If SBRB relaxation is used, the canonical grid transfer operators ($m_h^u = 1$) are sufficient: the high frequencies are so efficiently smoothed that they don't cause any problems. Here we see that the choice of the grid transfer operators is not only determined by the order of the differential equations, but is also influenced by the relaxation scheme.

The scaled norm of $\hat{\mathbf{M}}_h^{11}(\omega)$ only indicates what happens in a single two-grid cycle; the convergence rate after many cycles is estimated by the two-level convergence factor

$$\lambda_\nu = \sup_{0 \leq |\theta| \leq \frac{\pi}{2}} \rho(\hat{\mathbf{M}}_h^{\nu_1 \nu_2}),$$

with $\nu = \nu_1 + \nu_2$. In table 6.2 we show λ_ν for $\kappa = 0$ and for different values of ν. The combination of the transfer operators $\tilde{P}_h$ and $\tilde{R}_h$, and SBRB relaxation leads to a fast converging algorithm. In table 6.3 we show λ_ν for $\kappa = \frac{1}{6}$. We see that the introduction of a damping parameter α in SBGS relaxation indeed leads to faster convergence, but the best convergence factors are again obtained by using the combination of SBRB relaxation and the transfer operators $\tilde{P}_h$ and $\tilde{R}_H$.

7. Conclusions

By local mode analysis we have shown that symmetric block Gauss-Seidel relaxation is an efficient smoother indeed. Although lumping of the discrete equations spoils the Galerkin property of the coarse grid operator it generally leads to faster converging two-grid algorithms. The Fourier transform of the two-grid error amplification operator shows that the canonical grid transfer operators are insufficiently accurate in the 1D case if a lexicographical ordering of the gridpoints is used in the relaxation procedure; however they suffice if a red-black ordering is used.

References

[1] Molenaar, J., and P.W. Hemker (1990). *A multigrid approach for the solution of the 2D semiconductor equations*, IMPACT, to appear.

[2] Vanka, S.P. (1986). *Block-Implicit Multigrid Solution of Navier-Stokes Equations in Primitive Variables*, J. Comput. Phys., 65, 138-158.

[3] Polak, S.J. W.H.A. Schilders and H.D. Couperus (1988). *A finite element method with current conservation*, in Simulation of Semiconductor Devices and Processes, Vol. 3, Bologna, Technoprint.

[4] Brandt, A. (1982). *Guide to multigrid development*, in Multigrid Methods, Springer Verlag, Lecture Notes in Mathematics 960.

[5] Hemker, P.W. (1990). *On the order of prolongations and restrictions in multigrid procedures*, J. Appl. Math., to appear.

MULTIGRID APPLIED TO MIXED FINITE ELEMENT SCHEMES FOR CURRENT CONTINUITY EQUATIONS

by

Arnold Reusken

Abstract: *Mixed finite element discretizations for current continuity equations are presented in [5,6,8]. We consider the resulting system of equations and develop a multigrid method for such a system. Our method is based on a connection between the given system and nonconforming Crouzeix-Raviart finite elements.*

1. Introduction

Recently (new) mixed finite element discretizations for current continuity equations have been presented in [5,6,8]. These schemes have nice properties: They provide an M-matrix, there is current conservation, and a good approximation of sharp shapes. Such a scheme results in a large sparse system of equations for the unknowns. A "standard" multigrid solver cannot be used for this system due to the presence of (extremely) large convection in part of the domain and the use of mixed finite elements. In this paper we present a suitable multigrid method for solving this system. The method is based on a connection between the system resulting from the mixed FE discretization and nonconforming Crouzeix-Raviart FE. Using this connection together with multigrid theory for nonconforming FE then leads to a multigrid method for the given system. These ideas underlying our method are an important subject of this paper and might be useful in other situations too.

We consider the continuous problem as in [6]:

$$(1.1)\quad \begin{cases} \text{find } u \in H^1(\Omega) \text{ such that} \\ \operatorname{div}(\nabla u + u\nabla\psi) = f \quad \text{in } \Omega \subset \mathbb{R}^2 \\ u = g \quad \text{on } \Gamma_0 \subset \partial\Omega \\ \dfrac{\partial u}{\partial n} + u\,\dfrac{\partial \psi}{\partial n} = 0 \quad \text{on } \Gamma_1 = \partial\Omega \setminus \Gamma_0 \end{cases}$$

In this current continuity equation (the current is defined by $\mathbf{J} = \boldsymbol{\nabla} u + u \boldsymbol{\nabla} \psi$) we assume that ψ is a given function.

In §2 we collect some results from [6] concerning the mixed FE discretization of (1.1). In §3 we discuss a connection between the system resulting from §2 and a suitable variational problem in the Crouzeix-Raviart nonconforming FE space. In §4 we present our multigrid method. In §5 we give some numerical results.

2. Mixed finite element schemes for current continuity equations

In this section we collect some results from [5,6].

2.1. Continuous problem. The problem we want to solve is given in (1.1). Using the *Slotboom variable* $\rho := e^{\psi} u$ this results in the following problem

$$
(2.2)\qquad \begin{cases} \text{find } \rho \in H^1(\Omega) \text{ such that} \\ \operatorname{div}(e^{-\psi} \boldsymbol{\nabla} \rho) = f \quad \text{in } \Omega \\ \rho = \chi := e^{\psi} g \quad \text{on } \Gamma_0 \\ \dfrac{\partial \rho}{\partial n} = 0 \quad \text{on } \Gamma_1 \end{cases}
$$

2.2. Discretization. For ease we assume that Ω is a polygonal domain. Let $\{T_k\}_{k\geq 0}$ be a regular sequence of triangulations of Ω into triangles T. The set of edges of T_k is denoted by E_k; edges are denoted by e_i and $E_k = \{e_i\}_{i\in I_0 \cup I}$ with I_0 the index set of edges $e_i \subset \Gamma_0$ and I the index set of edges $e_i \subset \bar{\Omega} \setminus \Gamma_0$. Midpoints of edges are denoted by m_i.
We use the lowest order *Raviart-Thomas mixed finite element method* to discretize (2.2).
Set

$$
RT(T) = \{\boldsymbol{\tau} = (\tau_1, \tau_2) \mid \tau_1 = \alpha + \beta x \ , \quad \tau_2 = \gamma + \beta y \ , \quad \alpha, \beta, \gamma \in \mathbb{R}\} \qquad (T \in T_k)
$$

and define

$$
V_k = \{\boldsymbol{\tau} \in (L^2(\Omega))^2 \mid \operatorname{div}\boldsymbol{\tau} \in L^2(\Omega) \ , \quad \boldsymbol{\tau} \cdot \mathrm{n} = 0 \text{ on } \Gamma_1, \ \boldsymbol{\tau}_{|T} \in RT(T) \quad \text{for all} \quad T \in T_k\} \ ,
$$

$$
W_k = \{\varphi \in L^2(\Omega) \mid \varphi_{|T} \in P_0(T) \quad \text{for all} \quad T \in T_k\} \ .
$$

Then the discretization of (2.2) is as follows

$$
(2.3)\qquad \begin{cases} \text{find } \mathbf{J}_k \in V_k \text{ and } \rho_k \in W_k \text{ such that} \\ \displaystyle\int_\Omega e^{\psi} \mathbf{J}_k \cdot \boldsymbol{\tau} \, d\mathrm{x} + \int_\Omega \rho_k \operatorname{div}\boldsymbol{\tau} \, d\mathrm{x} = \int_{\Gamma_0} \chi \boldsymbol{\tau} \cdot \mathrm{n} \, d\Gamma \qquad \forall \boldsymbol{\tau} \in V_k \\ \displaystyle\int_\Omega \varphi \operatorname{div}\mathbf{J}_k \, d\mathrm{x} = \int_\Omega f \varphi \, d\mathrm{x} \qquad \forall \varphi \in W_k \ . \end{cases}
$$

The matrix associated with (2.3) is not positive definite. To circumvent this *Lagrange multipliers* are used.
Define $\tilde{V}_k = \{\boldsymbol{\tau} \in (L^2(\Omega))^2 \mid \boldsymbol{\tau}_{|T} \in RT(T) \quad \text{for all} \quad T \in T_k\}$, and for $\zeta \in L^2(\Gamma_0)$
$\Lambda_{k,\zeta} = \{\mu \in L^2(E_k) \mid \mu_{|e} \in P_0(e) \quad \text{for all} \quad e \in E_k, \ \int_{e_i}(\mu - \zeta)\, ds = 0 \quad \text{for all} \quad i \in I_0\}$.
Instead of (2.3) we now consider the discretization

$$
(2.4) \qquad \begin{cases}
\text{find } \tilde{\mathbf{J}}_k \in \tilde{V}_k \ , \quad \tilde{\rho}_k \in W_k \ , \quad \tilde{\lambda}_k \in \Lambda_{k,\chi} \quad \text{such that} \\[2mm]
\displaystyle\int_\Omega e^{\psi}\, \tilde{\mathbf{J}}_k \cdot \boldsymbol{\tau}\, d\mathrm{x} + \sum_T \int_T \tilde{\rho}_k \operatorname{div}\boldsymbol{\tau}\, d\mathrm{x} - \sum_T \int_{\partial T} \tilde{\lambda}_k\, \boldsymbol{\tau} \cdot \mathrm{n}\, ds = 0 \qquad \forall \boldsymbol{\tau} \in \tilde{V}_k \\[2mm]
\displaystyle\sum_T \int_T \varphi \operatorname{div}\tilde{\mathbf{J}}_k\, d\mathrm{x} = \int_\Omega f\, \varphi\, d\mathrm{x} \qquad \forall \varphi \in W_k \\[2mm]
\displaystyle\sum_T \int_{\partial T} \mu\, \tilde{\mathbf{J}}_k \cdot \mathrm{n}\, ds = 0 \qquad \forall \mu \in \Lambda_{k,0} \ .
\end{cases}
$$

The problem (2.4) has a unique solution and $\tilde{\mathbf{J}}_k \equiv \mathbf{J}_k$, $\tilde{\rho}_k \equiv \rho_k$ holds. Moreover $\tilde{\lambda}_k$ is a good approximation of ρ at the interelements (see [1]). In the resulting matrix-vector problem the unknowns corresponding to $\tilde{\mathbf{J}}_k$ and $\tilde{\rho}_k$ can be eliminated by static condensation. Lemma 2.5 below shows that in (2.4) $\tilde{\mathbf{J}}_k$ and $\tilde{\rho}_k$ can be eliminated a-priori. The proof of this lemma is straightforward using the arguments concerning static condensation in [6].

Notation. In the remainder we use the following notation: for $h \in L^2(A)$ we set $\bar{h}_{|A} := \frac{1}{|A|} \int_A h(x)\, dx$.

Lemma 2.5. *Define* $\mathbf{J}^f$ *by* $\mathbf{J}^f_{|T}(\mathrm{x}) = \frac{1}{2}\, \bar{f}_{|T}\, \mathrm{x}$. *Define a symmetric bilinear form* $b_k : L^2(E_k) \times L^2(E_k) \to \mathbb{R}$ *and a linear functional* $F_k : L^2(E_k) \to \mathbb{R}$ *as follows:*

$$
b_k(\lambda, \mu) = \sum_T (|T|\, \overline{e^{\psi}}_{|T})^{-1} \int_{\partial T} \lambda\, \mathrm{n}\, ds \cdot \int_{\partial T} \mu\, \mathrm{n}\, ds
$$

$$
F_k(\mu) = \sum_T (|T|\, \overline{e^{\psi}}_{|T})^{-1} \int_T e^{\psi}\, \mathbf{J}^f\, d\mathrm{x} \cdot \int_{\partial T} \mu\, \mathrm{n}\, ds - \sum_T \int_{\partial T} \mu\, \mathbf{J}^f \cdot \mathrm{n}\, ds \ .
$$

Then the solution $\tilde{\lambda}_k$ *of* (2.4) *is also the unique solution of the following problem*

$$
(2.6) \qquad \begin{cases}
\textit{find } \tilde{\lambda}_k \in \Lambda_{k,\chi} \quad \textit{such that} \\[2mm]
b_k(\tilde{\lambda}_k, \mu) = F_k(\mu) \quad \textit{for all} \quad \mu \in \Lambda_{k,0} \ .
\end{cases}
$$

2.3. Rescaling. The Lagrange multiplier $\tilde{\lambda}_k$ is an approximation of $\rho = e^{\psi}\, u$, and is not suited for actual computation if the range of ψ is large (which often happens in semiconductor

problems); moreover we are interested in approximating u instead of ρ. So we rescale $\tilde{\lambda}_k$ to get an approximation $\tilde{\mu}_k$ of u (at the interelements).
In addition to $\Lambda_{k,\zeta}$ we will also use the space $\Lambda_k := \{\mu \in L^2(E_k) \mid \mu_{|e} \in P_0(e) \text{ for all } e \in E_k\}$ with standard basis functions (1 on one edge and 0 on all other edges) denoted by $\{\mu_i\}_{i\in I_0 \cup I}$.

We define the isomorphism $Q_k : \Lambda_{k,g} \to \Lambda_{k,\chi}$ (g, χ as in (2.2)) as follows. Take $\mu \in \Lambda_{k,g}$ then:

$$\text{for } e \subset \Gamma_0 \qquad (Q_k\,\mu)_{|e} = \bar{\chi}_{|e}$$

$$\text{for } e \subset \bar{\Omega}\setminus\Gamma_0 \quad (Q_k\,\mu)_{|e} = \overline{e^{\psi}}_{|e}\,\mu_{|e}\ .$$

Using this isomorphism we can rewrite (2.6):

$$\text{(2.7)} \qquad \begin{cases} \text{find } \tilde{\mu}_k \in \Lambda_{k,g} \text{ such that} \\ b_k(Q_k\,\tilde{\mu}_k, \mu) = F_k(\mu) \quad \text{for all} \quad \mu \in \Lambda_{k,0}\ . \end{cases}$$

The problem (2.7) is the final one, which we actually want to solve.
Rewriting (2.7) as a matrix-vector problem using the basis $\{\mu_i\}$ of Λ_k yields the following system of equations for the unknowns $\{\alpha_j\}_{j\in I}$ with $\tilde{\mu}_k = \sum_{j\in I} \alpha_j\,\mu_j + \sum_{j\in I_0} \bar{g}_{|e_j}\,\mu_j$

$$\text{(2.8)} \qquad \sum_{j\in I} b_k(Q_k\,\mu_j, \mu_i)\ \alpha_j = -\sum_{j\in I_0} b_k(Q_k\,\mu_j, \mu_i)\ \bar{g}_{|e_j} + F_k(\mu_i) \quad \text{for } i \in I\ .$$

Remark 2.9. Expressions for $b_k(Q_k\,\mu_j, \mu_i)$ and $F_k(\mu_i)$ can be found in [6]. The resulting (non-symmetric) matrix is an M-matrix if the triangulation is weakly acute (no angle $> \frac{\pi}{2}$). The above discretization has upwinding features for strong convection. For a discussion of this upwinding effect to refer to [6].

3. Connection with nonconforming finite elements

In this section we show that the system (2.8) corresponds to a variational problem in the (nonconforming) $P1$ Crouzeix-Raviart finite element space.

Consider the *Crouzeix-Raviart* ($P1$) *space* corresponding to T_k:
$S_k = \{v \in L^2(\Omega) \mid v_{|T} \text{ is linear for all } T \in T_k,\ v \text{ is continuous at midpoints of edges}\}$. The standard basis of S_k is denoted by $\{\varphi_i\}_{i\in I\cup I_0}$ (I, I_0 as in §2). For $\zeta \in L^2(\Gamma_0)$ we define $S_{k,\zeta} := \{v \in S_k \mid v(m_i) = \bar{\zeta}_{|e_i}\ \ \forall i \in I_0\}$.
We also use the space $\tilde{S}_k := \{v \in L^2(\Omega) \mid v_{|T} \text{ is linear for all } T \in T_k\}$.

Let $R_k : S_k \to \tilde{S}_k$ be the linear operator which satisfies

$$\text{(3.1)} \qquad (R_k\,\varphi_i)_{|T} = (\overline{e^{\psi}}_{|T})^{-1}\,\overline{e^{\psi}}_{|e_i}\,\varphi_i \quad \text{for all basis functions } \varphi_i \text{ and all } T \in T_k.$$

Note that $(R_k v)_{|T} = v_{|T}$ if ψ is constant on T.

Theorem 3.2. *Let f be as in (1.1) and define the function $G_k(f) \in L^2(\Omega)$ by $G_k(f)_{|T} = \bar{f}_{|T}(1\frac{1}{2} - \frac{1}{2}(\overline{e^\psi}_{|T})^{-1} e^\psi)$. Consider the problem*

$$(3.3) \qquad \begin{cases} \text{find } \tilde{\eta}_k \in S_{k,g} \text{ such that} \\ \sum_T \int_T \nabla(R_k \tilde{\eta}_k) \cdot \nabla\varphi \, dx = -\int_\Omega G_k(f)\varphi \, dx \quad \text{for all } \varphi \in S_{k,0} . \end{cases}$$

Write $\tilde{\eta}_k = \sum_{j \in I} \alpha_j \varphi_j + \sum_{j \in I_0} \bar{g}_{|e_j} \varphi_j$. Taking $\varphi = \varphi_i$ $(i \in I)$ in (3.3) results in a system of equations for the unknowns $\{\alpha_j\}_{j \in I}$ that is given in (2.8).

Proof. Use the notation $a_k(\eta, \varphi) = \sum_T \int_T \nabla(R_k \eta) \cdot \nabla\varphi \, dx$. Taking $\varphi = \varphi_i$ in (3.3) yields

$$(3.4) \qquad \sum_{j \in I} a_k(\varphi_j, \varphi_i)\alpha_j = -\sum_{j \in I_0} a_k(\varphi_j, \varphi_i)\bar{g}_{|e_j} - \int_\Omega G_k(f)\varphi_i \, dx \qquad (i \in I) .$$

By comparing (3.4) with (2.8) it follows that we only have to show:

$$\text{(a)} \qquad b_k(Q_k \mu_j, \mu_i) = a_k(\varphi_j, \varphi_i)$$

$$\text{(b)} \qquad F_k(\mu_i) = -\int_\Omega G_k(f)\varphi_i \, dx .$$

Using the definitions and checking per triangle it is clear that it is sufficient to prove:

$$\text{(a')} \qquad |T|^{-1} \int_{\partial T} \mu_j \mathbf{n} \, ds \cdot \int_{\partial T} \mu_i \mathbf{n} \, ds = \int_T \nabla\varphi_j \cdot \nabla\varphi_i \, dx$$

$$\text{(b')} \qquad |T| \, (\overline{e^\psi}_{|T})^{-1} \int_T e^\psi \mathbf{J}^f \, dx \cdot \int_{\partial T} \mu_i \mathbf{n} \, ds - \int_{\partial T} \mu_i \mathbf{J}^f \cdot \mathbf{n} \, ds = -\int_T G_k(f)\varphi_i \, dx .$$

Let T be a given triangle with edges e_1, e_2, e_3, midpoints of edges m_1, m_2, m_3 (with corresponding coordinate vectors $\mathbf{M}_1, \mathbf{M}_2, \mathbf{M}_3$) and unit outward normals $\mathbf{n}^{(1)}, \mathbf{n}^{(2)}, \mathbf{n}^{(3)}$. Define $\boldsymbol{\nu}^{(i)} = |e_i| \, \mathbf{n}^{(i)}$. Now consider a basis function φ_i which is 1 in m_i $(i \in \{1,2,3\})$ and 0 in all m_j with $j \neq i$. One easily verifies that $\varphi_{i|T}$ can be represented as

$$(3.5) \qquad \varphi_{i|T}(\mathbf{x}) = |T|^{-1} \boldsymbol{\nu}^{(i)} \cdot (\mathbf{x} - \mathbf{M}_j) \quad \text{for } j \in \{1,2,3\} \setminus \{i\} .$$

Below we also use the following well-known result

(3.6) for $p \in P_2(T)$ we have $\displaystyle\int_T p(\mathbf{x})\,d\mathbf{x} = \frac{1}{3}|T|\,(p(m_1)+p(m_2)+p(m_3))\,.$

The lefthand side of (a') equals $|T|^{-1}\boldsymbol{\nu}^{(j)}\cdot\boldsymbol{\nu}^{(i)}$. Using (3.5) it follows that $\int_T \boldsymbol{\nabla}\varphi_j\cdot\boldsymbol{\nabla}\varphi_i\,d\mathbf{x} = |T|\,(|T|^{-1}\boldsymbol{\nu}^{(j)})\cdot(|T|^{-1}\boldsymbol{\nu}^{(i)}) = |T|^{-1}\boldsymbol{\nu}^{(j)}\cdot\boldsymbol{\nu}^{(i)}$. So (a') holds.
The proof of (b') runs as follows

$$(|T|\overline{e^{\psi}}_{|T})^{-1}\int_T e^{\psi}\mathbf{J}^f\,d\mathbf{x}\cdot\int_{\partial T}\mu_i\,\mathbf{n}\,ds - \int_{\partial T}\mu_i\,\mathbf{J}^f\cdot\mathbf{n}\,ds$$

$$= \tfrac{1}{2}\bar{f}_{|T}\,\{|T|^{-1}(\overline{e^{\psi}}_{|T})^{-1}\int_T e^{\psi}\,\mathbf{x}\cdot\boldsymbol{\nu}^{(i)}\,d\mathbf{x} - \frac{1}{|e_i|}\int_{e_i}\mathbf{x}\cdot\boldsymbol{\nu}^{(i)}\,ds\}$$

$$= \tfrac{1}{2}\bar{f}_{|T}\,\{|T|^{-1}(\overline{e^{\psi}}_{|T})^{-1}\int_T e^{\psi}(|T|\,\varphi_i+\boldsymbol{\nu}^{(i)}\cdot\mathbf{M}_j)\,d\mathbf{x} - \frac{1}{|e_i|}\int_{e_i}|T|\,\varphi_i+\boldsymbol{\nu}^{(i)}\cdot\mathbf{M}_j\,ds\}\quad\text{(use (3.5))}$$

$$= \tfrac{1}{2}\bar{f}_{|T}\,\{\int_T(\overline{e^{\psi}}_{|T})^{-1}e^{\psi}\,\varphi_i\,d\mathbf{x}+\boldsymbol{\nu}^{(i)}\cdot\mathbf{M}_j - |T| - \boldsymbol{\nu}^{(i)}\cdot\mathbf{M}_j\}\quad(\text{use } \varphi_i\equiv 1 \text{ on } e_i)$$

$$= \tfrac{1}{2}\bar{f}_{|T}\,\{\int_T(\overline{e^{\psi}}_{|T})^{-1}e^{\psi}\,\varphi_i\,d\mathbf{x} - 3\int_T\varphi_i\,d\mathbf{x}\}\quad\text{(use (3.6))}$$

$$= -\int_T \bar{f}_{|T}(1\tfrac{1}{2}-\tfrac{1}{2}(\overline{e^{\psi}}_{|T})^{-1}e^{\psi})\,\varphi_i\,d\mathbf{x}$$

$$= -\int_T G_k(f)\,\varphi_i\,d\mathbf{x}\,.$$

□

Remark 3.7. The system (2.8) corresponds to the variational problem (2.7), for the rescaled Lagrange multipliers $\tilde{\mu}_k$, which results in a natural way by using the Slotboom variable and mixed finite elements. From Theorem 3.2 we conclude that the system (2.8) also corresponds to the rather special nonconforming finite element discretization in (3.3). The discretization in (3.3) can be related to the given continuous problem in (1.1) as follows. Consider (1.1) as a variational problem in H^1 with a bilinear form $a_\Omega(u,v) = \int_\Omega (\boldsymbol{\nabla}u + u\boldsymbol{\nabla}\psi)\cdot\boldsymbol{\nabla}v\,d\mathbf{x} = \int_\Omega e^{-\psi}\,\boldsymbol{\nabla}(e^{\psi}u)\cdot\boldsymbol{\nabla}v\,d\mathbf{x}$ and a righthand side functional $v \to -\int_\Omega f\,v\,d\mathbf{x}$. The problem in (3.3) results from this by a modified discretization in the Crouzeix-Raviart P_1 space. The modification concerns the following:

(a) in the righthand side functional, f is replaced by $\tilde{f}$ with $\tilde{f}_{|T} = \bar{f}_{|T}(1\tfrac{1}{2}-\tfrac{1}{2}(\overline{e^{\psi}}_{|T})^{-1}e^{\psi})$

(b) $a_\Omega(u,v)$ is replaced by $\tilde{a}_\Omega(u,v)$ with $\tilde{a}_T(\varphi_i,\varphi_j) = \int_T (\overline{e^\psi}_{|T})^{-1}\, \nabla(\overline{e^\psi}_{|e_i}\,\varphi_i)\cdot\nabla\varphi_j\, d\mathbf{x}$ (so $e^{-\psi}{}_{|T}$ is replaced by the harmonic average $(\overline{e^\psi}_{|T})^{-1}$ and $e^\psi u = \Sigma\,\alpha_i\, e^\psi\,\varphi_i$ is replaced by $\Sigma\,\alpha_i\,\overline{e^\psi}_{|e_i}\,\varphi_i$).

Other relations between mixed and nonconforming finite elements are discussed in [1].

4. Multigrid method

In this section we develop a multigrid method that can be used to solve the system (2.8). The method is based on the equivalence between (2.8) and (3.3). Through this equivalence we are led to multigrid methods for nonconforming finite elements (as in [2], [3,4]). Before we specify the multigrid method we first discuss

- triangulation
- sequence of bilinear forms
- prolongation
- smoother.

4.1. Triangulation. In the remainder we assume a regular sequence of triangulations $\{T_k\}_{k\geq 0}$ in which T_k ("finer triangulation") is obtained from T_{k-1} ("coarser triangulation") by connecting the midpoints of the edges of the triangles of T_{k-1}. We also assume that T_0 is weakly acute (all triangles have angles $\leq \frac{\pi}{2}$). In the numerical experiments in §5 we will use $\Omega = [0,1]\times[0,1]$ and a triangulation as in Figure 1.

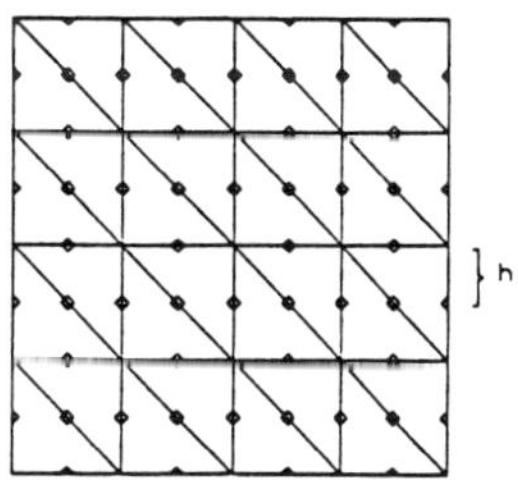

Figure 1.

4.2. Sequence of bilinear forms. We assume a given $k = k_{\max}$ for which we want to solve (2.8). We treat this problem in the (equivalent) form (3.3). For ease we use the notation $a_k(\eta,\varphi) = \sum_T \int_T \nabla(R_k\,\eta)\cdot\nabla\,\varphi\, d\mathbf{x}$ $(\eta,\varphi\in S_{k,0})$ and $\tilde{\eta}_k^d := \sum_{j\in I_0} \bar{g}_{|e_j}\,\varphi_j$; then we can rewrite (3.3) as follows:

$$
(4.1)\qquad \begin{cases} \text{find } \eta_k^* \in S_{k,0} \text{ such that} \\[2ex] a_k(\eta_k^*,\varphi) = -a_k(\tilde{\eta}_k^d,\varphi) - \displaystyle\int_\Omega G_k(f)\,\varphi\, dx \quad \text{for all } \varphi\in S_{k,0}\,. \end{cases}
$$

If we take $k = k_{\max}$ then (4.1) results in the problem we want to solve. We make an obvious choice for approximation on coarser grids: We take the discrete operators induced by the bilinear form $a_k(\cdot,\cdot)$ $(0 \leq k < k_{\max})$. We note, however, that it is not clear if a suitable "approximation property" (cf. [7]) holds for these coarse grid operators.

4.3. Prolongation. In multigrid methods we need mappings between the finite element spaces. In conforming finite elements the spaces are nested, so there is a natural imbedding of the coarse grid functions in the fine grid function space. In nonconforming finite elements this is not the case, so we need a suitable prolongation $P_k : S_{k-1,0} \to S_{k,0}$. Such prolongations are given in [2], [3,4]. We use the prolongation as proposed in [2]: For P_k we take the orthogonal projection w.r.t. the L^2-inner product of $S_{k-1,0}$ on $S_{k,0}$. More information concerning this prolongation is given in Lemma 4.2 below. We omit the proof because it is straightforward.

Lemma 4.2. *Define $W_k := S_{k,0} \oplus S_{k-1,0}$. Let P_k be the orthogonal projection w.r.t. the L^2-inner product of W_k on $S_{k,0}$. Take $u \in W_k$ and take the midpoint m_i of an edge $e_i \in E_k$.*
If $e_i \subset \partial\Omega$ then $(P_k u)(m_i) = u(m_i)$.
If $e_i \not\subset \partial\Omega$ then $(P_k u)(m_i) = (|T^L| + |T^R|)^{-1}\ (|T^L|\, u_{|T^L}(m_i) + |T^R|\, u_{|T^R}(m_i))$, where $T^L,\ T^R \in T_k$ are the two triangles with edge e_i.

From Lemma 4.2 it is clear that if u is continuous at m_i then $(P_k u)(m_i) = u(m_i)$. For this P_k formulas suited for computation can be found in [2].

4.4. Smoother. Systematic research concerning a suitable smoother for a system as in (2.8) has not been done yet. Two obvious candidates are ILU (which is a popular method for problems with strong convection) and Gauss-Seidel. Here we restrict ourselves to (a variant of) Gauss-Seidel. Numerical experiments for problems with a triangulation as in Fig. 1 show that lexicographic and red-black GS give grid independent convergence, however, with error reduction factors that are rather bad (even for the Poisson equation, i.e. $\psi = 0$). This is due to the fact that, in the special situation with right triangles, for the unknowns on horizontal and vertical lines we have only 3-point difference stars which is not very favourable for smoothing in $2D$. In view of this we use a variant of Gauss-Seidel in which we use a decoupling of unknowns. This variant, which is denoted by GSD (GS with decoupling), has much better smoothing properties for our system. The following explains the GSD method. If we collect the unknowns on diagonal lines in a vector $\mathbf{x}_d$ and the unknowns on horizontal or vertical lines in a vector $\mathbf{x}_{hv}$, then the system we want to solve can be written in the form

$$\begin{pmatrix} \mathbf{D}_1 & \mathbf{B} \\ \mathbf{C} & \mathbf{D}_2 \end{pmatrix} \begin{pmatrix} \mathbf{x}_d \\ \mathbf{x}_{hv} \end{pmatrix} = \begin{pmatrix} \mathbf{b}_d \\ \mathbf{b}_{hv} \end{pmatrix} \quad \text{with } \mathbf{D}_1,\ \mathbf{D}_2 \text{ diagonal matrices .}$$

This system is equivalent with

$$\begin{pmatrix} \mathbf{D}_1 - \mathbf{B}\,\mathbf{D}_2^{-1}\,\mathbf{C} & \emptyset \\ \mathbf{C} & \mathbf{D}_2 \end{pmatrix} \begin{pmatrix} \mathbf{x}_d \\ \mathbf{x}_{hv} \end{pmatrix} = \begin{pmatrix} \mathbf{b}_d - \mathbf{B}\,\mathbf{D}_2^{-1}\,\mathbf{b}_{hv} \\ \mathbf{b}_{hv} \end{pmatrix} .$$

The 5-point stencil of the matrix $\mathbf{K} := \mathbf{D}_1 - \mathbf{B}\,\mathbf{D}_2^{-1}\,\mathbf{C}$ and the 4-point stencil of $\mathbf{L} := \mathbf{B}\,\mathbf{D}_2^{-1}$ can be given a-priori. In GSD we first apply a number (say σ) of lexicographic GS iterations

to the system $\mathbf{K}\,\mathbf{x}_d = \mathbf{b}_d - \mathbf{L}\,\mathbf{b}_{hv}$, which results in $\mathbf{x}_d^{(\sigma)}$, and then we calculate an approximation of $\mathbf{x}_{hv}$ by replacing $\mathbf{x}_d$ by $\mathbf{x}_d^{(\sigma)}$. So one iteration of GSD for approximating $(\mathbf{x}_d, \mathbf{x}_{hv})$ consists of the computation of $(\mathbf{x}_d^{(\sigma)},\, \mathbf{D}_2^{-1}(\mathbf{b}_{hv} - \mathbf{C}\,\mathbf{x}_d^{(\sigma)}))$.
If $\sigma = 1$ then GSD is just a variant of collective Gauss-Seidel. In our experiments $\sigma = 2$ turned out to be a good choice (but also $\sigma = 1$ is acceptable). If $\sigma = 2$ then the cost of one GSD iteration is comparable with the cost of two usual Gauss-Seidel iterations.

4.5. MG algorithm. Using the bilinear forms $a_k(\cdot,\cdot)$, prolongations P_k and GSD smoother from above we now specify one iteration of the multigrid algorithm on level k $(1 \le k \le k_{\mathbf{max}})$ for approximating $u_k^* \in S_{k,0}$ which satisfies $a_k(u_k^*, \varphi) = l_k(\varphi)$ for all $\varphi \in S_{k,0}$ (l_k a given functional on $S_{k,0}$).

Algorithm (level k).
Step 1. (Pre-smoothing). Apply ν_1 iterations of GSD, resulting in $u_k^{(\nu_1)}$.

Step 2. (Coarse grid correction). Let $u_{k-1}^* \in S_{k-1,0}$ be such that

$$(4.3) \qquad a_{k-1}(u_{k-1}^*, \varphi) = l_k(P_k\,\varphi) - a_k(u_k^{(\nu_1)},\, P_k\varphi) \quad \text{for all} \quad \phi \subset S_{k-1,0}\,.$$

If $k = 1$ then $\tilde{u}_{k-1} := u_{k-1}^*$. If $k > 1$ then compute an approximation $\tilde{u}_{k-1}$ of u_{k-1}^* by using $\mu = 1$ or $\mu = 2$ iterations of the Algorithm on level $k-1$, with starting vector 0, applied to the problem (4.3).
Now put $u_k^{(new)} := u_k^{(\nu_1)} + P_k\,\tilde{u}_{k-1}$.

Step 3. (Post-smoothing). Apply ν_2 iterations of GSD.

Remark 4.4. There is no convergence proof for the algorithm above. The convergence analysis of [2], [3,4] cannot easily be modified for the situation here because the bilinear form $a_k(\cdot,\cdot)$ is nonsymmetric.

5. Numerical results

In this section we apply the algorithm of §4 to some model problems. In all experiments we use a triangulation as in Fig. 1 and a coarsest grid with $h = \frac{1}{4}$. On the finest grid we have $k_{\max}$ and $h_{\min}$ related by $h_{\min} = 2^{-(2+k_{\max})}$. We always take one pre- and one post-smoothing ($\nu_1 = \nu_2 = 1$). The model problems we consider are taken from the papers of Brezzi, Marini, Pietra [5,6]. First we consider a problem with (strong) convection in the whole domain (Experiments 1A, 1B) and then we consider a problem with (strong) convection in part of the domain (Experiment 2). In all our experiments we measure the performance of a method by way of the average reduction factor (arf) which results by taking an arbitrary starting vector (which is the same in all experiments) and then computing the average of the norm reduction of the defect in the first 15 iterations.

Experiment 1A. We take $f = 0$, $\Gamma_0 = \partial\Omega$ with $g(x,y) = 1$ if $x = 0$ or $y = 0$ and $g(x,y) = 0$ if $x = 1$ or $y = 1$. We take $\psi(x,y) = l(-2x - y)$ with $l = 10^3$.
We compare four different methods applied to (2.8) with $k_{max} \in \{1,2,3,4,5\}$:

1: Algorithm of §4 with $\mu = 0$ (so only 2 GSD iterations on the finest grid)

2: Algorithm of §4 with $\mu = 2$ (so MG W-cycle)

3: Algorithm of §4 with $\mu = 0$ and GSD replaced by the usual lexicographic GS

4: Algorithm of §4 with $\mu = 2$ and GSD replaced by the usual lexicographic GS.

The results are shown in Fig. 2.

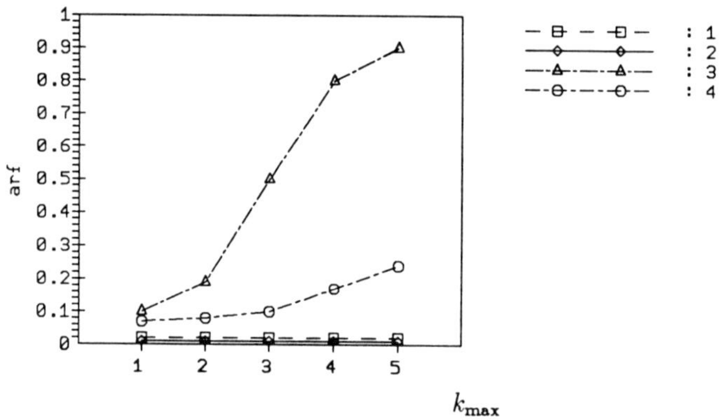

Figure 2.

In this problem with very strong convection and thus a strong upwinding effect in the discretization we expect a suitable GS iteration (without MG) to give good results. In this respect GSD is preferable to lexicographic GS. We also see that the good performance of GSD is not spoiled by going to (very) coarse grids and that lexicographic GS is significantly improved by going to coarser grids.

Experiment 1B. We consider the problem as in 1A but now with $l = 40$ (less convection). The solution of this problem on a 16×16 grid is shown in Fig. 3. We compare the methods 1 and 2 as described in Experiment 1A. The average reduction factors are given in Table 1.

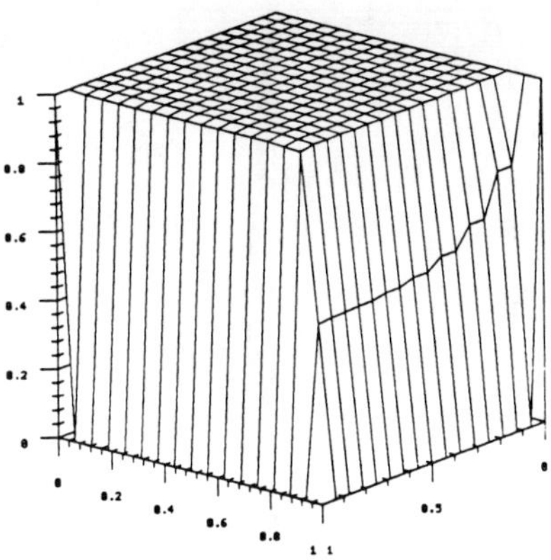

Figure 3.

k_{max}	1	2	3	4	5
method 1	exact	10^{-8}	$5\ 10^{-4}$	0.17	0.90
method 2	exact	10^{-8}	10^{-4}	0.020	0.058

Table 1.

As expected, for k_{max} large the *GSD* deteriorates due to the diffusion. Using coarser grids then gives a significant improvement.

Experiment 2. From the multigrid point of view this experiment is more interesting than Experiments 1A,B because here we consider a problem in which convection dominates in part of the domain and diffusion dominates in another part of the domain.
Again we take $f = 0$. On part of the boundary we have Neumann boundary conditions: $\Gamma_1 = \{(x,y) \mid ((x=1) \text{ and } (y<0.75)) \text{ or } ((y=1) \text{ and } (x<0.75))\}$. On Γ_0 we have $g(x,y)=0$ if $x=0$ or $y=0$ and $g(x,y)=1$ otherwise. The convection is determined by the function $\psi = l\,\psi_0$ with

$$\psi_0(\rho) = \begin{cases} 0 & \text{if } 0 \le \rho \le 0.8 \\ \rho - 0.8 & \text{if } 0.8 \le \rho \le 0.9 \\ 0.1 & \text{if } 0.9 \le \rho \le 1 \end{cases} \qquad \text{with } \rho := ((x-1)^2 + (y-1)^2)^{\frac{1}{2}}$$

So $\|\nabla \psi(\rho)\|_2 = 0$ if $0<\rho<0.8$ or $0.9<\rho<1$ and $\|\nabla \psi(\rho)\|_2 = l$ if $0.8<\rho<0.9$. For $l = 10^3$ the solution of (2.8) on a 16×16 grid is shown in Fig. 4. We consider the algorithm of §4 with different values of μ ($\mu = 0,1,2$) for problems with varying k_{max} ($k_{max} = 1,2,3,4,5$). For different values of μ the results for $l = 10$ and $l = 10^3$ are given in Fig. 5 and Fig. 6 respectively.

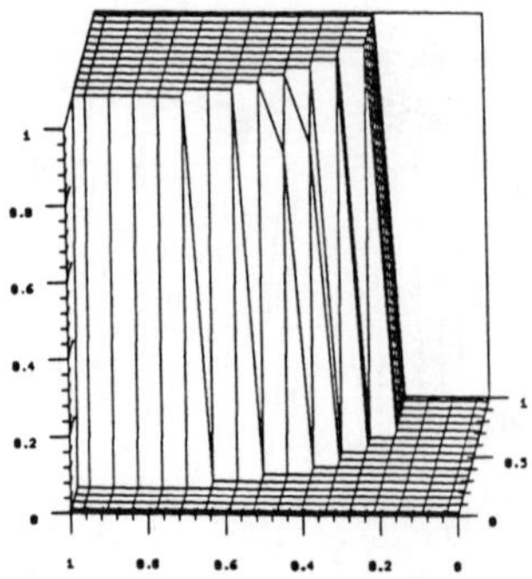

Figure 4.

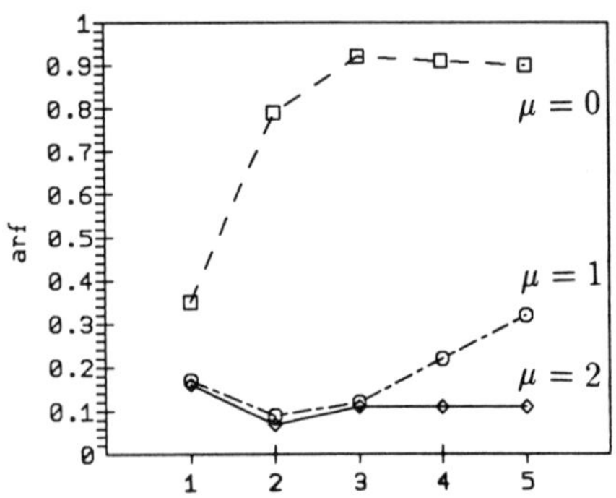

Figure 5.

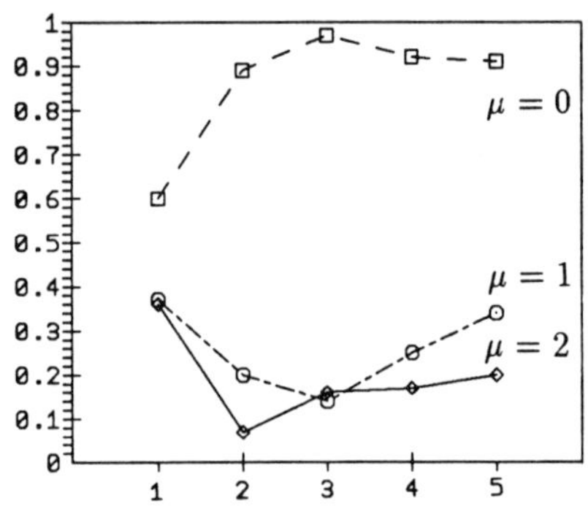

$k_{\mathbf{max}}$

Figure 6.

In Figure 7 we show the results for $k_{\mathbf{max}} = 4$ and with varying μ and l.

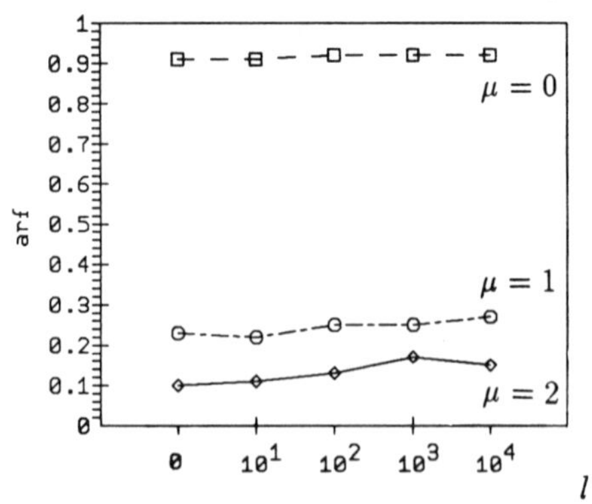

Figure 7.

These results show the typical behaviour one expects from a reasonable multigrid method: "small" (< 0.2) error reduction factors even for (very) fine meshes. Also our method seems to be rather robust with respect to the degree of convection present in the problem. Moreover in the multigrid method very coarse grids can be used.

References

1. Arnold, D.N. and Brezzi, F., Mixed and nonconforming finite element methods, postprocessing and error estimates, M^2AN **19** (1985), 7-32.

2. Braess, D. and Verfürth, R., Multi-grid methods for non-conforming finite element methods, SIAM J. Numer. Anal. **27** (1990), 979-986.

3. Brenner, S.C., Multigrid Methods for Nonconforming Finite Elements, Dissertation, The University of Michigan (1988).

4. Brenner, S.C., An optimal-order multigrid method for $P1$ nonconforming finite elements, Math. Comp. **52** (1989), 1-15.

5. Brezzi, F., Marini, L.D. and Pietra, P., Two-dimensional exponential fitting and application to drift-diffusion models, SIAM J. Numer. Anal. **26** (1989), 1342-1355.

6. Brezzi, F., Marini, L.D. and Pietra, P., Numerical simulation of semiconductor devices, Comp. Meths. Appl. Mech. and Engr. **75** (1989), 493-513.

7. Hackbusch, W., Multi-grid methods and applications, Springer, Berlin-Heidelberg (1985).

8. Marini, L.D. and Pietra, P., New mixed finite element schemes for current continuity equations, Publ. N. 695 Ist. di An. Num. Pavia (1989).

A. Reusken
Technical University Eindhoven
Department of Mathematics and Computer Science
P.O. Box 513
5600 MB Eindhoven
The Netherlands

Adaptive Higher Order Multigrid Methods

U. Rüde *
Institut für Informatik
Technische Universität München
Arcisstr. 21, D-8000 München 2
e-mail: ruede@lan.informatik.tu-muenchen.dbp.de

October 31, 1990

Abstract

Robust, efficient and reliable elliptic solvers require a combination of advanced techniques: multilevel solvers, adaptive mesh refinement, and higher order discretizations. In this paper current research on developing suitable concepts is discussed and is illustrated by numerical experiments. Furthermore an efficient adaptive relaxation and cycling strategy is introduced.

1 Introduction

For the truly effective solution of partial differential equations three basic algorithmic ingredients will be necessary: *Multi-leveling* to provide efficient solvers, *adaptive mesh refinement* to save work where the solution is well-behaved, and *higher order discretizations* to exploit the local smoothness properties of the solution. In this paper attempts to integrate the three concepts will be discussed.

Our work is based on finite elements on unstructured triangular meshes, thus allowing full flexibility in the discretization. Higher order is obtained through a technique called *energy-extrapolation* that will be shown to be equivalent to the use of higher order finite elements. This is achieved implicitly, having the advantage that the higher order stiffness matrix need not be explicitly formed. The extrapolation is motivated by an hierarchical basis analysis, but can also be understood as special form of multigrid τ-extrapolation. It depends on strictly local techniques and is thus applicable on arbitrary, unstructured meshes. We will demonstrate that the energy-extrapolation can improve the accuracy on adaptively refined finite element meshes as generated in our algorithm.

In analogy to adaptive refinement, *adaptive relaxation* will be introduced. This is a generalization of *local relaxation* (see Brandt [5]) and leads to particularly robust and

*partially supported by Deutsche Forschungsgemeinschaft

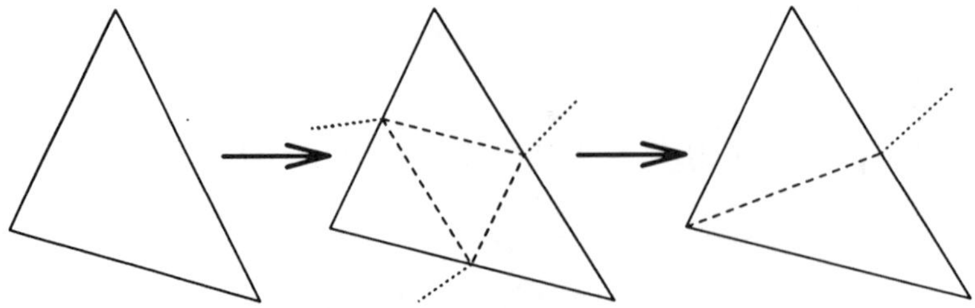

Figure 1: Local refinement in two steps

efficient algorithms. This technique will be extended further to provide switching criteria between relaxation, coarsening and refinement and results in an *adaptive multi-level cycling strategy.*

The paper will be concluded with a presentation and discussion of numerical experiments performed with a new adaptive multi-level finite element code that is being developed to implement these concepts.

2 Refinement

In this section we will briefly describe our refinement algorithms. They are closely related to standard techniques that have been described previously (see Bank [1]), so that only a brief summary will be given.

We assume that the solution process is started with a user supplied, coarse triangulation of the solution domain. Starting from this *primary* triangulation finer meshes are generated by refinement.

The refinement process is performed in two steps. In the first step a *regular* refinement is constructed by halving all edges and connecting the midpoints to form four congruent smaller triangles per coarse triangle (see fig. 1). This step provides the basis for our present error indicator and for the energy extrapolation that requires *one* step of regular refinement.

In a second step points may be eliminated from the fully refined mesh of step one. In this second step care is taken to maintain a consistent triangulation and that the quality of the triangulation does not degrade, by introducing too large obtuse angles. Details of this algorithm are described in Rüde [11]. The final result of this process is quite similar to that of the algorithms of Bank et. al (see [1]) with the difference that our algorithm yields triangulations (and finite element spaces) that are nested.

There remains the question of the error indicator. If a-priory information about the solution behavior cannot, we presently use a simple error indicator similar to Mitchell [8].

Nodes that are discarded in this second step on one level, will be re-introduced in the first step of refining the next level, where they may be discarded again, then they are re-introduced, discarded, and so on.

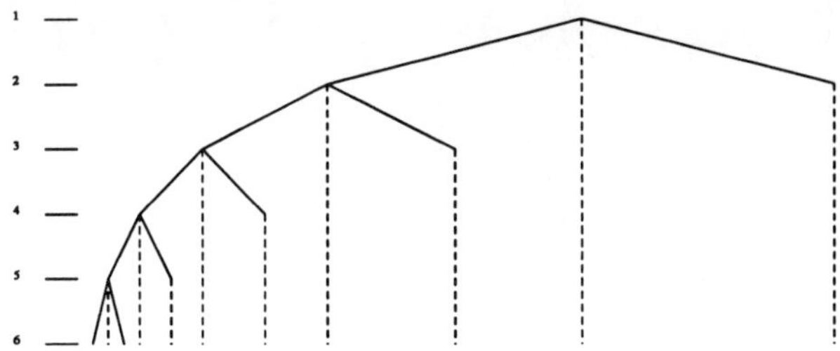

Figure 2: Hierarchical structuring of locally refined mesh

3 Algebraic Solution

In this section we will focus on how to obtain approximate solutions on the increasingly finer meshes. The triangulations generated by the refinement process of the previous section provide the basis for a *nested iteration.* On each level (except the coarsest), a starting guess is obtained by interpolation, which must then be improved by some iteration. We will consider two techniques: The hierarchical basis method (see Yserentant [13]), or multigrid. Both methods are closely related.

The hierarchical basis algorithm has the advantage of having a work estimate that is insensitive to the organization of levels. Nodes are uniquely assigned to one level, and a complete hierarchical basis step consists of relaxing each node exactly once. This clearly limits the work to $O(n)$ per cycle, no matter how the points are distributed across different levels. The price for this nice aspect is a slowly deteriorating convergence rate when the number of levels grows — $O(\log n)$ iterations are necessary to reduce the error by a prescribed factor (in 2-D).

Multigrid usually shows faster convergence. In many cases the rates are bounded away from one, independent of the number of nodes. This is well established (theoretically as well as in experiments) for expensive *cycling strategies*, like W-cycles. For the simpler V-cycles with just one visit to a coarser grid per level (which is more similar to the hierarchical basis method), the convergence may deteriorate, however, it is still faster than in the hierarchical basis algorithm.

The price of better performance is higher cost. A naive implementation of multigrid in a local refinement setting will not have a work estimate proportional to the number of grid points. Assuming, the number of points grows by a factor of two per level only, a V-cycle will still cost $O(n)$ operations but a W-cycle will have work-estimate $O(n \log n)$. If the number of points grows only linearly with the number of levels, even V-cycles cannot be used.

This problem is caused by an unbalanced hierarchical structure as shown for an one-dimensional example in figure 2, where only two new nodes are added per level. In this case the levels must be regrouped for the purpose of multigrid iteration if (asymptotically) optimal work-estimates should be maintained. This corresponds to switching from *locally uniform* patches as basic processing unit to processing in terms of *composite grids.* The

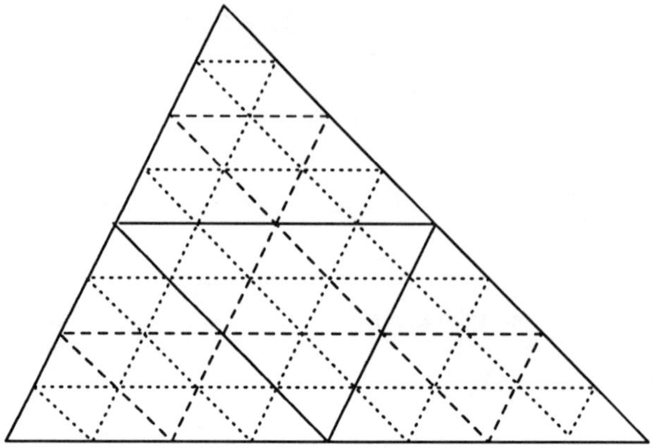

Figure 3: Regular refinement of a triangle

goal must be to group the points into "levels" such that there is a geometric progression in the number of nodes from level to level. This must be accomplished without affecting convergence too seriously.

4 Energy-Extrapolation

Elliptic problems typically have solutions that are smooth in large parts of the solution domain. This (local) smoothness should be exploited by higher order discretizations.

In this section we will briefly introduce the *energy extrapolation* that allows to construct higher order element equations from lower order systems. This technique has been introduced in Rüde [11] [12], see also McCormick and Rüde [7].

The theoretical foundation of extrapolation is based on asymptotic error expansions. In the classical analysis, however, the existence of an error expansion for an approximate solutions depends on the use of regular grids so that extrapolation seems to be impossible for finite element solutions on unstructured meshes. General finite element analysis typically only provides *error bounds* for the solution, not *error expansions*, as required for extrapolation.

The key for developing a suitable extrapolation technique on unstructured meshes is based on the variational principle. In [12] an analysis is developed showing that only one step of regular refinement is necessary per step of extrapolation. The primary mesh may be nonuniform. This analysis will now be briefly summarized.

Consider a single triangle K^0 of the mesh and a sequence of regular refinements $K^n, n = 1, 2, \ldots$ of this particular triangle (see fig. 3). Assume $E[f]$ denotes the *energy* of a function $f : K^0 \to \mathcal{R}$:

$$E = \int_{K^0} a(x, y)(\nabla f)^2 dx dy,$$

We introduce $h = 2^{-n}h_0$ with $h_0 = \text{diam}K^0$. Denote by $P^h f$ the linearly interpolating function of f on the mesh K^n such that $P^h f$ is linear on each triangle of K^n and coincides with f on the nodes of K^n. By calculating $E^n[f] := E[P^h f]$ we obtain a sequence of numerical approximations to $E[f]$. In Rüde [12] it has been shown, that these approximations have an h^2-expansion up to order $2N$ if $f \in C^{2N+3}(K^0)$, and if $a \in C^{2N+2}(K^0)$.

$$E^h[f] - E^0[f] = h^2 e_1 + h^4 e_2 + \cdots + h^{2N} e_N + h^{2N+2} R_{N+1}.$$

Here e_k are constants independent of h and R_{N+1} is a remainder term that satisfies $|R_{N+1}| \leq c$ where c is a constant independent of h.

Using extrapolation, we can thus calculate the energy of any sufficiently smooth function to $O(h^4)$ by one step of extrapolation, to $O(h^6)$ by two steps, etc. The complete analysis can be found in Rüde [12].

For constant coefficients ($a = const$), the remainder term depends on the derivatives of u such that it vanishes if the derivatives of u of order $N + 2$ vanish. In particular, if u is a polynomial of degree 1, (linear), then R_1 vanishes, which means that the error expansion degenerates: linear functions are represented correctly on all refinement levels.

Similarly, when u is polynomial of degree 2, then R_2 vanishes, so that the expansion has exactly one term. This means we can integrate the energy of quadratics correctly by the extrapolation formula:

$$E^0[f] = 4/3E^{h/2}[f] - 1/3E^h[f].$$

This linear combination gives us the element-stiffness-matrix for quadratic functions implicitly. Observe that this carries over from a single (macro-)element to the collection of several elements, providing us with an alternative method to generate the stiffness matrix for quadratic elements by taking a linear combination of the stiffness matrices for linear elements. In this process it is only required that each element is refined regularly *once*.

For the algebraic analogue, this becomes even simpler, when applied to a representation in hierarchical bases. Following the formalism of Rüde [11] we partition the nodal vectors u^k, $k > 1$:

$$u^k = \begin{bmatrix} u_C^k \\ u_F^k \end{bmatrix}, \tag{1}$$

where u_C^k is associated with the nodes of $K^{k-1} \subseteq K^k$ and u_F^k is associated with the nodes that are in K^k but not in K^{k-1}. Assume L^{k-1} is the stiffness matrix for a triangulation with linear elements, and L^k is the stiffness matrix for an associated regular (full) refinement. An approximation for the energy of any smooth function represented by the nodal values is given by $u_C^T L^{k-1} u_C$ and $\left[u_C^T u_F^T\right] L^k \begin{bmatrix} u_C \\ u_F \end{bmatrix}$. In this notation our results on asymptotic expansions of the energy become

$$\left[u_C^T u_F^T\right] L^k \begin{bmatrix} u_C \\ u_F \end{bmatrix} = E[u] + h^2 e,$$

provided u_C are the nodal values of a function that is piecewise quadratic on the coarser triangulation. For the same function u

$$u_C^T L^{k-1} u_C = E[u] + 4h^2 e.$$

such that extrapolation gives

$$\frac{4}{3}\left[u_C^T u_F^T\right] L^k \begin{bmatrix} u_C \\ u_F \end{bmatrix} - \frac{1}{3} u_C^T L^{k-1} u_C = E[u]. \tag{2}$$

As shown above, this is an *exact* representation of the energy for quadratic elements (if $a = const$). With hierarchical bases, the extrapolation of the stiffness equations can be written in an especially elegant form (see Rüde [11]):

$$\begin{bmatrix} L^{k-1} & \frac{4}{3}\hat{X}^k \\ \frac{4}{3}(\hat{X}^k)^T & \frac{4}{3}F \end{bmatrix} \begin{bmatrix} u_C \\ \hat{u}_F \end{bmatrix} = \begin{bmatrix} \frac{4}{3}(P_{k-1}^k)^T f_F \\ 4/3 f_F \end{bmatrix}. \tag{3}$$

In this equation, additionally the discrete analogues of the scalar product (f, u) have been extrapolated, such that the right hand side is also represented with improved accuracy.

The above technique can also be interpreted as multigrid with τ-extrapolation. As it has been shown by McCormick and Rüde [7] the coarse grid part of (3) is equivalent to an extrapolation of the τ-term (see also Brandt [4] or Hackbusch [6]).

The extrapolated system (3) will exhibit improved accuracy, provided quadratic elements have better accuracy for the problem considered. Note that this kind of analysis need not be concerned with regularity assumptions on the solution u. The only thing shown is an equivalence of energy-extrapolation with the use of higher order finite elements. Whether this provides improved solutions can then be decided by finite element theory.

If multigrid is applied to (3), where *all* nodes are relaxed on each level, a problem arises, because the smoothing will be performed on the "wrong" system of equations. The analysis in Hackbusch [6], chapter 14.1, shows that one step of extrapolation is still uncritical and that errors induced by the wrong smoother are negligible.

Extensions to higher order (by applying more steps of extrapolation) are also possible, as demonstrated in the example of section 6. Here the difficulty arises that conventional smoothing destroys the potential gain in accuracy. Unfortunately this is also true for hierarchical basis type smoothing. The reason is the existence of spurious basis functions, whose accuracy is not improved by the extrapolation process. This can be repaired by binding these basis functions by other means, e.g. by using high order interpolation or by using special smoothers as suggested in section 6 and in Rüde [9],

A suitable smoother is simultaneous Kaczmarz-under-relaxation, see [9]. This smoother is *slow* for low frequency modes, so that the order can be raised up to 6 on uniform grids. Unfortunately, the algebraic convergence rates are not satisfactory for practical applications, and along the boundaries a complicated extra treatment is necessary.

5 Adaptive Relaxation and Adaptive Cycling

In this section we introduce more self-adaptive strategies into multigrid algorithms. When one monitors the change in the solution during a relaxation sweep, one often observes that regions of significant changes are strongly localized. Just as in the case of mesh refinement, it is then tempting to save work by eliminating nodes with little significance from further relaxation sweeps.

This idea has already been presented in Rüde [10]. Here it will be adapted to the finite element equations and extended to adaptive cycling strategies.

When a point is relaxed, the associated residual becomes zero. Therefore a further relaxation sweep will only change this point, if in the meantime neighboring points are changed such that a significant residual is generated. This criterion can be used for deciding, whether points should be relaxed or not.

In a relaxation sweep we distinguish between *active* and *inactive* nodes. Active points are relaxed and then set to inactive. If the change due to relaxation was large, we assume that the residuals at *neighboring* points have changed (and possibly grown), so that we mark them as active, unless they already are.

Of course, a change at one point need not always *increase* the residual of its neighbors, but here a conservative strategy is quite suitable. Relaxing a point is computationally not more expensive than determining the precise size of the residual.

Each state of the relaxation is characterized by a *set of active points* and can be continued until no active points remain. Of course the relaxation must be suitably initialized by finding an initial set of active nodes. If nothing else is known, one can start with all nodes set active.

Up to now we have not considered the efficiency of possible implementations. It can be easily seen, that simply augmenting each node with a flag for marking its active/inactive-state is not sufficient. Finding the active nodes must then be implemented by sweeping through *all* nodes. This is unsatisfactory, if the relaxation is strongly localized, and a small subregion is relaxed several times. In this case the majority of the CPU-time might be spent on testing flags without doing any numerical work.

The solution is to introduce *lists of active nodes.* At each time a list contains the nodes to be relaxed in the present sweep, and a new list is constructed, pointing to all the nodes that need to be relaxed in the next sweep. Such a list of active nodes must contain each node at most once (essentially data structures for *sets* must be implemented). Therefore flags for each node are still useful in order to avoid the need to search the list, for finding multiple occurrences. Based on these lists an adaptive smoothing can be implemented with cost proportional to the actual number of points relaxed.

Note that this processing in separate *sweeps* implements a slightly modified algorithm. It is quite possible in this version that a point is entered into the active list for a future relaxation sweep, though it will still be relaxed in the present sweep and get a zero residual. The changes of the present sweep have only effect on the nodes for the next sweep, not on the current one.

This could again be repaired, however, the experiments in section 6 show that the

```
g_new= empty_list
FOR_ALL_ACTIVE_NODES p OF_GRID g
        SET active_flag OF p TO false
END_ALL_NODES
FOR_ALL_ACTIVE_NODES p OF_GRID g
        Real t= relax(p)
        IF | p - t | > limit THEN
                FOR_ALL_NEIGHBORS pn OF p
                        IF NOT active_flag OF pn THEN
                                APPEND pn TO g_new
                                SET active_flag OF pn TO true
                        END_IF
                END_ALL_NEIGHBORS
        END_IF
        p= t
END_ALL_NODES
```

Figure 4: Adaptive relaxation algorithm

present implementation already behaves quite satisfactory. A careful systematic examination of the algorithmic variants is not available at present. Figure 4 shows a pseudo-code fragment for one sweep of dynamic relaxation. In the algorithm of Rüde [10], the the adaptive relaxation idea hs been implemented for uniform grids using gridlines as basic processing unit.

Using dynamic relaxation in this form has great advantages. It has been shown (see Brandt [5]) that multigrid algorithms perform with good convergence rates even in cases where the usual regularity assumptions are violated, provided that local relaxations are added along the boundary. Our dynamic smoothing will automatically add the correct amount of extra relaxation where necessary.

Next we will extend the above ideas to construct a self-adaptive multigrid-cycling strategy. We start our derivation from a hierarchical basis algorithm, which differs from multigrid by relaxing only points that are not also represented on coarser levels. Our dynamic relaxation algorithm will now frequently mark points which cannot occur in the relaxation on the present level. We take this into account by appending those points to the active-list of their (coarser) level. When now the relaxation on the present level terminates, usually some points on the next coarser levels will have been marked. This constitutes our switching criterion: We proceed to the next coarser grid, if there are active points, otherwise we go to processing the next finer grid. On the coarser grid the dynamic relaxation is initialized with the list of active nodes that has been determined from relaxation on all finer grids.

When a coarse grid correction terminates and the algorithm returns to a finer grid, a

	Extrapolation steps		
Level	0	1	2
3	1.295e-02	1.751e-03	4.178e-04
4	3.219e-03	1.010e-04	9.718e-06
5	8.036e-04	6.311e-06	1.572e-07
6	2.008e-04	3.982e-07	2.624e-09
Convergence rate	2.001e+00	3.986e+00	5.905e+00

Table 1: Results for higher order multigrid algorithms

new set of active nodes for the finer grid is constructed by examining the change in the solution by the coarse grid correction.

These ideas lead to fully adaptive multilevel algorithms that choose their relaxation ordering dynamically, choose the cycling strategy dynamically, and are furthermore capable of performing fully *localized* cycles. For example in the presence of singularities, the iteration on the global solution may have become stable, while a neighborhood of the singularity needs further improvement. The dynamic algorithm may then implicitly perform a separate V-cycle for such a subdomain.

Though at present there is no analysis available for these methods, the potential efficiency will be illustrated by a numerical example in the following section.

6 Numerical Experiments

In this section we will illustrate the above techniques with numerical experiments. Our first model problem is

$$\begin{aligned} \Delta u &= 2\pi^2 \sin(\pi x)\sin(\pi y) && \text{on } (-1,1)^2, \\ u &= \sin(\pi x)\sin(\pi y) && \text{on } \partial[-1,1]^2. \end{aligned}$$

with analytical solution $u = \sin(\pi x)\sin(\pi y)$. First, we compare τ-extrapolation algorithm of different order on uniform grids. The numerical solution is obtained by applying a sufficient number of (static) V-cycles, because we are only interested in the approximation accuracy of our implicit discretization processes. The error norm displayed in table 1 is a (discrete) L_2-norm. As explained in section 4, for higher than fourth order the smoother becomes a problem. Here we use a Kaczmarz-Jacobi underrelaxation, because it preserves up to sixth order on uniform grids. The effectivity of the method is quite poor and at this stage the experiment only gives an indication that higher order extrapolation based multilevel algorithms are possible.

In another experiment we demonstrate the combination of local refinement with extrapolation. Our second example is

$$\begin{aligned} \Delta u &= 0 && \text{on } (-1,1)^2, \\ u &= \tfrac{\cos(2\pi(x-y))\sinh(2\pi(x+y+2))}{\sinh 8\pi} && \text{on } \partial[-1,1]^2. \end{aligned}$$

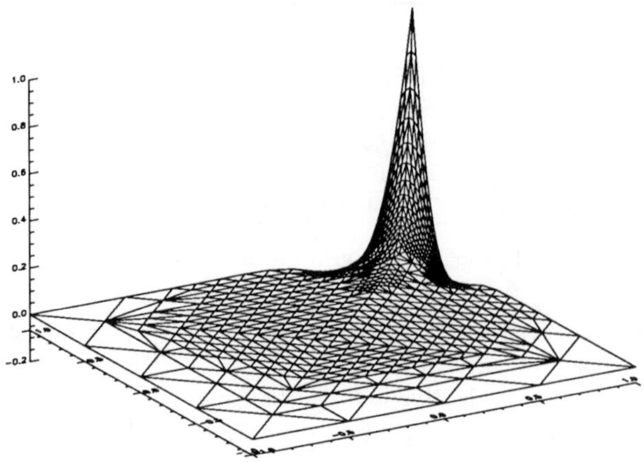

Figure 5: Locally refined finite element mesh for example 2

The solution on a locally refined grid is depicted in figure 5. For this problem we compare two algorithms, both using local refinement, but the first applied without, the second with energy extrapolation. In each of these algorithms a step of full refinement is performed. On the new grid two V-cycles are are performed. In algorithm 2, the *extrapolated* system is used. After that, nodes that have not changed significantly are deleted, before another step of full refinement is performed. The results are summarized in table 2. The given figures show that the method with extrapolation is more efficient for the finer grids: Comparable accuracy is obtained with much fewer nodes when extrapolation is used. Remember that the algorithm is driven by a prescribed accuracy, that is decreased by a factor two from level to level. Thus the better algorithm shows its superiority not by providing lower errors, but by needing fewer nodes to reach its goal.

Finally, in a third example problem, we demonstrate the operation of adaptive refinement, in combination with adaptive relaxation and cycling. The model problem has a singularity and is given by:

$$\begin{aligned} \Delta u = & \quad 0 & \text{on } \Omega = (-1,1)^2 \setminus (-1,0)^2, \\ u = & \quad r^{2/3}\sin(2/3\phi) & \text{on } \partial\Omega. \end{aligned}$$

The algorithm proceeds from coarser to finer levels. Similar to the previous example, only points are kept, whose values change by more than a prescribed limit in a first (hierarchical) relaxation sweep. From level to level of the nested iteration, this limit value is reduced by a factor of two. As described in section 5, the adaptive relaxation (and cycling) is characterized by the limiting number, when (neighboring) nodes are made active. This number is determined from which level the cycling was started, and is calculated as a fourth of the refinement limit parameter introduced above. This is

	without extrapolation		with extrapolation	
Level	# nodes	L_2-error	# nodes	L_2-error
2	41	7.827e-2	41	6.897e-2
3	145	1.141e-2	112	1.803e-2
4	329	2.801e-3	329	7.804e-4
5	726	6.781e-4	418	7.394e-5
6	1681	1.702e-4	567	8.338e-5
7	4134	4.390e-5	828	4.877e-5
8	10993	1.229e-5	1447	2.528e-5

Table 2: Local refinement with and without extrapolation

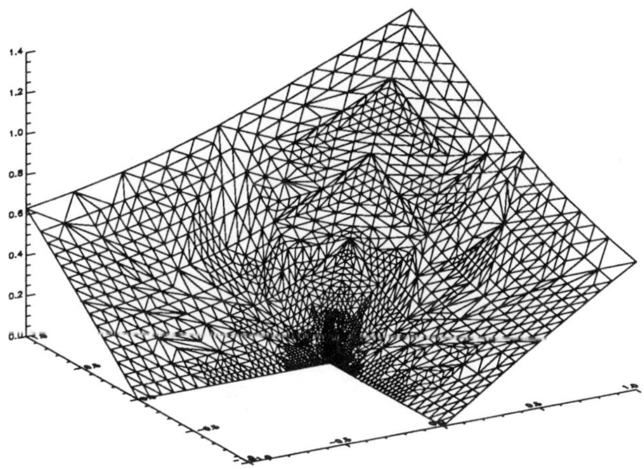

Figure 6: Locally refined finite element mesh on L-region

Level	Number nodes	L_2-Error	Number relax.	Relax./Node
1	15	3.654e-02	9	0.60
2	24	1.678e-02	58	2.42
3	41	1.036e-02	151	3.68
4	87	5.941e-03	402	4.62
5	197	2.646e-03	1082	5.49
6	353	1.457e-03	2325	6.59
7	756	7.415e-04	5364	7.10
8	1450	3.990e-04	11161	7.70
9	2830	1.946e-04	22446	7.93
10	5709	1.019e-04	45815	8.03
11	10773	5.243e-05	91996	8.54
rate	0.99	-0.96	1.06	

Table 3: Performance of adaptive multigrid

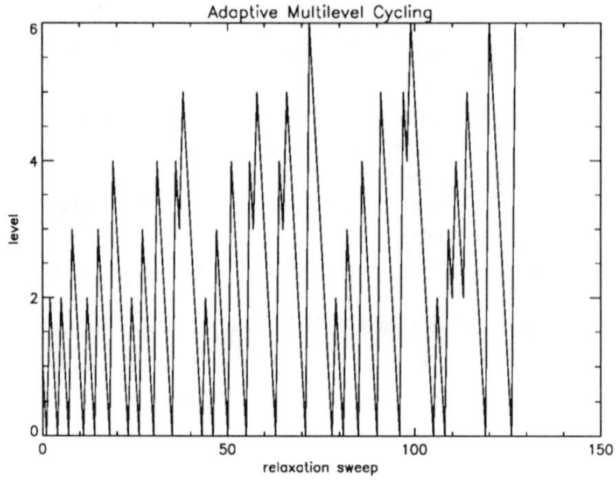

Figure 7: Cycling Strategy

motivated by the desire to have the algebraic errors somewhat (but not too much) smaller than the differential errors. A numerical solution on a locally refined finite-element mesh is shown in figure 6. The performance of the algorithm is summarized in table 3. The columns show the number of nodes, the L_2-error after an adaptive multilevel cycle on the level, the total number of relaxations to obtain this accuracy, and the relation number of nodes vs. number relaxations. The last row gives $\log_2$ of the progression, averaged over the last five levels. The refinement algorithm is driven by the requirement to reduce the error by a factor of 2 per level. Note that this is reflected in the number L_2-error in column two of table 3. Furthermore, the number of nodes grows with the same factor, as does (approximately) the computational work, measured in points relaxed.

On the other hand the number of relaxations per point is still slowly increasing with the number of levels. From the given figures it is impossible to decide whether this relation has an upper bound, or will continue to grow according to some logarithmic dependence on the number of nodes.

Another quite interesting aspect is the particular cycling generated by the adaptive strategy. This is depicted in figure 7. On each level only the hierarchical nodes are relaxed, so that a V-cycle has a work estimate comparable to single relaxation sweep of a conventional multigrid algorithm. Thus the sawtooth-shape that is prominent in fig. 7 corresponds to a conventional multigrid V-cycle.

7 Acknowledgement

The author is grateful for the hospitality of the Computational Mathematics Group at University of Colorado at Denver.

References

[1] R. Bank: *PLTMG Users' Guide, Edition 4.0*, Department of Mathematics, University of California at San Diego, 1985

[2] D. Bai and A. Brandt: *Local Mesh Refinement Multilevel Techniques*, SIAM J. Sci. Math. Comp. 8 (1987), pp. 109-134

[3] R. Bank, T. Dupont and H. Yserentant: *The Hierarchical Basis Multigrid Method*, Konrad Zuse Zentrum Preprint, SC-87-2, April 1987, Berlin

[4] A. Brandt: *Multigrid Techniques: 1984 Guide with Applications to Fluid Dynamics*, GMD Studien 85, 1984

[5] A. Brandt: *Rigorous Local Mode Analysis of Multigrid*, Preliminary Proceedings of the Fourth Copper Mountain Conference on Multigrid Methods, April 9-14, 1989

[6] W. Hackbusch: *Multigrid Methods and Applications*, Springer Verlag, Berlin, 1985

[7] S. McCormick and U. Rüde: *On Local Refinement Higher Order Methods for Elliptic Partial Differential Equations*, submitted to the Journal of High Speed Computing.

[8] W. F. Mitchell, *Optimal Multilevel Iterative Methods for Adaptive Grids*, to be published in the proceedings of the First Copper Mt. Conference on Iterative Methods, April 1-5, 1990.

[9] U. Rüde, *Multiple tau-Extrapolation for Multigrid Methods*, Institut für Informatik, Technische Universität München, I-8701, 1987

[10] U. Rüde: *Local Corrections for Eliminating the Pollution Effect of Reentrant Corners*, Proceedings of the Fourth Copper Mt. Conference on Multigrid Methods, April 9-14, 1989.

[11] U. Rüde: *The Hierarchical Basis Extrapolation Method*, to be published in the proceedings of the First Copper Mt. Conference on Iterative Methods, April 1-5, 1990.

[12] U. Rüde: *Extrapolation Techniques for Constructing Higher Order Finite Element Methods*, to be published

[13] H. Yserentant, *On the Multi-level Splitting of Finite Element Spaces*, Numer. Math. 49, pp 379-412, 1986

International Series of Numerical Mathematics, Vol. 98, © 1991 Birkhäuser Verlag Basel

Shape from Shading using Parallel Multigrid Relaxation

by

Wim Sweldens and Dirk Roose

Abstract: *The aim of a shape from shading algorithm is to reconstruct the shape of an object from a two-dimensional image of that object. This can be formulated as a variational problem that leads to a system of non-linear partial differential equations. In order to find a solution, accurate information about the occluding boundary was required in early methods. This is often not available or difficult to obtain. Omitting occluding boundary information for scenes lighted from overhead leads to ambiguous problems that cannot be solved with standard relaxation methods. This difficulty can be overcome by using the full multigrid scheme. Since most of the computations are local, grid partitioning and parallel processing can be exploited successfully. We report on results for synthetic and camera images, obtained on the Intel iPSC/2, a distributed memory parallel computer.*

1. Problem definition

1.1 Introduction

The purpose of the shape from shading problem is to reconstruct the three-dimensional shape of an object from a two-dimensional image of this object. Applications of shape from shading can be found in robot vision and relief reconstruction from aerial photos. Recently the most important papers on this subject are reprinted in a book [8]. It is important to distinguish shape from shading from *stereo matching* that tries to determine the spatial position of several objects using two slightly different images.

First we take a closer look at the image forming process. We consider a scene containing an eye (human eye or camera), one light source, and one object (Figure 1). Light rays from the source reflect on the surface of the object and reach the eye where the image is formed.

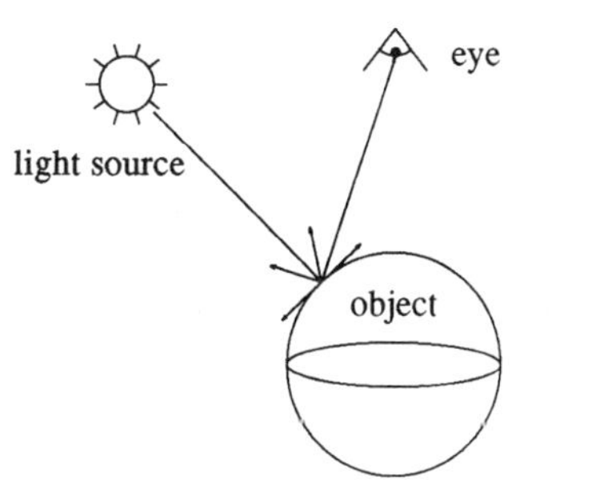

Figure 1. Scene geometry.

The resulting image depends on scene geometry, surface reflectance properties, scene illumination and local surface orientation. If the light source and eye are sufficiently distant, light rays can be assumed to be parallel and the position of the source and the eye can be given by a direction. The Z-axis is chosen parallel to the viewing direction. The image is given by an irradiance function $I(x,y)$ over a domain $\Omega \subset \mathbb{R}^2$ while the shape of the object is given by a height function $z(x,y)$ over Ω. So the shape from shading problem can be formulated as: given an image of an object $I(x,y)$, reconstruct its shape $z(x,y)$.

1.2 The reflectance map and image irradiance equation

Since essential information is lost during the image forming process, it is necessary to make the following assumptions:

- the direction of the light source is known,
- the shape of the image is smooth,
- the reflectance properties of the surface are homogeneous,
- the illumination is uniform.

Under these assumptions there remains a relationship between surface orientation and image brightness [10]. Surface orientation is given by means of the first partial derivatives of the height function $z(x,y)$. We use the notation p and q:

$$p(x,y) = \frac{\partial z}{\partial x}(x,y) \quad \text{and} \quad q(x,y) = \frac{\partial z}{\partial y}(x,y). \tag{1}$$

The pq-space is referred to as the *gradient space*. Every point (p,q) of the gradient space corresponds to a particular surface gradient. If the surface belongs to C^2, p and q are dependent since cross derivatives have to be equal:

$$\frac{\partial p}{\partial y}(x,y) = \frac{\partial q}{\partial x}(x,y). \tag{2}$$

This equation is called the *integrability constraint* [9]. If it holds, it ensures the existence of a smooth surface z that satisfies (1).

The relationship between image brightness $I(x,y)$ and surface orientation (p,q) is expressed by a function $R(p,q)$, referred to as the *reflectance map* [9]:

$$I(x,y) = R(p(x,y), q(x,y)). \tag{3}$$

This equation is called the *image irradiance equation*.

1.3 The Lambertian reflectance model

A simple reflectance map exists for objects with a Lambertian surface. A patch of a Lambertian surface looks equally bright from all viewing directions and its brightness is proportional to the light flux falling on it. Thus the reflection is proportional to the cosine of the angle α between surface normal and incident light ray. The direction of the surface normal is given by $(1,-p,-q)$ and let the direction of the light source be $(1,-p_s,-q_s)$. Then the following expression holds for the Lambertian reflectance map R_L:

$$R_L(p,q) = \cos\alpha = \frac{1 + p\,p_s + q\,q_s}{(1+p^2+q^2)^{1/2}\,(1+p_s^2+q_s^2)^{1/2}}. \tag{4}$$

We consider scenes that are lighted from overhead. This means that the direction of the light source is parallel to the viewing direction (Z-axis) so p_s and q_s are equal to zero:

$$R_L^o(p,q) = \frac{1}{\sqrt{1+p^2+q^2}}. \tag{5}$$

Notice that $R_L^o(p,q) = R_L^o(-p,-q)$. This means that if there exists a solution z of the image irradiance equation, also $-z$ will satisfy this equation. Those two solutions correspond to two different interpretations of the image. If we look at the synthetic image of a Lambertian sphere (Figure 2), we can impossibly distinguish whether this is made from either a convex or a concave hemisphere. This ambiguity is typical for scenes lighted from overhead.

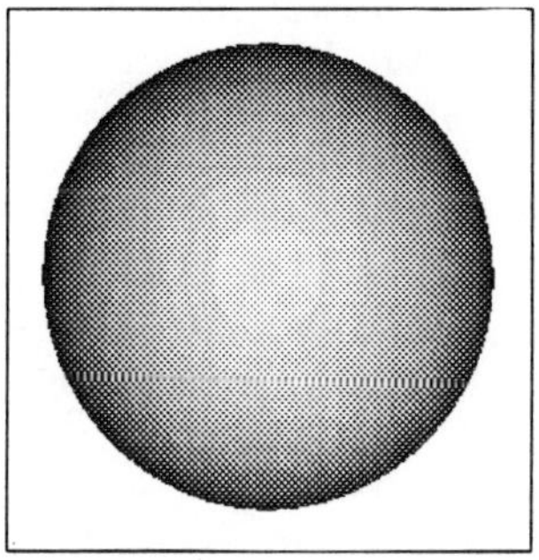

Figure 2. Lambertian sphere.

2. Variational approach to shape from shading

2.1 *Solution method*

The problem is traditionally solved in two steps [9]. In a first step the partial derivatives p and q of the height are calculated and in a second step the height z is reconstructed from p and q. The partial derivatives p and q can be found by solving the system consisting of the image irradiance equation (3) and the integrability constraint (2). Since images can be noisy and the assumptions made are sometimes not perfectly satisfied, there might be no theoretical solution to this system. Therefore a solution can best be found by minimising the residual of these equations. We use one of the variational solution schemes proposed by Horn and Brooks [9]. For the first step we solve:

$$\min_{p,q} \Gamma_1(p,q) \quad \text{with} \quad \Gamma_1(p,q) = \iint_\Omega \left[(R(p,q) - I(x,y))^2 + \lambda (p_y - q_x)^2 \right] dxdy. \tag{6}$$

The weight of the integrability constraint is given by the fixed parameter λ. If λ is chosen too big or too small, the problem will be ill-posed.

Since the solution of the first step might not perfectly satisfy the integrability constraint, we also use a variational approach in the second step:

$$\min_{z} \Gamma_2(z) \quad \text{with} \quad \Gamma_2(z) = \iint_\Omega \left[(z_x - p)^2 + (z_y - q)^2 \right] dxdy. \tag{7}$$

The solution of a variational problem has to satisfy a corresponding Euler equation [4]. The Euler equations of the first variational problem (6) form a system of two non-linear partial differential equations:

$$\begin{cases} (I(x,y)-R(p,q))\,R_p(p,q)+\lambda(p_{yy}-q_{xy})=0 \\ (I(x,y)-R(p,q))R_q(p,q)+\lambda(q_{xx}-p_{yx})=0. \end{cases} \tag{8}$$

The Euler equation of the second variational problem (7) is a Poisson partial differential equation:

$$z_{xx}+z_{yy}=\Delta z=p_x+q_y. \tag{9}$$

2.2 Boundary information

These Euler equations cannot be solved without appropriate boundary conditions. There are two possibilities: one can impose boundary conditions either at the *image boundary* or at the *occluding boundary* (Figure 3). The occluding boundary can be defined as the set of points $(x,y)\in\Omega$ for which the tangent plane at the point $(x,y,z(x,y))$ of the object is parallel to the viewing direction (Z-axis). Consequently, the surface normal $(-p,-q,1)$ is perpendicular to the Z-axis. This is only possible if at least one of the parameters p and q becomes unbounded.

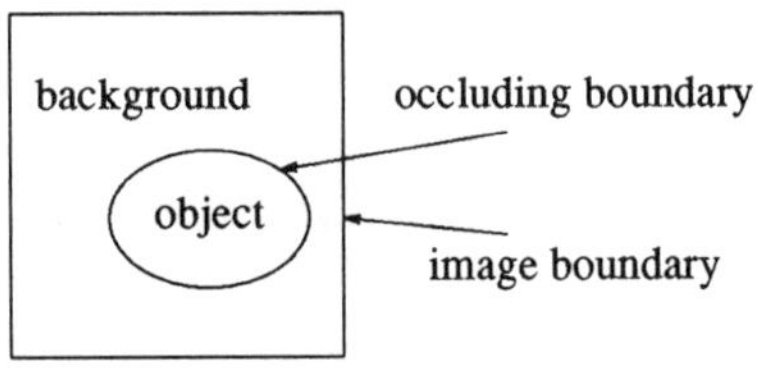

Figure 3. Image structure.

Ikeuchi and Horn [10] propose to provide Dirichlet boundary information on the occluding boundary. This is not possible in the gradient space since p or q become unbounded at this occluding boundary. Therefore the problem is solved in a transformed gradient space. This method however has three drawbacks:

- It becomes too difficult to express the integrability constraint in this transformed gradient space. The integrability constraint in the variational expression is therefore replaced by a general smoothness term. This however essentially changes the problem, since if there exists a exact solution to the equations, this will not be a minimum of the integral.
- When working with images taken by a camera (in contrast to synthetic ones) it is difficult to find the precise position of the occluding boundary and even more difficult to provide the boundary information since this requires the calculation of the normal to the occluding boundary. Even a small error in the position of the occluding boundary can yield substantial errors in the calculation of the occluding boundary normal. Incorrect boundary information inevitably leads to incorrect solutions.
- This solution technique can lead to irregular domains.

With these boundary conditions, Ikeuchi and Horn use a Picard iterative scheme to solve the Euler equations. This scheme can also be used as a smoother in a multigrid procedure as proposed by Terzopoulos [11].

The second possibility, proposed by Battiti, is to provide boundary information on the image boundary [1]. This approach, which we follow, does not bother with the theoretically infinite values of p and q on the occluding boundary and trusts that they will automatically be replaced by "large" values. This has three major consequences:

- The domain Ω where the solution has to be sought, does not correspond only to the irradiance of the object but also to the background irradiance. The assumption that the object is surrounded with a flat plane, leads to boundary conditions $p = q = 0$ for the first step and $z = 0$ for the second.
- Due to the homogeneous boundary conditions, two solutions z and $-z$ remain possible. In the previous case it was possible to avoid this ambiguity through the non-homogeneous boundary conditions.
- Since we assume a smooth solution, the accuracy will be low near the occluding boundary where the theoretical solution has an orientation discontinuity.

2.3 Discretisation and iterative scheme

We will not discuss solutions methods for the second step since this is a simple Poisson equation. Discretisation of the equations of the first step (8) using central differences yields a nine point stencil (h is the grid distance in both directions):

$$\begin{cases} (I_{ij} - R(p_{ij},q_{ij}))R_p(p_{ij},q_{ij}) + \dfrac{\lambda}{h^2}[2(\bar{p}_{ij} - p_{ij}) - \hat{q}_{ij}] = 0 \\ (I_{ij} - R(p_{ij},q_{ij}))R_q(p_{ij},q_{ij}) + \dfrac{\lambda}{h^2}[2(\bar{q}_{ij} - q_{ij}) - \hat{p}_{ij}] = 0, \end{cases} \tag{10}$$

with (analogous for q)

$$\hat{p}_{ij} = \frac{p_{i+1,j+1} + p_{i-1,j-1} - p_{i+1,j-1} - p_{i-1,j+1}}{4} \quad \text{and} \quad \bar{p}_{ij} = \frac{p_{i,j+1} + p_{i,j-1}}{2}.$$

Isolating the center linear terms yields a Picard iterative scheme [9]. Updating can be done in a Gauss-Seidel fashion:

$$\begin{cases} p_{ij}^{k+1} = (I_{ij} - R(p_{ij}^k,q_{ij}^k))R_p(p_{ij}^k,q_{ij}^k)\dfrac{h^2}{2\lambda} + \bar{p}_{ij}^k - \hat{q}_{ij}^k/2 \\ q_{ij}^{k+1} = (I_{ij} - R(p_{ij}^k,q_{ij}^k))R_q(p_{ij}^k,q_{ij}^k)\dfrac{h^2}{2\lambda} + \bar{q}_{ij}^k - \hat{p}_{ij}^k/2, \end{cases} \tag{11}$$

with (analogous for q)

$$\hat{p}_{ij}^k = \frac{p_{i+1,j+1}^l + p_{i-1,j-1}^l - p_{i+1,j-1}^l - p_{i-1,j+1}^l}{4} \quad \text{and} \quad \bar{p}_{ij}^k = \frac{p_{i,j+1}^l + p_{i,j-1}^l}{2}.$$

with $l = k$ or $k+1$ depending on the updating order (lexicographic, red-black,...).

Another possibility is to use Newton-Raphson to solve (10).

2.4 Convergence of standard relaxation methods

Whether standard relaxation schemes will lead to a solution and which of the two possible solutions will be found, depends on the initial approximation. No convergence problems occur if initial values having the right sign for one of the two solutions are used. In the case of the sphere, an initial approximation with p negative on the right half and positive on the left half, will ensure convergence to the concave hemisphere. But in general the sign of p and q is not known in advance. Taking zero as initial approximation is impossible since this is a local maximum of the integral and thus satisfies the Euler equations. In our experiments we use small ($O(10^{-6})$) randomly generated initial values. This technique is known as symmetry breaking.

For the image of a Lambertian sphere (Figure 2), the result after 200 standard relaxation steps (11) starting from this random initial approximation is shown in Figure 4. No solution is found. The reason for this has to be sought in the ambiguity. As indicated before two solutions are possible: a convex and concave hemisphere. Starting from these initial values, some solution patches choose for one interpretation of the image and others for the other one. Each patch tries to extend its interpretation. This struggle continues and possibly no solution is found even after a large number of relaxation steps. Analogous results were obtained by Battiti [1].

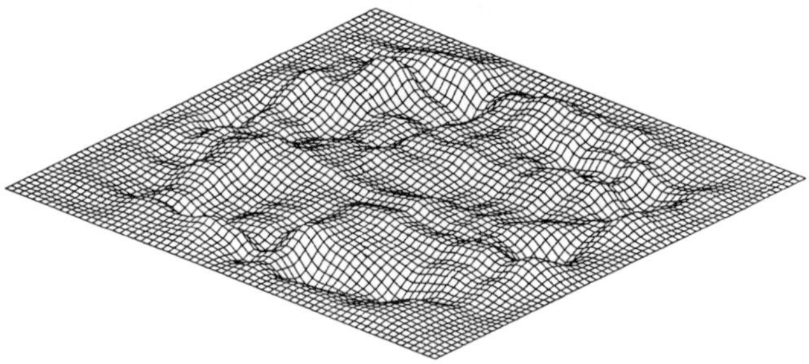

Figure 4. Result after 200 relaxation steps.

2.5 Multigrid

The full multigrid scheme can help to solve the ambiguity problem. In this scheme the equations are first solved on the coarsest grid. On this grid, information can propagate in a few relaxation steps over the whole area. After some relaxations, the whole area chooses for one interpretation: either the convex or concave one. This decision is then passed to the finer grids so that everywhere the same interpretation is followed. We report on results obtained with full multigrid using the following parameters:

- 4 grids (63×63, 31×31, 15×15 and 7×7 grid points),
- 2 pre and 2 post smoothing steps,
- 10 relaxations steps on the coarsest grid (7×7 grid points),
- full weighting as restriction,
- biquadratic interpolation as prolongation,
- V-cycle,
- 4 multigrid steps on each level.

For the Lambertian sphere, we obtained convergence for $\lambda \in [0.005, 0.02]$. If the parameter λ is chosen outside this interval, the problem will become ill-posed and convergence is lost. The result obtained after the full multigrid cycle for $\lambda = 0.01$ is shown in Figure 5.

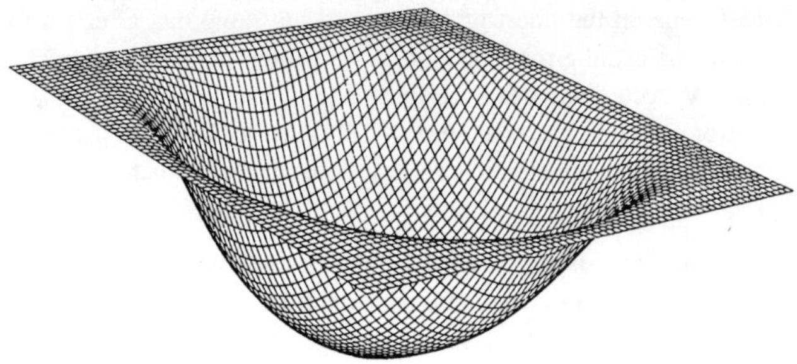

Figure 5. Result obtained after the full multigrid cycle.

The accuracy of the solution is good in the inner region. The error on the computed height in the middle of the sphere is less than one percent. On the other hand the accuracy is, as predicted, low near the occluding boundary.

Table 1

Convergence history of the full multigrid scheme for the Lambertian sphere.

Level	Iteration	Residual L_∞ norm	Residual L_2 norm	Convergence factor L_∞ norm	Convergence factor L_2 norm
0	0	1.32e-06	6.02e-07	-	-
0	1	7.17e-06	1.84e-06	5.4	3.1
0	2	2.33e-04	5.88e-05	32.5	31.8
0	3	8.11e-03	2.11e-03	34.7	35.9
0	4	1.42e-01	5.09e-02	17.6	24.0
1	0	2.22e-01	7.17e-02	-	-
1	1	9.52e-02	2.76e-02	0.42	0.38
1	2	3.77e-02	1.03e-02	0.39	0.37
1	3	2.31e-02	6.07e-03	0.61	0.58
1	4	1.68e-02	4.28e-03	0.72	0.70
2	0	1.61e-01	4.24e-02	-	-
2	1	1.28e-01	1.92e-02	0.79	0.45
2	2	8.42e-02	1.34e-02	0.65	0.69
2	3	5.85e-02	1.00e-02	0.69	0.74
2	4	4.23e-02	7.88e-03	0.72	0.78
3	0	2.64e-01	4.45e-02	-	-
3	1	2.13e-01	2.47e-02	0.80	0.55
3	2	1.55e-01	1.97e-02	0.72	0.79
3	3	1.43e-01	1.64e-02	0.92	0.83
3	4	1.34e-01	1.39e-02	0.93	0.85

The convergence history of the full multigrid scheme is given in Table 1. On the coarsest grid (level 0) the residual increases since computations start from an initial approximation close to zero that is also an (unstable) solution of the Euler equations. On the other levels, the convergence factor is 0.4 ... 0.5 for the first step, but convergence slows down during the other three steps. At the end of the full multigrid cycle the residual is still relatively high. It is possible to reduce the residual further with additional V-cycles or additional Gauss-Seidel steps on the finest grid. Experiments show that after the full multigrid cycle 19 additional V-cycles yield the same error reduction as 400 additional Gauss-Seidel relaxation steps. The convergence factor for a V-cycle increases even further to ≈ 0.98. Applying these additional steps is not very useful since the error on the image is at least $O(10^{-3})$ due to the discretisation with 256 gray values. Moreover, experiments show that the solution does not change very much when the residuals are further reduced.

The poor convergence rate is not caused by the discontinuity of the problem at the occluding boundary, since examples without a discontinuity show similar convergence rates. The non-ellipticity of the problem can cause the slow convergence, see also the results of Carter and Asher [3].

The convergence factors of the Newton-Raphson iterative scheme applied to (10) with one Newton step in each iteration, are somewhat better. However, this scheme requires more calculations than the Picard scheme. Our experiments indicate that a Newton scheme does not lead to shorter execution times to achieve a certain accuracy.

3. Parallel implementation

3.1 The Intel iPSC/2

The Intel iPSC/2 is a distributed memory parallel computer with nodes connected in hypercube topology. Each node consists of a 80386/387 processor, memory and a separate communication processor. Communication is done with an asynchronous message passing system. Information on distributed memory parallel computers and parallel algorithms can be found in e.g. [6], [13], [14].

3.2 The parallel solution method

The image domain is divided into as many rectangular subdomains as there are processors available. Each node is assigned to one subdomain and is responsible for the computations on the grid points in this subdomain for all multigrid levels. In order to perform the calculations (restriction, relaxation, etc) for a grid point on the border of a subdomain, values of grid points assigned to other nodes are required. An overlap area of one grid line is stored in each node to avoid separate communication for each point (Figure 6). The actual image data are distributed over the nodes by a parallel data distribution algorithm [5].

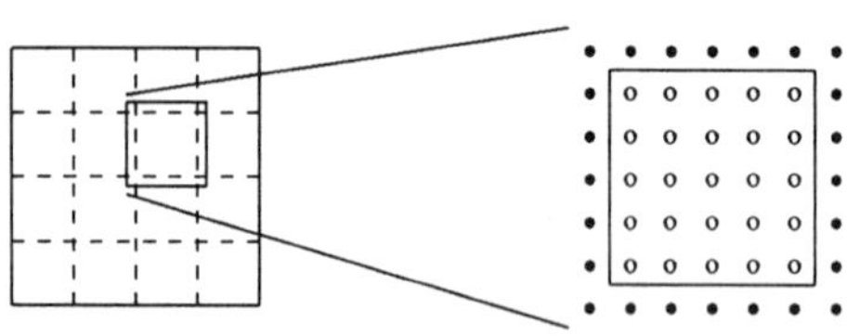

Figure 6. Division of an image between the nodes.

In order to perform a parallel Gauss-Seidel iteration with a nine point stencil, the unknowns are divided into four subsets or colours (Figure 7). Within a subset the order of updating is arbitrary and all calculations can be performed in parallel. Communication of overlap areas is needed each time one subset is updated.

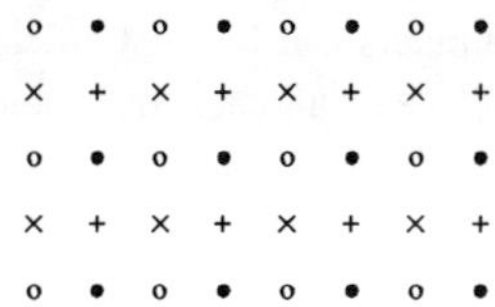

Figure 7. Division of the unknowns into four colours.

3.3 Parallel efficiency

To measure the efficiency of a parallel implementation of an algorithm, one can use the following characteristics [6]:

1. The *speedup* S_p is the execution time on one node divided by the parallel execution time on p nodes: $S_p = T_1 / T_p$.
2. The *efficiency* ε_p is the actual speedup divided by the number of processors: $\varepsilon_p = S_p / p$.

Speedups and efficiencies are given for images of a Lambertian sphere using the parameters mentioned in paragraph 2.5, but for several problem sizes (Figure 8). For each problem size the coarsest level consists of 7×7 grid points.

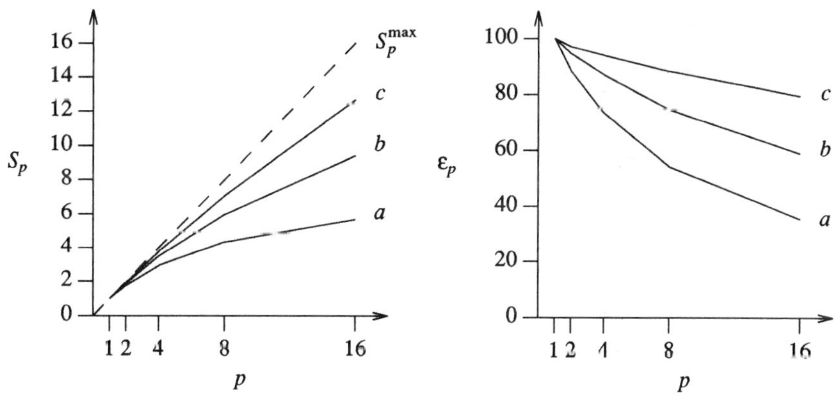

Figure 8. Speedup and efficiency : $a = 63{\times}63$, $b = 127{\times}127$, $c = 255{\times}255$.

Speedup and efficiency are not maximal due to:

- Imbalanced load distribution: The number of grid lines is always odd so the load balance will never be perfect. The imbalance increases as the grids become coarser.
- Communication time: We can separate the parallel execution time in two parts: the calculation time (t_{calc}) and the communication time (t_{comm}). It can easily be shown that in case of a perfect load balance $\varepsilon_p = (1 + t_{comm}/t_{calc})^{-1}$ [6]. The t_{comm}/t_{calc} ratio depends on the size of the problem solved in each node (say $n \times n$ grid points): t_{calc} grows quadratic with n and t_{comm} linear. This phenomenon is called the perimeter effect [6]. When the same problem is solved by using an increasing number of nodes, n will decrease and so will ε_p. On the other hand, when the problem grows larger and is solved with a fixed number of nodes, n and thus ε_p will increase.

The results show that parallel computers can be used efficiently for shape from shading, even for problems of moderate size, due to the large calculation to communication ratio.

4. Results for camera images

We also used this method with the Lambertian reflectance map (5) on a camera image of a sphere. In order to reduce the influence of noise the original image (512×512 pixels) was reduced to 63×63 pixels using full weighting. Most of the assumptions we made in the mathematical deduction now are only partly correct. The obtained results were poor. The algorithm was not capable of reconstructing the shape of the sphere. Instead we obtained a sort of pyramid. The reason for this can be found in the background brightness. According to our model, the background corresponds to a plane ($p = q = 0$) so, following (5), background brightness should be maximal (white). However, the background of this image is rather grey, due to the illumination and the nonhomogeneous reflectance properties of object and background. This grey background is interpreted, according to the reflectance map as a leaning plane instead of a flat one.

This problem can be overcome by processing the background in order to satisfy the image irradiance equation. This essentially means withening the background (Figure 9). The occluding boundary can be found by checking the grey values or by using an edge detector. The accuracy of this edge detection is much less important than the one needed in the method of Ikeuchi and Horn. Notice that in the latter approach not only the position of the occluding boundary is needed but also its derivative and that this information is *imposed* as boundary condition.

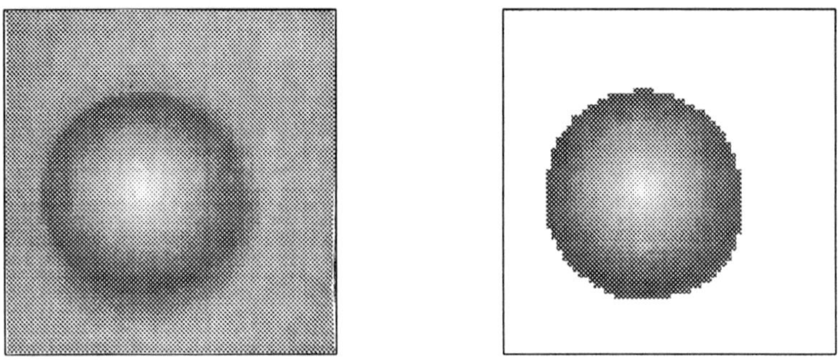

Figure 9. Camera image before and after processing the background.

The result obtained with the same parameters as in paragraph 2.5 and with $\lambda = 0.02$ is shown in Figure 10. The accuracy of the solution is rather low. The computed height in the middle of the sphere is not equal to the diameter and the orientation discontinuity at the occluding boundary is seriously smoothed. The main reason for this is that the Lambertian reflection model is only an approximation of the real situation.

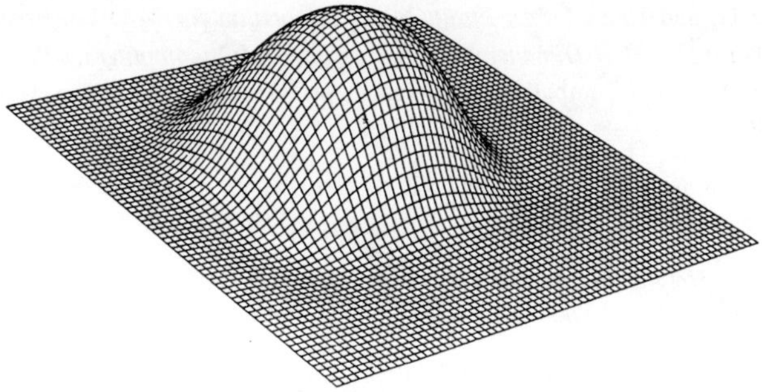

Figure 10. Result for camera image.

5. Conclusion

In contrast to earlier approaches, the method proposed by Battiti [1] for the shape from shading problem does not require accurate occluding boundary information. This results in an ambiguous problem for which the use of the full multigrid scheme is essential to obtain convergence.

In this paper we showed that this method can be applied to camera images after processing the background. Only a simple algorithm is needed to detect the occluding boundary. However our experiments indicate that the convergence is rather slow and a sufficient accuracy is only obtained if the image strictly satisfies the conditions of the theoretical deduction. Possible ways to achieve this are the use of more than one image as proposed by Carter and Asher [3] and the use of more realistic reflectance models.

We also showed that parallel computers can be used efficiently to solve the shape from shading problem.

Acknowledgement

We thank Stefan Vandewalle for the use of GRIDLIB, a library for solving partial differential equations on distributed memory parallel computers, and Hugo Embrechts for the data distribution software.

References

[1] Battiti R., *Surface reconstruction and discontinuity detection: a fast hierarchal approach on a two-dimensional mesh*, Proceedings of the Fifth Distributed Memory Computing Conference, pp 184-193, Charleston, USA, D. Walker and Q. Stout, 1990

[2] Briggs W.L., *A Multigrid Tutorial*, SIAM, Philadelphia, Pennsylvania, 1987

[3] Carter P. and Asher U., *Multigrid Solution for Shape from Shading*, these proceedings.

[4] Courant R. and Hilbert D., *Methods of Mathematical Physics vol.1*, Interscience, London, 1953

[5] Embrechts H. and Jones J., *An Input/Output Algorithm for M-Dimensional Rectangular Domain Decompositions on N-Dimensional Hypercube Multicomputers*, Proceedings of the Fifth Distributed Memory Computing Conference, pp 876-882, Charleston, U.S.A., D. Walker and Q. Stout, 1990

[6] Fox G.C., Johnson M.A., Lyzenga G.A., Otto S.W., Salmon J.K. en Walker D.W., *Solving Problems on Concurrent Processors*, Prentice Hall, 1988

[7] Horn B.K.P., *Height and gradient from shading*, International Journal of Computer Vision 5, pp 37-75, 1990

[8] Horn B.K.P., *Shape from Shading*, MIT Press, Cambridge, 1989

[9] Horn B.K.P. and Brooks M.J., *The variational approach to shape from shading*, Computer vision, Graphics and Image Processing 33, pp 174-208,1986

[10] Ikeuchi K. and Horn B.K.P., *Numerical shape from shading and occluding boundaries*, Artificial Intelligence 17, pp. 141-184, 1981

[11] Stüben W. and Trottenberg U., *Multigrid methods: fundamental algorithms, model problem analysis and applications*, Lecture Notes in Mathematics 960, pp 1-176, Springer-Verlag, Berlin, 1982

[12] Terzopoulos D., *Image analysis using multigrid relaxation methods*, IEEE Trans. Pattern Analysis Machine Intelligence 8, pp. 129-139, 1986

[13] Vandewalle S., De Keyser J. and Piessens R., *The numerical solution of elliptic partial differential equation solvers on a hypercube multiprocessor*, Scientific computing on supercomputers, pp 69-97, J.T. Devreese en P.E. Van Camp, Plenum Press, New York, 1989

[14] *iPSC/2 Users guide*, Intel Scientific Computers, Beaverton, 1987

Wim Sweldens* and Dirk Roose
Katholieke Universiteit Leuven
Departement of Computer Science
Celestijnenlaan 200 A
B 3001 Heverlee Leuven
Belgium

e-mail: wim@cs.kuleuven.ac.be
dirkr@cs.kuleuven.ac.be

* Research Assistant of the National Fund of Scientific Research Belgium

Cell-centered Multigrid Methods in Porous Media Flow

R. Teigland, G.E. Fladmark,
Dep. of Applied Math., University of Bergen,
Allegt.55, N5007 Bergen, Norway.

Abstract

A cell-centered multigrid method is presented for the solution of the system of differential equations used to model multiphase flow in a porous medium. We have applied the cell-centered multigrid method as a linear solver both to the symmetric set of equations arising from an IMPES formulation and to the unsymmetric coupled set of equations that comes from a fully implicit formulation.

In the fully implicit case the system of nonlinear equations is solved by Newton iterations. The cell-centered multigrid method works well even in the case of jump discontinuities in the coefficients. This method avoids using special interpolation operators for multigrid transfers based on the discretized differential equation. Useful properties as conservation of mass is conserved by this scheme. Results of numerical experiments are presented.

1 Introduction

Conservation of mass is an important and desirable property of a discrete scheme. The conservation laws can be written in the form of balance equations approximated over grid cells. This cell-centered finite difference approach is widely used in reservoir simulation. A cell-centered multigrid method is presented for the solution of the equations arising in reservoir simulation. The model that we consider is 2-phase, 2-D compressible flow in a porous media.

We investigate the use of the cell-centered approach proposed by Wesseling and Khalil [15] for problems with discontinuous coefficients. Simulation of large reservoirs or entire fields containing several thousand grid blocks entails solution of very large sets of linear equations. The most time consuming part of a reservoir simulator is solving the set of linear equations. Therefore it is clear that fast iterative methods that can handle jump discontinuities in the coefficients are needed. Preliminary tests with a cell-centered multigrid method applied to flow in porous media have been reported in [12]. Cell-centered two-level methods have been widely used in nuclear reactor calculations. The method used was a two-step method based on a finite dimensional Galerkin technique. The Galerkin technique was first applied by E. L. Wachspress in 1965 for acceleration of the iterative solution of the equations in nuclear reactor calculations [14]. In reservoir simulation discontinuities in permeability occur between different layers having different geological structure. The transition zone between layers is of the order of milimeters. The finite difference approximation of the flow equations contain discontinuities on the scale of practical finite difference cells. These jump discontinuities may be several orders in magnitude. Similar problems with jump discontinuities also arise in neutron diffusion problems [1,14].

In the next section we set up the coupled set of non-linear partial differential equations to be solved and discretize them using cell-centered differencing. Two common solution procedures are used in solving the equations, namely the IMPES and fully implicit method. In the fully implicit scheme water saturation and oil pressure as unknowns are used in the solution of the system. The discrete set of non-linear equations is solved by Newton iterations. In section 3 we discuss the use of a linear multigrid scheme in order to solve the set of linear equations. Straightforward application of multigrid methods to problems arising in reservoir simulation will generally fail if there are large differences in transmissibilities. Techniques have been developed for problems with jump discontinuities in equation coefficients in the vertex-centered case and recently in the cell-centered case [1,3,5,15]. The technique used in this paper is similar to the one used by Wesseling & Khalil [15].

In section 4 we present numerical results.

2 Model equations and discretization

The model equations are derived by combining the mass conservation equations for the two components and Darcy's law. (In general three phases are present in the black-oil equations.)

The mass conservation equations are given below.

$$\frac{\partial}{\partial t}(\phi\rho_w S_w) + \tilde{q}_w = -\nabla\cdot(\rho_w\vec{u}_w) \tag{1}$$

$$\frac{\partial}{\partial t}(\phi\rho_o S_o) + \tilde{q}_o = -\nabla\cdot(\rho_o\vec{u}_o) \tag{2}$$

$\tilde{q}_o$ and $\tilde{q}_w$ are source terms, production \ injection rates at reservoir conditions. In addition to the equations of mass conservation, a relationship between the flow rate and pressure gradient in each phase is required. In fluid mechanics this is given by the momentum equation. For laminar flow through a porous medium it is given by an empirical relationship, Darcy's law:

$$\vec{u}_l = -\frac{Kk_{r_l}}{\mu_l}(\nabla P_l - \gamma_l\nabla H) \tag{3}$$

where the subscript l refers to the oil, gas or water phase and k_{r_l} is the relative permeability of phase l. The mass conservation equations and Darcy's law can be combined to give the flow equations. After dividing through by $\rho_{l_{stc}}$ we get the standard model equations.

$$\nabla\cdot(\lambda_w(\nabla P_o - \nabla P_c - \gamma_w\nabla H)) = \frac{\partial}{\partial t}(\phi b_w S_w) + q_w \tag{4}$$

$$\nabla\cdot(\lambda_o(\nabla P_o - \gamma_o\nabla H)) = \frac{\partial}{\partial t}(\phi b_o S_o) + q_o \tag{5}$$

where

$$\lambda_l = \frac{Kk_{r_l}b_l}{\mu_l}$$

$l = o, w$ and the b_l's are the formation volume factors defined as:

$$b_l = \frac{V_{l_{stc}}}{V_{l_{rc}}} \tag{6}$$

rc stands for reservoir conditions and stc stands for stock tank conditions. V is the volume of a fixed mass. Initial conditions are established by specification of pressure at a point, capillary pressure at water-oil contact, and calculated throughout the reservoir based on the hypothesis of gravitational equilibrium. At the reservoir boundary we assume no net current of fluids. This implies that the component of the Darcy velocity normal to the boundary surface is zero for each fluid.

A box integration technique or cell-centered technique is used in the finite difference formulation. Cell-centered differencing is one of the simplest ways to approach the discretization of the flow equations. Equations (4) and (5) are integrated over a cell. After applying Gauss divergence theorem to the volume integrals and discretizing the equations reads as follows.

$$\triangle T_w(\triangle(P_o - P_c - \gamma_w \triangle h)_{ij} = \frac{V_{p_{ij}}}{\Delta t} \Delta_t (S_w b_w)_{ij} + Q_{w_{ij}} \tag{7}$$

$$\triangle T_o(\triangle(P_o - \gamma_o \triangle h)_{ij} = \frac{V_{p_{ij}}}{\Delta t} \Delta_t (S_o b_o)_{ij} + Q_{o_{ij}} \tag{8}$$

where

$$\Delta_t(S_l b_l) = (S_l b_l)^{n+1} - (S_l b_l)^n, \; l = (o, w)$$

and

$$V_{p_{ij}} = \phi \Delta x_i \Delta y_j$$

The difference operator $\triangle$ is such that

$$[\Delta T \triangle P]_{ij_l} = (T_{x_{i+\frac{1}{2}j}}(P_{i+1j} - P_{ij}) - T_{x_{i-\frac{1}{2}j}}(P_{ij} - P_{i-1j})$$

$$+T_{y_{ij+\frac{1}{2}}}(P_{ij+1} - P_{ij}) - T_{y_{ij-\frac{1}{2}}}(P_{ij} - P_{ij-1}))_l$$

The transmissibility term for phase l is given by

$$T_{x_{i+\frac{1}{2}}} = \Lambda_{x_{i+\frac{1}{2}}} \frac{\Delta y_j \Delta z_{i+\frac{1}{2}j}}{\Delta x_{i+\frac{1}{2}}}$$

and $\Lambda_{x_{i+\frac{1}{2}}}$ is the discretized form of λ defined above. The other transmissibility terms are defined analogously.

$$T_{y_{j+\frac{1}{2}}} = \Lambda_{y_{j+\frac{1}{2}}} \frac{\Delta x_i \Delta z_{ij+\frac{1}{2}}}{\Delta y_{j+\frac{1}{2}}}$$

The absolute permeabilities at the block boundaries are defined using a harmonic average. For relative permeabilities k_{r_l} we use one-point upstream weighting. Note that the cell-centered grid does not adversely affect the convergence properties of the discretization. It was thought for some time that cell-centered grids were only $\mathcal{O}(\delta x)$, while point-distributed grids where $\mathcal{O}(\delta x^2)$, (See Aziz and Settari, [2]). In a point-distributed grid, nodes are selected first and cells second. In a cell-centered (or block-centered) grid cells are selected first and nodes are placed at cell centers. The latter is obviously more useful in modelling physical problems, where cell edges can be aligned with physical discontinuities in the problem. It is now known that the $\mathcal{O}(\delta x^2)$ convergence property applies to both uniform and non-uniform cell-centered grids (see Weiser and Wheeler [17]). There is therefore no reason to favour the point-distributed (or vertex-centered) construction over the more physically reasonable cell-centered approach.

If we expand the terms where

$$\Delta_t(S_l b_l) = (S_l b_l)^{n+1} - (S_l b_l)^n, \; l = (o, w)$$

in terms of oil pressure P_o and water saturation S_w and define

$$d_{11} = \frac{V_p}{\Delta t}(S_w^n b_w')$$

$$d_{12} = \frac{V_p}{\Delta t}(b_w^{n+1} - S_w^n b_w' P_c')$$

$$d_{21} = \frac{V_p}{\Delta t}((1 - S_w^n) b_o')$$

$$d_{22} = \frac{V_p}{\Delta t}(-b_o^{n+1})$$

the set of equations (7) and (8) becomes

$$\Delta T_w(\Delta(P_o - P_c - \gamma_w \Delta h)_{ij} = d_{11} \Delta_t P_o + d_{12} \Delta_t S_w + Q_{w_{ij}} \tag{9}$$

$$\Delta T_o(\Delta(P_o - \gamma_o \Delta h)_{ij} = d_{21} \Delta_t P_o + d_{22} \Delta_t S_w + Q_{o_{ij}} \tag{10}$$

The IMPES solution consists of two steps:

(a) With P_c evaluated explicitly, solve for P_o^{n+1} implicitly from:

$$\begin{aligned} -(\frac{d_{22}}{d_{12}}) \Delta T_w \Delta P_o^{n+1} + \Delta T_o \Delta P_o^{n+1} = \\ (-(\frac{d_{22}}{d_{12}}) d_{11} + d_{21}) \Delta_t P_o - (\frac{d_{22}}{d_{12}}) Q_w + Q_o \\ -(\frac{d_{22}}{d_{12}} \Delta (T_w(\gamma_w \Delta h + \Delta P_c^n)) + \Delta T_o \gamma_o \Delta h \end{aligned} \tag{11}$$

(b) Solve explicitly for $\Delta_t S_w$, from the water equation:

$$\Delta_t S_w = \frac{1}{d_{12}}(-d_{11} \Delta_t P_o - Q_w + \Delta T_w(\Delta P_o^{n+1} - \gamma_w \Delta h - \Delta P_c^n)) \tag{12}$$

If we assume incompressibility we get an elliptic pressure equation to solve for. This is particularly well suited to multigrid solution techniques. Because of the explicit treatment of saturation, however, there is a severe restriction on the size of the time step if the mesh sizes are very small or P_c' is very large.

If the terms on the left of equations (7) - (8) are to be evaluated at the new time level $n + 1$, the resulting formulation is termed a fully implicit formulation. The fully implicit formulation has become widely used in reservoir simulation because it is a stable and robust method. The fully implicit method produces a set of $N \times n_{ph}$ nonlinear algebraic equations (where n_{ph} is the number of coupled equations per grid node and N is the number of grid nodes). The system of equations is solved using Newton iterations.

Equations (7) and (8) are put in a residual form where the ith cell in the grid is a two component vector

$$R_i = \begin{bmatrix} R_o \\ R_w \end{bmatrix}_i$$

The ith element of the solution vector, U, to be evaluated at the end of the timestep is

$$U_i = \begin{bmatrix} P_o \\ S_w \end{bmatrix}_i$$

such that the equations (7) and (8) are of the form

$$R(U) = 0 \tag{13}$$

In Newtons method, the nonlinear equation is linearized about the current estimate of the solution vector, U,

$$R(U + \Delta U) = R(U) + \mathcal{J} \cdot \Delta U \tag{14}$$

where the ij^{th} element of the Jacobian matrix takes the form

$$\mathcal{J} = \begin{bmatrix} \frac{\partial R_o}{\partial P_o} & \frac{\partial R_o}{\partial S_w} \\ \frac{\partial R_w}{\partial P_o} & \frac{\partial R_w}{\partial S_w} \end{bmatrix}_{ij} \tag{15}$$

The resulting system of linear equations can be written as

$$\mathcal{J} \cdot \Delta U = -R(U) \tag{16}$$

If we order the unknowns in a vector

$$U^h = [(\Delta P_o)_1, (\Delta S_w)_1, \ldots, (\Delta P_o)_k, (\Delta S_w)_k, \ldots, (\Delta P_o)_N, (\Delta S_w)_N]^\top \tag{17}$$

we can write the difference equations (7) - (8) in matrix form as follows

$$A^h U^h = F^h \tag{18}$$

at each grid point (x_i, y_j) it can be written as

$$A_{ij}^h U_{ij}^h = W_{ij}^h U_{i-1j}^h + S_{ij}^h U_{ij-1}^h + C_{ij}^h U_{ij}^h + E_{ij}^h U_{i+1j}^h + N_{ij}^h U_{ij+1}^h \tag{19}$$

where the terms in front of the U's are $n_{ph} \times n_{ph}$ matrices and U_{ij}^h are the values of U^h at (x_i, y_j). The matrix A^h is highly unsymmetric.

3 The Linear Multigrid Scheme.

Multigrid methods have been developed for problems with jump discontinuities in equation coefficients in the vertex-centered case and recently in the cell-centered case [1,3,5,15]. In the vertex-centered case the multigrid method can be expensive in terms of preliminary work and storage, especially in the three-dimensional case. Equation dependent interpolation operators are needed in this case [1,3,5,6]. Because of its greater simplicity, the recently developed cell-centered multigrid method [15] promises greater efficiency.

The idea of using a coarser grid in order to get a correction to the fine grid equations has been used by researchers in the field of reactor physics where they termed the method "coarse mesh rebalancing technique" [14,10]. Their construction of the matrices on coarser levels is similar to the automatic prescription or Galerkin approximation suggested by Nicolaides [11] and Alcouffe et al [1]. We refer to [4] and [7] for a general discussion of basic multigrid ideas and the Galerkin approach.

Let a sequence of grids G^k (k=1(1)M) be defined with $h_1 > h_2 > \ldots > h_M$, where h_k is the mesh size on grid G^k. Denote the finite difference operator to the differential operator A on G^M by A^M. The coarse-grid operator A^{k-1} is defined in terms of the fine-grid operator A^k and prolongation and restriction operators P_{k-1}^k and R_k^{k-1}:

$$A^{k-1} = R_k^{k-1} A^k P_{k-1}^k$$

This definition is called the Galerkin approximation. It seems natural to use a variational formulation in the construction of the coarse-grid operator in the case of discontinuous coefficients. In reservoir simulation using this automatic prescription would only require the representation of the differential operator (or stencil) on the finest grid, the rest is automatically set up by the Multigrid solver. Such schemes seem attractive in large reservoir simulators where one could use the Multigrid scheme essencially as a "black-box" solver [5].

The so-called full-approximation scheme (FAS) that does not require linearity could be applied directly to the nonlinear equations, requiring only local linearization. One would like to use a Galerkin approach on this nonlinear MG scheme. The major limitation with the Galerkin approach for nonlinear schemes (as noted by Mc. Cormick [8]) stems from the fact that explicit use of the definition of $A^{k-1}(u^k)$ (now a nonlinear operator) involves grid h^k computations.For the defect computations on grid G^{k-1}, $A^{k-1}(u^{k-1})$ may be evaluated by interpolating u^{k-1} to grid G^k, adding u^k, applying A^k, and transferring the result to grid h^k.

Since we usually have a good initial guess when using values from the previous time step, Newton iterations will usually converge in a few iterations. This led us to use the linear multilevel scheme and not the nonlinear FAS approach.

A simple well-structured FORTRAN description, using only one goto, of the fundamental MG algorithm due to Wesseling [16], is given below.

```
Subroutine MG(φ, f)
ic(1) = maxit
k = 1
if (V-cycle) then γ = 1 else γ = 2
  10 if (ic(1).gt.0) then
       if (ic(k).eq.0.or.k.eq.M) then
         if (k.eq.M) then
           call S(φ^M, f^M)
           ic(M) = ic(M) - 1
           if (F-cycle) γ = 1
         endif
         k = k - 1
         φ^k = φ^k + P^k φ^{k+1}
         call S(φ^k, f^k)
         ic(k) = ic(k) - 1
       else
         call S(φ^k, f^k)
         f^{k+1} = R^{k+1}(f^k - A^k φ^k)
         k = k + 1
         φ^k = 0
         ic(k) = γ
       endif
       if (k.eq.1.and. F-cycle) γ = 2
       go to 10
  endif
end.
```

$k = 1$	: finest grid
maxit	: number of MG iterations desired
ic(k)	: counter of MG cycles on grid k
k = M	: coarsest grid
S	: smoothing subroutine
f^k	: right-hand side on grid k
ϕ^k	: correction ($k \geq 2$) or solution ($k = 1$)

In multigrid methods P^k_{k-1} and R^{k-1}_k must satisfy the following requirements. Let m_P-1, m_R-1 be the maximum degree of polynomials that are interpolated exactly by sP or tR^* respectively for some real value of s or t. Then we must have $m_P + m_R > 2m$ with $2m$ the order of the partial differential equation to be solved. It is sufficient here to take $m_P = 1$ and $m_R = 2$ as suggested by Wesseling & Khalil [15]. The multigrid treatment of the system is not different from the scalar case. The degree of the interpolation m_{p_α} has to be chosen according to the αth differential equation. Prolongation is chosen as constant interpolation, and restriction as scaled adjoint of linear interpolation in triangles. This preserves 7-point stencils. We refer to [15] for further details.

The weights for the restriction are such that for inner points we have

$$\begin{aligned} u^c_{i,j} = u^f_{2i-2,2j+1} + u^f_{2i-1,2j+1} + u^f_{2i-2,2j} \\ 3u^f_{2i-1,2j} + 2u^f_{2i,2j} + 2u^f_{2i-1,2j-1} + 3u^f_{2i,2j-1} \\ u^f_{2i,2j-2} + u^f_{2i+1,2j-2} + u^f_{2i+1,2j-1} \end{aligned} \tag{20}$$

At the boundary the restriction operator has to be modified. Wesseling and Khalil have suggested modifications based on Dirichlet or Neumann conditions at the boundary. We are using the modification based on a Neumann condition since that is the usual condition at the boundary that we encounter in reservoir simulation. It also gives the best convergence rates for general boundary conditions. We refer to Wesseling and Khalil [15] for details about this restriction operator. We have also experimented with restriction as adjoint of bilinear interpolation. In that case the stencil size on the coarser grids will be a nine-point one. The convergence rates in that case where comparable to the ones obtained by restriction operator defined by equation (20).

In the construction of the coarse-grid matrix $\tilde{A}$ we are using the Galerkin-approximation. We will describe this construction considering a fine level and a coarse level. We also assume to simplify discussion that $m_P + m_R = 2$, i.e. simplest restriction and interpolation.

Consider a system of $N = n \times n$ linear equations

$$A\vec{u} = \vec{f} \tag{21}$$

Consider a grid system as below, we impose on this a coarse system:

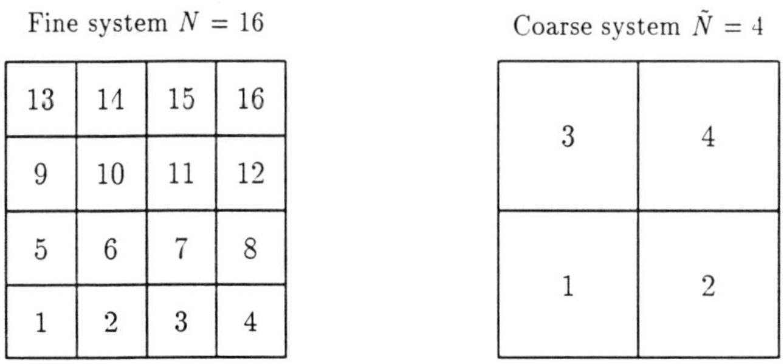

Figure 1:

Let $M_{R_{\tilde{i}}}$ denote the set of fine cells that are defined in the weighted sum for the restriction operation centered on coarse cell $\tilde{i}$. In case $m_R = 1$ we have for coarse cell number $\tilde{1}$: $M_{R_{\tilde{1}}} = (1,2,5,6)$, see Figure 1. Since $P = R^*$, we have $M_{P_{\tilde{i}}} = M_{R_{\tilde{i}}}$.

Let $\vec{u}$ be the iterative solution before the coarse-grid correction step. Let $\vec{\Psi}_{\tilde{i}}$ denote a vector of order N consisting of only zero-elements except for unit-elements at all $i \in M_{R_{\tilde{i}}}$. For $\tilde{i} = \tilde{1}$ we have:

$$\vec{\Psi}_{\tilde{i}} = [1,1,0,0,1,1,0,\ldots,0]^{\top}$$

Define

$$\vec{z} = \sum_{\tilde{i}=1}^{\tilde{N}} \tilde{u} \cdot \vec{\Psi}_{\tilde{i}} + \vec{u} \tag{22}$$

By using $\vec{z}$ as a base function in a Galerkin method, we get that for each coarse cell $\tilde{j}$ the following equation should be satisfied:

$$< \vec{\Psi}_{\tilde{j}}, A\vec{z} >=< \vec{\Psi}_{\tilde{j}}, \vec{y} > \tag{23}$$

The unknown vector is then given by:

$$\tilde{A}\vec{\tilde{u}} = \vec{\tilde{f}} \tag{24}$$

where

$$\tilde{a}_{\tilde{i},\tilde{j}} = \sum_{i \in M_{R_{\tilde{i}}}} \sum_{j \in M_{P_{\tilde{j}}}} a_{i,j} \tag{25}$$

$$\tilde{f}_{\tilde{i}} = \sum_{i \in M_{R_{\tilde{i}}}} r_i \tag{26}$$

and r_i is the fine grid residual. The new estimate for the solution of (21) is then:

$$\vec{u} = \vec{u} + \sum_{\tilde{i}=1}^{\tilde{N}} \tilde{u} \cdot \vec{\Psi}_{\tilde{i}} \tag{27}$$

In the more general case where $P \neq R^*$ the sets $M_{P_{\tilde{i}}}$ and $M_{R_{\tilde{i}}}$ are different.

In the homogeneous case where K_x and K_y are of the same order we use the point Gauss-Seidel scheme (In the fully implicit case we use collective Gauss-Seidel schemes since we are treating the system of equations as a coupled system). In the anisotropic case we have three options in our code:

- x-line Gauss-Seidel for $K_x \gg K_y$ denoted XLGS.
- y-line Gauss-Seidel for $K_y \gg K_x$ denoted YLGS.
- Alternating line Gauss-Seidel if both kinds of anisotropies appear, denoted ALGS.

We have also tried to accelerate the scheme using the multigrid scheme as a preconditioner to ORTHOMIN.

4 NUMERICAL EXPERIMENTS

We have tested the multilevel linear solver with our black-oil simulator. The following cases were considered. Ω is a square of constant thickness, $\Omega = (x,y) \mid 0 \leq x, y \leq 1000\,ft$. Water is injected in the lower left corner of Ω, and oil is produced in the upper right corner.

Pb 1. $K_x = K_y = 100\, mD\ \forall(x,y) \in \Omega$
Pb 2. $K_x = 1000\, mD, K_y = 10\, mD\ \forall(x,y) \in \Omega$
Pb 3. $K_x = 10\, mD \quad 0 \leq x \leq 200.\ ft,$
$K_x = 1000\, mD\ 200.0 < x \leq 1000.0\, ft$
$0 \leq y \leq 1000\, ft$
$K_y = 10\, mD \quad \forall(x,y) \in \Omega$
Pb 4. $K_x = K_y = 10.0\, mD, 0 \leq x \leq 500\, ft$
$K_x = K_y = 500.0\, mD, 500.0 < x \leq 1000\, ft$
$0 \leq y \leq 1000\, ft$
Pb 5. $K_x = K_y = 0.001\, mD, 0 \leq x \leq 950\, ft$
$120 \leq y \leq 130\, ft$
$K_x = K_y = 1000.0\, mD$, for other values of x and y
Pb 6. $K_x = 1.0\, mD\ 0 \leq x \leq 200.\ ft,$
$K_x = 10000.0\, mD\ 200.0 < x \leq 1000.0\, ft$
$0 \leq y \leq 1000\, ft$
$K_y = 10.0\, mD \forall(x,y) \in \Omega$

Let ν_1 and ν_2 denote the number of pre- and post-smoothing steps respectively. We define the average reduction factor by

$$\bar{\kappa} = \left(\frac{\|r^\nu\|_2}{\|r^0\|_2}\right)^{\frac{1}{\nu}}$$

where r^ν is the residual at the iteration ν , $r^\nu = f - Au^\nu$. The tabulated average reduction factors are averages over 20 timesteps.

Results for fully implicit case:
Smoother for test problems 1.2 and 3 is XLGS, $m_P + m_R = 3$.

h	$\nu = 10$		
	$pb.1$	$pb.2$	$pb.3$
1/16	.09	.07	.08
1/32	.13	.09	.09
1/64	.13	.09	.11

Table 1: results for $V(2,1)$-cycles.

h	$\nu = 10$		
	$pb.1$	$pb.2$	$pb.3$
1/16	.07	.06	.06
1/32	.09	.08	.07
1/64	.10	.09	.09

Table 2: results for $F(2,1)$-cycles.

h	$\nu = 10$
	$F(2,1)$
1/16	.07
1/32	.07
1/64	.06
1/128	.06
1/256	.06

Table 3: Results for pb. 5, smoother is ALGS.

In the next table we have tabulated results for test problem 4. Average reduction factors with a W(2,1)-cycle and the same cycle as preconditioner to ORTHOMIN (labeled orto-mg).

h	$\nu = 7$	
	mg	$ortho - mg$
1/16	.06	.009
1/32	.06	.01
1/64	.08	.03
1/128	.08	.03

Table 4: results for test problem 4, smoother is ALGS.

Results for IMPES case:

h	$\nu = 10$	
	(1,1)	(2,1)
1/8	.12	.07
1/16	.13	.08
1/32	.15	.08
1/64	.15	.09

Table 5: Results for pb. 2, smoother is XLGS.

h	$\nu = 10$	
	(1,1)	(2,1)
1/8	.09	.04
1/16	.09	.04
1/32	.07	.04
1/64	.07	.05

Table 6: Results for pb. 4, smoother is ALGS.

h	$\nu = 10$	
	(1,1)	(2,1)
1/8	.18	.07
1/12	.16	.07
1/16	.19	.07
1/24	.21	.08
1/32	.19	.07
1/48	.20	.09
1/64	.19	.07

Table 7: Results for pb. 6, smoother is ALGS.

5 CONCLUDING REMARKS

We have implemented a cell-centered multigrid linear solver in a reservoir simulator. The method uses geometric restriction and prolongation operators. The multigrid solver can easily be implemented on existing simulators, requiring only stencils on the finest grid. Numerical experiments show that W- and F-cycles seem marginally better than V-cycles. Implementation in three dimensions is easier in the cell-centered case than in the vertex-centered case. The convergence rates that we achieved are typical for those found when the multilevel method is used.

6 Acknowledgements

This work has been supported by VISTA, a research cooperation between the Norwegian Academy of Science and Letters and Den Norske Stats Oljeselskap A.S. (Statoil), and Norsk Hydro Research Centre.

References

[1] R. E. Alcouffe, R. E. Brandt, J. E. Dendy, J. W. Painter. The multi-grid method for the diffusion equation with strongly discontinuous coefficients.
SIAM J. Sci. Statist. Comp. 2, pp 430-454, 1981.

[2] K. Aziz, A. Settari. Petroleum reservoir simulation.
Elsevier Applied Science Publishers 1979.

[3] A. Behie, P. Forsyth Jr. Comparisons of fast iterative methods for symmetric systems.
IMA Journal of Numerical Analysis, **3**, pp 41-63, 1983.

[4] A. Brandt. Multi-level adaptive solutions to boundary-value problems.
Math. of Comp. **31**, pp 333-390, 1977.

[5] J. E. Dendy. Black Box Multigrid.
Journal of Computational Physics **48**, pp 366-386, 1982.

[6] T. W. Fogwell and F. Brakhagen. Multigrid Methods for the Solution of Porous Media Multiphase Flow Equations.
Nonlinear Hyperbolic Equations-Theory, Computation Methods, and Applications. Vol. 24, Notes on numerical Fluid Mechanics,ed. Josef Ballmann and Rolf Jeltsch, 139-148. Braunschweig, Vieweg, 1989.

[7] W. Hackbush. Multi-grid Methods and Applications.
Springer series in Computational Mathematics.
Springer-Verlag, 1985.

[8] S. F. McCormick. Multilevel Adaptive Methods for Partial Differential Equations.
Frontiers in applied mathematics. vol. 6. SIAM. Philadelphia 1989.

[9] J. A. Meijerink and H. A. van der Vorst. An iterative solution method for linear systems of wich the coefficient matrix is a symmetric M-matrix.
Math. of Comp., vol. 31. pp 148-162, 1977.

[10] S. Nakamura and T. Onishi. The Iterative solutions for the finite element method.
IAEA Report SM-154/57, 1972.

[11] R. A. Nicolaides. On some theoretical and practical aspects of multi-grid methods. Math. of Comp., vol. 33, 933-952, 1979.

[12] R. Teigland, G. E. Fladmark. Multilevel Methods in Porous Media Flow.
Proceedings of the 2. European Conference on the Mathematics of Oil Recovery.
Sept. 11-14, 1990, Arles, France.
Edited by D. Guerillot and O. Guillon.

[13] P. W. Vinsome. Orthomin, an iterative method for solving sparse sets of simultaneous linear equations.
Proceedings of the Fourth SPE Symposium on Reservoir performance, SPE of AIME, 1976.
Paper SPE 5729.

[14] E. L. Wachspress. Iterative Solution of Elliptic Systems and Applications to the Neutron Diffusion Equations of Reactor Physics.
Prentice-Hall, Englewood Cliffs, New Jersey 1966.

[15] M. Khalil, P. Wesseling. Vertex-centered and cell-centered multigrid methods for interface problems.
Report 88-42, Delft University of Technology.

[16] P. Wesseling. Proceedings of th fourth GAMM Seminar, Kiel, West-Germany,
January 1988, edited by W. Hackbush (Vieweg, Braunschweig, 1988).

[17] A. Weiser, M. F. Wheeler. On convergence of block-centered finite differences for elliptic problems.
SIAM j. Numer. Anal. vol. 25, No. 2, April 1988.

International Series of Numerical Mathematics, Vol. 98, © 1991 Birkhäuser Verlag Basel

Multigrid Waveform Relaxation for Solving Parabolic Partial Differential Equations

by

Stefan Vandewalle* and Robert Piessens

Abstract. *The numerical solution of a parabolic partial differential equation is usually calculated by a time stepping method. This precludes the efficient use of vectorization and parallelism if the problem to be solved on each time level is not very large. In this paper we present an algorithm based on the waveform relaxation technique, which overcomes the limitations of the standard marching schemes. The method is applicable to linear and nonlinear problems on arbitrary domains. It can be used to solve initial-boundary value problems as well as time-periodic equations. We have implemented the method on an Intel iPSC/2 hypercube. By several numerical examples we illustrate the properties of the method and we compare its performance to that of the best standard solvers.*

1. Introduction

In a previous study we analyzed the parallel characteristics of several standard parabolic time stepping techniques, [15]. As it was shown, standard methods can only be parallelized efficiently for problems that are *large* enough, i.e. the number of unknowns (and consequently the arithmetic complexity) per processor at each time level is large enough to outweigh communication. New algorithms are needed for solving relatively *small* parabolic problems, i.e. problems with few unknowns per time level or problems solved on large scale parallel machines.

A direction taken by several researchers is to improve the parallel efficiency of the implicit methods by operating on several or on all time levels at once. With the *windowed relaxation* methods proposed by J. Saltz and V. Naik the standard Jacobi and SOR relaxations are extended to operate on a window of time levels, [11]. In the *parallel time stepping* method of D. Womble one or more processors are assigned to successive time levels, [18]. While the solution is being calculated on one time level other processors update the approximation on the subsequent time levels. In [4] the multigrid technique was extended by W. Hackbush to solve a parabolic equation on several time levels simultaneously. A parallel implementation of this *parabolic multigrid* method was presented by P. Bastian, J. Burmeister and G. Horton in [1].

In this paper we present a technique which is different from the previous approaches in that it can be defined without explicit reference to any time discretization technique or to any time levels. The resulting algorithm was first published by Ch. Lubich and A. Ostermann in [6] and rediscovered independently by

---

* Senior research assistant of the National Fund for Scientific Research (Belgium)

the current authors, [13,14,16]. The method is based on *waveform relaxation*, which is presented in section 2. We discuss the application of waveform relaxation for solving initial-boundary value problems in section 3. In section 4 we consider the solution of time-periodic parabolic equations. Some implementation aspects are discussed in section 5. In section 6 we illustrate the method by several examples and compare its performance to that of a parallel implementation of the best standard methods.

2. The waveform relaxation method

Waveform relaxation, also called dynamic iteration or Picard-Lindelöf iteration, is an iterative solution technique for systems of ordinary differential equations, [7,8,17]. Its application has proven to be very effective in solving the systems of differential equations that describe the behaviour of large, complex VLSI devices, [17]. Further applications are found in the literature on the simulation of chemical distillation processes, [12].

Consider the following general system of N first order differential equations,

$$\frac{d}{dt} y_i = f_i(t, y_1, \ldots, y_N) \quad \text{for } t \in [\, t_0, t_f\,], \quad \text{with initial condition } y_i(t_0) = y_{i0},\ i=1,\ldots,N. \tag{2.1}$$

The Jacobi variant of the waveform relaxation algorithm for solving (2.1) can be stated as follows.

ALGORITHM Jacobi waveform relaxation:

n := 0

choose $y_i^{(0)}(t)$ for $t \in [t_0, t_f]$, $i=1,\ldots,N$

repeat

for all i:

solve $\frac{d}{dt} y_i^{(n+1)} = f_i(t, y_1^{(n)}, \ldots, y_{i-1}^{(n)}, y_i^{(n+1)}, y_{i+1}^{(n)}, \ldots, y_N^{(n)})$ *with* $y_i^{(n+1)}(t_0) = y_{i0}$ *on* $[t_0, t_f]$

n := n+1

until convergence.

The idea is first to make an initial guess of the solution of the system and then to solve each equation in sequence as part of a relaxation loop until convergence is achieved. In the iteration step each differential equation is solved as an equation in one variable. The adaptation of the algorithm to obtain a Gauss-Seidel type iteration is straightforward. Often it is advantageous to solve for the variables in groups of unknowns. This leads to a waveform extension of the familiar block-Jacobi and block-Gauss-Seidel procedures. In the case of a nonlinear problem each differential equation may be solved approximately, e.g. by applying a Newton linearization and solving the resulting linear ordinary differential equation. This approach is similar to the approach often found in relaxation methods for calculating the solution of nonlinear algebraic equations. The resulting technique is usually referred to as the Waveform Newton Relaxation method.

Waveform relaxation is known to converge superlinearly on finite time intervals, i.e. the spectral radius vanishes. An in depth convergence analysis for linear systems on unbounded time intervals is given by Miekkala and Nevanlinna in [7]. The effect of discretization is studied in [8]. In [17], one of the basic papers on waveform relaxation in the electrical engineering literature, uniform convergence is proven for nonlinear systems. Further convergence results are given in [5], in which the relation is established between the number of iterations and the accuracy order, i.e. the number of correct terms in the Taylor expansion of a partially converged solution.

3. Waveform relaxation applied to initial-boundary value problems

3.1 Standard waveform relaxation

We consider the following general parabolic initial-boundary value problem

$$\frac{\partial u}{\partial t} + \mathrm{L}(u) = f_1 \quad (x,t) \in \Omega \, x \, [t_0, t_f] \tag{3.1.a}$$

$$B(u) = f_2 \qquad (x,t) \in \partial\Omega \, x \, [t_0, t_f] \tag{3.1.b}$$

$$u(x,t_0) = u_0 \qquad x \in \Omega \tag{3.1.c}$$

where $\Omega \subset \mathbb{R}^n$, L is an elliptic, possibly non-linear operator and B is the boundary operator. After spatial discretization and incorporation of the boundary conditions, the parabolic problem is transformed into a system of ordinary differential equations with one equation defined at each grid point,

$$\frac{dU}{dt} + L(U) = F \quad with \quad U(t_0) = U_0. \tag{3.2}$$

In equation (3.2) U is the vector of unknown functions defined at the grid points, L is the operator derived from L and B by discretization and F is the vector of functions determined by f_1 and f_2.

The waveform method discussed in section 2 may be applied to solve (3.2). However, this does not lead to a satisfactory algorithm due to slow convergence. As was shown in [7], the convergence rates for the Jacobi and Gauss-Seidel scheme in the case of the discretized heat equation are of order $1-O(h^2)$. The convergence is thus rapidly deteriorating with an increasing number of grid lines. In contrast to the case of a linear system of equations arising from a discretized elliptic equation, overrelaxation in a SOR fashion does not lead to significantly improved convergence characteristics, [7].

3.2 Multigrid Waveform Relaxation

The convergence can be accelerated if the waveform algorithm is combined with the multigrid idea. The multigrid method can be extended to time dependent problems in essentially the same way as the classical relaxation methods. Each of the standard multigrid operations is replaced by a similar operation defined to operate on functions.

- **Smoothing** is performed by applying one or more Gauss-Seidel or damped Jacobi waveform relaxations. In the case of a nonlinear problem a Newton waveform relaxation technique may be used. Smoothing rates for these relaxations have been calculated for solving the heat equation and are published in [6].

- The **defect** of an approximation $\overline{U}$ is defined as

$$D = \frac{d}{dt}\overline{U} + L(\overline{U}) - F. \tag{3.3}$$

- **Restriction** and **prolongation** are calculated with identical formulae as in the elliptic case. However, these formulae now operate on functions.

We may now state the equivalent waveform algorithm of the *multigrid full approximation scheme*, which is used for solving nonlinear problems. Let G_i, $i = 0, 1, \cdots, k$, be the hierarchy of grids with G_k the finest grid and G_0 the coarsest grid. Equation (3.2) or equivalently,

$$\frac{dU_k}{dt} + L_k(U_k) = F_k \quad with \quad U_k(t_0) = U_{k,0} \tag{3.4}$$

is solved by iteratively applying the following algorithm to an initial approximation of U_k.

PROCEDURE fas (k, F_k, U_k)

if $k = 0$ *solve the coarse grid problem* $\frac{d}{dt} U_0 + L_0(U_0) = F_0$ *exactly*

else

- *perform* ν_1 *smoothing operations (e.g. by red/black waveform relaxation)*
- *project* U_k *onto* G_{k-1}*:* $\bar{U}_{k-1} := \bar{I}_k^{k-1} U_k$
- *calculate the coarse problem right hand side:* $F_{k-1} := \frac{d\bar{U}_{k-1}}{dt} + L_{k-1}(\bar{U}_{k-1}) - I_k^{k-1}(\frac{dU_k}{dt} + L_k(U_k) - F_k)$
- *solve on* G_{k-1}*:* $\frac{d}{dt} U_{k-1} + L_{k-1}(U_{k-1}) = F_{k-1}$ *with* $U_{k-1}(t_0) = \bar{I}_k^{k-1} U_{k,0}$

  repeat γ_k *times* *fas* $(k-1, F_{k-1}, U_{k-1})$, *starting with* $U_{k-1} := \bar{U}_{k-1}$.
- *interpolate the correction to* G_k *and correct* U_k*:* $U_k := U_k + I_{k-1}^k (U_{k-1} - \bar{U}_{k-1})$
- *perform* ν_2 *smoothing operations (e.g. by red/black waveform relaxation)*

endif

The algorithm is completely defined by specifying the grid sequence G_i, $i=0,...,k$, the discretized operators L_i, the inter-grid transfer operations I_{i-1}^i and I_i^{i-1}, the nature of the smoothing relaxations, and by assigning a value to the constants ν_1, ν_2 and γ_i. A similar extension of the multigrid correction scheme is discussed in our papers [14,16]. Note that the derivative calculation in the formula for the determination of the coarse grid problem right hand side can be avoided. Indeed, when the two restriction operators I_i^{i-1} and $\bar{I}_i^{i-1}$ are equal, the two derivatives cancel. The right hand side of the problem on the coarse grid may then be calculated by the following formula,

$$F_{k-1} := L_{k-1}(\bar{U}_{k-1}) - I_k^{k-1}(L_k(U_k) - F_k). \tag{3.5}$$

The algorithm can be combined with the idea of nested iteration. The initial approximate solution to the problem on G_i is then derived from the solution obtained on G_{i-1}. This leads to the waveform equivalent of the *full multigrid method.*

4. Waveform relaxation applied to time-periodic parabolic problems

4.1 The standard algorithms

In this section we consider the non-autonomous parabolic problem (3.1a-c) where the initial condition (3.1.c) is replaced by the periodicity condition

$$u(x, t_0) = u(x, t_f). \tag{4.1}$$

This problem is of considerable importance in various areas of practical interest, such as wing flutter, ferro-conductor eddy currents, chemical reactor theory, pulsating stars, and fluid dynamics. Various algorithms have been proposed to compute the stable periodic solutions. One approach is a time-integration of the studied system, starting from an arbitrary initial condition until a stable periodic orbit is reached. This *brute force* method however may be prohibitively expensive in the case of slowly decaying transients. A second approach uses difference methods where a large system of nonlinear algebraic equations is obtained after discretization. This system may be solved with direct or iterative sparse solvers. A third and commonly used approach is based on the shooting method. The major computational cost of this method lays in the determination of a large Jacobian matrix, needed in a Newton-Raphson procedure. This matrix may be approximated numerically by difference methods and requires the calculation of a possibly very large number of initial value problems. We have applied this technique for the determination of the periodic solutions of autonomous parabolic problems in [10]. Finally, a very fast algorithm was presented by W. Hackbush, in which the periodic problem is reformulated as an integral equation and solved by the multigrid method of the second kind, [3]. One iteration of this method is based on a multigrid W-cycle, where in each smoothing and defect calculation step an initial-boundary value problem is to be solved.

4.2 A waveform relaxation algorithm

Spatial discretization of (3.1.a-b) and (4.1) leads to the following system of ordinary equations,

$$\frac{dU}{dt} + L(U) = F \quad with \quad U(t_0) = U(t_f). \tag{4.2}$$

This system may be solved with a waveform relaxation algorithm that is only slightly different from the algorithm discussed in section 2. Instead of repeatedly solving an equation of initial value type at each grid point, one repeatedly solves the following *periodic differential equation*

$$\frac{du_{ij}}{dt} + (L(U))_{ij} = f_{ij} \quad with \quad u_{ij}(t_0) = u_{ij}(t_f). \tag{4.3}$$

This periodic boundary value problem may be solved e.g. by a discretization method, which results in a sparse matrix equation. Application of an implicit one-step discretization method, such as the trapezoidal rule, leads to an easily solvable, cyclic bidiagonal system.

The modified waveform relaxation can be used as such, or it can be integrated as a smoother into the multigrid waveform scheme of section 3. The only algorithmic change is then in core of the ODE solver. The other multigrid operations remain totally unchanged. The changes in an implementation comprise maybe only ten lines of code. Numerical evidence shows that the new method, *periodic multigrid waveform relaxation,* leads to a rapidly converging iteration, with typical multigrid convergence rates. As will be shown by examples in section 6, the arithmetic complexity of solving a time-periodic problem with the new method is similar to the complexity of solving one initial-boundary value problem.

5. Implementation aspects

5.1 Time discretization

Any stiff integrator can be used to solve the ordinary differential equations that arise in the smoothing step of the algorithm. A sophisticated variable time step variable order integrator from the ODEPACK library was used in our early experiments, [13]. However, the overheads of step size and order selection, reoccurring at each grid point, badly deteriorated the execution times. Consequently, our experience now advocates the use of a simple but fast ODE integrator. For the examples in the next section we used the second-order accurate trapezoidal rule (or Crank-Nicolson method) or the backward differentiation method, both with fixed time step, Δt. The same time step was used for the integrations on the fine grid and on the coarse grids. This choice of time integration precludes taking advantage of waveform relaxation as a multirate integration method. However it simplifies implementation and, as will be explained in section 5.3, it allows for vectorization.

In the waveform method several functions are associated with each grid point. For these functions, an effective and compact computer representation should be chosen. It should allow for efficient evaluation of the functions and, more importantly, for efficient algebraic manipulation (mainly the summation of functions and multiplication with a scalar). For the latter reason we have represented each function as a *vector* of values evaluated at equidistant time levels. We denote the vector length by n_t (number of time levels). The time increment is chosen equal to the time increment used by the integration formula.

In the case of a time stepping procedure the unknowns need to be stored in the computer memory on one or two time levels only. In the case of waveform relaxation the values at every time level need to be available. Storage requirements are therefore very high. However, if storage is a problem, the time interval of interest may be partitioned into several smaller *windows*, which can be treated in sequence.

5.2 Parallelization

The implementation of the waveform algorithm and the parallelization of the standard time stepping methods is discussed in detail in our studies [14,15]. In this section we only review some of the main issues. A classical data decomposition is used to evenly distribute the computational workload. The processors are arranged in a rectangular array and are mapped onto the domain of the partial differential equation, which, in our implementation, is restricted to be a rectangle. Each processor is responsible for doing all computations on the grid points in its subdomain. During the computation, communication with neighbouring processors is needed mainly to update local boundary values. Some other communication strategies are further used to improve the parallel performance. We'd like to mention in particular the use of an *agglomeration* strategy to reduce the communication complexity of the coarse grid operations, [15].

The arithmetic complexity of the waveform algorithms increases linearly with the value of n_t. In the same way, the total length of the messages exchanged during the computation is proportional to n_t. The number of messages, however, and the sequential overhead due to program control are independent of n_t. From the high message start-up time on most parallel machines it is clear that the communication time to calculation time ratio will decrease with increasing function length. The larger the number of time levels calculated simultaneously, the better the parallel efficiency will be.

5.3 Vectorization

The use of vector processors for executing the multigrid waveform algorithm may result in a substantial reduction of computing time. Indeed, most of the operators can be expressed as simple arithmetic operations on functions, i.e. on *vectors*. In contrast to the standard approach *vectorization is in the time direction* and not in the spatial direction ! The vector speedup of the arithmetic part of the computation will mainly depend on the value of n_t. It will be virtually independent of the size of the spatial grid, the number of multigrid levels, the multigrid cycle used and the number of processors. This contrasts sharply to standard vectorization, which only leads to a performance improvement when the number of grid points per processor is very large.

The only operation which is not perfectly vectorizable is the core of the ODE integrator used in the smoothing step. It consists of one or two first order recurrence formulae of length n_t (one in the case of an initial value problem and two in the case of a time-periodic problem solved by discretization with a one step formula). This reduces the possible gain through vectorization. It can be shown, however, that the cost of the recurrence relation makes out only a small fraction of the total computation cost, typically 5 to 10%. This leads to a possible vector speedup of 10, or more.

6. Numerical examples

We illustrate in this section the numerical and parallel characteristics of the waveform relaxation methods discussed in sections 3 and 4. We have implemented the algorithms on an Intel iPSC/2 hypercube, without vector processors. For a description of this multiprocessor, its hardware characteristics and various performance benchmarks, we refer to [2].

6.1 A nonlinear initial-boundary value problem

We consider the numerical solution of the Brusselator, a well-known nonlinear system of parabolic partial differential equations which models a chemical reaction-diffusion process, [9]. The two-dimensional Brusselator model is described by the following equations defined over the unit square,

$$\frac{\partial X}{\partial t} = \frac{D_X}{L^2}\left[\frac{\partial^2 X}{\partial x^2} + \frac{\partial^2 X}{\partial y^2}\right] - (B+1)X + X^2 Y + A \tag{6.1}$$

$$\frac{\partial Y}{\partial t} = \frac{D_Y}{L^2}\left[\frac{\partial^2 Y}{\partial x^2} + \frac{\partial^2 Y}{\partial y^2}\right] + X^2 Y + B X$$

The functions $X(t,x,y)$ and $Y(t,x,y)$ denote chemical concentrations of intermediate reaction products. $A(t,x,y)$ and $B(t,x,y)$ are the concentrations of the input reagents, which, for simplicity, are considered constant and homogeneous, i.e. $A = 2.0$ and $B = 5.45$. The diffusion coefficients are given by $D_X = 0.004$ and D_Y=0.008. The value of the chemical reactor length L will be specified further. We consider Dirichlet boundary conditions $X(t,x,y) = A$, $Y(t,x,y) = B/A$, $(x,y) \in \partial\Omega$. Second order finite differences are used on an equidistant mesh, with mesh size $h = 1/16$ and $h = 1/32$. For $L = 0.9$ we have first calculated the time-evolution of the system for $t \in [0,3]$, starting from the following initial conditions,

$$X(t,x,y) = A + \sin(\pi x)\sin(\pi y) \quad \text{and} \quad Y(t,x,y) = B/A - \sin(\pi x)\sin(\pi y) \tag{6.2}$$

Fig. 1a illustrates the convergence behaviour of red/black Gauss-Seidel waveform relaxation. For all grid points ($h = 1/16$ in this case) a constant profile in time was used as the starting approximation. The figure shows the successive approximations for the function $Y(t, ½,½)$. It clearly illustrates the typical waveform relaxation phenomenon that the largest error occurs at the end of the integration interval. Corresponding results obtained with multigrid waveform relaxation are shown in fig. 1b. Multigrid V-cycles are used with one red/black WR pre-smoothing step and one similar post-smoothing step. Standard coarsening is used down to a coarse level with mesh size $h = 1/2$. In fig. 1c the results obtained with the full multigrid approach are shown.

In fig.2 the performance of the multigrid waveform relaxation (MWR) and the Crank-Nicolson (C-N) methods are compared. To make a fair comparison possible we also use the Crank-Nicolson rule to integrate the ODEs in the waveform method, with the same time step as the one used by C-N. This guarantees that the discrete equations of both techniques are similar and that the solutions are identical. The accuracy, i.e. the largest residual w.r.t. the discrete equations, is plotted versus the execution time on 1 processor and on 16 processors. The figures show *smooth curves* for the waveform method. The residual of the initial approximation decreases as more and more multigrid cycles are applied. The C-N results show up as *discrete points*. The C-N solution process is advanced time step per time step in a total of t seconds. The residual of the result is represented by a "+" -sign at position (t,error) in the figure. The results are annotated as follows: with "WR V(1,1) with FMG" we denote "waveform relaxation using V-cycles with 1 pre- and 1 post-smoothing step and the full multigrid technique with 1 cycle at each grid level"; with "C-N, 2 V(1,1)" we denote "Crank-Nicolson method with 2 V(1,1) cycles per time step". The dashed line indicates the residual of the 'true solution', i.e. the solution obtained with the mesh parameters $h = 1/64$ and $\Delta t = 1/128$.

On one processor MWR turns out to be as efficient as C-N if the problem is to be solved to discretization accuracy, indicated by the dashed line. The method is less efficient if the discrete equations are to be solved to a high accuracy. On 16 processors however MWR clearly outperforms C-N. This is due to its superior parallel characteristics, i.e. the much lower communication to computation ratio. The use of vectorization would make the difference even larger in favour of MWR.

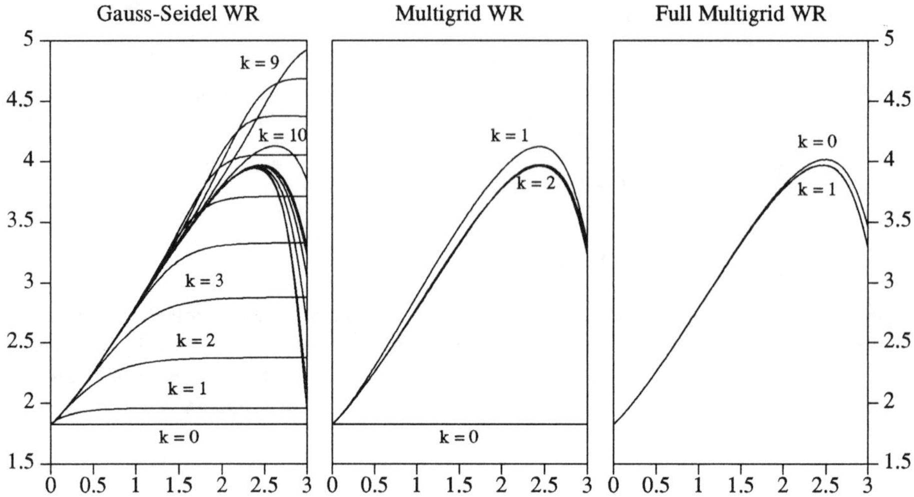

Figure 1. Waveform relaxation for time integration of (6.1) with L=0.9 and h=1/16. Approximations $Y^{(k)}(t, \frac{1}{2}, \frac{1}{2})$ after k iteration steps.

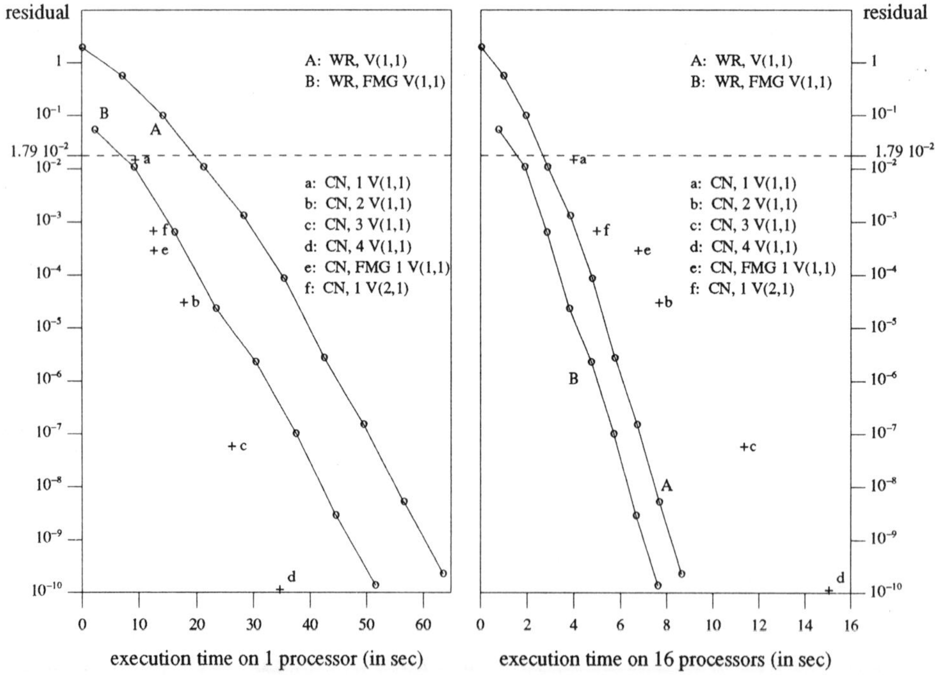

Figure 2. Comparison of multigrid waveform relaxation and Crank-Nicolson method for time-integration of (6.1), $t \in [0,1]$, with L=0.75, h=1/16 and Δt=1/32.

6.2 A nonlinear time-periodic problem

We will determine the stable periodic solution of the Brusselator model in the case of a periodically varying concentration of the input reagent,

$$A(t,x,y) = 2 + sin(4t), \text{ (i.e. with period } T = \pi/2 \text{)} \tag{6.3}$$

To this end the problem is discretized on an equidistant grid with $h = 1/32$. The problem is solved with the periodic multigrid waveform relaxation method (PMWR), using 5 multigrid levels. Smoothing is performed by block-wise periodic red/black waveform relaxation. The system of two periodic differential equations at each grid point is solved with one Newton linearization step. The linear problem is then solved by discretization with the trapezoidal rule (100 time steps, or $\Delta t = \pi/200$) and by direct solution of the resulting cyclic block-bidiagonal matrix. The convergence is illustrated in fig. 3, in which the discrete l_2-norm of the residual (w.r.t. the set of discrete equations) is plotted for consecutive approximations. The results indicated by solid lines are obtained by starting the iterative algorithm with a constant time-profile in each grid point. The dashed lines represent the results with the full multigrid procedure. In addition we tabulate in table 1 the experimentally observed convergence factors, averaged over 10 iterations.

table 1: averaged convergence factor, periodic brusselator problem					
PMWR V(1,1)	PMWR V(2,1)	PMWR W(1,1)	PMWR W(2,1)	PMWR F(1,1)	PMWR F(2,1)
0.163	0.163	0.063	0.041	0.064	0.041

The profile of the stable periodic solution at the mid-point of the chemical reactor is plotted in fig. 4. The figure to the left shows the time behaviour, while the figure to the right illustrates the limit cycle in the state diagram. This solution can also be determined by a straightforward time integration of the equations starting from some initial condition. The resulting process is illustrated in fig. 5. The periodic solution is obtained when the transient part of the solution has decayed. Integration over a long time interval may be required when the periodic solution is not strongly attracting. Furthermore it is difficult to detect in an automatic way that the periodic behaviour is attained.

6.3 A linear time-periodic problem

We consider the diffusion equation on the unit square, with periodic Dirichlet boundary conditions,

$$\frac{\partial u}{\partial t} = \frac{\partial^2 u}{\partial x^2} + \frac{\partial^2 u}{\partial y^2}, \ \textit{with} \ \ u(t,0,y) = u(t,x,1) = u(t,1,y) = 0 \ \text{ and } \ u(t,x,0) = sin(\pi x) sin(10\pi t). \tag{6.4}$$

We calculate the stable periodic solution, with period $T = 0.2$, on an equidistant grid with mesh size $h = 1/32$ and with the time interval, [0,0.2], being divided into 64 time levels.

Three methods are compared. The first method is the multigrid method of the second kind with second order backward differentiation (BDF(2)) used for time integration. In this method a number of initial boundary value problems are to be solved on each multigrid level. The spatial mesh size on each grid is determined by standard coarsening. The use of BDF(2) allows the time increment on each grid level to be chosen equal to the mesh size, [3]. Thanks to its $O(h^2)$ convergence rate, one iteration of the multigrid method of the second kind is sufficient to solve the problem to discretization accuracy. The arithmetic complexity of the above algorithm is equal to 2.5 times the complexity of solving an initial boundary value problem on the fine mesh, [16].

A related method is obtained if MWR is used as the initial-boundary value solver inside the multigrid method of the second kind. In that case the arithmetic complexity will not change significantly when the same discretization parameters are used. The third algorithm is the periodic multigrid waveform relaxation method. Its complexity is similar to the complexity of solving only one initial boundary value problem.

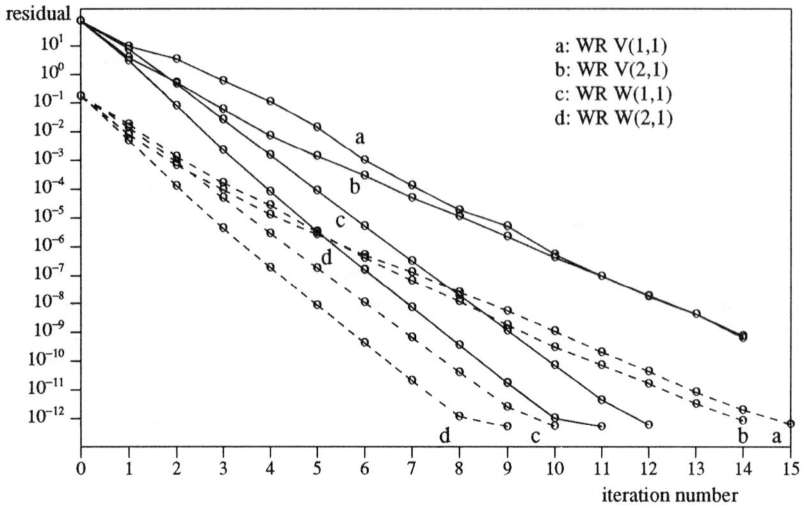

Figure 3. PMWR convergence for solving (6.2) with (6.3), L=0.15, h=1/32, Δt=π/200.

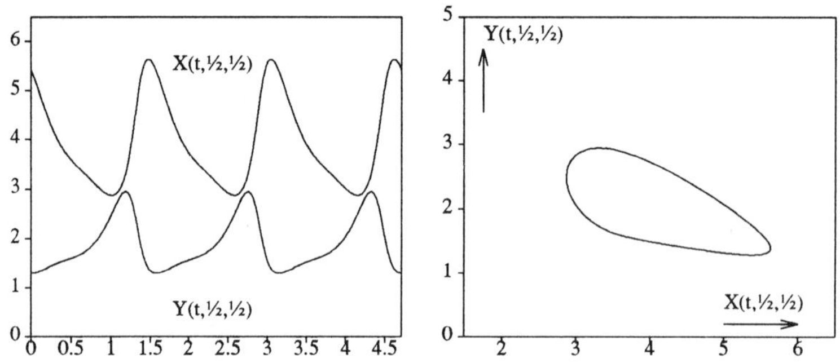

Figure 4. Periodic solution obtained by PMWR (time domain - state space)

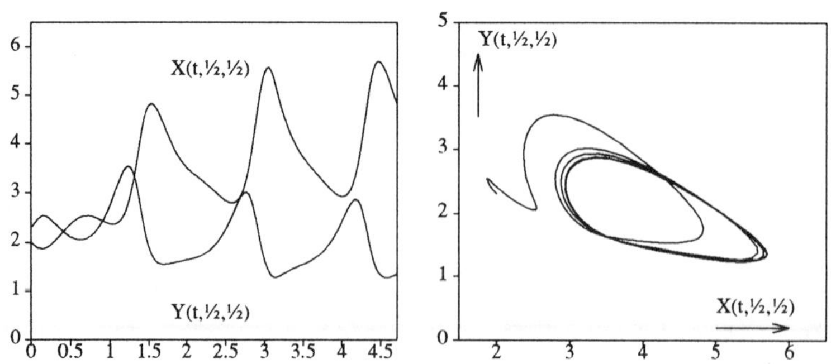

Figure 5. Periodic solution obtained by brute force time integration (time domain - state space)

The timing results are presented in fig. 6. We have plotted the execution time on 1 and on 16 processors versus the accuracy. The latter is defined as the maximum difference at the grid points on time level $t = 0$, between the approximation and the 'true' solution, obtained with $h = 1/64$ and $\Delta t = 1/128$.

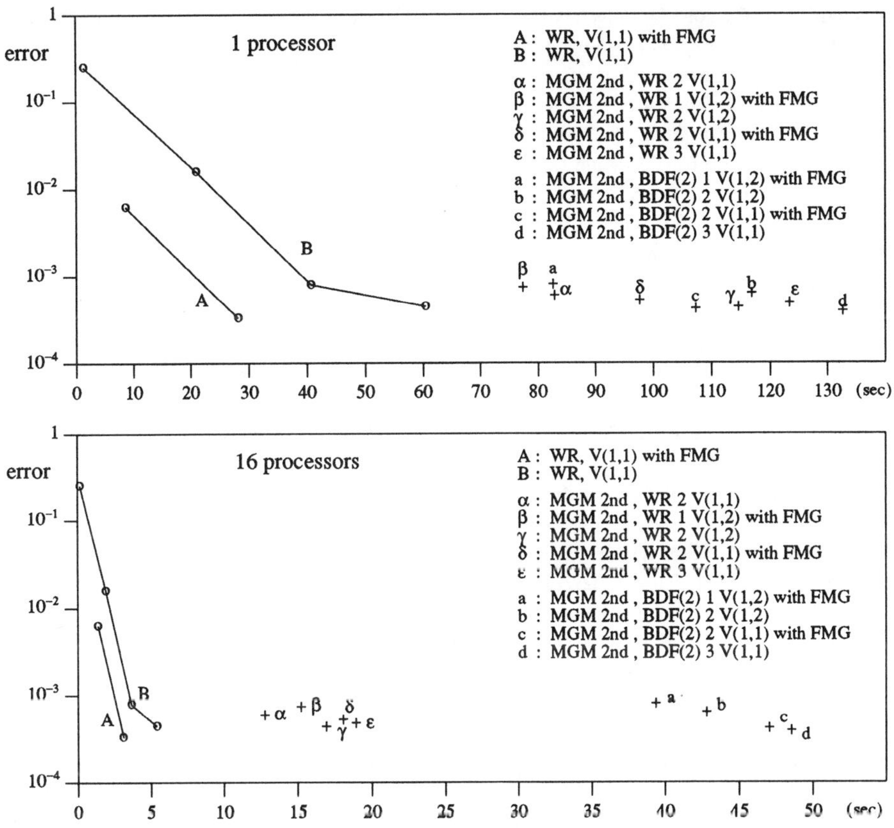

Figure 6. Executions times for solving (6.4) with PMWR and the multigrid method of the second kind

The one-processor results are interesting in their own right as they confirm the arithmetic complexities mentioned above. As it shows, the periodic multigrid waveform method outperforms the 'best' standard method, multigrid of the second kind, with a factor of 2.5. The timings on 16 processors illustrate the parallel characteristics. A very low parallel efficiency is obtained with the multigrid method of the second kind (although we have applied various techniques to optimize its parallel performance). This is due to the fact that the multigrid method of the second kind visits the coarse grids very frequently. As coarse grid operations are difficult to parallelize efficiently, the algorithm is not well-suited for parallel implementation. The multigrid method of the second kind combined with waveform relaxation achieves a speedup of about 6. As such it outperforms the parallel implementation of the standard method by a factor of 3. The periodic multigrid waveform method is faster than the multigrid method of the second kind by a factor of 12. This is due to its lower arithmetic complexity and its much better better parallel characteristics. In addition, we should note again that vectorization will lead to an additional speedup in the case of the waveform relaxation algorithm.

References

[1] Bastian P., Burmeister J. and Horton G., *Implementation of a parallel multigrid method for parabolic partial differential equations*, in: Proceedings of the Sixth GAMM - Seminar Kiel, (to appear).

[2] Bomans L., and Roose D., *Benchmarking the iPSC/2 hypercube multiprocessor*, Concurrency: Practice and Experience, 1 (1) 1989, pp. 3-18.

[3] Hackbusch W., *Fast numerical solution of time-periodic parabolic problems by a multigrid method,* SIAM J. Sci. Stat. Comput., Vol. 2., No. 2, 1981, pp. 198-206.

[4] Hackbusch W., *Parabolic multigrid methods,* in: Computing Methods in Applied Sciences and Engineering VI, R. Glowinski and J.-L. Lions (eds), North-Holland, Amsterdam, 1984, pp. 189-197.

[5] Juang F., *Accuracy increase in waveform relaxation*, Report No. UIUCDCS-R-88-1466, department of computer science, university of Illinois at Urbana-Champaign, Urbana, Illinois, Oct. 1988.

[6] Lubich Ch. and Ostermann A., *Multi-Grid Dynamic Iteration for Parabolic Equations,* BIT, 27 (1987), pp. 216-234.

[7] Miekkala U. and Nevanlinna O., *Convergence of Dynamic Iteration Methods for Initial Value Problems,* SIAM J. Sci. Stat. Comput., Vol. 8, No. 4, 1987, pp. 459-482.

[8] Miekkala U. and Nevanlinna O., *Sets of convergence and stability regions*, BIT 27, 1987, pp. 554-584.

[9] Prigogine I. and Lefever R., *Symmetry breaking instabilities*, J. Chem. Phys. 48 (4), 1968, pp. 1695-1700.

[10] Roose D. and Vandewalle S., *Efficient Parallel Computation of Periodic Solutions of Parabolic Partial Differential Equations,* in: Proceedings of "Bifurcations and Chaos: Analysis, Algorithms, Applications", R. Seydel, F. Schneider, T. Küpper and H. Troger (eds), Birkhäuser Verlag (to appear).

[11] Saltz J. and Naik V., *Towards developing robust algorithms for solving partial differential equations on MIMD machines,* Parallel Computing, No. 6, 1988, pp.19-44.

[12] Skjellum A., *Concurrent dynamic simulation: multicomputer algorithms research applied to ordinary differential-algebraic process systems in chemical engineering*, PhD-thesis, California Institute of Technology, May 1990.

[13] Vandewalle S. and Roose D., *The Parallel Waveform Relaxation Multigrid Method*, in: Rodrigue G., (ed.), Parallel Processing for Scientific Computing, SIAM, Philadelphia, 1989, pp. 152-156.

[14] Vandewalle S. and Piessens R., *A Comparison of the Crank-Nicolson and Waveform Relaxation Multigrid Methods on the Intel Hypercube,* in: Proceedings of the Fourth Copper Mountain Conference on Multigrid Methods, SIAM, Philadelphia, 1990, pp. 417-434.

[15] Vandewalle S., Van Driessche R. and Piessens R., *The Parallel Implementation of Standard Parabolic Marching Schemes*, submitted to Int. J. of High Speed Computing.

[16] Vandewalle S. and Piessens R., *Efficient Parallel Algorithms for Solving Initial-Boundary Value and Time-Periodic Parabolic Partial Differential Equations,* Report TW 139 Katholieke Universiteit Leuven, November 1990, submitted to SIAM J. Sci. Stat. Comput.

[17] White J., Odeh F., Sangiovanni-Vincentelli A.S. and Ruehli A., *Waveform Relaxation: Theory and Practice,* Memorandum No. UCB/ERL M85/65, Electronics Research Laboratory, College of Engineering, University of California, Berkeley, 1985.

[18] Womble D., *A time-stepping algorithm for parallel computers,* SIAM J. Sci. Stat. Comput., Vol. 11, No. 5, 1990, pp 824-837.

Stefan Vandewalle, dept Computer Science, Katholieke Universiteit Leuven
Celestijnenlaan 200 A, B-3001 Leuven, Belgium.

CONTENT OF GMD–STUDIEN NR. 189

(ISSN 0170 8120, *available from:* Gesellschaft für Mathematik und Datenverarbeitung mbH (GMD), Schloß Birlinghoven, Postfach 1240, W-5205 Sankt Augustin 1, Germany)

Amini, S.; Ke, Chen
Multigrid solution of boundary integral equations with non-compact operators – application to the exterior acoustic problem

Ascher, U.M.; Carter, P.M. Multigrid solution for shape from shading

Brakhagen, F.; Fogwell, T.W.
Multigrid for the fully implicit formulation of the equations for multiphase flow in porous media

Catalano, L.A.; Deconinck, H.
Two-dimensional optimization of smoothing properties of multi-stage schemes applied to hyperbolic equations

Constapel, R.
A multigrid approach for two-dimensional simulation of semiconductor devices

Douglas, C.C.
An almost assumptionless convergence theory for the full approximation scheme or multiple coarse space multigrid algorithms

Estivalezes, J.-L.; Rompteaux, A.
Fast mesh generation with non linear multigrid algorithms

Ferry, M.; Piquet, J.
A new fully-coupled method for the solution of Navier-Stokes equations

Finnemann, H.; Böer, R.; Müller, R.; Kim, Y.I.
Adaptive multi-level techniques for the solution of nodal transport equations

Fujino, S.; Takeuchi, T.; Kuwahara, K.
Visualization of convergence behavior of multigrid methods

Gaskell, P.H.; Sleigh, P.A.; Wright, N.G.
Multigrids and their role in the solution of practical CFD problems

Heinrichs, W.
Spectral multigrid techniques for the Stokes problem in streamfunction formulation

Joppich, W.; Lorentz, R.A.
High-order monotone and convex multi-grid interpolations for a VLSI model problem

Kanarachos, A.; Pantelesis, N.; Provatides, Chr.: An algebraic multiblock-multigrid solver for the Euler equations

Lemke, M.; Quinlan, D.
Local refinement based fast adaptive composite grid methods on SUPRENUM

Melson, N.D.; Cannizzaro, F.E.; Elmiligui, A.; von Lavante, E.
A multiblock multigrid three-dimensional Euler method

Mijalkovic, S.
An adaptive multigrid simulation of coupled multiparticle diffusion in semiconductor device fabrication

Mikulinsky, V.; Brandt, A. Multigrid treatment of free-boundary conditions

Morano, E.; Lallemand, M.-H.; Leclercq, M.-P.; Steve, H.; Stoufflet, B.; Dervieux, A.
Local iterative upwind methods for steady compressible flows

Nowak, Z.P.
Fine and coarse grid discretizations for a high-order integro-differential formulation of the potential flow problem in three dimensions

Pardhanani, A.; Carey, G.F.
Adaptive redistribution and multilevel techniques for time-dependent and steady problems

Popa, C.
ILU decomposition for coarse grid correction step on algebraic multigrid

van Rienen, U.
A multigrid-algorithm for an indefinite, non-hermitian linear system in the calculation of electromagnetic fields

Sauter, St.; Wittum, G.
On the computation of the eigenmodes of Lake Constance by means of a multi-grid method

Sbosny, H.
Domain decomposition methods and parallel multigrid algorithms

Schmidt, G.H.; de Sturler, E.
Multigrid for porous medium flow on locally refined three dimensional grids

Schwichtenberg, H.; Winter, G.
Standard multigrid-components in a classical drift-diffusion model for MOS devices

Steffen, B.
Treatment of very small structures in finite difference multigrid

Tu, J.Y.; Fuchs, L.; Bai, X.S.
Finite volume and multi grid methods on 3-D zonal overlapping grids

Weinerfelt, P.
Multigrid technique applied to the Euler equations on a multiblock mesh

Weiss, R.; Schönauer, W.
Solution of partial differential equations by means of data reduction (DARE)

Weyand, C. A multigrid method for the Mindlin-Reissner plate equations

PARTICIPANTS

Akian, Marianne	INRIA Rocquencourt, France
Amini, S.	University of Salford , Great Britain
Ashby, J.V.	Rutherford Appelton Lab., Didcot, Great Britain
Balsara, Dinshaw S.	University of Illinois at Urbana, USA
Bey, Jürgen	Aachen, Germany
Biermann, Ralf	GMD-F1, St.Augustin, Germany
Blazek, Jiri	DLR, Braunschweig, Germany
Bloß, Martina	Technische Universität Berlin, Germany
Böer, Rainer	Siemens AG, Erlangen, Germany
Brakhagen, Franz	GMD-F1, St.Augustin, Germany
Brandt, Achi	The Weizmann Institute of Science, Rehovot, Israel
Brenner, Susanne C.	Clarkson University, Potsdam, NY, USA
Bünten, Rolf	RWTH Aachen, Germany
Büsse, Joachim	GMD-F1, St.Augustin, Germany
Burg, Jan Wim Van der	Enschede, The Netherlands
Burmeister, Jens	Universität Kiel, Germany
Callies, Ulrich	GKSS-Forschungzentrum, Geesthacht, Germany
Carter, Paul M.	The University of British Columbia, Vancouver, Canada
Catalano, Luciano A.	Istituto di Macchine ed Energetica, Bari, Italy
Constapel, Rainer	Fraunhofer-Institut, Duisburg, Germany
Cote, Owen R.	EOARD/LDG, London, Great Britain
De Keyser, Johan	Katolieke Universiteit Leuven, Belgium
Dendy, Joel E.	Los Alamos National Laboratory, NM, USA
Dervieux, Alain	INRIA Sophia Antipolis, France
De Sturler, E.	TU-Delft-TWI, Delft, The Netherlands
Deuflhard, Peter	Konrad-Zuse-Zentrum, Berlin, Germany
Dick, Erik	State University of Gent, Belgium
Douglas, Craig C.	IBM Research Center,Yorktown Heights, NY, USA
Dreyer, Thomas	Universität Heidelberg, Germany
Eppel, Dieter P.	GKSS-Forschungszentrum, Geesthacht, Germany
Epple, B.	Universität Stuttgart, Germany
Ernst, Roland	John-Andrews-Entwicklungszentrum, Köln, Germany
Esposito, Pier Giorgio	INSEAN, Roma, Italy
Findling, Andreas	DLR, Braunschweig, Germany
Finnemann, H.	Siemens AG, Erlangen, Germany
Fogwell, T.W.	GMD-F1, St.Augustin, Germany
Frese, Hans-Hermann	Konrad-Zuse-Zentrum, Berlin, Germany
Frohn, Claudia	GMD-F1, St.Augustin, Germany
Fuchs, Laszlo	The Royal Institute of Technology, Stockholm, Sweden
Fuhrmann, Jürgen	Akademie der Wissenschaft, Berlin, Germany
Fujino, Seiji	Institute of Computational Fluid Dynamics, Tokyo, Japan

Gärtel, Ute	GMD-F1, St.Augustin, Germany
Galbas, Hans Georg	GMD-F1, St.Augustin, Germany
Geiger, Alfred	Rechenzentrum der Universität Stuttgart, Germany
Ginzel, Christian	MPI für Quantenoptik, Garching, Germany
Goergens, Günter	GMD-F1, St.Augustin, Germany
Greenwald, Joseph	The Weizmann Institute of Science, Rehovot, Israel
Griebel, Michael	Technische Universität München, Germany
Hackbusch, Wolfgang	Universität Kiel, Germany
Hahn, Harro H.	Technische Universität Braunschweig, Germany
Hebeker, Friedrich-Karl	IBM Deutschland GmbH, Heidelberg, Germany
Heinrichs, Wilhelm	Universität Düsseldorf, Germany
Hemker, Pieter W.	CWI, Amsterdam, The Netherlands
Hempel, Rolf	GMD-F1, St.Augustin, Germany
Hengst, Sabine	Akademie der Wissenschaft, Berlin, Germany
Holzbecher, Ekkehard	Technische Universität Berlin, Germany
Hoppe, Hans-Christian	GMD-F1, St.Augustin, Germany
Hoppe, Ronald H.W.	Technische Universität München, Germany
Horton, Graham	Universität Erlangen-Nürnberg, Germany
Jameson, Anthony	Princeton University, USA
Jarausch, Helmut	RWTH Aachen, Germany
Joppich, Wolfgang	GMD-F1, St.Augustin, Germany
Katzer, Edgar	Universität Kiel, Germany
Klinkhammer, Ralf	GMD-F1, St.Augustin, Germany
Koren, Barry	CWI, Amsterdam, The Netherlands
Kornhuber, Ralf	Berlin, Germany
Krechel, Arnold	GMD-F1, St.Augustin, Germany
Kröner, Dietmar	Universität Heidelberg, Germany
Küster, Uwe	Rechenzentrum der Universität Stuttgart, Germany
Kuhfuß, Rudolf	Rechenzentrum der Universität Stuttgart, Germany
Lacor, Chris	Vrije Universiteit Brüssel, Belgium
Lauwers, Paul G.	Universität Bonn, Germany
Lavante, E. von	Gesamthochschule Essen, Germany
Lemke, Max	SUPRENUM GmbH, Bonn, Germany
Lien, Fue-Sang	University of Manchester, Great Britain
Linden, Johannes	GMD-F1, St.Augustin, Germany
Lötstedt, Per	SAAB-SCANIA, Linköping, Sweden
Lonsdale, Guy	GMD-F1, St.Augustin, Germany
Lorentz, Rudolph	GMD-F1, St.Augustin, Germany
Luh, Yvonne	GMD-F1, St.Augustin, Germany
Maitre, Jeans-Francois	Ecole Centrale de Lyon, Ecully Cedex, France
Mayers, David F.	Oxford University, Great Britain
McBryan, Oliver	University of Colorado at Boulder, USA
Meinke, M.	RWTH Aachen, Germany
Michelsen, Jess	Technical University of Denmark, Lyngby, Denmark
Mijalcovic, Slobodan	Faculty of Electronic Eng., Beogradska, Jugoslavia

Mikulinsky, Vladimir	The Weizmann Institute of Science, Rehovot, Israel
Molenaar, J.	CWI, Amsterdam, The Netherlands
Morano, Eric	INRIA Rocquencourt, France
Müller, Markus	GMD-F1, St.Augustin, Germany
Münzer, Bernhard	Ges. für Strahlen- u. Umweltforschung, Neuherberg, Germany
Munthe-Kaas, Hans	CERFACS, Toulouse Cedex, France
Niessner, Herbert	ABB Informatik AG, Baden, Switzerland
Niestegge, Anton	GMD-F1, St.Augustin, Germany
Nurmann, Jens	Aachen, Germany
Oertel, Dieter	SUPRENUM GmbH, Bonn, Germany
Oosterlef, C.W.	TU-Delft-TWI, Delft, The Netherlands
Pantelelis, Nikos	National Technical University of Athens, Greece
Parhanani, Anand	University of Texas at Austin, USA
Piquet, J.	ENSM CFD Group, Nantes-Cedex, France
Popa, Constantin	Polytechnic Institute, Bucharest, Rumania
Reichelt, Claudia	GMD-F1, St.Augustin, Germany
Ressel, Klaus	GMD-F1, St.Augustin, Germany
Reusken, A.	Technische Universiteit Eindhoven, The Netherlands
Rienen, Ursula van	DESY, Hamburg, Germany
Ries, Manfred	Universität Trier, Germany
Ritzdorf, Hubert	GMD-F1, St.Augustin, Germany
Rogers, C.F.	BP Research Centre, Sunbury-on-Thames, Great Britain
Rompteaux, Arnaud	ONERA, Toulouse, France
Roose, Dirk	Katholieke Universiteit Leuven, Belgium
Rüde, Ulrich	Technische Universität München, Germany
Ruge, John W.	University of Denver at Boulder, USA
Rust, Wilhelm	Universität Hannover, Germany
Sauter, Stefan	Universität Kiel, Germany
Sbosny, Hannes	Universität Düsseldorf, Germany
Schmatzer, Franz-Karl	München, Germany
Schmidt, Geert H.	Shell Exploratie en Produktie Lab., Rijswijk, The Netherlands
Schrauf, Geza	Deutsche Airbus GmbH, Bremen, Germany
Schüller, Anton	GMD-F1, St.Augustin, Germany
Schwamborn, D.	DLR, Göttingen, Germany
Schwichtenberg, Horst	GMD-F1, St.Augustin, Germany
Simon, Matthias	Universität -GH- Paderborn, Germany
Singh, J.P.	DLR, Göttingen, Germany
Sleigh, Andrew	The University of Leeds, Great Britain
Solchenbach, Karl	SUPRENUM GmbH, Bonn, Germany
Stamp, Jochen	Rechenzentrum der Universität Hannover, Germany
Steckel, Barbara	GMD-F1, St.Augustin, Germany
Steffen, Bernhard	Forschungszentrum Jülich, Germany
Ströder, Thomas	GMD-F1, St.Augustin, Germany
Stüben, Klaus	GMD-F1, St.Augustin, Germany
Sweldens, Wim	Katholieke Universiteit Leuven, Belgium

Ta'asan, Shlomo	The Weizmann Institute of Science, Rehovot, Israel
Teigland, Rune	University of Bergen, Norway
Thole, Clemens-August	SUPRENUM GmbH, Bonn, Germany
Tiero, Alessandro	II Universita di Roma 'Tor Vergata', Roma, Italy
Tolsma, Bob	Eindhoven University of Technology, The Netherlands
Trottenberg, Ulrich	GMD-F1, St.Augustin, Germany
Tu, Jiyuan	The Royal Institute of Technology, Stockholm, Sweden
Turek, Stefan	Universität Heidelberg, Germany
Vandewalle, Stefan	Katholieke Universiteit Leuven, Belgium
Vournas, Ioannis	National Technical University of Athens, Greece
Wang, Dieqian	FFA The Aeronautical Research, Bromma, Sweden
Weinerfelt, Per	CERFACS, Toulouse Cedex, France
Weiss, Rüdiger	Rechenzentrum der Universität Karlsruhe, Germany
Werner, Klaus-Dieter	GKSS-Forschungszentrum, Geesthacht, Germany
Wesseling, Pieter	Delft University of Technology, Delft, The Netherlands
Weyand, Clemens	Universität Düsseldorf, Germany
Winter, Gerd	GMD-F1, St.Augustin, Germany
Witsch, Kristian	Universität Düsseldorf, Germany
Wittum, Gabriel	Universität Heidelberg, Germany
Womble, David E.	Sandia National Laboratories, Albuquerque, NM, USA
Wright, Nigel G.	The University of Leeds, Great Britain
Wunderlich, Peter	Prime Computer GmbH, München, Germany
Yserentant, Harry	Universität Tübingen, Germany